Complexity and Complex Thermo-Economic Systems

Complexity and Complex Thermo-Economic Systems

Stanisław Sieniutycz
Warsaw University of Technology,
Faculty of Chemical and Process Engineering,
Warsaw, Poland

Elsevier
Radarweg 29, PO Box 211, 1000 AE Amsterdam, Netherlands
The Boulevard, Langford Lane, Kidlington, Oxford OX5 1GB, United Kingdom
50 Hampshire Street, 5th Floor, Cambridge, MA 02139, United States

Library of Congress Cataloging-in-Publication Data
A catalog record for this book is available from the Library of Congress

British Library Cataloguing-in-Publication Data
A catalogue record for this book is available from the British Library

ISBN: 978-0-12-818594-0

For information on all Elsevier publications visit
our website at https://www.elsevier.com/books-and-journals

Publisher: Susan Dennis
Acquisition Editor: Anita Koch
Editorial Project Manager: Emerald Li
Production Project Manager: Prem Kumar Kaliamoorthi
Cover Designer: Miles Hitchen

Typeset by SPi Global, India

Contents

Preface

The title of this book at the time when it was planned was *Complexity and Complex Thermodynamic Systems*, i.e., a narrowed treatise was intended, associated only with thermodynamic aspects of complexity. However, as stressed by Reynolds and Holwell in their book *System Approaches to Managing Change* (Springer 2010; ISBN 978-1-84882-808-7) we live and work in a highly complex and interconnected world, thus even small enlargement of the decision manifold may have an impact on organizations, societies, environment, etc. Therefore, it was natural to extend in the present volume the class of complex thermodynamic systems to the class of complex thermoeconomic systems, thus increasing the applicative contents of the book (even at the cost of occasional violation of the book coherence). Keeping this change in mind we will, nevertheless, preserve the priority of thermodynamics, "a queen of macroscopic sciences which will never be overthrown." In fact, thermodynamics is invariably the basis and the inevitable component in modeling, analysis, synthesis, and optimization of most practical and industrial systems, in particular those considered in this book.

Thermodynamic models consider energy and matter balances, the results of invariance principles of physics. These balances incorporate limitations on working parameters of real processes, which should be considered when formulating constraints on the performance of practical and industrial systems. Thermodynamics and kinetic theory provide data of static and transport properties needed in calculations. These data are necessary to express balance or conservation laws for energy and substances in terms of variables used in systems modeling, analysis, synthesis, the stages terminating sometimes at an optimization problem (after making a selection of state variables, controls, and parameters). In fact, thermodynamic variables are frequently state and/or control coordinates in systemic optimizations; moreover, some thermodynamic variables or their functions may constitute performance criteria of optimization. Thermodynamics may also help to set selling prices of chemical substances and thus influence the formulation of paraeconomic optimization criteria (thermoeconomic optimization of systems).

Briefly, the basic purpose of this book is to achieve and investigate formulations, solving tools, and solutions describing the performance of selected complex systems, including those considered from the standpoint of a preassigned optimization criterion (chemical conversion, thermoeconomical or economical profit, or a technical index).

In Chapters 1–3, cybernetics and system theory are defined as two sister disciplines governed by general system methodology, and introduce cyberneticians and system experts: researchers working in systems theory, scientists who by assumption apply their basic objective of building general, domain-independent systemic models, in opposition to traditional scientists who focus on applications coming from various specific disciplines (chemistry, economics, thermodynamics, pattern recognition, etc.).

Further chapters investigate various complex systems as they differ in their structure and properties. Chapter 4 is devoted to systems of neural networks and their application for emission prediction of pollutants. Chapter 5 characterizes in detail components of system design which involves, in sequence, modeling, analysis, synthesis, and optimization, often with adaptive controls. Chapter 6 focuses on systems of energy engineering with a few excursions to ecology. Chapter 7 offers a rigorous mathematical treatment of optimal dynamics in both continuous and discrete systems governed by ordinary differential equations, and discusses nonclassical optimal solutions stemming from the use of the dynamic programming method. Chapter 8 focuses on optimization of chemically complex systems with catalyst decay and regeneration. Chapter 9 deals with diverse cases of complex dynamics and thermodynamics, and includes decision-making processes within economic systems analyzed in terms of Simon's conception of rationality and Morecroft's bounded rationality, both cases portraying situations when decisions are made by individuals.

System optimization, described briefly with mathematics in Chapters 3–6 and 8, offers a formal way of assuring the best intervention into a complicated reality, either by providing limiting values of certain quantities (extremum power, minimum cost, maximum yield, etc.) or by finding economically optimal solutions (optimal trajectories and optimal controls). These solutions ensure optimal feasible profits or costs attributed to economical or exergy-economical models. Optimal solutions, also called "best solutions," are obtained by calculations applying suitable computational algorithms. The corresponding computer programs use these algorithms, each leading to optimal solution in its own way. Results of optimization calculations include optimal trajectories and controls usually in discrete forms (discrete representation of the optimization solution).

This book is intended to supplement the volume written by Sieniutycz and Jeżowski titled *Energy Optimization in Process Systems*, Elsevier, Oxford, 2009 and its second and third editions both titled *Energy Optimization in Process Systems and Fuel Cells*, Elsevier, Oxford, 2013 and 2018. In these books, the authors defined and investigated limits of energy and power for various deterministic, power-yield and mechanical-energy-yield systems, important in chemical and mechanical engineering.

The contents of the book extend the previously treated contents by inclusion of self-organization issues, emergent properties, economic and thermoeconomic solutions, deactivation and regeneration of catalysts, hierarchical decision models, and other factors characteristic of complex systems. Of special interests are limiting solutions obtained with performance criteria involving thermoeconomic, environmental, and, occasionally, ecological terms. Thermodynamic behavior of systems is investigated in terms of finite-rate processes with heat, work, and mass transfer, occurring in apparatuses of finite dimension. Analyses of complex chemical systems focus on heterogeneous catalytic reactors, deactivating catalysts and catalyst regenerators. Optimization approaches, occasionally in the spirit of finite-rate thermodynamics, are based on the condition that in order to make the results of thermodynamic analyses usable in engineering it is the thermodynamic limit or a thermoeconomic limit, not the maximum of thermodynamic efficiency, which must be overcome for prescribed process requirements.

Not surprisingly, systems requiring determination of power or energy limits, i.e., practical power generators, treated in detail in the previous books of the present author, are only briefly mentioned here as systems in which the optimal control theory serves to determine optimal inputs of driving energy or driving substance. They are control systems of heat pump type, engine type, or a separator type which are assumed to utilize the low-grade heat or exergy of transferred heat and mass. Links with exergy definition in reversible systems and classical problems of extremum work are rediscovered in these systems in the context of reversible problems of maximum power. On the other hand, more attention is devoted in the book to optimization of multilevel control systems with decaying catalysts and catalyst regenerators.

The book has some unique features of related benefit to the reader, in view of still demanded literature information on solutions of practical optimization problems. The book extends the family of previously treated optimization problems by inclusion of economic and environmental issues, deactivation and regeneration of catalysts, and other factors appropriate for using thermodynamic models and methods of thermoeconomic optimization.

The literature of optimizations in chemical and economical worlds keeps growing, so that efficient literature searches are still necessary. The book is intended as a collection of chapters addressed to actively working scientists and students (both undergraduates and graduates). Associated professionals are: chemical engineers, economists, environmental engineers, control engineers, thermal engineers, solar system engineers, and some others. The authors offer basically a textbook equipped in the feature of a reference book. Because of abundance of literature discussion and citation of numerous references, the book should constitute a valuable and helpful reference volume for any reader, including readers employed in industries and universities. Because of the textbook property of the proposed volume, names and levels of appropriate courses are listed below. The book can be used in schools, libraries, industries, and internet centers as a directory of guided tour described in the book chapters, each tour representing a research problem.

One can ask: What type of knowledge do readers need before reading the book? The book applies analytical reasoning, and transfers to the reader a reasonable amount of analytical mathematics.

Therefore, the knowledge of elementary and differential calculus is mandatory. Basic equations of optimal control (Pontryagin's maximum principle and the corresponding optimization theory) should be the requirement only when reading Chapters 7 and 8. Since, however, the satisfaction of this condition might be difficult to the general reader, Chapter 12 of Sieniutycz's and Szwast's 2018 book *Optimizing Thermal, Chemical and Environmental Systems* offers a simple, synthesizing text on Pontryagin's maximum principle and related criteria of dynamic optimization. Because of the presence of the above theoretical information, it is not assumed that using other sources will be necessary. In the most advanced examples, complete analyses are developed to achieve solutions enabling verbal descriptions and subsuming of the discussed problems.

Another question is: What should the readers gain (academically/professionally) from the reading of the book? When the volume is used as a textbook, it can constitute a basic or supplementary text in the following courses conducted in engineering departments of technical universities:

- Technical thermodynamics and industrial energetics (undergraduate)
- Chemical reaction engineering (undergraduate)
- Alternative and unconventional energy sources (graduate)
- Catalyst decay and optimal catalyst regeneration (graduate)
- Complexity and complex systems theory (graduate)
- Economic models in engineering (graduate)

Having read the book, the readers will gain the necessary information on what has been achieved to date in the field of complexity and complex systems; what new research problems could be stated; or what kind of further studies should be developed within specialized optimizations of the discussed cases. It is expected that the information contained in the book will help to improve technical skills of the reader. This book is especially intended to attract graduate students and researchers in engineering departments (especially in chemical, mechanical, material, and environmental engineering). The author hopes that the book will also be a helpful source to actively working economists, engineers, and students.

Finally, we like to list other related books that can target a similar audience with this type of content. Elsevier has recently published the following thermodynamically oriented books:

Variational and Extremum Principles in Macroscopic Systems (Ed. by S. Sieniutycz and H. Farkas, Elsevier, Oxford, 2000).
Energy Optimization in Process Systems (by S. Sieniutycz and J. Jeżowski, Elsevier, Oxford, 2009).
Energy Optimization in Process Systems and Fuel Cells (by S. Sieniutycz and J. Jeżowski, Elsevier, Oxford, 2013 (2nd ed.)).
Thermodynamic Approaches in Engineering Systems (by S. Sieniutycz, Elsevier, Oxford, 2016).

Optimizing Thermal, Chemical and Environmental Systems (by S. Sieniutycz and Z. Szwast, Elsevier, Oxford 2018).
Energy Optimization in Process Systems and Fuel Cells (by S. Sieniutycz and J. Jeżowski, Elsevier, Oxford, 2018 (3rd ed.)).

Reading this book on *Complexity and Complex Thermoeconomic Systems* will provide the opportunity for treating all the above-mentioned books because of their unity of teaching style as a teaching cluster.

Our future plans involve the publication of the new title: *Complexity and Complex Chemoelectric Systems*, the manuscript submission predicted in December 2021.

Acknowledgments

The author started his research and collecting scientific materials on system theory and complexity principles during his 2001 stay at the Chemistry Department of The University of Chicago and then while lecturing on a course on the systems theory for students of Faculty of Chemical Engineering at the Warsaw University of Technology (Warsaw TU). A part of suitable materials was obtained in the framework of two national grants, namely Grant 3 T09C 02426 from the Polish Committee of National Research (KBN) and the Hungarian OTKA Grant T-42708, the latter in cooperation with Henrik Farkas of the Department of Physics at the Budapest University of Technology and Economics. In preparing this volume, the author received help and guidance from Marek Berezowski (Faculty of Engineering and Chemical Technology, Cracow University of Technology), Andrzej Ziębik, (Silesian University of Technology, Gliwice), Andrzej B. Jarzębski (Institute of Chemical Engineering of Polish Academy of Science and Faculty of Chemistry at the Silesian TU, Gliwice), Elżbieta Sieniutycz (University of Warsaw), Lingen Chen (Naval University of Engineering, Wuhan, PR China), Zbigniew Szwast, and Piotr Kuran (Faculty of Chemical and Process Engineering at the Warsaw University of Technology). The author is also sincerely grateful to Piotr Juszczyk (Warsaw TU) for his careful and creative work in making all necessary artworks for this book. An important part of preparing any book is the process of reviewing, thus the author is very much obliged to all researchers who patiently helped him to read through subsequent chapters and who made valuable suggestions. The author, furthermore, owes a debt of gratitude to his students who participated and listened to his lectures on systems engineering in the period 2010–16. Finally, appreciations also go to Anita Koch, Elsevier's Acquisition Editor, and the whole book's production team in Elsevier for their cooperation, help, patience, and courtesy.

CHAPTER 1

Systems science vs cybernetics

1.1 General systems theory

General systems theory (GST) is a science investigating general laws for arbitrarily complex arrangements—"systems"—which constitute functional integrities. The origin of the general systems theory is associated with the publication in 1928 of a seminal book (Von Bertalanffy, 1928, 1968) titled *Kritische Theorie der Formbildung*, authored by eminent Austrian biologist and philosopher Ludwig von Bertalanffy (1901–71).

General systems theory (GST) is linked with cybernetics and information theory. Rigorous (axiomatic) mathematical formulation for GST was provided by Mesarović and his collaborators (Mesarović, 1964; Mesarović and Macko, 1970; Mesarović and Takahara, 1975, 1988), whereas important contributions to developments of systems theory were made by G.J. Klir in the field of computer methods of solving various systemic problems (Klir, 1969, 1972, 1981, 1985, 1992; Klir and Folger, 1987) and J.G. Miller in the field of living systems theory (Miller, 1978).

The most typical ingredients (components) of the systems theory include: basic definitions, system thinking, system topologies, life cycles, system performance, conceptual design, current state evaluations, related sciences, solving methods, creative solutions, system synthesis, system analysis, optimization, solution assessment, virtual optimizing, system engineering, and evaluation of knowledge in economy and society. Contents of university courses usually involve: the significance of proper modeling and the role of identification and control aspects for systems of various structure. These issues are analyzed and the importance of computers in applications is evaluated.

1.2 Cybernetics and systems science in academics

The basic concepts of cybernetics have proven to be very powerful in a variety of disciplines: computer science, management, sociology, biology, thermodynamics, etc. Cybernetics and systems science link the abstraction of philosophy and mathematics with the concreteness of (dealing with) the theory and modeling of real world systems. As exemplarily interdisciplinary sciences, cybernetics and systems science work between and among rudimentary disciplines, usually pairwise, but occasionally across more than two kinds of systems.

Complexity and Complex Thermo-Economic Systems. https://doi.org/10.1016/B978-0-12-818594-0.00001-5

Some recent approaches have their origins in ideas and concepts proposed by scientists many years ago: for example, neural networks, artificial intelligence, human-machine interfaces, organized therapies, etc. In these approaches, the majority of basic concepts and problems have already been formulated, mainly in the period of years 1940–80, by cyberneticians such as Von Bertalanffy (1928, 1968), Wiener (1948), Ashby (1952), Boulding (1978), Forrester (1961), Von Foerster (1979), Von Neumann (1958), McCulloch (1965), and Pattee (1973).

Since their founding, cybernetics and systems science have struggled to achieve a level of respectability in the academic society. Cyberneticians especially have found it difficult to find their homes in academic institutions or to create their own. Presently, a number, perhaps still insufficient, of academic programs in cybernetics and systems science (CSS) exist, and those working in the new disciplines described seem not to always remember their cybernetic predecessors.

What is the reason that the progress in the popularity of cybernetics is so slow? The difference between cyberneticians and researchers in the previously mentioned areas is that "*the former stubbornly stick to their objective of building general, domain independent theories, whereas the latter focus on very specific applications, such as: expert systems, psychotherapy, thermodynamics, pattern recognition, etc. General integration of the former researchers is quite abstract, hence it is not sufficiently easy to be really appreciated*" (Joslyn and Heylighen, 1992).

As a common interdisciplinary field, cybernetics and systems science (CSS) offers common concepts used in multiple traditional disciplines and attempts to achieve a logically consistent unification by finding common terms for similar concepts in these multiple disciplines. Thus, CSS unifies individual concepts, theories, and terminologies in a specific discipline directed toward general, and perhaps idiosyncratic tools (usages). These new conceptual categories may not be recognizable to the traditional researchers, or they may find no reason in the use of the general concepts (Joslyn and Heylighen, 1992).

Clearly, the problem of building a global theory is much more complex than any of the more down-to-earth goals of the fashionable approaches (Joslyn and Heylighen, 1992). But we may also say that the generality of the problem is dangerous in itself if it leads to being "stuck" in abstractions which are so far removed from everyday world that it is difficult to use them, interact with them, or test them on concrete problems; in other words, "to get a feel for how they behave and what their strengths and weaknesses are" (Joslyn and Heylighen, 1992).

Although there are many exceptions, researchers in cybernetics and systems science tend to be trained in a traditional specialty (like biology, management, or psychology) and then come to apply themselves to problems in other areas, perhaps a single other area. Thus, their exposure to cybernetics and systems science concepts and theory tends to be somewhat *ad hoc* and specific to the two or three fields they apply themselves to (Joslyn and Heylighen, 1992). However, this

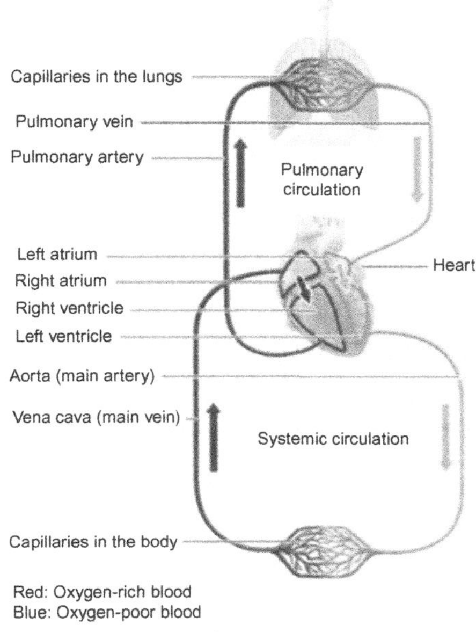

Fig. 1.1
The human cardiovascular system.

doesn't mean that the tremendous research record and great achievements of last years (especially those in biology after the discovery of DNA) are not appreciated enough. Rather, a much longer period is expected for inclusion of basic biological and biochemical systems into the general scheme of cybernetics and systems science (Voet and Voet, 1995; Menche, 2016; Schmidt et al., 2017). Fig. 1.1, which schematizes the human cardiovascular system, is an example of a very large number of complex biological subsystems found in living organisms.

The blood circulatory system (cardiovascular system) consists of the heart and the blood vessels running through the entire body. The arteries carry blood away from the heart; the veins carry it back to the heart. There isn't only one blood circulatory system in the human body, but two, which are connected: The systemic circulation provides organs, tissues, and cells with blood so that they get oxygen and other vital substances. The pulmonary circulation is where the fresh oxygen we breathe in enters the blood. At the same time, carbon dioxide is released from the blood (Menche, 2016; Schmidt et al., 2017).

1.3 Relation to other disciplines

Ideas related to the domain of cybernetics and systems are firstly used in the emerging "sciences of complexity," also called "complex adaptive systems," studying selforganization and heterogeneous networks of interacting entities (e.g., the work of the Santa Fe Institute,

Joslyn and Heylighen, 1992), and, secondly, in an associated research in the natural sciences such as far-from equilibrium thermodynamics, stability theory, catastrophe theory, chaos, and dynamical systems. An example is a bifurcation diagram of the stability theory, Fig. 1.2, which shows how the change of the control parameter C influences the change of the stability properties of the system.

A third strand is constituted by different high-level computing applications such as artificial intelligence, neural networks, man-machine interaction, and computer modeling and simulation.

Unfortunately, few practitioners in these recent disciplines seem to be aware that many of their concepts and methods have been proposed and used by cyberneticians for many years (Joslyn and Heylighen, 1992). Subjects like complexity, selforganization, connectionism, and

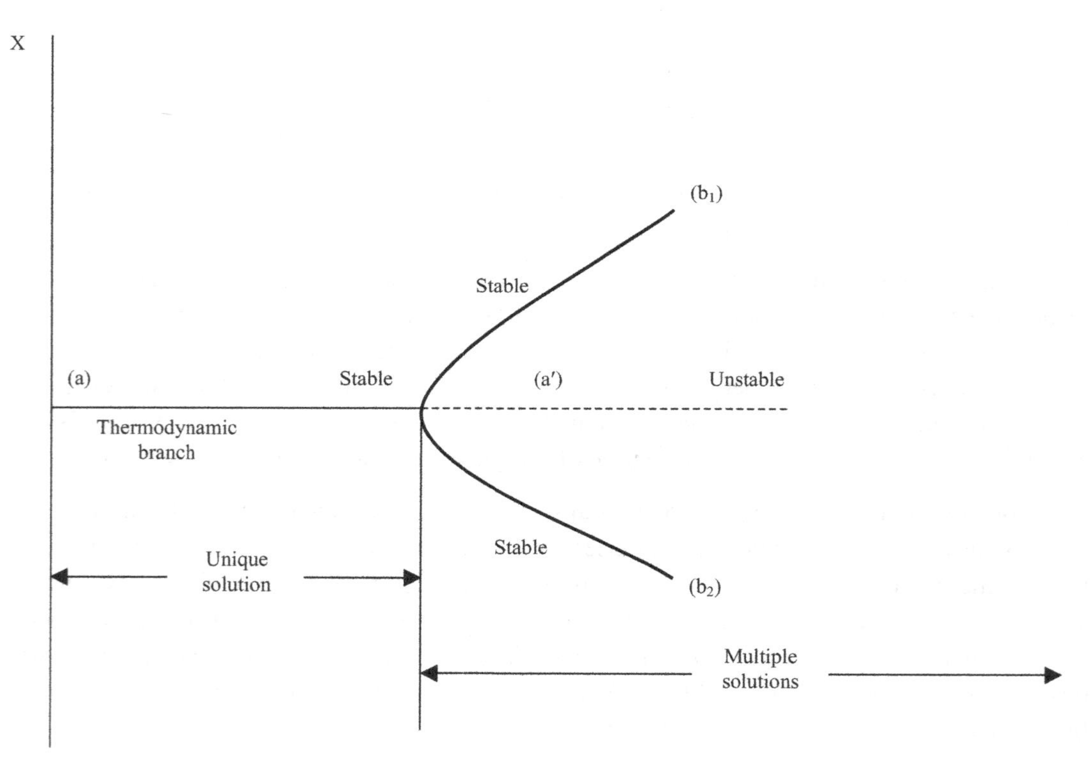

Fig. 1.2
Bifurcation diagram showing how the state variable X is affected when an arbitrary control parameter varies along the horizontal direction: the unique solution (a), the thermodynamic branch loses its stability at the point where the curvilinear boundary line touches the vertical straight line separating a unique solution and multiple solutions. At this value of control parameter, new branches of solutions (b_1, b_2), which are stable as implied by this example, are generated. *Based on Nicolis, G., Prigogine, I., 1989. Exploring Complexity. W.H. Freeman. p. 72.*

adaptive systems have already been extensively studied, in the 1940s and 1950s, by researchers like Wiener, Ashby, von Neumann, and von Foerster, and in discussion forums like the famous Josiah Macy meetings on cybernetics (Heims, 1991). Some popularizing books on "the sciences of complexity," e.g., Waldroop (1993), seem to ignore the fact, creating the false impression that work on complex adaptive systems only started in earnest with the creation of the Santa Fe Institute in the 1980s (Joslyn and Heylighen, 1992).

1.4 What are cybernetics and systems science?

Cybernetics and systems science (also "general system theory" or "system's research") constitutes a somewhat fuzzily defined academic domain which touches upon virtually all traditional disciplines, from mathematics, technology, and biology to philosophy and social sciences. It is more specifically related to the recently developing "sciences of complexity," including artificial intelligence (AI), neural networks (NN), dynamical systems, chaos, and complex adaptive systems. The history of "sciences of complexity" dates back to the 1940s and 1950s when thinkers such as Wiener, von Bertalanffy, Ashby, and von Foerster founded the domain through a series of interdisciplinary meetings.

Systems theory or system science argues that however complex or diverse the world that we experience is, we will always find different types of organizations in it, and such organizations can be described by concepts and principles which are *independent from the specific domain* at which we are looking. Hence, if we were to uncover these general laws, we would be able to analyze and solve problems in any domain, pertaining to any type of system. The system approach distinguishes itself from the more traditional analytical approach by emphasizing the interactions and connectedness of different components of a system.

Although the system approach in principle considers all types of systems, in practice, it focuses on more complex adaptive, self-regulating systems which we might call *cybernetic* (Heylighen et al., 1999).

Many concepts used by system scientists come from closely related approach of cybernetics: information, control, feedback, communication, etc. Cybernetics, derived from the Greek word for steersman (*kybernetes*), was first introduced by the mathematician Wiener, as the science of communication and control in the animal and the machine (to which we now might add: in social society and in individual human beings). It grew out of Shannon's information theory, which was designed to optimize the transmission of information through communication channels and the feedback concept used in engineering control systems. See various schemes of a communication channels in Shannon and Weaver (1949) or other books on the information theory. In the present incarnation of "second order cybernetics," the emphasis is on how observers construct models of the systems with which they interact (c.f. constructivism).

In fact, cybernetics and systems theory study essentially the same problem. The problem is organization independent of the substrate in which it is embodied. Insofar as it is meaningful to make a distinction, systems theory is focused more on the structure of systems and their models, whereas cybernetics is focused on how systems function, that is to say how they control their actions and how they communicate with other systems or with their own components. Since structure and function of a system cannot be understood in separation, it is clear that cybernetics and systems theory should be viewed as two facets of a single approach (Heylighen et al., 1999).

This insight has resulted in the two domains almost merging in practice: many, if not most, of the central associations, journals, and conferences in the field include both terms, systems and cybernetics, in their title. In spite of the absence of strict subdivisions, though, the field is rather fragmented with many different approaches, similar in some respects, different in others, existing side-by-side (Heylighen et al., 1999).

Many "schools," such as autopoietic systems, anticipatory systems, living systems, viable systems, or soft systems, are associated with a particular theorist or thinker, respectively Maturana and Varela (1987), Rosen (1970), Miller (1978), Beer (1975), Turchin (1997). As a result, occasionally, the cybernetics and systems domain still lacks clear foundations. Yet, the authors of the Principia Cybernetica Project (Heylighen et al., 1999), believe that "the commonalities are much larger than the differences," and, therefore, it is worthwhile attempting to integrate different approaches in a common conceptual network.

While the present chapter cites many statements discussed in the text of Heylighen et al. (1999), some important issues contained therein were omitted for brevity. This pertains, in particular, to information about outside internet links. Some very good, readable introductory books on cybernetics and systems can be downloaded from the Principia Cybernetica Library (Heylighen et al., 1999). Together with a dictionary and bibliography of basic books and papers, this should be sufficient for at least an introductory course in the domain.

Optimization of energy systems is covered in three editions of the book written by the present author with Jacek Jeżowski (Sieniutycz and Jeżowski, 2009, 2013, 2018); for a broader perspective, a comprehensive review of various thermodynamic approaches to practical systems should be helpful (Sieniutycz, 2016). Occasionally useful are sources and links to diverse articles, books, and websites, which provide further information and references. In particular, diverse sources are essential for understanding basic problems of selforganization.

Summing up, in this introduction, typical ingredients of the systems theory are specified and its link with cybernetics is pointed out. General systems theory (GST) is defined as a science investigating general laws for topologically complicated arrangements—"systems"—which constitute the functional integrities. This introductory chapter provides the reader with both a historical outline and brief state-of-art of the GST a well as the list of the scientists who contributed to the birth of the field and to the development of its mathematical backgrounds (in particular to the development of computer methods and indication of their applicative potential).

References

Ashby, R., 1952. Design for a Brain. Wiley, New York.
Beer, S., 1975. Platform for Change. Wiley, Chichester.
Boulding, K., 1978. Ecodynamics. Sage, Beverly Hills.
Forrester, J.W., 1961. Industrial Dynamics. Appendix E. MIT Press, Cambridge, MA.
Heims, S., 1991. The Cybernetics Group. MIT Press, Cambridge, MA.
Heylighen, F., Joslyn, C., Turchin, 1999. What are Cybernetics and Systems Science? http://pespmc1.vub.ac.be/CYBSWHAT.html.
Joslyn, C., Heylighen, F., 1992. Cybernetics and Systems Science in Academics. http://pespmc1.vub.ac.be/CYBSACAD.html.
Klir, G., 1969. An Approach to General Systems Theory. Van Nostrand, New York.
Klir, G. (Ed.), 1972. Trends in General Systems Theory. Wiley, New York.
Klir, G., 1981. Special issue on reconstructibility analysis. Int. J. Gen. Syst. 7 (1), 1–107.
Klir, G., 1985. Architecture of Systems Problem Solving. Plenum, New York.
Klir, G., 1992. Facets of Systems Science. Plenum, New York.
Klir, G., Folger, T., 1987. Fuzzy Sets, Uncertainty, and Information. Prentice Hall, Englewood Cliffs.
Maturana, H.R., Varela, F., 1987. The Tree of Knowledge—The Biological Roots of Human Understanding. Shambala Publications, Boston.
McCulloch, W., 1965. Embodiments of Mind. MIT Press, Cambridge, MA.
Menche, N., 2016. Biologie Anatomie Physiologie. Urban und Fischer, Munich (refer to Fig. 1.1).
Mesarović, M.D., 1964. Views of General Systems Theory. Wiley, New York.
Mesarović, M.D., Macko, D., 1970. Theory of Hierarchical Multi-Level Systems. Academic Press, New York.
Mesarović, M.D., Takahara, Y., 1975. General Systems Theory: Mathematical Foundations. Academic Press, New York.
Mesarović, M.D., Takahara, Y., 1988. Abstract Systems Theory. Springer-Verlag, Berlin.
Miller, J.G., 1978. Living Systems. McGraw Hill, New York.
Pattee, H. (Ed.), 1973. Hierarchy Theory. George Braziller, New York.
Rosen, R., 1970. Dynamical Systems Theory in Biology. Wiley, New York.
Schmidt, R., Lang, F., Heckmann, M., 2017. Physiologie des Menschen: Mit Pathophysiologie. Springer, Berlin (refer to Fig. 1.1).
Shannon, C.E., Weaver, W., 1949. The Mathematical Theory of Communication. The University of Illinois Press, Urbana.
Sieniutycz, S., 2016. Thermodynamic Approaches in Engineering Systems. Elsevier, Oxford.
Sieniutycz, S., Jeżowski, J., 2009. Energy Optimization in Process Systems, first ed. Elsevier, Oxford.
Sieniutycz, S., Jeżowski, J., 2013. Energy Optimization in Process Systems and Fuel cells, second ed. Elsevier, Oxford.
Sieniutycz, S., Jeżowski, J., 2018. Energy Optimization in Process Systems, third ed. Elsevier, Oxford.
Turchin, V., 1997. Phenomenon of Science. Columbia University, New York.
Voet, D., Voet, J.G., 1995. Biochemistry, second ed. Wiley, New York.
Von Bertalanffy, L., 1928. Kritische Theorie der Formbildung. Gebrüder Borntraeger, Berlin.
Von Bertalanffy, L., 1968. General Systems Theory. George Braziller, New York.
Von Foerster, H., 1979. In: Kripendorf, K. (Ed.), Cybernetics of Cybernetics. Gordon and Breach, New York.
Von Neumann, J., 1958. Computer and the Brain. Yale University, New Haven.
Waldroop, M.M., 1993. Complexity: The Emerging Science at the Edge of Order and Chaos. (A Popularizing Book on "The Sciences of Complexity"). Viking Books, London.
Wiener, N., 1948. Cybernetics or Control and Communication in the Animal and Machine. Massachusetts Institute of Technology Press, Cambridge, MA.

Further reading

Mittleton-Kelly, E., 2003. Ten principles of complexity. In: Complex Systems and Evolutionary Perspectives on Organizations. Elsevier, Oxford, ISBN: 0-08-043957-8.
Nicolis, G., Prigogine, I., 1989. Exploring Complexity. W.H. Freeman. p. 72.
Pschyrembel, 2017. Klinisches Wörterbuch. De Gruyter, Berlin.

CHAPTER 2

Selforganization and complexity

2.1 What is selforganization

Selforganization is a particular process in which the organization of a system, or its susceptibility to a constraining action, spontaneously increases, and this increase follows without control of any external forces of the environment. One may regard selforganization as a sort of evolution process in which the effect of the external world (environment) is so small that it may be ignored. The development of new, complex structures takes place primarily in and through the system itself. In the text below, the following problems which accompany selforganization are considered: definition of selforganization, significance of selforganization and complexity in the natural sciences, selforganization principle as a possible fourth law of thermodynamics, the role of selforganization in the emergence and the architecture of complexity, and a possible link with maximum power or energy dissipation principles in thermodynamics. Currently, due to the effective use of computers, it is possible to obtain many numerical solutions to complex selforganization problems. However, in order to ensure explicit selforganization effects, a suitable process requires subtle assumptions for rate functions and constraining sets. Nondifferentiable solutions to the equations governing selforganization may occur. In effect, selforganization can be understood on the basis of the same variation and natural selection processes as other, environmentally propelled evolution processes. Yet, selforganization of a complex system often represents "holistic dynamics" (Cross and Hohenberg, 1993; Heylighen, 1997; Cempel and Natke, 1996; Cempel, 2008). Rayleigh-Benard convection provides a simple, classical example of selforganization.

The increase of selforganization can be measured more objectively by a decrease of the system's statistical entropy. This property can be briefly summed up by the so-called Principle of Asymmetric Transitions (Heylighen, 1992), which states as follows: A transition from an unstable configuration to a stable one is possible, but the converse is not. This principle implies a fundamental asymmetry in evolution: one direction of change (from unstable to stable) is more likely than the opposite direction. The generalized, "continuous" version of the principle is the following: The probability of transition from a less stable configuration A to a more stable one B is larger than the probability for the inverse transition: $P(A \rightarrow B) > P(B \rightarrow A)$ (holds under the condition $P(A \rightarrow B) \neq 0$).

Complexity and Complex Thermo-Economic Systems. https://doi.org/10.1016/B978-0-12-818594-0.00002-7

A similar principle was proposed by Ashby (1962) in his "principles of self-organizing systems": "We start with the fact that systems in general go to equilibrium. Now, most of a system's states are nonequilibria. So in going from any state to one of the equilibria, the system is going from a larger to a smaller number of states. In this way, it is performing a selection, in the purely objective sense that it rejects some states, by leaving them, and retains some other state, by sticking to it."

Selforganization is usually triggered by small perturbations accompanying internal variation processes, usually called "fluctuations" or "noise." The phenomenon of these processes yielding a selective retained ordered configuration is called the "order from noise" principle by von Foerster (1949, 1958, 1961, 2002), von Foerster and Pask (1961), von Foerster et al. (1960), and by Prigogine (1947, 1955) and Glansdorff and Prigogine (1971) the "order through fluctuations mechanism." Both these terms may be regarded as special cases of what Heylighen proposes to call the *principle of selective variety* (Heylighen, 1997).

This reduction in the number of reachable states signifies that the variety, and hence the statistical entropy, of the system diminishes. It is because of this increase in neg-entropy or organization that Ashby (1962) calls the process selforganization. But how does this fit in with the 2nd law of thermodynamics which states that entropy in closed systems cannot decrease? The easy way out is to conclude that such a self-organizing system cannot be closed and must lose entropy to its environment (von Foerster et al., 1960).

A deeper understanding can be reached by going back from the statistical definition of entropy to the thermodynamic one, in terms of energy or heat. Energy is defined as the capacity to do work, and working means making changes, that is to say exerting variation. Hence, energy input can be viewed as potential variation. A stable configuration is not subject to variation. In order to destroy a stable equilibrium, you need to add energy, and the more stable the configuration, the more energy you will need. Therefore, stability is traditionally equated with minimal energy.

The 1st law of thermodynamics states that energy is conserved. A simple interpretation of that law would conclude that the principle of asymmetric transitions cannot be valid, since it postulates a transition from an unstable (high energy) to a stable (low energy) configuration. If energy is absolutely conserved, then an unstable configuration can only be followed by another unstable configuration. This is the picture used in classical mechanics, where evolution is reversible, that is to say symmetric. Incidentally, this shows that the principle of asymmetric transitions is not tautological—though it may appear self-evident—since a perfectly consistent theory (classical mechanics) can be built on its negation (Heylighen, 1997).

This principle is again equivalent to an increase in redundancy, information, or constraint: after the process is self-organized, there is less ambiguity about which state the system is in. A self-organizing system, which also decreases its thermodynamic entropy, must necessarily (because of the second law of thermodynamics) export ("dissipate") such entropy to its surroundings, e.g., to the ambient, as explained by von Foerster (2002) and Prigogine and others. Prigogine (1947, 1955) and Glansdorff and Prigogine (1971) used the term "dissipative structures" to name systems which continuously export entropy in order to maintain their organization.

2.2 Selforganization and complexity in the natural sciences

According to Heylingen, selforganization is usually associated with nonlinear phenomena of increased complexity rather than with a relatively simple process of structure maintenance by diffusion (Heylighen, 1997). All the intricate phenomena associated with nonlinearity (limit cycles, chaos, sensitivity to initial conditions, dissipative structuration, etc.) can be understood through the interplay of positive and negative feedback cycles in which some variations or perturbations tend to reinforce themselves.

An important strand of work leading to the analysis of complex evolutions is laid down by thermodynamics. Ilya Prigogine received the Nobel Prize for his work in collaboration with other members of "Brussels School" after showing that physical and chemical systems far from thermodynamic equilibrium tend to self-organize by exporting entropy and to form disequilibrium systems called dissipative structures. Both Prigogine's philosophical considerations about the new world view implied by selforganization and irreversible change (Prigogine and Stengers, 1984) and his scientific research (Nicolis and Prigogine, 1977, 1989; Prigogine, 1980) on bifurcations and *order through fluctuations* remain classics, cited in very diverse contexts.

Lotka-Volterra oscillations (Fig. 2.1), Lorentz attractor (Fig. 2.2), and the roll streamlines in Rayleigh-Benard convection (Fig. 2.3), were crucial for the development of the theory of self-organizing systems. These models showed that the competition of dissipation and damping with restoring forces leads to oscillations and limit cycles. The early history and the debate surrounding oscillating reactions can be found in numerous references such as Glansdorff and Prigogine (1971), Nicolis and Portnow (1973), Gray and Scott (1975, 1990), Nicolis and Prigogine (1977, 1979), Nicolis and Rouvas-Nicolis (2007a,b), and many others.

The same basic approach may be used for molecular collisions, hydrodynamics, and chemical or electrochemical processes. Nonlinear molecular mechanism gives rise to critical points, instabilities, bifurcations, oscillations, chaotic behavior, and limit cycles (Figs. 2.1 and 2.2). It leads, in general, to various forms of organization (Ebeling, 1985; Ebeling and Klimontovich, 1984; Klimontovich, 1986, 1991, 1982; Feistel and Ebeling, 1989; Ebeling and Feistel, 1982, 2011).

Feistel and Ebeling (1989) write: One of the most (or even the most at all) striking features in nature is the phenomenon of natural evolution. Since a long time, it has attracted the interest of people and especially of scientists, and up to our days, it has not lost this attractivity. It is impressive to observe how in the course of scientific history the evolution concept influenced

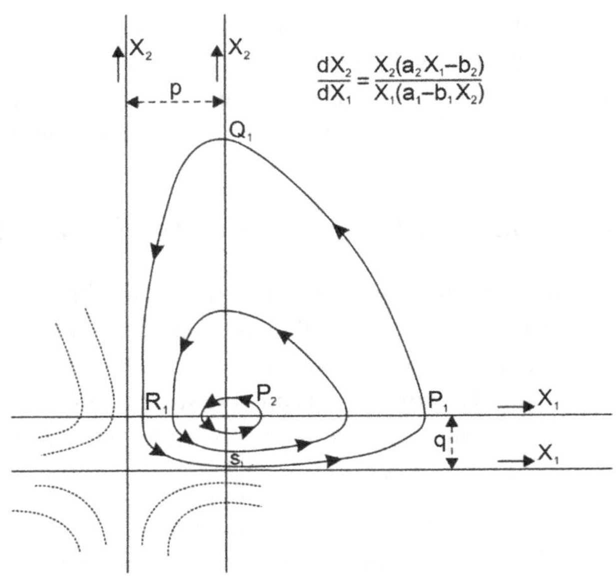

Fig. 2.1

Lotka-Volterra model is a classical scheme describing interactions between populations of predators and preys. The time variability of the populations of predators (foxes) and preys (rabbits) provides the ecological interpretation of the model. The Lotka diagram shows the general character of integral curves of his equation which describes ratio dX_2/dX_1 in terms of X_1 and X_2. In the positive quadrant of X_1X_2, the curves are closed, contained entirely within the quadrant, and intersecting the axes of x_1x_2 orthogonally. Near origin curves are very nearly elliptical (Lotka, 1920).

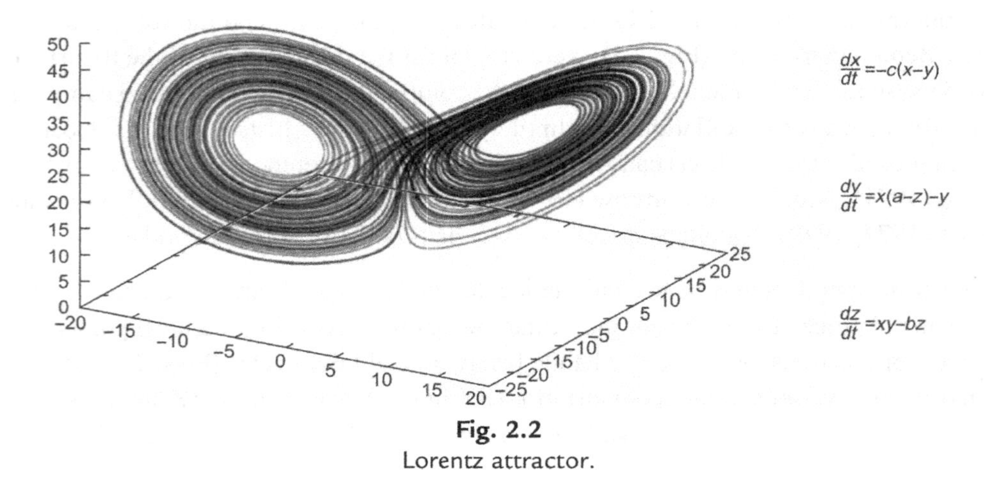

Fig. 2.2
Lorentz attractor.

practically all branches of science, replacing finally the old view of the world as a static (or strictly periodic) building.

The biological world with fixed and unchangeable species could not stand the overwhelming number of facts supporting Darwin's picture of a flexible, continuously changing and adapting

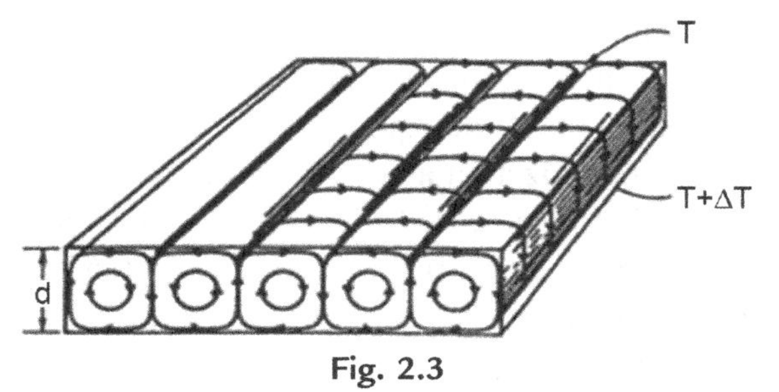

Fig. 2.3
Schematic picture of Rayleigh-Benard convection with streamlines in an ideal roll state. *Based on Cross, M.C., Hohenberg, P.C., 1993. Pattern formation outside of equilibrium. Rev. Mod. Phys. 65(3), 851–1125.*

biosphere under the selection pressure of competition and changing environmental conditions. The crystal sphere of heaven with twinkling stars pinned on it was replaced by an expanding universe filled with evolving galaxies and solar systems. And even the seemingly rigid earth crust with continents and oceans has nowadays to be seen as a complex system of drifting plates with continued exchange of matter between the solid earth surface and the fluid mantle. Interestingly enough, the most fundamental natural science, physics, dealing with the first principles that govern the motion of the world, was for a long time untouched by the evolution ideas that occupied many other sciences. This situation is changing now. It was a breakthrough to understand that evolving systems are essentially systems being far from thermodynamic equilibrium (Glansdorff and Prigogine, 1971; Haken, 1984; Feistel and Ebeling, 1989) and that, under these conditions, the second law of thermodynamics not at all states the destruction of any existing structure.

Ebeling (1983, 1985) corroborates by experiments the theory of hydrodynamic turbulence of Klimontovich (1982), which yields effective turbulent viscosity as a linear function of the Reynolds number. This theory seems to be able to describe the observed effective viscosity of vortices and, after modifications, the fluid flow in tubes up to very high Reynolds numbers.

Klimontovich (1986, 1991) evaluates the entropies of laminar and turbulent motions with respect to the velocity of the laminar flow and the velocity of the averaged turbulent flow, for the same values of the average kinetic energies. He shows that, when transition occurs from laminar flow to turbulent flow, the entropy production and entropy itself decrease. This indicates that the disequilibrium phase transition from laminar flow to turbulent flow transfers the system to a more ordered state. He also proves that the turbulent motion should be thought as having a lower temperature than laminar motion. Applying his theory of turbulence, Klimontovich (1991) examines the link between the turbulent motion and the structure of chaos.

Klimontovich (1999) considers entropy, information, and relative criteria of order for states of open systems. He points out that two definitions of the "information" concept are known in the theory of communication. The first definition coincides in form with the Boltzmann entropy.

The second one refers to the difference between unconditional and conditional entropies. Klimontovich investigates two kinds of open systems. The systems of the first kind, with zero value of a controlling parameter, are in the equilibrium state. Those of the second kind are disequilibrium systems. In selforganization associated with the process of escaping from the equilibrium state, the information is increased. For open systems and all values of the controlling parameter, a conservation law is proved for the sum of information and entropy with all values of controlling parameter. Klimontovich's (1999) treatment is carried out on a number of (classical and quantum) examples of physical systems. An example of a medical-biological system is also introduced.

Ebeling (1983, 1985) compares with experiments the theory of hydrodynamic turbulence of Klimontovich (1982), which yields effective turbulent viscosity as a linear function of the Reynolds number. This theory describes the observed effective viscosity of vortices and, after modifications, the fluid flow in tubes up to very high Reynolds numbers.

Ebeling (1985, 2002) studies a special class of nonequilibrium systems which allows to develop an ensemble theory of some generality. For these so-called canonical-dissipative systems, the driving terms are determined only by invariants of motion. First, Ebeling constructs systems which are ergodic on certain surfaces of the phase plane. These systems may be described by a nonequilibrium microcanonical ensemble, corresponding to an equal distribution on the target surface. Next, he constructs and solves Fokker-Planck equations. In the last part, he discusses a special realization: systems of active Brownian particles. In conclusion, his work is devoted to the study of "canonical" dissipative systems which are pumped from external sources with free energy. Ebeling started his work from the Hamiltonian theory for conservative mechanical systems. In order to extend the known solutions for conservative systems to dissipative systems, he used a general theory of canonical-dissipative systems. He constructed special canonical-dissipative systems whose solution converges to the solution of the conservative system with given energy, or other prescribed invariants of motion. In this way, he was able to generate nonequilibrium states characterized by certain prescribed invariants of mechanical motion. He postulated special distributions which are analogues of the microcanonical ensemble in equilibrium. Further on, he found solutions of the Fokker-Planck equation which may be considered as analogues of the canonical equilibrium ensembles. Ebeling proposed calling these distributions canonical-dissipative ensembles. With the help of explicit nonequilibrium solutions, he was able to construct for canonical-dissipative systems a complete statistical thermodynamics which includes an evolution criterion for the nonequilibrium.

Ebeling and Feistel's (2011) updated book on selforganization and evolution includes a review of more recent literature. It retains the original fascination surrounding thermodynamic systems far from equilibrium, synergetics, and the origin of life. While focusing on the physical and theoretical modeling of natural selection and evolution processes, the book covers in detail experimental and theoretical fundamentals of self-organizing systems as well as such selected features as random processes, structural networks, and multistable systems. The authors take examples from physics, chemistry, biology, and social sciences. The result is a resource suitable for students and scientists in physics and related interdisciplinary fields, including mathematical physics, biophysics, information science, and nanotechnology.

Ideas of Nicolis and Prigogine (1977) diffused quickly to the ecology and soon received its own characteristic picture. Prigogine's name is best known for extending the second law of thermodynamics to systems that are far from equilibrium and implying that new forms of ordered structures could exist under such conditions. Prigogine called these forms "dissipative structures," pointing out that they cannot exist independently of their environment. Nicolis and Prigogine (1977) showed that the formation of dissipative structures allows order to be created from disorder in nonequilibrium systems. These structures have since been used to describe many phenomena of biology and ecology.

Entropy and entropy lowering, in comparison to their equilibrium values, are good indicators of disequilibrium. Ebeling and Engel-Herbert (1986) prove that entropy decreases with excitation under the condition of fixed energy. They also show that entropy lowering refers to the contraction of the occupied part of the phase space due to the formation of attractors. Excitation of oscillations in solids and turbulence in liquid flows serve as examples. The entropy statement formulated by Klimontovich is discussed in the turbulence context. Two examples are given: (1) selfoscillations of nonlinear oscillators, and (2) laminar and turbulent flows in tubes.

Inspired by Prigogine's theories, Erich Jantsh made an ambitious attempt to synthesize everything which was known at the time (Jantsh, 1979) about self-organizing processes, from the Big Bang to the evolution of society, into an encompassing world view. The physicist Herman Haken has suggested in 1978 the label of synergetics for the field which investigates the collective patterns emerging from many interacting components, as they are found in chemical reactions, crystal formations, or lasers. A Nobel laureate, Eigen (1992), has focused on the origins of life, the domain where chemical selforganization meets biological evolution. He introduced the concept of *hypercycle*, which is an autocatalytic cycle of chemical reactions containing other cycles, and of *quasispecies*, the fuzzy distribution of genotypes characterizing a population of quickly mutating organisms or molecules (Eigen and Schuster, 1979).

The modeling of nonlinear systems in physics has led to the concept of *chaos*, the behavior of a deterministic process characterized by extreme sensitivity to the initial conditions of trajectories (Crutchfield et al., 1986). Although chaotic dynamics is not strictly a form of evolution, it is an important aspect of the behavior of complex systems. The science journalist J. Gleick has written a popular history and evolution of the field in the book on cellular automata. Mathematical models of distributed dynamical processes characterized by a discrete space and time have widely been used to study phenomena such as chaos, attractors, and the analogy between dynamics and computation through computer simulation. Wolfram (2017) has made a fundamental classification of the critical behavior of continuous mappings. Catastrophe theory was developed by Thom (1975). His results served, in particular, the purpose of modeling the continuous development of discontinuous forms of organisms, thus extending the much older work by the biologist Thompson (1917).

Another French mathematician, Mandelbrot (1983), has founded the field of fractal geometry, which models the recurrence of similar patterns at different scales characterizing most natural systems. Such self-similar structures exhibit power laws like the famous Zipf's law governing the frequency of words. By studying processes such as avalanches and earthquakes, Bak and Chen (1989a,b) have shown that many complex systems will freely evolve to the critical edge between stability and chaos, where the size of disturbances obeys a power law and large disturbances are less frequent than small ones. The phenomenon, which he called self-organized criticality, may also provide an explanation for the so-called punctuated equilibrium dynamics observed in biological evolution.

2.3 Selforganization: The fourth law of thermodynamics?

The text below follows with small changes the opinion of Charles Gurtin and Drew Kerkhoff expressed in 1996 at the Ecological Complexity Seminar, held in the Department of Biology at the University of New Mexico.

The following criteria for evaluation of new theoretical constructs in ecology are suggested by O'Neill et al. (1986):

1. *The theory must be internally consistent*
2. *The theory must not be adopted simply because of success in other fields*
3. *The theory must agree with known properties of ecosystems*
4. *The theory must be capable of producing new and testable hypotheses*

The above criteria are a basis for the evaluation of "complexity theory" as we have come to understand it over the course of numerous discussions. It may be argued that selforganization lies at the basis of complexity as we have come to interpret it. For example, it is the self-organized order of Bernard cells (Chapter 10) which renders the system complex. It is well-known that, below a critical temperature, the systems exhibit only uniform conduction. The system becomes more attractive when selforganization occurs, and selforganization requires that a system be away from the thermodynamic equilibrium. Morowitz (1992) provides the following intuitive explanation:

> Indeed, there is a statement, sometimes called the fourth law of thermodynamics which states that the flow of energy from a source to a sink through an intermediate system orders that system. Here the word order must be taken as increased complexity.

Bak and Chen (1989a,b) extend this notion of the fourth law by suggestion that "slowly driven systems naturally self-organize into a critical state." This implies that, in disequilibrium systems, criticality and complexity may be a rule rather than an exception (Bak and Chen,

1989a,b). Moreover, O'Neill et al. (1986) anticipate Bak's and Chen's conclusion with their proposal that the hierarchical structure of an ecosystem is a consequence of a fourth law which drives the evolution of an increasingly complex nonequilibrium system from the prebiotic, molecular level to the level of ecosystem. O'Neill et al. (1986) present the above criteria as a challenge to their notion of hierarchical structure in ecosystems, but perhaps these criteria can be even more usefully applied to the construct that underlies the proposed origin of order.

2.4 Emergence and the architecture of complexity

Heylighen (1996, 1997, 2009) argues that the problems of emergence and the architecture of complexity can be solved by analyzing self-organized evolution of complex systems. He proposes a generalized, distributed variation-selection model in which internal and external aspects of selection and variation are contrasted. "Relational closure" is introduced as an internal selection criterion. He outlines a possible application of the theory in the form of a pattern-directed computer system for supporting complex problem-solving.

2.4.1 Emergent selforganization

Emergence is a classical concept in systems theory, where it denotes the principle that the global properties defining higher order systems or "wholes" (e.g., boundaries, organization, control, etc.) cannot be reduced to the properties of the lower order systems or "parts." Such irreducible properties are called emergent. Until now, there is no satisfactory theory explaining what characterizes emergent properties or what are the conditions for their existence. Therefore, Heylighen (1997, 2009) proposes to look at this question not from the traditional static viewpoint, but from a dynamic, evolutionary viewpoint, thus replacing the question "How can a property be emergent?" by "How can a property become emergent? (i.e., how can it emerge?)." This should also lead us to answer the question "Where do *wholes* or systems come from?"

A promising approach to the problem of dynamical emergence is provided by the recently developed models of selforganization. Selforganization may be defined as a spontaneous (i.e., not steered or directed by an external system) process of organization, i.e., of the development of an organized structure. This spontaneous creation of an "organized whole" out of a "disordered" collection of interacting parts, as witnessed in self-organizing system in physics, chemistry, biology, sociology etc., is the basic part of dynamical emergence.

However, another essential characteristic of emergence as understood in systems theory is its hierarchical or multilevel nature: an emergent whole at one level is merely a component of an emergent system at the next higher level. Until now, the most popular paradigms used for explaining selforganization (e.g., attractors, synergetics, catastrophes, etc.) are characterized by a mere two-level structure: the "microscopic" level where a multitude of building blocks or elements (e.g., molecules, individual organisms) interact, and the "macroscopic" level where these interactions lead to certain global patterns of organization, e.g., a dissipative structure or a crystalline symmetry. The resulting systems as studied through these paradigms, e.g., an ideal crystal, a trail of ants carrying food back to the nest, or a regular pattern of fluid rolls in the Benard problem (Fig. 2.3), are usually simpler in their structures and treatments than multilevel hierarchical systems. See Schneider and Kay (1995), and Chapter 10 for more details on Benard phenomenon.

2.4.2 Emergence and the complexity

Realistically complex systems (e.g., organisms, societies, ecologies), however, are characterized by a multilevel structure. A classic explanation for this hierarchical "architecture" of a complex system was given by Simon (1962). His argument is based on a variation-and-selection of natural and artificial evolution: elements are connected and combined by natural interactions, or equivalently, by the trials of a problem solver, thus creating a variety of assemblies. Of these assemblies, only those will survive which are sufficiently stable. The other assemblies will be all apart before they can undergo any further evolution. The stable assemblies, forming "naturally selected wholes," can again function as building blocks to be combined into higher order assemblies, and so the process can be repeated at even higher levels, forming a set of hierarchically structured complexes.

Simon (1962) uses this model in order to show why multilevel systems are more probable to emerge than two-level systems of comparable complexity: in a two-level system, all the components must "fall in place" at once, otherwise the assembly will be unstable and fall apart before the missing components are added by the natural variation mechanism. In a multilevel system, on the other hand, it suffices that small subsets of components would "fall into place" forming stable subassemblies (modules), which can then again be recursively combined in small sets forming higher level modules. Clearly, the smaller the set of elements which must fall into place, the higher the probability that this will happen to be a random combination.

However, Simon (1962) acknowledges that there are exceptions to the rule that nonhierarchical complex systems are highly improbable. For example, most polymers are formed by a very simple linear, two-level assembly of a large number of molecules. One of the important achievements of the present day selforganization models is that they can explain the emergence of such nonmodular, two-level systems, which have a very large number of elements. Such processes are usually characterized by nonlinear autocatalytic mechanisms, whereby the presence of a small stable assembly (whose emergence is quite probable) enhances the probability that other elements will join the assembly, thus making it grow and become even more stable. In the formulation of Haken (1984), a stable mode controls the remaining unstable modes. No intermediate level of modules is needed in such a process with positive feedback. The emergent stable configuration can be thought of as an "attractor," exerting a force on the configurations in its neighborhood, so that the configurations which are close enough to the attractor will automatically move closer and closer toward this stable configuration.

It is clear that both the hierarchical model of Simon and the "nonlinear model of selforganization" only describe part of features of emergence. A real complex system, e.g., the human body, has as well hierarchical, multilevel aspects (e.g., the organelle being a subsystem of the cell, being a subsystem of the organ, being a subsystem of...) as nonlinear, two-level aspects (e.g., the system of blood vessels as a coordinated closed circuit consisting of billions of blood cells).

However, in general, there is not just one global hierarchy or nonlinear organization, but a multitude of inextricably entwined suborganizations and subsystems. If we wish to understand the architecture of such complexity, we will need a more general integrating theory of emergence and selforganization. Heylighen (1996, 2009) proposes some basic principles on which such a theory should be founded.

2.4.3 Toward generalized dynamics of variation-and-selection

The theory of natural selection as it is used to describe biological evolution can be generalized to any kind of systemic evolution. It suffices to consider a system susceptible to variations and an environment exerting a "selective pressure" on the system: only those configurations of the system will maintain (or grow) which are "fit" or adapted to the environment. The evolution system can be likened to a problem solver, generating possible solutions by trial (variation) to a problem posed by the environment: how to be optimally adapted? A problem arises as soon as adaptation is not optimal, i.e., the system is not completely stable or invariant with respect to the environment. The larger the instability, the more serious the problem, and the more variation the system must undergo before it reaches a new equilibrium. It does not suffice to blindly try out possibilities, in the hope that accidentally one of them might prove to be the optimal solution: the chances that this would succeed are very small. You can enhance your chances by looking for intermediate steps, i.e., relatively easy-to-find problem states or configurations, which are not final solutions but which are somehow "closer" to the goal than the configuration you started with. This is what also happens during natural selection. For the continuation of this problem, see Heylighen (1996, 1997, 2009).

2.5 Maximum power vs energy dissipation

The maximum power principle (Lotka, 1922) can be stated as follows: During selforganization system, designs develop and prevail which maximize power intake and energy transformation, and those uses reinforce production and efficiency. This sort of "fourth law of thermodynamics" constraints and guides the selforganization of open systems. Such an approach has been largely advocated by H.T. Odum and his coworkers because it provides a coherent framework for the physical structure of natural systems (Odum, 1971, 1983, 1988).

In contrast, work of Prigogine and his coworkers uses an approach based on dissipative structures, which focuses on the rate of entropy generation. While concepts related to entropy generation have been widely applied to ecological problems (O'Neill et al., 1989), the approach has been assessed as controversial. Primary criticisms involved concern that biological systems are far from equilibrium constructs where either thermodynamic laws don't apply, or they become increasingly uncertain as the system moves away from equilibrium. In contrast, the maximum power approach which integrates kinetics and thermodynamics may have more utility because kinetic forces increase with distance away from equilibrium (Berry, 1995).

The utility of these approaches is that they provide a null-model for addressing biological systems (Turner, 1979). The principles derived from thermal physics are related to complexity by their link with a minimal energy dissipation or a maximum power during driving the system away from equilibrium. It is these increases in complexity which order the system.

It has been shown that the competition of dissipation and damping with restoring forces leads to oscillations and limit cycles. In numerous references, the reader can study the early history and the debate surrounding oscillating reactions (Glansdorff and Prigogine, 1971; Nicolis and Prigogine, 1977, 1979; Nicolis and Rouvas-Nicolis, 2007a,b; and many others). In an original PhD thesis, Zalewski (2005) provides a unified picture of chaos and oscillations in chemical reactors.

Erikson et al. (1987) investigate the role of the transformed information in creation processes by showing how destructive processes of entropy production are related to the creative processes of formation. Zainetdinov (1999) develops an approach which attempts to attribute the informational neg-entropy to selforganization processes in open systems.

2.6 Self-organized heterogeneous systems

Zalewski and Szwast (2007) find self-organized oscillations and chaos in selected biosystems of prey-predator type in presence of deactivating catalysts. To identify and interpret the phenomena, they test physiochemical models of these systems and investigate the dependence of average concentrations on models parameters. In particular, they determine areas of maximal values of rate constants in these models.

Zalewski and Szwast (2008) consider the chaotic behavior in a stirred tank reactor with first-order exothermic chemical reaction ($A \rightarrow B$) running in a mixture with deactivating catalyst particles. The reactor is continuous with respect to reacting mixture and periodic (batch) with respect to catalyst particles. The process yield is determined by the chemical reaction subject to the Arrhenius-type catalyst deactivation. The movement of the areas with self-organized oscillations and chaos is influenced by the catalyst deactivation. This effect is illustrated in Figs. 2.4 and 2.5.

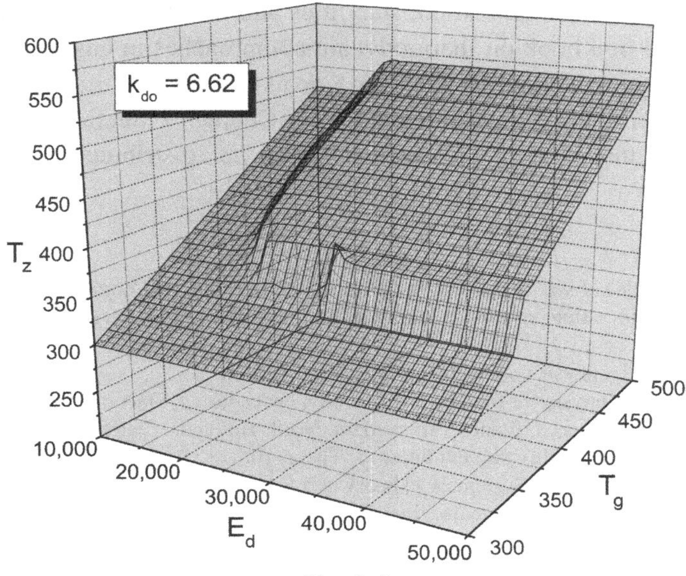

Fig. 2.4
Catalyst's temperature T as a function of the inlet mixture temperature, T_g, and the activation energy of catalyst deactivation E_d, after time $t = 100$ min, for $k_{do} = 6.62$.

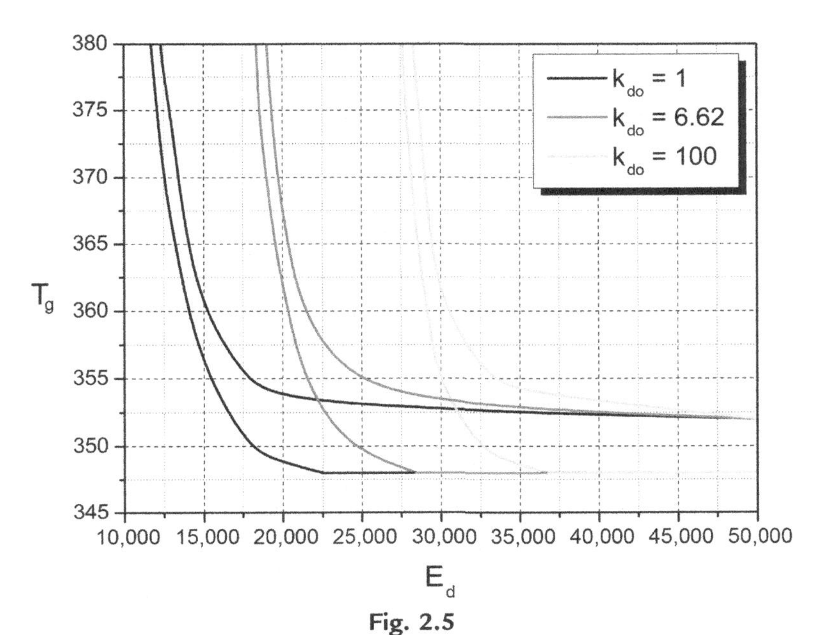

Fig. 2.5
Areas $T_g - E$ with chaotic behavior over them, for various values of k_{do}, after time $t = 100$ min.

More information about selforganization, stabilities and instabilities, and chaos in chemical systems can be found in a book on thermodynamic approaches in engineering systems (Sieniutycz, 2016, Chapters 5 and 9). Chapter 5 deals mainly with Lapunov functions and stability of paths rather than fields, whereas a part of Chapter 9 (Sections 9.7–9.9) provides an account of stabilities, instabilities, and chaos in fields, i.e., distributed chemical systems.

References

Ashby, W.R., 1962. Principles of self-organizing systems. In: von Foerster, H., Zopf Jr., G.W. (Eds.), Principles of Self-Organization: Transactions of the University of Illinois Symposium. Pergamon Press, London, UK, pp. 255–278.

Bak, P., Chen, K., 1989a. The physics of fractals. Physica D 38, 5–12.

Bak, P., Chen, K., 1989b. The physics of fractals. In: Aharony, Feder (Eds.), Fractals in Physics. Proceedings of the International Conference, Vance, France, Amsterdam, North-Holland, (Chapter 2).

Berry, S., 1995. Entropy, irreversibility and evolution. J. Theor. Biol. 175, 197–202.

Cempel, C., 2008. Teoria i Inżynieria Systemów (Theory and Engineering of Systems). vol. 1 (2). Wydawnictwo Naukowe Instytutu Technologii Eksploatacyjnej ITE-PIB, Radom, pp. 1–291.

Cempel, C., Natke, H.G., 1996. Holistic dynamics of systems. J. Syst. Eng. 6 (1), 33–45.

Cross, M.C., Hohenberg, P.C., 1993. Pattern formation outside of equilibrium. Rev. Mod. Phys. 65 (3), 851–1125.

Crutchfield, J.P., Doyne Farmer, J., Packard, N.H., Shaw, R.S., 1986. Chaos. Sci. Am. 54 (12), 46–57.

Ebeling, W., 1983. Discussion of the Klimontovich theory of hydrodynamic turbulence. Ann. Phys. 40 (1), 25–33.

Ebeling, W., 1985. Thermodynamics of self-organization and evolution. Biomed. Biochim. Acta 44, 831–838.

Ebeling, W., Engel-Herbert, H., 1986. Entropy lowering and attractors in phase space. Acta Phys. Hungarica 66 (1–4), 339–348.

Ebeling, W., Feistel, R., 1982. Physik der Selbstorganisation und Evolution. Akademie-Verlag, Berlin.

Ebeling, W., Feistel, R., 2011. Physics of Self-Organization and Evolution. Wiley-VCH, Berlin, ISBN: 978-3-527-40963-1.

Ebeling, W., Klimontovich, Y.L., 1984. Self-organization and Turbulence in Liquids. Teubner Verlagsgesellschaft, Leipzig.

Eigen, M., 1992. Steps Towards Life: A Perspective on Evolution. Oxford University Press, Oxford.

Eigen, M., Schuster, P., 1979. The Hypercycle: A Principle of Natural Self-Organization. Springer-Verlag, Berlin.

Erikson, K.-E., Lindgren, K., Månsson, B.A., 1987. Structure, Context, Complexity, Organization. World Scientific, Singapore.

Feistel, R., Ebeling, W., 1989. Evolution of Complex Systems: Self-Organization, Entropy and Development. vol. 30. Kluwer Academic Publishers/Springer, Wiley-VCH, Dordrecht/Berlin, ISBN: 90-277-2666-3.

Glansdorff, P., Prigogine, I., 1971. Thermodynamic Theory of Structure Stability and Fluctuations. Wiley, New York.

Gray, P., Scott, O.K., 1975. Kinetics of oscillatory reactions. In: Reaction Kinetics. The Chemical Society, London, Burlington House (Chapter 8).

Gray, P., Scott, O.K., 1990. Chemical Oscillations and Instabilities. Clarendon Press, Oxford.

Haken, H., 1984. Advanced Synergetics. Springer-Verlag, Berlin.

Heylighen, F., 1992. Principles of systems and cybernetics: an evolutionary perspective. In: Trappl, R. (Ed.), Cybernetics and Systems '92. World Science, Singapore, pp. 3–10.

Heylighen, F., 1996. Self-Organization and Complexity in the Natural Sciences. Principia Cybernetica Web.http://pespmc1.vub.ac.be/COMPNATS.html.

Heylighen, F., 1997. Self-Organization. Principia Cybernetica Web.http://pespmc1.vub.ac.be/SELFORG.html.

Heylighen, F., 2009. Self-organization. Emergence and the architecture of complexity. ftp://ftp.vub.ac.be/pubprojects/Principia_Cybernetica/P.../Self-Organization_Complexity.tx.

Jantsh, E., 1979. Gauthier Lectures in Systems Science at Berkeley. www.eocht.info/page/ErichJantsh.
Klimontovich, Y.L., 1982. Kinetic Theory of Nonideal Gases and Nonideal Plasmas. Pergamon, Oxford.
Klimontovich, Y.L., 1986. Statistical Physics. Harwood Academic Publishers, Chur.
Klimontovich, Y.L., 1991. Turbulent Motion and the Structure of Chaos. Kluwer Academics, Dordrecht.
Klimontovich, Y.L., 1999. Entropy, information, and criteria of order in open systems. Nonlinear Phenomena Complex Syst. 2 (4), 1–25.
Lotka, A.J., 1920. Undamped oscillations derived from the law of mass action. J. Am. Chem. Soc. 42, 1595–1599.
Lotka, A.J., 1922. Contribution to the energetics of evolution. Proc. Natl. Acad. Sci. U. S. A. 8, 147–151.
Mandelbrot, B.B., 1983. The Fractal Geometry of Nature. Henry Holt and Co., New York, NY.
Morowitz, H., 1992. The Thermodynamics of Pizza: Essays on Science and Everyday Life Paperback – January 1, 1992.
Nicolis, G., Portnow, J.I., 1973. Chemical oscillations. Chem. Rev. 73 (4), 365–384.
Nicolis, G., Prigogine, I., 1977. Self-Organization in Nonequilibrium Systems. Wiley, New York.
Nicolis, G., Prigogine, I., 1979. Irreversible processes at nonequilibrium steady states and Lyapounov functions. Proc. Natl. Acad. Sci. U. S. A. 76 (12), 6060–6061.
Nicolis, G., Prigogine, I., 1989. Exploring Complexity: An Introduction. W.H. Freeman, New York, NY. ISBN: 0-7167-1859-6.
Nicolis, G., Rouvas-Nicolis, C., 2007a. Foundations of Complex Systems. World Scientific, Singapore.
Nicolis, G., Rouvas-Nicolis, C., 2007b. Complex systems. Scholarpedia 2 (11), 1473.
O'Neill, R.V., DeAnzelis, D.L., Waide, J.B., Allen, T.F.H., 1986. Hierarchical structure as the consequence of evolution in open, dissipative systems. In: O'Neill, R.V., DeAnzelis, D.L., Waide, J.B., Allen, T.F.H. (Eds.), A Hierarchical Concept of Ecosystems. Princeton University Press, Princeton, USA, p. 1986.
O'Neill, R.V., Johnson, A.R., King, A.W., 1989. A hierarchical framework for the analysis of scale. Landsc. Ecol. 3, 193–206.
Odum, H.T., 1971. Environment, Power, and Society. Wiley, New York.
Odum, H.T., 1983. Systems Ecology. Wiley, New York.
Odum, H.T., 1988. Self-organization, transformity, and information. Science 242, 1132–1139.
Prigogine, I., 1947. Etude Thermodynamique des Phenomenes Irreversibles. Liege Desoer.
Prigogine, I., 1955. Thermodynamics of Irreversible Processes. John Wiley, New York.
Prigogine, I., 1980. From Being To Becoming. Freeman, New York ISBN: 0-7167-1107-9.
Prigogine, I., Stengers, I., 1984. Order out of Chaos. University of Michigan: Bantam Books, Ann Arbor, ISBN: 0-00-654115-1. Published also by Flamingo Books, Temecula, CA, USA.
Schneider, E.D., Kay, J.J., 1995. Order from disorder: the thermodynamics of complexity in biology. In: Murphy, M.P., O'Neill, L.A.J. (Eds.), What Is Life: The Next Fifty Years. Reflections on the Future of Biology. Cambridge University Press, Cambridge, UK, pp. 161–172.
Sieniutycz, S., 2016. Thermodynamic Approaches in Engineering Systems. Elsevier, Oxford.
Simon, H.A., 1962. The architecture of complexity. Proc. Am. Philos. Soc. 106 (6), 467–482.
Thom, R., 1975. Structural Stability and Morphogenesis. English Translation of 1973 Stabilite Structuree et Morphogenese, Translated by D. H. Fowler. W. A. Benjamin, Inc., 1976, Reading, MA.
Thompson, D.W., 1917. On Growth and Form. http://www.darcythompson.org/index.html.
Turner, J.S., 1979. Nonequilibrium thermodynamics, dissipative structures and self-organization: some implications for biomedical research. In: Scott, G.P., Ames, J.M. (Eds.), Dissipative Structures and Spatiotemporal Organization Studies in Biomedical Research. Iowa State University Press, Ames.
von Foerster, H. (Ed.), 1949. Cybernetics: Transactions of the Sixth Conference. Josiah Macy Jr. Foundation, New York.
von Foerster, H., 1958. Basic concepts of homeostasis. In: Homeostatic Mechanisms. Brookhaven National Laboratory, Upton, New York, pp. 216–242.
von Foerster, H., 1961. A Predictive Model for Self-Organizing Systems, Part I. vol. 3. Cybernetica, pp. 258–300.
von Foerster, H., 2002. Understanding Understanding, A Volume of von Foerster's Papers. Springer-Verlag, Belin.

von Foerster, H., Pask, G., 1961. A Predictive Model for Self-Organizing Systems, Part II. vol. 4. Cybernetica, pp. 20–55.

von Foerster, H., Mora, P.M., Amiot, L.W., 1960. Doomsday: Friday, November 13, AD 2026. Science 132, 1291–1295.

Wolfram, S., 2017. A New Kind of Science: A 15-Year View. https://blog.stephenwolfram.com/2017/05/a-new-kind-of-science-a-15-year-view/.

Zainetdinov, R.I., 1999. Dynamics of informational negentropy associated with self-organization process in open system. Chaos Soliton. Fract. 10, 1425–1435.

Zalewski, M., 2005. Chaos i Oscylacje w Reaktorach Chemicznych (Chaos and Oscillations in Chemical Reactors) (Ph.D. thesis). Faculty of Chemical and Process Engineering, Warsaw University of Technology Press, Warszawa.

Zalewski, M., Szwast, Z., 2007. Chaos and oscillations in chosen biological arrangements of the prey–predator type. Chem. Process. Eng. 28, 929–939.

Zalewski, M., Szwast, Z., 2008. Chaotic behaviour of chemical reaction in a deactivating catalyst particles. In: Proceedings of the 18th International Congress of Chemical and Process Engineering CHISA August 24–28, 2008, Praha, Abstr. P7.22, pp. 1–10. on CD, http://www.chisa.cz/2008/.

Further reading

Composition Book – Biology/Research, Dr. Morowitz/Biogenesis, Anne Mitchell, 1991–1992. https://scrc.gmu.edu/finding_aids/morowitz.html.

Eigen, M., 1971. Self-organization of matter and the evolution of biological macro molecules. Naturwissenschaften 58 (10), 465–523.

Joslyn, C., 1990. Basic Books on Cybernetics and Systems Science. Materials of Course SS-501: Introduction to Systems Science. Systems Science Department of SUNY, Binghampton.

Mynarski, S., 1979. Elementy Teorii Systemów i Cybernetyki. PWN, Warszawa.

Prigogine, I., 1949. On the domain of validity of the local equilibrium hypothesis. Physica 15, 1942.

Simon, H.A., 1947. Administrative Behavior: A Study of Decision-Making Processes in Administrative Organization, first ed. The Macmillan Company, New York.

Simon, H.A., 1955. A behavioral model of rational choice. Q. J. Econ. 69 (1), 99–118 (compiled by Barros in 2010, and quoted after Simon (1957: 241–260)).

Simon, H.A., 1957. Models of Man, Social and Rational: Mathematical Essays on Rational Human Behavior in a Social Setting. John Wiley and Sons, New York.

Simon, H.A., 1976. Administrative Behavior, third ed. Free Press, New York.

Simon, H.H., 1979. Rational decision making in business organizations. Am. Econ. Rev. 69 (4), 493–513.

Thom, R., Brookes, W.M., Rand, D., Bell, G.M., 1985. Mathematical models of morphogenesis. In: Ferson, S. (Ed.), The Quarterly Review of Biology, vol. 60, no. 2 (June), pp. 216–217. https://doi.org/10.1086/414350.

Thompson, D.W., 1992. On Growth and Form: The Complete Revised Edition – June 23, 1992. http://www.darcythompson.org/index.html.

Zalewski, M., Szwast, Z., 2006. Obszary chaosu dla heterogenicznej katalitycznej reakcji chemicznej. Inżynieria i Aparatura Chemiczna 6s, 251–252.

CHAPTER 3

Properties of complex systems

3.1 Introduction

In this chapter, we describe several basic properties of complex systems resulting from their structure and properties of participating media. These systems may exhibit nonlinearities, instabilities, chaos, and emergent properties. The difference between "simple" and "complex" systems may sometimes be associated with more complex control structure in the latter systems. Complexity may also appear as a purely mathematical effect when the system is described in terms of certain "improper" variables. Our main interest is the complexity arising as the consequence of system links (interdependences) and complicated topology. Examples considered here serve to introduce complex systems possessing the simplest topologies (sequential structures); more complicated topologies are considered in Chapter 9.

3.1.1 Difference with simple arrangements

Many systems in the natural sciences and humanities fall under the heading "Complex Systems." Many are multiagent systems, the agents being living organisms, human societies, industrial firms, animal groups, diverse technologies, complex molecules, special cells, etc. But what is a complex system?

It follows from the observation that, when a set of evolving agents interacts, the resulting (overall) system exhibits collective properties which look qualitatively different from a superposition of their elementary behaviors. The common abstract features of such systems are:

- Nonlinearity
- Interdependence
- Emergence

These features will be described in detail as long as we proceed.

3.1.2 Nonlinearities, interdependence, and emergence

Nonlinearity means that the dynamical behavior of the system cannot be viewed as a superposition of the elementary effects of its components, nor reconstructed from elementary "modes." Speaking in simple terms, doubling the input does not necessarily double the output.

Complexity and Complex Thermo-Economic Systems. https://doi.org/10.1016/B978-0-12-818594-0.00003-9

Interdependence means that the response of each of the agents or parts from which the system is built depends on the evolution of the others, in fact in a self-consistent way.

Emergence means the creation of collective properties qualitatively different from the individual behavior.

The above abstract features are associated with the following material or physiochemical properties of complex systems:

- Open to substances and energy flows.
- Exist in various state spaces as disequilibrium structures or manifolds, in steady or quasisteady states.
- Behave as a whole, characteristic for "a system." This means that because of emergence properties the systems represent (or are able to create) collective behavior which is entirely different than the individual behavior of their components; consequently, they cannot be understood by decomposing into pieces which are simply added together.

3.1.3 Thermodynamics of selforganization and development

In a synthesizing paper, Ebeling (1985) considers a group of physical processes which play the main role in the development of selforganization and evolution. His special attention is devoted to the thermodynamic aspects of selforganization and evolution, associated with (bio) energetics, turbulence, and creation of dissipative structures.

Selforganization is defined as an irreversible process which is going on far from equilibrium, beyond a critical distance from equilibrium, leading to the creation of order by the cooperative action of subsystems (Ebeling and Feistel, 1992). Interestingly, evolution is understood as an unlimited sequence of selforganization steps (Ebeling, 1985). This is similar to the approach in Section 3.2, where the complexity increase is a consequence of the process in which a self-organizing system optimizes its organization with respect to a locally defined potential. Having selforganization steps, we have to define order.

Evidently, there is no unique measure of order. However, there are several special measures of order, such as:

- Entropy lowering with respect to its equilibrium value, S_{eq}, taken at the same energy, volume, and composition (Ebeling and Klimontovich, 1984; Ebeling and Feistel, 1992).
- Breaking of the symmetry. If any symmetry of a system is broken, e.g., with respect to time or space, the order of the system is increased by definition.
- Decrease of the length of programs. If the length of a program necessary for the description of the internal structure decreases, this will also mean the increasing order.

All these measures of order are of relative nature; they inform us about the increase or decrease of order with respect to a reference state. However, we may note that each of these measures covers only a special aspect of the order concept which has many faces. Ebeling (1985) is concerned only with the thermodynamic measure of order according to Boltzmann. For Ebeling, entropy is not only a measure of (dis)order, but also a measure of the value of the energy contained in the body (Ebeling and Klimontovich, 1984). Therefore, selforganization and evolution are necessarily connected with the creation of high-valued forms of energy (Ebeling and Klimontovich, 1984). The structures created in the selforganization process are often called "dissipative structures." Apart from this term which stresses the aspect of dissipation, Ebeling advices to also use "autostructures" which is a generalization of the more special terms "autooscillations" and "autowaves." It also seems that the term self-organizing oscillations could also be applied (Zalewski, 2005).

Nowadays, it is well-understood that autostructures (dissipative structures) are of much importance for many biological phenomena, such as morphogenesis, excitation waves in nerves, the brain, the heart, and in muscles. In nerve excitation, the basic phenomenon is an electrical autostructure. Several ions, especially K^+, are pumped by active (enzyme-containing) membranes to the other side, producing a charged capacitor representing high-valued electrical energy. This electrical autostructure is an energy-transforming engine which changes chemical energy into electrical one and stores it in the electrical field energy. Other examples are also available (Ebeling, 1985).

3.1.4 Inclusion of information fluxes

Any contemporary human activity usually requires transferring a certain amount of information. Information within a system's objects or between different systems may be transferred in time and space. Transfer of information in time is usually named recalling or remembering. Transfer of information in space is called communication. In large systems, where information transmission is essential, the role of information channels becomes increasingly important because of the need for the preservation of good quality of transferred signals. This need increases with the length of the information channel. Possession of suitable information enables us, among others, to save (minimize) useful energy input or to maximize useful energy yield.

3.1.5 State of art

Because systems share some common structural properties, it is tempting to ask: Is there a unified theory of complex systems?

The answer is: probably not. Unifying ideas like "the edge of chaos," self-organized criticality, etc. have at times been proposed as general paradigms for practically all complex adaptive systems. But in all cases, examples are found of systems that do not follow these general schemes, and nevertheless, look fairly complex to the normal eye. And even if all of complex systems arise from the same basic principles, the principles themselves might be of limited utility. (The notion of forest may be useful to understand plant epidemics, but is of limited utility to treat a specific tree disease). Complex systems have their specific properties which deserve specialized scientific effort, and for such different corpora, different strategies need to be applied in order to advance the operative knowledge in each particular field. As somebody writes: "For complex systems, to learn from their differences is an even more exciting challenge than to find a grand unified theory."

To reach a better understanding of each field and to transfer techniques and insights from one discipline to another requires the development of a common language and an open forum engaging a broad community to perform an efficient cross-fertilization of research. This is why, in a complexity project, mathematicians, physicists, biologists, medical doctors, economists, sociologists, linguists, and engineers attempt to share their experiences, models, and the many questions on what role the complexity notion may play in their fields.

By joint cooperative researches and seminars or workshops, phenomena and paradigms of fields other than their own became familiar to the scientists who shared their experiences in complexity projects. Because "the simple is what becomes familiar" and, by broadening their knowledge on how complexity unfolds in its many forms, scientists and engineers might hopefully be able, in the future, to approach the study of complexity through simplicity.

Based in part on http://www.uni-bielefleld.de/ZIF/FG/2000Complexity//introduction.html.

3.2 A complexity-based approach to evolution of living systems

The physical contents of the next part of this chapter are focused mainly on selected thermodynamic aspects of development and bistability in complex multistage systems. We consider extremum properties of entropy and/or its associates in complex systems. In particular, we transform the well-known extremum principle for a thermodynamic potential of a physical system into the extremum principle for an entropy-based complexity in a biological system with variable number of states, thus making it possible to investigate processes of biological development and evolution. The results imply that the increase in a number of states is governed by gradient of the complexity function based on information-theoretic entropy. Tensor form of associated dynamics can be obtained. For the accepted complexity model, some developmental processes may progress in a relatively undisturbed manner, whereas others may terminate in a rapid way due to the inherent instabilities. We show that these features can be predicted when describing complex evolutions in terms of variational principles for shortest paths along with suitable transversality conditions. Reversible modification of states appears as a basic reference frame.

One of the essential properties of evolution is the observed increase of complexity. Saunders and Ho (1976, 1981) decisively argue that complexity, rather than fitness or organization, is the quantity whose increase fixes a direction to evolutionary processes. They argue that the complexity increase is a consequence of the process by which a self-organizing system optimizes somehow its organization with respect to a locally defined fitness potential. Their basic premise is that a system which is not only organized but also capable of undergoing a continual selforganization or modification optimizes its structure with respect to a criterion and tends to permit the addition of components more readily than their removal. It is this asymmetry which, in Saunders and Ho's opinion, is the "chief cause of the observed increase of complexity during evolution."

Following thermodynamicists, selforganization is the spontaneous creation of order in open entropy-exporting systems operating beyond a critical distance from equilibrium. Introduction of the second law of thermodynamics into the argument does seem to suggest that there should be a complementary criterion which could increase during evolution of complex systems, such as living systems. As a living organism is not an isolated system, its entropy need not increase either during the single-organism development or during the species evolution. On the other hand, in effect of considering organization as a structure on a set of components (analogously as topology on a set of points), complexity emerges as the basic quantity in evolution. For biological systems, a reasonable measure of complexity is the different number of components they contain, so as to be consistent with the well-known Williston's law which predicts that very similar components will either merge or specialize. The increase in organization seems a secondary effect which emerges because the more complex system requires more organization in order to survive.

An example is provided by large, randomly connected linear systems, whose likelihood to be stable is very small. To ensure their survival, components must be added to them in a nonrandom way, which increases their organization. Staying on the ground of the idea of increasing complexity, the system's organization acts as a force which prevents loss of components and allows complexity increase by the integration of new ones.

When speaking about complexity as a quantity related to information or entropy, one issue is particularly important: a nonequilibrium entropy has to be applied, as, in any sufficiently general description of a complex system, the researcher must deal with an inherently nonequilibrium system. Sometimes, for special paths, they can achieve an equilibrium and derive equilibrium conditions, but generally transition between two nonequilibrium states is of interest. This state of affairs implies an analogy with nonequilibrium statistical physics. There, even in the special case of terminating the process at (isoenergetic) equilibrium, the equilibrium conditions along with Boltzmann distribution are derived from the entropy maximum condition in a closed system, and extremizing method uses entropy in an arbitrary macroscopic state, not in the equilibrium state (Landau and Lifshitz, 1975). Generally, the idea of maximum of a potential function, which may even be more general than complexity or nonequilibrium entropy, subject to given side conditions or constraints, is used.

In this chapter, we consider the idea of the extremal development of complexity or complexity-related criterion of evolution in the context of some recent information-theoretic models for multistage living systems characterized by sequentially evolving states. The information concept is not only appropriate to complex systems, but is also well-quantitatively defined. As natural selection cannot act on a variation which does not exist, whether or not a particular feature is observed in evolution largely depends on the probability that it occurs in the first place. For the probability p_i, the occurrence of the character increases the complexity by $\delta X = \log_2(1/p_i)$, i.e., larger values of p imply smaller values of δX. Consistently, evolution should proceed in accordance with a principle of minimum increase in complexity (Saunders and Ho, 1976). This resembles the thermodynamic principle of minimum of entropy production (Prigogine, 1955), yet is free of any equilibrium assumptions, including that of the local thermal equilibrium. The principle is essentially "a postulate in gradualism" which explains the two well-known trends in evolution, Williston's law and parallelism, and which helps distinguish between inherent and accidental properties. For example, the principle explains that the number of different type of components (states) is a more remarkable marker in evolution than is the total number of components (states). Changes in the number of some components, state coordinates, etc. are relatively common in evolution because "change in number (repetition) is a relatively easy alternation to make, and not because decrease or, for that matter, increase - is in itself inherently advantageous" (Saunders and Ho, 1976). Yet, in view of large difficulties in applying statistical mechanics to even relatively simple thermodynamic systems, these authors doubt if it will ever be possible to effectively use the information-theoretic expression for entropy for complex living systems especially because these are fairly more complex than thermodynamic ones. Also, they doubt if other measures of complexity, the information content of the genome or that based on genetic code, which are complexities measured by a molecular biologist, are appropriate and operative for complex-system modeling.

However, there may be two different formulations of the principle of increase in complexity, and the distinction between these two settings is not sufficiently well-stressed in the current literature. If complexity is considered as a function of variables characterizing current state of the system or related parameters (e.g., entropy, as described in Section 3.2), then, since the complexity increases, its value will always be greater at the end of the process stage than that at any one of the previous states. In this sense, we can say that the final complexity is a maximum with respect to changes of state or an indicator of this state. An example of application of this idea is described in Section 3.2 when discussing behavior of complexity with respect to entropy. The approach which is applied there is called in economics "comparative statics" as it is associated with seeking *maximum* of Γ at a given distance of states J. In a dual problem (Section 3.3), we seek for min J at a given $\Delta\Gamma$.

On the other hand, the principle of *minimum* increase in complexity means that when process rates are taken as controls, then the complexity increase in the natural evolution will be minimum in comparison with that calculated for any of virtual evolutions subject to the same constraints. Of these two formulations, the first involves the mathematical programming for static maximization of Γ, whereas the second uses the variational calculus for dynamic minimization of Γ. They should not be confused. Each of them has definite advantages and disadvantages. The virtue of the first is its simplicity, whereas that of the second is the information about dynamics generated in terms of the extremum conditions of the problem (Euler-Lagrange equations, Hamiltonian equations, or geodesic equations).

Complexity and selforganization can be increased in various ways (Ebeling and Klimontovich, 1984; Ebeling and Feistel, 1992; Saunders and Ho, 1976). Development proceeds by various interrelated physiochemical rate processes which can result in remarkably complicated structures without or with very little intervention from any controlling mechanisms. Recently, significant progress has been made toward explaining how the forms which are observed in nature are produced. Relatively simple differential equations were shown to produce bifurcations and singularities even without discontinuities in their coefficients or boundary conditions. A particularly large number of researchers have shown bifurcations emerging out of the reaction-diffusion interactions (Orlik, 1996; Malchow and Schimansky-Geier, 1985). Complex living systems have developed various strategies to manipulate their selforganization in order to satisfy the principle of minimum complexity increase, ultimately, however, the physical laws set limits to their size, functioning, and rate of development. For example, the physical law of thermal conduction sets the size of warm-blooded aquatic animals which require a minimum diameter (of c. 15 cm) in order to survive in cold oceans (Ahlborn and Blake, 1999; Ahlborn, 1999). Species which survive in ecosystems are those that funnel energy into their own production and reproduction and contribute to autocatalytic processes in the ecosystem. Also, there are data which show that poorly developed ecosystems degrade the incoming solar energy less effectively than more mature ecosystems (Schneider and Kay, 1994). The cornerstone here is to view living systems as stable structures increasing the degradation of the incoming solar energy, while surviving in a changing and sometimes unpredictable environment. All these structures have one feature in common: they increase the system's ability to dissipate (degrade) the applied gradient in accordance with the reformulated second law of thermodynamics (Schneider and Kay, 1994 and Chapter 10). In all these situations, the second law imposes constraints which are necessary but not a sufficient cause for life itself. In fact, reexamination of thermodynamics proves that the second law underlines and determines the direction of many processes observed in the development of living systems. As an ecosystem develops, it becomes more effective in removing the exergy part in the energy it captures, and this exergy is utilized to build and support organization and structure (Schneider and Kay, 1994). Time and its derivative cycling play a key role in the evolution of these complex systems. Evolution itself is a time-dependent process and the understanding of cycling is of great methodological and cognitive importance. In particular, an optimization theory of pulsating physiological processes can shed some light on the basic understanding of development and evolution (Bejan, 1997). Optimal strategies of streets' tree networks and urban growth can mimic development of living systems (Bejan and Ledezma, 1998).

One way in which complexity of evolving system can be increased is the repetitions of elements, the particular interest of this chapter. Saunders and Ho (1976, 1981) and others point out that repetition can result in an apparent increase in the information content of an organism during development and tend to quantify repeated and homologous structures. The common occurrence of sequences of similar structures proves that a form of loop exists, as in a computer program; a number of diverse biochemical processes accounts for this fact. And a gradient of some criterion seems to affect the subsequent stages in such a way as to produce similar, rather than identical structures.

Keeping in mind that repetition is not the only way in which the complexity can be increased, we refer the reader in Section 3.2 to a work of Szwast (1997), which analyses evolution of living organisms treated as multistage systems by using the complexity criterion based on information-theoretic entropy. In this section, we also propose a modified entropy-like criterion of complexity. Next, Section 3.3 addresses an extremality principle for a potential as a driving factor in the evolution dynamics, the potential being entropy itself or an entropy-like complexity function. Classical thermodynamic quantities do not appear in this approach, yet its statistical model is governed by an extremum principle which, as in thermodynamics, implies extremal properties for a potential. Section 3.4 discusses discrete and nonlinear generalizations of dynamics in metric spaces which may be curvilinear, whereas the final Section 3.5 concludes that many features of living organisms can be predicted when describing their complex evolutions in terms of variational principles for shortest paths along with suitable transversality conditions. More information about complexity-related criteria of evolution can be found in complementary papers (Sieniutycz, 2002; Szwast et al., 2002).

A basic claim of this section is that a bridge can be constructed to link purely thermodynamic approaches with those which must abandon thermodynamics and rely on methods of statistical physics and information theory. In fact, the main theoretical tools for all disequilibrium approaches are provided by the union of information theory (Tribus, 1970) and thermodynamic optimization (Berry et al., 2000). With the help of these approaches, some recent findings in the mathematical theory of species selection, e.g., Kimura's maximum principle for increase in average fitness, can not only be explained, but also properly restricted to their own realm of validity (Hofbauer and Sigmund, 1992).

Here, we shall analyze these aspects of modeling in the context of multistage systems which can successfully mimic some processes of biological development and evolution. In this modeling, the information-theoretic entropy can successfully be used to measure changes in the system's statistical properties and provide a suitable complexity criterion.

The so-called Williston's law is often quoted in the evolution literature, which states that, for an organism possessing many of the same or similar elements, a tendency appears to reduce the number of these elements along with the simultaneous modification (specialization) of these elements which are saved by the organism.

Saunders and Ho (1976, 1981) recall that the evolution leading from trylobite to crab is a relevant example illustrating this tendency. It should be kept in mind that annelids, ancestors of trilobite, live today, and crabs, their posteriorities, live today as well. Trilobites became extinct in perm (270–220 millions of years ago) in camber, and the problem of their extinction calls for explanation. Saunders and Ho (1981) and Szwast (1997) address the evolution leading from the trylobite, which possesses a large number of segments and pairs of legs, to the crab, with far fewer number of these segments and pairs of legs along with the single pair of legs being modified (specialized) to the pair of claws. These researchers next ask: what are causes of such path of evolution?

Considering their understanding of this particular evolution, the researchers state that, even if the energy benefits could be shown, the three following questions would still remain unanswered. The first is: Why, if there are definite benefits of a small number of segments and pairs of legs to an organism, extra segments and pairs of legs do evolve? The second is: Why the majority of recent evolutions show the tendency to increase rather than decrease the number of elements? The third is: Why the same natural selection enables evolution in various parallel directions?

Szwast (1997) has investigated evolution from trilobite—as an organism with many states: segments and pairs of legs—to crab, with far fewer of each and with one pair of legs modified (specialized) into claws. The basic problem was: why extra states (pairs of legs) evolved and what was the cause of trilobites extinction. His mathematical model of evolution implied that the increase of the number of states (pairs of legs) is governed by extremum gradient of a potential which may possibly be either the informational entropy or a complexity function (each of these quantities increases with the number of states in the system). Perhaps, a decrease of number of states (pairs of legs) seemed to Szwast to result from catastrophes on the evolution path (in fact, he was first to show that a state modification or specialization may lead evolution to a catastrophe region). What evolution criterion should actually be preferred is still the subject of investigations; according to Saunders and Ho, a principle of minimum increase holds for a potential function which includes the effect of environment rather than for complexity or entropy. However, as shown in Section 3.3, a related extremality analysis can be developed for an arbitrary potential function.

Szwast's (1997) description of evolution involves calculations of information entropy, $S = -\Sigma p_i \ln p_i$, where the summation is over the number of states. Here p_i is the probability of finding an element in the state i among n states possible and the sum $\Sigma p_i = 1$. Maximum of unconstrained entropy appears for the total randomness. In this case, $p_i = 1/n$, for n states, and $S_{\max} = -n\ (1/n)\ln(1/n) = \ln n$. In the trilobite analysis, however, the number of states is $2n+1$ ($2n$ legs plus the rest part of the body); thus $S_{\max} = \ln(2n+1)$.

However, the well-known Schrödinger's definition of disorder is not used in the analysis in question (Schrödinger, 1967). In fact, as explained by Landsberg, entropy itself is not a direct measure of disorder for growing systems as it is an extensive quantity (Landsberg, 1984a,b). In other words, for growing systems, the classical definitions of disorder $D = e^S$ and order $\Omega = e^{-S}$ proposed by Schrödinger are inapplicable. Their flaws are not shared by Landsberg's definitions of disorder and order (Landsberg, 1984a,b, 1994).

According to Landsberg, disorder $D = S/S_{max}$ and order $\Omega = 1 - D$. As S_{max} depends on n, in Landsberg's definition, both disorder D and order Ω are functions of entropy S and number of states in the system, n; i.e., $D = D(S, n)$ and $\Omega = \Omega(S, n)$. For evolution processes, the particularly important role is played by the complexity Γ, which is a function of disorder D and order Ω, i.e., $\Gamma = f(D, \Omega)$. In the literature, various functions f are advocated (Shiner et al., 1999; Shiner, 2000). A popular form has the simple structure $\Gamma = 4D\Omega = 4D(S, n)[1 - D(S, n)]$, which was used by Szwast (1997). In this case, the maximum of Γ is attained for $D = 0.5$ and equals the unity. In Szwast's (1997) analysis, two statements of Saunders and Ho (1981) are confirmed: (1) "While the increase of complexity Γ is more likely and the trend will be in this direction, we must also expect occasional decrease of complexity" and (2) "Only completely reversible changes are processes which develop without change of complexity (isocomplex processes)."

This reversible application requires constancy of complexity or entropy, the requirement which restricts motions to those along "entropy isolines" when modifications or specializations occur at a constant number of states, n. This means that the decrease of states (pairs of legs) is a possible process which is compensated during the evolution by the creation of modified states (claws). This process occurs along an isoline of entropy which is not a maximum entropy for a given n; it is rather the entropy which maximizes complexity Γ (in fact, probabilities which would maximize entropy would rather minimize complexity). Therefore, it is the set of conditions $d\Gamma/dS = 0$ and $d^2\Gamma/dS^2 < 0$, which defines an entropy isoline along which modification or specialization of an organism is possible. When the maximum of entropy for the n-state system at complete randomness is $S_{n(\max)}$, then, in that model, the complexity-maximizing entropy follows as S_n^{opt} or $\hat{S}_n = (1/2)S_{n(\max)}$, and the corresponding complexity $\hat{\Gamma}_n = 4S_n/S_{n(\max)}(1 - S_n/S_{n(\max)}) = 1$. Both the maximal entropy (at the complete randomness) and the one which maximizes complexity ($\hat{S}_n$ or one half of $S_{n(\max)}$) are increasing functions of the number of states, n. A slight disadvantage contained in the normalized structure of complexity $\Gamma = 4D\Omega = 4D(S, n)[1 - D(S, n)]$ is that the extremum complexity Γ_n follows the same regardless the number of states, n. Therefore, we propose here to apply a modified (nonnormalized) definition of Γ, say $\Gamma = 4S_{n(\max)}D\Omega = 4S_{n(\max)}D(S, n)[1 - D(S, n)]$, whose upper-bound values increase with n (Fig. 3.1).

As the result of the redefinition, the complexity profile associated with a system in which the number of states grows during its evolution is the zigzag line whose upper edges are constrained by the values $S_{n(\max)}$ which increase with n, as illustrated in Fig. 3.1.

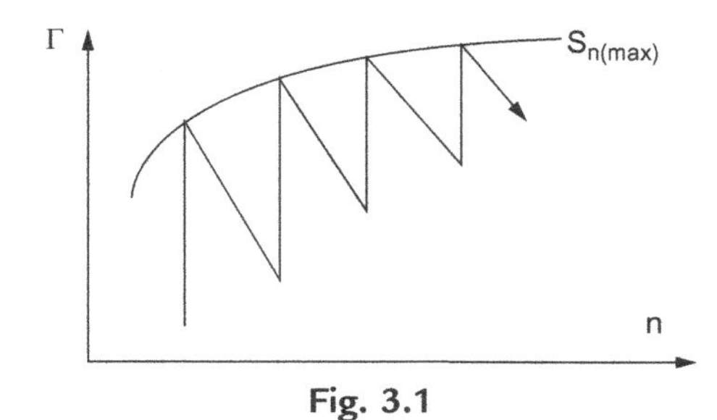

Fig. 3.1
A sketch of complexity profile for a system with growing number of states, n.

Now, we can return to the trilobite problem. For the purpose of analysis, it is sufficient to distinguish one pair of legs, of the probability $2p_1$, from the remaining parts of the trilobite, of the overall probability p_0. The condition $2np_1 + p_0 = 1$ holds in this case. In the analysis of a crab, claws should be distinguished with an additional probability $2p_2$, subject to the condition $2p_2 + 2(n-1)p_1 + p_0 = 1$, which holds for the case when the modification (speciation) accompanies the trilobite's evolution to a crab without change in number of pairs of legs and claws. With values of $\hat{S}_n$ calculated for various n, values of independent probabilities p_1 and p_0 can be calculated (also in terms of n) which correspond with optimal $\hat{S}_n$. For the trilobite case, results of these calculations (Szwast, 1997) show the decrease of both independent probabilities p_1 and p_0 with n. With various values of the entropy $\hat{S}_n = (1/2)S_{n(\max)}$ for organisms which have various pairs of legs, n, data of order, disorder, and complexity can be found on the basis of their definitions. Yet, additional observation that has to be incorporated takes into account the fact that as long as an embryo of new pair of legs (a new life) is created in the organism of n pair of legs, the optimal entropy $\hat{S}_n$ should be referred to the maximum entropy $S_{n+1(\max)}$ rather than to $S_{n(\max)}$, that is, to the entropy of an organism with $n+1$ pair of legs. Consequently, an inequality $\Omega_{n+1} = 1 - \hat{S}_n/S_{n+1(\max)} > 1 - \hat{S}_n/S_{n(\max)} = \Omega_n$ is valid which describes the order increase with the increase of the number of states in the system (here the number of legs, n) due to a new life creation during the evolution.

A map of possible states was obtained in the form of "entropy isolines" in Szwast's (1997) evolution description. Yet, according to the principle of complexity increase (Saunders and Ho, 1976), we should rather deal with "complexity isolines." Evolution of states toward a crab refers to a situation in which an organism with one pair of states (legs) is modified (specialized) along a reversible path. With the normalization condition $\Sigma p_i = 1$ incorporated to eliminate p_2, the probability of one pair of states (legs), the complexity isoline along which the specialization proceeds is $\Gamma_n(p_0, p_1) = \hat{\Gamma}_n$.

Fig. 3.2 gives the qualitative description of trajectories of reversible modification (specialization) of one pair of original states (legs) for various values of variable n, which constitutes the sum of $n-1$ identical pairs of states (legs) and one pair being modified. Dots on trajectories mark findings for organisms without specialization of states (legs); thus they refer

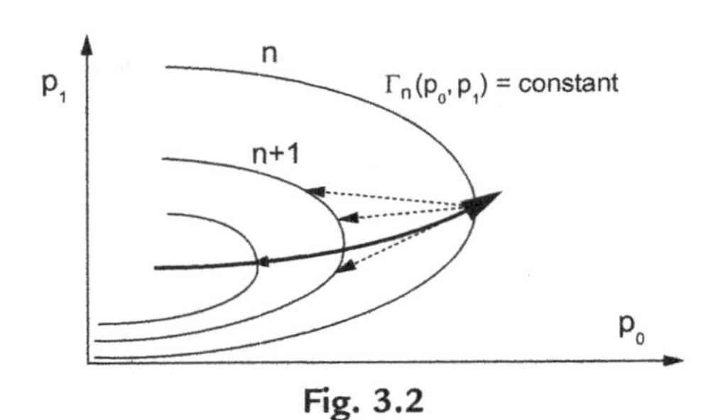

Fig. 3.2
A minimum distance problem between a dot point on isoline n and the isoline $n+1$. In evolution analysis, isolines are paths of the reversible modification of single pair of states (legs) for various n, where n is the sum of $n-1$ pairs of the same states (legs) and one modified pair.

to organisms possessing n pair of identical states (legs). Constant attributed to an isoline Γ_n in Fig. 3.2 refers to extremum values $\hat{\Gamma}_n$. The figure also sketches a minimum distance problem between a dot point on isoline n and the isoline $n+1$, as the variational problem of minimum length with left free end of an extremal (see below).

The following question arises: why an organism possessing n of identical pair of states (legs) tends to the state associated with $n+1$ pair of legs during the evolution, or, why the evolution proceeds in the direction of an increase of identical states? The results below imply that it is the shortest line between the dot point on the curve n and the curve $n+1$, which determines the location of the dot point on the curve $n+1$. Of course, we need to take into account that each curve in Fig. 3.2 refers to a fixed value of complexity $\Gamma_n = \hat{\Gamma}_n$.

3.3 Extremality principle for complexity potential in evolution dynamics

The analysis involved is essentially the classical variational problem of minimum length along with the use of a transversality condition. It can be developed for an arbitrary potential function (Sieniutycz, 2000; Szwast et al., 2002).

Let us introduce the function

$$F_n(p_0, p_1) = \Gamma_n(p_0, p_1) - \hat{\Gamma}_n = 0 \tag{3.1}$$

where $\hat{\Gamma}_n$ is the constant value of Γ_n along the isoline of complexity potential Γ. The numerical value of F_n equals zero for all states consistent with the complexity-maximizing entropy S. They are located on "reversible isolines" of Fig. 3.2, for various n. Eq. (3.1) states that the total differential of F is that of potential Γ restricted to values $\hat{\Gamma}_n$.

$$dF_n(p_0, p_1) = d\Gamma_n(p_0, p_1) = \frac{\partial \Gamma_n(p_0, p_1)}{\partial p_0} dp_0 + \frac{\partial \Gamma_n(p_0, p_1)}{\partial p_1} dp_1 = 0 \tag{3.2}$$

The direction coefficient of the tangent to each of the lines $\Gamma_n(p_0, p_1)$ is the derivative $v = (dp_1/dp_0)_n$. With Eq. (3.2), this derivative can be determined in terms of the partial derivatives of Γ_n or F_n as

$$v_n = \left(\frac{dp_1}{dp_0}\right)_n = -\frac{\partial\Gamma_n(p_0, p_1)}{\partial p_0} \bigg/ \frac{\partial\Gamma_n(p_0, p_1)}{\partial p_1} \tag{3.3}$$

To predict the location of the dot point on the curve $n+1$ when the dot point on the curve n is given, consider the variational problem of the shortest line between the dot point on the curve n and the curve $n+1$. This corresponds with the minimum of a length functional

$$J = \int_{p_0^n}^{p_0^{n+1}} a(p_0, p_1)\sqrt{1 + \left(\frac{dp_1}{dp_0}\right)^2} dp_0 \tag{3.4}$$

where a "conformal constant" $a(p_0, p_1)$ subsumes the deviation from the Euclidean measure of length. It depends on p_0 only if it is possible to assume that only basic part of the body (with probability p_0) influences the geometry properties of the system. The integrand of this Lagrangian functional is designated by L.

In terms of Lagrangian L, the general Euler-Lagrange equation for the minimum of J is

$$\partial L/\partial p_1 - \frac{d}{dp_0}(\partial L/\partial(dp_1/dp_0)) = 0 \tag{3.5}$$

and, for the considered model, it has the form

$$\frac{\partial a(p_0, p_1)}{\partial p_1}\sqrt{1 + \left(\frac{dp_1}{dp_0}\right)^2} - \frac{d}{dp_0}\left(\frac{a(p_0, p_1)}{\sqrt{1 + \left(\frac{dp_1}{dp_0}\right)^2}}\right)\left(\frac{dp_1}{dp_0}\right) = 0 \tag{3.6}$$

Eq. (3.6) incorporates generally a non-Euclidean metrics but, if $a = 1$ or $a = a(p_0)$, it refers to the Euclidean distance. In the general case, when the coefficient a in Eq. (3.4) depends on all probabilities, the problem is referred to a geodesics.

Consider first the case of a constant a. Then the solution of Eq. (3.6) describes the family of the straight lines

$$p_1 = C_1 p_0 + C_2 \tag{3.7}$$

which are extremals in the flat space. Any extremal which starts from the dot point on the curve n and terminates on the curve $n+1$ should satisfy a transversality condition (Sieniutycz, 1991)

$$\delta J = \left\{L - \frac{\delta L}{\delta(dp_1/dp_0)}(dp_1/dp_0)\right\}\delta p_0 + \left(\frac{\delta L}{\delta(dp_1/dp_0)}\right)\delta p_1 = 0 \tag{3.8}$$

where variations δp_0 and δp_1 are linked by condition (3.3) applied for the curve $n+1$

$$(dp_1/dp_0)_{n+1} = -\frac{\delta\Gamma_{n+1}(p_0, p_1)}{dp_0} \Big/ \frac{\delta\Gamma_{n+1}(p_0, p_1)}{dp_1} \tag{3.9}$$

This equation assures optimal location of the final point on the curve $n+1$. For any J which has meaning of a length, Eq. (3.6) describes the condition for the length extremum of the extremal which starts from the dot point on the curve n and terminates on the curve $n+1$. Substituting variation dp_1 from Eq. (3.9) into Eq. (3.8) yields the condition

$$\delta J = \left\{ L - \frac{\delta L}{\delta(dp_1/dp_0)}(dp_1/dp_0)_{n+1} + \frac{\delta\Gamma_{n+1}(p_0, p_1)}{dp_0} \Big/ \frac{\delta\Gamma_{n+1}(p_0, p_1)}{dp_1} \right\} \delta p_0 = 0 \tag{3.10}$$

In view of arbitrary variations of p_0, this is equivalent with the condition

$$\begin{aligned} L - \frac{\delta L}{\delta(dp_1/dp_0)}(dp_1/dp_0)_{n+1} + \frac{\delta\Gamma_{n+1}(p_0, p_1)}{dp_0} \Big/ \frac{\delta\Gamma_{n+1}(p_0, p_1)}{dp_1} \\ \equiv L - \frac{\delta L}{\delta(dp_1/dp_0)}(u_{n+1} - v_{n+1}) = 0 \end{aligned} \tag{3.11}$$

Condition (3.11) links two slope coefficients, $u_{n+1} = dp_1/dp_0$ or the tangent to the extremal of integral J and v_{n+1} or the tangent to the isoline $\Gamma_{n+1}(p_0, p_1) = \hat{\Gamma}_{n+1}$, Eq. (3.3), at the optimal end point where the extremal reaches the curve $n+1$. For our Lagrangian L, the above condition takes the form

$$a(p_0, p_1)\left\{ \sqrt{1 + (u_{n+1} + 1)^2} - \frac{u_{n+1}}{\sqrt{1 + (u_{n+1} + 1)^2}}(u_{n+1} - v_{n+1}) \right\} = 0 \tag{3.12}$$

We observe here the consequence of Euclidean geometry assumed by restricting ourselves to straight-line extremals: conformal coefficient $a(p_0, p_1)$ does not change the transversality condition, which is the same in the case when $a = 1$ and when a is arbitrary function, $a(p_0, p_1)$. In fact, however, extremals not need to be straight lines, the fact contained in Eq. (3.6).

Eq. (3.12) can be simplified to the form

$$\frac{1 + (u_{n+1} + 1)^2 - u_{n+1}(u_{n+1} - v_{n+1})}{\sqrt{1 + (u_{n+1} + 1)^2}} = \frac{1 + u_{n+1}v_{n+1}}{\sqrt{1 + (u_{n+1})^2}} = 0 \tag{3.13}$$

or

$$u_{n+1} = -1/v_{n+1} \tag{3.14}$$

which describes orthogonality of the slopes u_{n+1} and v_{n+1}. From Eqs. (3.3), (3.14),

$$(dp_1/dp_0)_{n+1} = -\frac{\partial\Gamma_{n+1}(p_0, p_1)}{dp_1} \Big/ \frac{\partial\Gamma_{n+1}(p_0, p_1)}{dp_0} \tag{3.15}$$

This equation implies the gradient dynamics for change of probabilities with time in the form

$$\left(\frac{dp_0}{dt}\right)_{n+1} = \omega \frac{\partial \Gamma_{n+1}(p_0, p_1)}{dp_1} \tag{3.16}$$

and

$$\left(\frac{dp_1}{dt}\right)_{n+1} = \omega \frac{\partial \Gamma_{n+1}(p_0, p_1)}{dp_1} \tag{3.17}$$

Indeed, by taking the ratio of Eqs. (3.17), (3.16), the orthogonality condition contained in Eq. (3.15) is recovered. The frequency-type coefficient ω has interpretation of a kinetic constant governing the state modification process.

3.4 Discrete and nonlinear generalizations of dynamics in metric spaces

The time interval needed for modification of two states (single pair of legs into claws) in the process which starts with n states and terminates at $n+1$ states is $\theta_{n+1} = t(n+1) - t(n)$, and it can be calculated from this model as

$$\theta(n+1) = \frac{p_0(n+1) - p_0(n)}{\omega \partial \Gamma_{n+1}(p_0, p_1)/\partial p_0} \tag{3.18}$$

The associated change of the nondimensional time is $\omega\theta(n+1)$. Should we wish to directly measure the increase of number of states (pair of legs) in terms of this nondimensional time, we must identify the increment $\omega\theta(n+1)$ with the unity. Then, an associated discrete dynamics is obtained which describes the passage of the system from the state $\mathbf{p}(n)$ to the state $\mathbf{p}(n+1)$

$$p_0(n+1) - p_0(n) = \frac{\partial \Gamma_{n+1}(p_0, p_1)}{\partial p_0} \tag{3.19}$$

$$p_1(n+1) - p_1(n) = \frac{\partial \Gamma_{n+1}(p_0, p_1)}{\partial p_1} \tag{3.20}$$

A characteristic feature of this model is worth stressing: each single stage number is identified with the change in the number of state coordinates (pair of legs) by one unit. This is characteristic of evolution processes with specializations, (speciations), or modifications of states, or those with births and deaths, and may not be admissible when no new organs or organisms are born. The applicability of this set can be extended by assumption that the starting point at the nth entropy isoline is not necessarily the dot point; with this extension the dynamics generated by Eqs. (3.16) and (3.17) or (3.19) and (3.20) covers the whole probability space in which it is represented by trajectories orthogonal to complexity isolines.

The results of calculations imply that the gradient of information entropy increases with number of pairs of legs, and that the modification (specialization) leads evolution to the region of catastrophes (Szwast, 1997). In this region, the entropy gradient tends to plus infinity, and then passes from plus to minus infinity, which is connected with the inversion of the flow direction of the complexity isolines with n. Consistently, the decrease of number of pairs of legs is observed which results from catastrophe on the trajectory of evolution. A more detailed explanation of the mechanism of this phenomenon is given in a publication (Szwast et al., 2002). These qualitative results suggest that bistability and Schlögl's mechanism are involved in these phenomena (Orlik, 1996; Malchow and Schimansky-Geier, 1985).

The results of trilobite analysis agree with the Williston's law which subsumes the results of observation and comparative analyses. This law states that if an organism possesses many of the same or similar elements, a tendency appears to reduce the number of these elements along with the simultaneous modification (specialization) of these elements which are saved by the organism. This is, in fact, observed for evolution leading from the primitive trilobite, which possesses a large number of segments and pairs of legs, to the crab, with far fewer number of these segments and pairs of legs along with the single pair of legs being modified (specialized) to the pair of claws.

While Eqs. (3.16) and (3.17) or (3.19) and (3.20) refer to the simplest metric $g^{ik}=\delta^{ik}$, their tensor generalization can easily be obtained for an arbitrary g^{ik} and, in particular, to an arbitrary function $a(p_0, p_1)$ in Eq. (3.4). Taking into account that the notation deals with probabilities indexed by subscripts, it will be the easiest to treat all independent probabilities as covariant coordinates. To attain the tensor generalization, we transform the set of Eqs. (3.16), (3.17) to another (primed) coordinate frame which may be curvilinear. Assuming that the state vector is s-dimensional we find

$$\frac{d\mathbf{p}}{dt}=\frac{\partial \mathbf{p}}{\partial \mathbf{p}'}\frac{\mathrm{d}\mathbf{p}'}{\mathrm{dt}}=\omega\frac{\partial \Gamma(p_0,..p_s)}{\partial \mathbf{p}}=\omega\frac{\partial \mathbf{p}'}{\partial \mathbf{p}}\frac{\partial \Gamma(p_0,..p_s)}{\partial \mathbf{p}'} \tag{3.21}$$

$$\frac{d\mathbf{p}'}{dt}=\omega\frac{\partial \mathbf{p}'}{\partial \mathbf{p}}\frac{\mathrm{d}\mathbf{p}'}{\mathrm{dt}}=\omega\frac{\partial \Gamma(p_0,..p_s)}{\partial \mathbf{p}'}=\omega\mathbf{g}^{-1}\frac{\partial \Gamma(p_0,..p_s)}{\partial \mathbf{p}'} \tag{3.22}$$

where $\mathbf{g}^{-1}$ is the matrix of the contravariant metric tensor. The corresponding matrix of covariant metric tensor is then $\mathbf{g}$. We can now ignore primes and write down the obtained result in the following tensor form

$$\frac{dp_i}{d\tau}=g_{ik}\frac{\partial \Gamma(p_0,..p_s)}{\partial p_i} \tag{3.23}$$

where the nondimensional tine $\tau=\omega t$. The product of the covariant metric tensor g_{ik} and the contravariant gradient vector $d\Gamma/dp_k$ is the covariant gradient vector, $d\Gamma/dp^k$. Thus, Eq. (3.23) states that the covariant rate $dp_i/d\tau$ is equal to the covariant gradient vector, $d\Gamma/dp^k$.

In the generalized case considered, an associated discrete dynamics, which describes the passage of the system from the state $\boldsymbol{p}(n)$ to the state $\boldsymbol{p}(n+1)$, can be written in the Onsager-like form

$$p_0(n+1)-p_0(n)=L_{11}\frac{\partial\Gamma_{n+1}(p_0,..p_s)}{\partial p_0}+L_{12}\frac{\partial\Gamma_{n+1}(p_0,..p_s)}{\partial p_1}....+L_{1s}\frac{\partial\Gamma_{n+1}(p_0,..p_s)}{\partial p_s} \tag{3.24}$$

$$p_s(n+1)-p_s(n)=L_{s1}\frac{\partial\Gamma_{n+1}(p_0,..p_s)}{\partial p_0}+L_{s2}\frac{\partial\Gamma_{n+1}(p_0,..p_s)}{\partial p_1}....+L_{ss}\frac{\partial\Gamma_{n+1}(p_0,..p_s)}{\partial p_s} \tag{3.25}$$

where $L_{ik} \equiv g_{ik}\Delta\tau$. In this model, each single stage number is identified with the change in the number of state coordinates (as, e.g., in Section 3.2) by one unit.

The discrete nature of dynamical equations in processes of biological development and evolution was postulated as a most relevant feature which is characteristic of evolution processes with births, deaths, or modifications (Gontar, 2000). The applicability of the above set can be extended by assumption that the starting point at the *n*th isoline is not necessarily the dot point; with this extension, the dynamics generated by Eqs. (3.16) and (3.17), (3.23) or (3.24) and (3.25) cover the whole probability space in which this dynamics is represented by trajectories orthogonal to the isolines of the complexity potential. An explanation of the mechanism of state catastrophes is based on this model and bistability of Schlögl's mechanism (Orlik, 1996; Schlögl, 1980; Schöll, 1986). While many results obtained in this context can only be qualitative in view of different definitions of complexity (Shiner et al., 1999), combining results of the contemporary theory of thermodynamic metric with analysis of some numerical tests of evolution can improve the situation.

3.5 Final remarks

The principle of minimum increase of complexity, while strongly postulated by some authors, should not be accepted without limitations (Saunders and Ho, 1976, 1981). The requirement for the principle to hold exactly is unchanged or very slowly variable fitness. Whether or not one accepts the neo-Darwinian ideas of evolution, no doubts are expressed that natural selection is an important factor. Read Polani's (2010) review on J.S. Avery's book "Information Theory and Evolution." While the idea of a constantly increasing fitness is unacceptable, a generalization of complexity definition seems necessary for those evolutions in which changes in the fitness are essential. In these cases, in view of tautological nature of the concept of the survival of the fittest, a measure of fitness rather than fitness itself, called homeostasis $\mathcal{H}$, could play a role (Saunders and Ho, 1976, 1981). When studying ecological systems which interact actively with environment (exchange energy and matter), fitness is suggested to play a role of energy-like potential, and for strongly interacting systems, the quantity $\mathcal{G}=\mathcal{H}-T\Gamma$ should be taken into considerations, similarly like Gibbs free energy in thermodynamics of systems interacting with a thermal reservoir. Yet, it should be assumed that, as in thermodynamics, modeling of this sort would be possible provided that a sort of local equilibrium situation would exist associated with rapid microscopic processes (quick collisions causing rapid averaging on microscale), whereas the global process still would be out of equilibrium, i.e., finite rates would exist. But with this reorientation, the properties of maximum of final entropy (complexity) and minimum change of entropy (complexity) should, respectively, be replaced by the properties of minimum of a final value of G and minimum change of G. This last property may also be expressed as the requirement that the final value of the potential G should attain a maximum with respect to rates of all related kinetic equations. With this definition of G, selforganization is associated with spontaneous creation of an order in systems exporting entropy or importing the energy of high quality; this agrees with majority of recent opinions on the subject. In such general systems, the entropy and complexity could decrease when selforganization of the system takes place. More research, however, is necessary to verify how much these ideas will prove their usefulness. In particular, the recognition of geometries governing evolutions described by Eqs. (3.16) and (3.17), (3.23), or (3.24) and (3.25) should be essential for further progress.

Geometrical measures for energy and entropy changes are well-known in thermodynamics: Weinhold (1975, 1976), Ingarden and coworkers (Ingarden, 1977; Ingarden et al., 1982; Antonelli et al., 1996), Schlögl (1980), Wootters (1981), Levine (1985), Ruppeiner (1979, 1985, 1990, 1995), Mrugala (1990, 1996), Mrugala et al. (1990), and coworkers of these authors. A review on geometrical methods in equilibrium thermodynamics appeared (Mrugala, 2000). Extensions to local-disequilibrium systems progressed (Casas-Vazquez and Jou, 1985; Chen, 1991; Sieniutycz and Berry, 1991). As the formalism of thermodynamic geometry can be derived from an arbitrary potential V and its Legendre transform (Sieniutycz and Berry, 1991), the transfer of geometric ideas to complexity-governed realms will allow to establish metrices for complex systems. For this purpose, equivalence of statistical and phenomenological metrices is an important issue (Salamon et al., 1985; Feldmann et al., 1986). As the Finslerian metric tensor can be associated with an arbitrary (inhomogeneous) variational problem, approaches which use Lagrangian equations, Hamilton's canonical sets, and related Fokker-Plank equations can provide further generalizations (Plastino et al., 1997; Janyszek and Mrugala, 1990; Ingarden, 1996).

Variational formulations have well-defined virtues. First, physical insight is gained when a single scalar quantity is found which generates the whole vector field represented by many equations of motion. Second, unification of various mechanisms within a complex process is often possible, for example: mechanical, electrical, and chemical mechanisms. Third, by applying the so-called direct variational methods with some trial functions used, approximate solutions can be obtained which are usually of good accuracy (Kantorovich and Krylow, 1958); they may be the only usable solutions when the analytical solution of the differential equations of interest cannot be found. Fourth, integrals of motion and conservation laws can be obtained from the symmetry principles (Sieniutycz, 1994). However, constructing variational principles for irreversible fields is since years associated with large difficulties. Most essential difficulty is caused by unknown Lagrangians, as there are hardly any systematic rules to obtain relevant Lagrangians. Equations of dissipative fluids are particularly difficult in this regard. Some results, however, suggest that the inclusion of thermodynamic irreversibility does not change either lagrangians or action functionals; it only complicates potential representations of physical fields in comparison with those describing the reversible evolution (Sieniutycz, 1994,

2000). Therefore, it may be expected that irreversible features of complex systems can be predicted when describing these systems in terms of variational principles for shortest paths along with suitable transversality conditions.

Some recent works on modeling of selforganization and life define processes of evolution as branching chains composed of the cycles of selforganization (Ebeling and Feistel, 1992; Rymarz, 1999a,b). We briefly recapitulate below especially the last two references which propose a novel definition of the living system where phenomena of selforganization and chaos may coexist. Contrary to the definition of life as a balanced coexistence of the homeostasis and adaptivity, which comes from biologists, Rymarz's definition expresses the attitude of a physicist. He characterizes the life as the branching chain of elementary links, where the elementary link comprises the triad: birth, evolution, and death. Each elementary link contains at least one bifurcation point and it is open for external action. The branching phenomena can take place during the process evolution and in the state of death. They can also occur when division of the living cell takes place. The new notion is the elementary link, which models the individual development in relation to the group evolution, the latter running through many parallel elementary links, in a way open to external action, often with increasing fitness. Various tree structures are suitable examples in this context. The birth is an instability or a phase transition, which influences the growth of order (coherence of states). The evolution of an individual (development) is a sequence of selforganization processes with both creative and destructive parts. If the creation prevails, coherence of states is preserved or increased, associated with branching phenomena such as bifurcations and reproductions. All these phenomena increase the number of states in the system; compare the trilobite evolution analyzed in previous sections. The death appears in this context as the significant decrease of coherence of the state of an organism. Well-organized, living systems are open for import of high-quality energy (e.g., solar energy) and information. Unstable modes, growing at the expense of other ones, lead systems to selforganization. Periodic changes of chemical components occur in living organisms. Examples from meteorology such as creation and evolution of Bernard's cells and those from chemical kinetics of oscillatory reactions show the parallelism between animate and inanimate matter (Rymarz, 1999a,b). In this logical structure, entropy creates the possibility of storing and processing the high-quality energy and information, and the information processing is the basic phenomenon for any form of "life," and mainly for biological one. Current achievements in the selforganization theory (Ebeling, 1985) and in systems theory (Auyang, 1998) improve our understanding of the depth and sense of these developments.

Appendix: Contribution of diversity and complexity to the structure of the terrestrial world (based on Auyang, 1998)

Once, the well-known physicist Richard Feynman closed his lecture on the relations among various sciences by contemplating a glass of wine. He said: if we look at the wine closely enough we see the entire universe: the optics of crystals, the dynamics of fluids, the array of chemicals, the life of fermentation, the sunshine, the rain, the mineral nutrient, the starry sky, the growing vine, the pleasure it gives us. "How vivid is the claret, pressing its existence into the consciousness that watches it! If our small minds, for some convenience, divide this glass of wine, this universe, into parts - physics, biology, geology, astronomy, psychology, and so on - remember that nature does not know it!" Feynman did not mention the price tag, the symbol of the market economy (Auyang, 1998). The division of intellectual labor, he highlighted, however, underlies the success of science just as the division of labor generates the prosperity in which common people can enjoy fine wines from around the world. How is the division of intellectual labor possible? What are the general conditions of the world that allow so many scientific disciplines to investigate the same glass of wine? What does the organization of science tell us about the general structure of theoretical reason?

Our scientific enterprise exhibits an interesting double-faced phenomenon (Auyang, 1998). On the one hand, scientists generally acknowledge that everything in the universe is made up of microscopic particles. On the other, various sciences investigate macroscopic entities with concepts of their own, which are drastically different from those of microscopic particles, with little if any reference to the latter. The phenomenon is often noticeable in a single science. For instance, hydrodynamics depicts a fluid as a continuous whole governed by its own dynamic equations with no mention of molecules, but when asked, hydrodynamic physicists readily admit that the fluid is made up of molecules (Auyang, 1998).

The division of science into many disciplines is not merely a matter of academic politics or intellectual fashion. It is objectively underwritten by the structure of the world. One can hardly imagine that our many sciences are supported by a world of perfect simplicity like that which prevailed shortly after the Big Bang or a world of utter randomness like that projected for the end of the universe. The plurality of science suggests a world with great diversity, but not so great as to overwhelm our cognitive capacity and be regarded as totally chaotic; a world that permits the discernment of phenomena at various levels of scale, organization, and complexity, each of which can be conceptualized and investigated with little reference to the others (Auyang, 1998).

A diverse world alone does not guarantee the plurality of science. We could have been one-track-minded and insisted on explaining everything by a single set of substantive concepts and principles, be they divine laws or the concepts of things observable by our naked senses or the laws of elementary particles. Then science would be like a centrally planned economy in which everyone labors to work out the consequences of the supreme directive. It is doubtful whether such a science can prosper. Fortunately, the human mind is more versatile. Actual science is more like a free-market economy. Various sciences find their special niches and open their domains of inquiry, each raising its own questions, positing its basic assumptions, adopting a suitable method, individuating a set of entities, and setting forth predicates to characterize a specific group of phenomena. The specialization of the sciences does not imply that they are totally disjoint and proceed regardless of each other. Just as players in the market observe legal and social rules that enforce contracts and promote cooperation, the sciences implicitly observe some general principles, which are revealed, among other things, in the acknowledgment of composition (Auyang, 1998).

Our sciences present the world as a hierarchical system with many branches, featuring individuals of all kinds. The individuals investigated by various sciences appear under different objective conditions. Despite the apparent independence of individuals in various levels, the hierarchy of individuals is not merely a matter of classification. It is underwritten by the composition of the individuals, although the idea of composition is tacit in many sciences. All individuals except the elementary particles are made up of smaller individuals, and most individuals are parts of larger individuals. Composition includes structures and is not merely aggregation. To say that something is made up of others is not to say that it is nothing but the others. A brick house is more than a heap of bricks; it has its architecture. A natural compound differs from a house in that its structure results from the spontaneous interactions among the elements that become its components (Auyang, 1998).

There are four known kinds of basic physical interaction: gravity, electromagnetism, and the two nuclear interactions (Auyang, 1998). The weak nuclear interaction plays little role in the formation of stable matter. The strong nuclear interaction is crucial below the nuclear level, but insignificant beyond, where the characteristic distances far exceed its range. Gravity, extremely weak but having purely cumulative effects, is responsible for the formation of astronomical objects and the large structures of the universe. The electromagnetic interaction produces the awesome diversity of medium-sized things we see around us. Like gravity, it has infinite range; unlike gravity, it is strong but its effects are not cumulative. When positively charged protons and negatively charged electrons combine to form an atom, they almost neutralize the electromagnetic interaction among them. The residue electromagnetic force among the atoms is weak and assumes various forms such as the hydrogen bond, the covalent bond, and the van der Waals potential, which are responsible for chemical bonds and reactions. Relations among organisms and humans take on forms that are drastically different from the forms of physical and chemical interactions. They have some kind of physical base. However, their connection to the physical interactions is best articulated in the negative sense that they will disappear if the electromagnetic force is turned off, in which case the organisms themselves disintegrate. We know from experience that the behavior of one organism or human affects the behaviors of others within a certain range, and the relations between the behaviors and effects have enough regularity to be characterized in general terms. Thus, we posit some kind of interaction among organisms and humans. Historically, the "power" one organism has on another was recognized before the physical interactions (Auyang, 1998).

Let us start with elementary-particle physics, which aims to study the elementary building blocks of the universe. It presents the elementary ontology as a set of quantum fields, three of which, the electron and the two quark fields, constitute all known stable matter of the universe. The quanta of excitation of the quantum fields are called particles, which are not immutable, but can easily transform into each other when fields interact. Up the ladder of complexity, we have nuclear and atomic physics. Nuclear physics investigates the structures of nucleons (neutrons and protons) and atomic nuclei made up of nucleons. The strong nuclear interaction, which binds quarks into nucleons and nucleons into nuclei, has a range of about 10–13 cm. Beyond this distance, it is unable to overcome the electric repulsion among the protons. The short range of the nuclear interaction limits the size of the nuclei, hence their variety (Auyang, 1998). Constrained by the variety of nuclei, there are only 109 kinds of atom, the heaviest of which, unnilennium, contains 266 nucleons and 109 electrons. A kind of atom can have several isotopes, which have the same number of electrons, but vary in the number of neutrons. They are the topics of atomic physics.

The diversity of composite systems on larger scales is helped by the availability of more than a hundred kinds of atom, instead of a few kinds of elementary particle as building blocks. More importantly, the electromagnetic and gravitational interactions both have infinite ranges and impose no basic limit on the size of the systems. Variety explodes, as the number of possible combinations of building blocks increases exponentially with the number of blocks. There are 1.5 million known kinds of stable chemical compound, even more kinds of organisms. Responding to the diversity, the sciences diverge. Very roughly, physics branches into cosmology, general relativity, the studies of solids, fluids, and plasmas in which atoms are decomposed into electrons and positively charged ions.

Chemistry studies the reactions among a great variety of molecules, organic and inorganic. It also enables us to synthesize long chains of molecules that constitute various plastics. Further up the slope of complexity, the miracle of life occurs in entities that grow and reproduce themselves. Biology divides into molecular biology, genetics, cytology for cells and cellular organelles, histology for tissues, ethology for organismic behaviors, physiology, ecology, evolution, and more (Auyang, 1998).

Within the diversity of life, the organisms of one species are conscious of their existence and have the capability to think in symbolic terms. They become the topic of the human sciences including psychology and cognitive research. People engage in meaningful communication, exchange, and conflict; they form groups and institutions that are meaningful to them. To study these phenomena, the social sciences spring up. Individuals in a certain level of scale and complexity interact most strongly among themselves. They are often posited independently and described without mentioning individuals in other levels. Concepts pertaining to one level bear little or no relation to concepts in remote levels, and most satisfactory explanations of an individual's behavior use concepts of its own level (Auyang, 1998).

Elementary-particle concepts are irrelevant to theories of solids, nuclear concepts are not pertinent in biology, and biological notions are beside the point in explaining many human actions. When a camper asks why the food takes so long to cook, physicists answer in terms of pressure at high altitudes. Physicists use macroscopic thermodynamic concepts, although they have microscopic theories in reserve, as the microscopic information is irrelevant and would obscure the explanation. The prominence of explanations on the same level secures a significant degree of autonomy for the sciences, for they investigate disparate entities and use logically independent concepts (Auyang, 1998).

Coherence and disruption. Adhesive interaction among individuals is the basis of composition and structure formation, but not all interactions are adhesive. The electromagnetic force among electrons is repulsive; hence, a group of negatively charged electrons will fly apart and never form a unit by themselves. They can participate in a system only if the system contains other ingredients, notably positively charged protons, which neutralize their antagonism. Besides antagonistic relations, excessive idiosyncrasy in individual behaviors also frustrates unity. No unity can be maintained if each individual goes its own way. A system can crystallize only if the individuals somehow harmonize their behaviors through agreeable relations, as a democracy exists only when a significant degree of consensus prevails among its citizens (Auyang, 1998). Furthermore, a system is endangered if it is subjected to external forces that attract the constituents, as villages dwindle when outside opportunities entice young people to leave. To succeed in forming a system, the cohesive interaction among the individuals in a collection must be strong enough to overcome internal disruption and external impediment. If the cohesion exceeds the forces that pull the individuals apart or attach them to outside elements, the collection acquires an integral structure and becomes a composite system on a larger scale. Thus, the criterion of cohesion is not absolute but relative; it is a tug-of-war between the ordering and disordering forces. The war can often be represented in terms of some variables suitable to the particular systems under study (Auyang, 1998).

In physics, the crucial variable is energy. The cohesiveness of a physical system is determined by the relative importance of two kinds of energy: the kinetic energy characterizing the agitation of each constituent and the binding energy characterizing the bond among the constituents. The kinetic energy is mainly determined by the ambient temperature. The higher the temperature, the higher the kinetic energy, and the more agitated the constituents become to go their own ways. When the kinetic energy matches or exceeds the binding energy of an entity, the constituents break free and the entity disintegrates (Auyang, 1998).

Thus, an entity is stable only in temperatures below its binding energy. Suitable data list composite entities together with their characteristic binding energies, which must be supplied to break an entity into its constituents. At the high temperatures found within split seconds of the Big Bang, quarks were free entities. Quarks interacted, but the interacting energy was small relative to the kinetic energies of individual quarks, so that each quark went its own way (Auyang, 1998). As the universe cooled, the kinetic energy of the quarks decreased. When the temperature dropped below the characteristic energy of the strong nuclear interaction, the quarks were bound into protons and neutrons.

At still lower temperatures, but no less than tens of billions of degrees Celsius, which are the temperatures found in the interior of stars, protons and neutrons fuse into nuclei. When a collection of elements coalesces into a stable composite entity such as a nucleus, it gives off an amount of energy equal to its binding energy. The energy released in nuclear fusion fuels the sun and the stars. Physicists are trying to harness it in nuclear fusion reactors. As the temperature falls further, nothing can happen to the nuclei. They are so stable and tightly bound that the motions of their components are negligible; thus, they are often approximately treated as simple units without internal structures (Auyang, 1998). New entities, the atoms, overcome the destructive force of thermal agitation and emerge to take over the center stage. In the low-energy environment typical of condensed-matter physics and chemistry, a large part of an atom is tightly bound and only electrons on the outermost atomic shell can break free, leaving a positively charged ion. The atoms, ions, and electrons become building blocks of molecules, chemical compounds, and solids (Auyang, 1998).

References

Ahlborn, B.K., 1999. Thermodynamic limits of body dimension of warm blooded animals. J. Non-Equilib. Thermodyn. 40, 407–504.

Ahlborn, B.K., Blake, R.W., 1999. Lower size limit of aquatic mammals. Am. J. Phys. **67**, 1–3.

Antonelli, P.L., Ingarden, R.S., Matsumoto, M., 1996. The Theory of Sprays and Finsler Spaces With Applications in Physics and Biology. Kluwer, Dordrecht.

Auyang, S.Y., 1998. Foundations of Complex-System Theories: In Economics, Evolutionary Biology, and Statistical Physics. Cambridge University Press, Cambridge, MA, USA.

Bejan, A., 1997. Theory of organization in nature: pulsating physiological processes. Int. J. Heat Mass Transf. 40, 2097–2104.

Bejan, A., Ledezma, G.A., 1998. Streets tree networks and urban growth: Optimal geometry for quickest access between a finite-size volume and one point. Physica A 255, 211–217.

Berry, R.S., Kazakov, V.A., Sieniutycz, S., Szwast, Z., Tsirlin, A.M., 2000. Thermodynamic Optimization of Finite Time Processes. Wiley.

Casas-Vazquez, J., Jou, D., 1985. On the thermodynamic curvature of nonequilibrium gases. J. Chem. Phys. 83, 4715–4716.

Chen, M., 1991. Dynamical stability and thermodynamic stability in irreversible thermodynamics. J. *Math. Phys.* 32, 744–748.

Ebeling, W., 1985. Thermodynamics of selforganization and evolution. Biomed. Biochim. Acta 44 (6), 831–838.

Ebeling, W., Feistel, R., 1992. Theory of selforganization and evolution: the role of entropy, value and information. J. Non-Equilib. Thermodyn. **17**, 303–332.

Ebeling, W., Klimontovich, Y.L., 1984. Selforganization and Turbulence. Teubner, Leipzig.

Feldmann, T., Levine, R.D., Salamon, P., 1986. A geometrical measure for entropy changes. J. Stat. Phys. 42, 1127–1133.

Gontar, V., 2000. Entropy principle of extremality as a driving force in the discrete dynamics of complex and living systems. Chaos Soliton. Fractals 11, 231–236.

Hofbauer, J., Sigmund, K., 1992. The Theory of Evolution and Dynamical Systems-Mathematical Aspects of Selection. Cambridge University Press.

Ingarden, R.S., 1977. Information thermodynamics and differential geometry. Tenso 31, 124–155.

Ingarden, R.S., 1996. On physical applications of Finsler geometry. Contemp. Math. 196, 213–222.

Ingarden, R.S., Kawaguchi, M., Sato, Y., 1982. Information geometry of classical thermodynamical systems. Tenso 39, 267–278.

Janyszek, H., Mrugala, R., 1990. Riemannian and Finslerian geometry and fluctuations of thermodynamic systems. Adv. Thermodyn. 3, 159–175.
Kantorovich, L.V., Krylow, V.I., 1958. Approximate Methods of Higher Analysis. Nordhoff, Groningen.
Landau, L., Lifshitz, E., 1975. Statistical Physics. Pergamon, Oxford.
Landsberg, P.T., 1984a. Two general problems in quantum biology. Int. J. Quantum Chem. Quantum Biol. Symp. 11, 55–61.
Landsberg, P.T., 1984b. Can entropy and "order" increase together. Phys. Lett. A 102, 171.
Landsberg, P.T., 1994. Self-organization, entropy and order. In: Misra, R.K., Maass, D., Zwierlein, E. (Eds.), On Self-Organization. Springer, Berlin., pp. 158–184.
Levine, R.D., 1985. Geometry in classical statistical thermodynamics. J. Chem. Phys. 84, 910–916.
Malchow, H., Schimansky-Geier, L., 1985. Noise and Diffusion in Bistable Nonequilibrium Systems. In: Teubner-Texte zur Physik. Teubner, Berlin. Band 5.
Mrugala, R., 1990. Contact and Metric Geometry in Thermodynamics (Thesis). Copernicus University, Institute of Physics, Torun.
Mrugala, R., 1996. On a Riemannian metric on contact thermodynamic spaces. Rep. Math. Phys. 38, 339–347.
Mrugala, R., 2000. Geometrical methods in thermodynamics. In: Sieniutycz, S., de Vos, A. (Eds.), Thermodynamics of Energy Conversion and Transport. Springer, New York, pp. 257–285.
Mrugala, R., Nulton, J.D., Schön, J.C., Salamon, P., 1990. A statistical approach to the geometric structure of thermodynamics. *Phys. Rev.* A 41, 3156–3165.
Orlik, M., 1996. Oscillatory Reactions: Order and Chaos. Wydawnictwa Naukowo Techniczne, Warszawa (in Polish).
Plastino, A.R., Miller, H.G., Plastino, A., 1997. Minimum Kullback entropy approach to the Fokker-Planck equation. *Phys. Rev.* E 56, 3927–3934.
Polani, D., 2010. On J. S. Avery's "Information theory and evolution": book review. Artif. Life 16 (4), 333–335.
Prigogine, I., 1955. Thermodynamics of Irreversible Processes. Wiley, New York.
Ruppeiner, G., 1979. Thermodynamics: a Riemannian geometric model. Phys. Rev. A 20, 1608–1613.
Ruppeiner, G., 1985. Comment on "length and curvature in the geometry of thermodynamics" Phys. Rev. A 20, 1608–1613.
Ruppeiner, G., 1990. Thermodynamic curvature: origin and meaning. In: Nonequilibrium Theory and Extremum Principles. Advances in Thermodynamics Series, vol. 3. Taylor and Francis, New York, pp. 129–159.
Ruppeiner, G., 1995. Riemannian geometry in thermodynamic fluctuation theory. Rev. Mod. Phys. 67, 605–659.
Rymarz, C., 1999a. Chaos and self-organization in living systems. J. Tech. Phys. 40, 407–504.
Rymarz, C., 1999b. Self-organization and chaos in atmosphere. Int. J. Bifurc. Chaos 9, 361–370.
Salamon, P., Nulton, J.D., Berry, R.S., 1985. Length in statistical thermodynamics. J. Chem. Phys. 82, 2433–2436.
Saunders, P.T., Ho, M.W., 1976. On the increase in complexity in evolution. J. Theor. Biol. 63, 375–384.
Saunders, P.T., Ho, M.W., 1981. On the increase in complexity in evolution. II. The relativity of complexity and the principle of minimum increase. J. Theor. Biol. 90, 515–530.
Schlögl, F., 1980. Stochastic measures in nonequilibrium thermodynamics. Phys. Rep. 62, 267.
Schneider, E., Kay, J., 1994. Complexity and thermodynamics: towards a new ecology. Future 26, 626–647.
Scholl, E., 1986. Influence of boundaries on dissipative structures in the Schlögl model. Z. Phys. B-Condensed Matter 62, 245–253.
Schrödinger, E., 1967. What Is Life? University Press, Cambridge (the first edition in 1944).
Shiner, J.S., 2000. Self-organized criticality: self organized complexity? The disorder and "simple complexities" of power law distributions. Open Syst. Inf. Dyn. 7, 131–137.
Shiner, J.S., Davison, M., Landsberg, P.T., 1999. Simple measure for complexity. Phys. Rev. E 59, 1459–1464.
Sieniutycz, S., 1991. Optimization in Process Engineering. Wydawnictwa Naukowo Techniczne (in Polish).
Sieniutycz, S., 1994. Conservation Laws in Variational Thermodynamics. Kluwer Academic Publishers, Dordrecht.
Sieniutycz, S., 2000. Action-type variational principles for hyperbolic and parabolic heat & mass transfer. Int. J. Appl. Thermodyn. 3 (2), 73–81.
Sieniutycz, S., 2002. Optimality of nonequilibrium systems and problems of statistical thermodynamics. Int. J. Heat Mass Transf. 45, 1545–1561.

Sieniutycz, S., Berry, R.S., 1991. Field thermodynamic potentials and geometric thermodynamics with heat transfer and fluid flow. *Phys. Rev.* A 43, 2807–2818.
Szwast, Z., 1997. An approach to the evolution of selected living organisms by information-theoretic entropy. In: Reports of Faculty of Chemical and Process Engineering at the Warsaw TU.vol. 24, pp. 123–143.
Szwast, Z., Sieniutycz, S., Shiner, J.S., 2002. Complexity principle of extremality in evolution of selected living organisms by information – Theoretic entropy. Chaos Soliton. Fractals 13 (9), 1871–1888.
Tribus, M., 1970. Thermostatics and Thermodynamics. Van Nostrand, Princeton.
Weinhold, F., 1975. The metric geometry of equilibrium thermodynamics. J. Chem. Phys. 63, 2479. ibid 2484, 2488 and 2496.
Weinhold, F., 1976. Metric geometry of equilibrium thermodynamics. V. Aspects of heterogeneous equilibrium. *J. Chem. Phys.* 65, 559.
Wootters, W.K., 1981. Statistical distance and Hilbert space. *Phys. Rev.* D 23, 357–362.
Zalewski, M., 2005. Chaos i Oscylacje w Reaktorach Chemicznych (Chaos and Oscillations in Chemical Reactors) (Ph.D. thesis). Faculty of Chemical and Process Engineering, Warsaw University of Technology Press, Warszawa.

Further reading

Lawton, B., Kingsburg, G., 1996. Transient Temperature in Engineering and Science. Oxford University Press.
Schneider, E., Kay, J., 1993. Energy degradation, thermodynamics and the development of ecosystems. In: Szargut, J., Kolenda, Z., Tsatsaronis, G., Ziębik, A. (Eds.), Proceedings of the International Conference ENSEC'93, Energy Systems and Ecology, vol., Cracov, Poland, July 5–9, pp. 33–42.

CHAPTER 4

Complex systems of neural networks

4.1 Introduction: General issues

We shall start this review with pointing out the role of ANN in fuel cells modeling and the development of low pollution vehicles. Increasing interest is, nowadays, devoted to the development of this field. Many related studies and implementations of fuel-cell-based vehicles are progressed. In the simplest form of a fuel-cell-based vehicle, a fuel-cell-based engine is fed with a stream of hydrogen, which reacts with air oxygen in a device called fuel cell stack, producing electric energy which can be exploited by an electric engine.

4.1.1 Review of some central notions of ANN theory

Van Hemmen and Kühn (1991) review some basic notions of the theory of neural networks. In so doing, they concentrate on collective aspects of the dynamics of large networks. The neurons are usually treated formally; formal neurons make the theory simpler and more lucid. The authors state that there are at least two ways of reading their review. It may be read as a self-contained introduction to the theory of neural networks. Alternatively, it may be regarded as a vademecum on models of neural networks. The article also contains some new and previously unpublished results.

4.1.2 ANN modeling of SOFC

Artificial-neural networks (ANN) are applied to model performance of fuel cell systems (Milewski and Świrski, 2009; Świrski and Milewski, 2009). The artificial-neural network is an arrangement that can be applied to simulate an object's behavior without an algorithmic solution merely by utilizing available experimental data. ANNs are used for modeling single cell behavior. The error back-propagation algorithm is used for an ANN training procedure. The ANN-based SOFC model has the following input parameters: current density, temperature, fuel volume flow density, and oxidant volume flow density. Based on these input parameters, cell voltage is predicted by the model. The results obtained show that the ANN can successfully be used for modeling a solid oxide fuel cell. The self-learning process of the ANN provides an opportunity to adapt the model to new situations (e.g., certain types of impurities at inlet streams, etc.). Based on the results of this study, it can be concluded that, by using the ANN, a SOFC system can be modeled with reasonable accuracy. In contrast to traditional models, the

Complexity and Complex Thermo-Economic Systems. https://doi.org/10.1016/B978-0-12-818594-0.00004-0

ANN model is able to predict the cell voltage without reference to numerous physical, chemical, and electrochemical factors.

In the neural network application described by Milewski and Świrski (2009), the optimal network architecture can be displayed, analyzed, and assessed. The ANN-based SOFC model has the following input parameters: current density, temperature, fuel volume flow density (mL/min/cm^2), and oxidant volume flow density. Based on these inputs, cell voltage is predicted by the model. The results obtained show that ANN can be successful for modeling of solid oxide fuel cells. The self-learning process of the ANN provides an opportunity to adapt a model to new situations (e.g., certain types of impurities at inlet streams, etc.). Many results show that, by using ANNs, an SOFC can be modeled with relatively high accuracy. In contrast to traditional models, the ANN is able to evaluate cell voltage without knowledge of numerous physical, chemical, and electrochemical factors.

4.2 Training, modeling, and simulation

4.2.1 NN-based dynamic controller of PEM FC

Hattia and Tioursi (2009) describe artificial intelligence techniques which may be applied to control a proton exchange membrane fuel cell (PEMFC system). In particular, they consider the use of dynamic neural networks. They work with a dynamic neural network control model obtained by introducing a delay line in the input of the neural network. A static production system including a PEMFC is subjected to variations of active and reactive power. Therefore, the goal is to produce a system that follows these imposed variations. The simulation requires modeling of the principal element (PEMFC) in a dynamic mode. The simulation results demonstrate that the model-based dynamic neural network control scheme is appropriate for controlling operations. The stability of the identification and the tracking error are analyzed, and some reasons for the usefulness of this methodology are given.

4.2.2 ANN role in development of low pollution vehicles

In the paper by Caponetto et al. (2005), a nonlinear dynamical model of a fuel cell stack is developed by means of artificial-neural networks. The model presented is a black-box model, based on a set of easily measurable exogenous inputs such as pressures and temperatures at the stack. The model is capable of predicting the output voltage of the fuel cell stack. The model is exploited as a component of complex control system able to manage the energy flows between fuel cell stack, battery pack, auxiliary systems, and electric engine in a zero-emission vehicle prototype. This paper involves topics on control engineering computing, fuel cell vehicles and fuel cells, multilayer perceptrons, nonlinear dynamical systems, and power engineering computing.

4.2.3 ANN to predict SOFC performance in residential spaces

Entchev and Yang (2007) apply the techniques of adaptive neuro-fuzzy inference system (ANFIS) and an artificial-neural network (ANN) to predict the performance of solid oxide fuel cell (SOFC), while supplying both heat and power to a residence. A microgeneration SOFC system installed at the Canadian Centre for Housing Technology (CCHT) is integrated with existing mechanical systems and connected in parallel to the grid. SOFC performance data are collected during the winter heating season and used for training both ANN and ANFIS models. The ANN model is built on back-propagation algorithm; as for ANFIS model, a combination of least squares method and back-propagation gradient decent method is developed and applied. Both models are trained with experimental data and used to predict the selective SOFC performance parameters, such as fuel cell stack current, stack voltage, etc.

The study reveals that both ANN and ANFIS models' predictions agree with a variety of experimental data sets representing steady state, start-up, and shut-down operations of the SOFC system. The initial data set is subjected to detailed sensitivity analysis, and statistically insignificant parameters are excluded from the training set. As a result, significant reduction of computational time is achieved without affecting the models' accuracy. The study shows that adaptive models can be applied with confidence during the design process and for performance optimization of existing and newly developed SOFC systems. It demonstrates that by using ANN and ANFIS techniques, SOFC microgeneration system's performance can be modeled with minimum time and with a high degree of accuracy.

4.2.4 Modeling the study and development of fuel cells

Sousa Jr. et al. (2006) and Sousa (2007) apply mathematical modeling to the study and development of fuel cells. Models of gas diffusion electrodes and proton exchange membrane biochemical fuel cells are developed. For typical physical systems, mechanistic models are obtained, based on available knowledge. However, for biochemical fuel cells, semiempirical and empirical models are used. In this work, there are three objectives: to characterize a phenomenological model for a Pt-air cathode and perform appropriate simulations; to characterize a semiempirical model to predict the performance of a Pt-H_2/H_2O_2-peroxidase fuel cell; and to evaluate the effectiveness of (empirical) neural networks to predict the performance of a Pt-H_2/O_2-peroxidase fuel cell. The developed mechanistic model of a Pt-air cathode is based on material balances, on Fick's law of diffusion, and on Tafel kinetics. A semiempirical model based on Michaelis-Menten kinetics, in turn, can predict the performance of a Pt-H_2/H_2O_2-peroxidase biochemical fuel cell. Artificial-neural networks are capable of fitting the potential/current relationship of a Pt-H_2/O_2-peroxidase biochemical fuel cell.

4.2.5 NN modeling of polymer electrolyte membrane FC

Lobato et al. (2009) investigate three types of neural networks which have as common characteristic the supervised learning control (multilayer perceptron, generalized feed-forward network, and Jordan and Elman network). These neural networks have been designed to model the performance of a polybenzimidazole-polymer electrolyte membrane fuel cells operating upon a temperature range of 100–175°C. The influence of temperature of two periods is studied: the temperature in the conditioning period and temperature when the fuel cell is operating. Three inputs variables, the conditioning temperature, the operating temperature, and current density, are taken into account in order to evaluate their influence upon the potential, the cathode resistance, and the ohmic resistance. The multilayer perceptron model provides good predictions for different values of operating temperatures and potential and, hence, it is the best choice among the study models, recommended to investigate the influence of process variables of PEMFCs.

4.3 Performance prediction, optimization, and related issues

4.3.1 GA-RBF neural networks modeling

Wu et al. (2007) build a nonlinear offline model of the solid oxide fuel cell (SOFC) by using a radial basis function (RBF) neural network based on a genetic algorithm (GA). During the process of modeling, the GA aims to optimize the parameters of RBF neural networks and the optimum values are regarded as the initial values of the RBF neural network parameters. Furthermore, we utilize the gradient descent learning algorithm to adjust the parameters. The validity and accuracy of modeling are tested by simulations. Besides, compared with the BP neural network approach, the simulation results show that the GA-RBF approach is superior to the conventional BP neural network in predicting the stack voltage with different temperature. Thus, it is feasible to establish the model of SOFC stack by using RBF neural networks identification based on the GA.

4.3.2 NN on-board FC power supplier

Proton exchange membranes are associated with one of the most promising fuel cell technologies for transportation applications. Considering the aim and importance of transportation applications, a simulation model of the whole fuel cell system may be a major milestone. The model (Jemeï et al., 2003) enables the optimization of the complete vehicle (including all ancillaries, output electrical converter, and their dedicated control laws). In a fuel cell system, there is a strong relationship between available electrical power and actual operating conditions: gas conditioning, membrane hydration state, temperature, current set point.... Thus, a "minimal behavioral model" of a fuel cell system able to evaluate output variables and their variations is highly interesting. Artificial-neural networks (NN) are a very efficient tool to reach such an aim. In this paper, a proton exchange membrane fuel cell (PEMFC) system neural network model is proposed. It is implemented on Matlab/Simulink® software and integrated in a complete vehicle powertrain. This makes it possible to carry out the development and a simulation of the control laws in order to drive energy transfers on-board fuel cell vehicles.

4.3.3 NN-based modeling of PEM FC and controller synthesis

Mammar and Chaker (2012) concentrate on possibilities of applying artificial-neural networks for creating an optimal model of the proton exchange membrane fuel cell (PEMFC). Their paper represents techniques based on neural networks to control the active power output from a fuel cell power generation system. Various ANN approaches are tested; the back-propagation feed-forward networks show satisfactory performance with regard to cell voltage prediction. The model is then used in a power system for residential application. This approach requires an ANN fuel cell stack model, reformer model, and DC/AC inverter model. Furthermore, a neural network controller (NNTC) and fuzzy logic controllers (FLC) are used to control active power of PEM fuel cell system. A controller modifies the hydrogen flow feedback from the terminal load. The validity of the controller is verified when the FC system is used jointly with the neural network controller NNTC to predict the response of the active power to: (a) computer-simulated step changes in the load active and reactive power demand, and (b) actual active and reactive load demand of a single family residence. Simulation results confirm the high performance capability of the NNTC to control power generation. Artificial-neural networks can be trained to simulate the performance of a fuel cell with great accuracy; consequently, the same concept could be extended to other components, and thus, bigger and more complex cycles can be simulated in reduced time.

4.3.4 NN model of drying and thermal degradation

Kamiński and Tomczak (1999) test the description of drying dynamics by an artificial-neural network. Drying is an evaporation process from solids with change of moisture and temperature in time. The whole operation is split into two separate stages, drying and degradation. Drying is modeled by a neural network which uses radial basis function, while degradation is modeled with the help of a multilayer perceptron. Final inputs to the neural model are drying agent parameters at the inlet and the initial moisture and temperature of the material, while the output is the quality index and it changes with time. Experimental data for drying and degradation of vitamin C for sliced potatoes, green peas, and silica gel saturated with ascorbic acid are used to prove the accuracy of the approach.

4.3.5 NN optimization of energy systems

Artificial-neural networks are useful both in designing and in optimization of energy systems. Investigations and models based on artificial-neural networks (ANN) have been made first for SOFC type of fuel cells. Since the obtained results are promising (Świrski and Milewski, 2009), the need emerges for making similar ANN investigations for one of the most developed high-temperature fuel cells, i.e., molten carbonate fuel cells (MCFCs). However, the difficulties with obtaining reliable models are particularly troubling in case of MCFCs. The complexity of FC processes causes complications in their mathematical description; therefore, it seems reasonable to expect more ANN FC applications in the near future.

4.3.6 NN model for fluidized bed

Palancar et al. (2001) explore the application of an artificial-neural network (ANN) to model a continuous fluidized bed dryer. The ANN predicts the moisture and temperature of the output solid. A three-layer network with sigmoid transfer function is used. The ANN learning is accomplished by using a set of data obtained for the operation simulation by a classical model of the dryer. The number of hidden nodes, the learning coefficient, the size of learning data set, and a number of iterations in the learning of the ANN are optimized. The optimal ANN has five input nodes and six hidden nodes. It is possible to predict, with an error less than 10%, the moisture and temperature of the output dried solid in a small pilot plant which can treat up to 5 kg/h of wet *alpeorujo*. This is identified as the wet solid waste generated in the two-phase decanters used to obtain olive oil.

4.3.7 NN control of power of a FC

El-Sharkh et al. (2004) present a neural networks (NN)-based active and reactive power controller of a stand-alone proton exchange membrane (PEM) fuel cell power plant (FCPP). The controller modifies the inverter modulation index and the phase angle of the ac output voltage to control the active and reactive power output from an FCPP to match the terminal load. The control actions are based on feedback signals from the terminal load, output voltage, and hydrogen input. The validity of the controller is verified when the FCPP model is used in conjunction with the NN controller to predict the response of the power plant to: (a) computer-simulated step changes in the load active and reactive power demand, and (b) actual active and reactive load demand of a single family residence. The response curves indicate the load following characteristics of the model and the predicted changes in the analytical parameters highlighted by the analysis.

4.3.8 Performance prediction and analysis of the cathode catalyst layer

Khajeh-Hosseini-Dalasm et al. (2011) highlight the following results of their research into using artificial-neural network in prediction and analysis of the cathode catalyst layer of proton exchange membrane fuel cells: (1) For the first time, artificial-neural networks together with statistical methods (ANOM and ANOVA methods) are employed for modeling, prediction, and analysis of an agglomerate cathode catalyst layer (CL) performance. (2) ANOM and ANOVA methods allow to extract physical explanations regarding the underlying system modeled using the ANN approach. Determined cathode CL thickness and membrane volume content in the catalyst layer are the most significant structural parameters affecting the catalyst layer performance.

4.3.9 Neural network model for a PEM FC

Saengrung et al. (2007) investigate performance prediction of a commercial proton exchange membrane (PEM) FC system by using artificial-neural networks (ANNs). Two artificial-neural networks including back-propagation (BP) and radial basis function (RBF) networks are constructed, tested, and compared. Experimental data as well as preprocess data are utilized to determine the accuracy and speed of several prediction algorithms. The performance of the BP network is investigated by varying error goals, the number of neurons, the number of layers, and training algorithms. The prediction performance of RBF network is also presented. The simulation results show that both the BP and RBF networks can successfully predict the stack voltage and current of a commercial PEM fuel cell system. Speed and accuracy of the prediction algorithms are quite satisfactory for the real-time control of this particular application.

4.3.10 ANN capillary transport characteristics of FC diffusion media

An appropriate choice of the fuel cell diffusion media design criteria is critical to achieving high fuel cell performance and durability, since the porous diffusion medium (DM) plays a deterministic role in establishing an effective micro-fluidic management in fuel cell operations. The study by Kumbur et al. (2008) addresses the development of a DM design tool using artificial-neural network and newly available direct benchmark data to describe the capillary pressure-saturation relationship in various fuel cell DMs. Direct drainage capillary pressure-saturation data have been generated for DMs coated with different loadings. Detailed descriptions of the experimental approach and major findings have been documented in a series of publications. The benchmark data compiled from these experiments have been integrated into the three-layered ANN which processes the feed-forward error back-propagation methodology. The network was systematically trained with the novel benchmark data, and then utilized to delineate the relative significance of PTFE content and compression on the capillary pressure of this class of DM materials, within the range of tested parameters. The artificial intelligence model presented in this research can be extended further into a more detailed design algorithm which couples the DM internal architecture with the cell performance, manufacturer variables, and operating conditions as more benchmark data are available. Such a tool can potentially reduce the number of required experiments and provide improved means of selecting the optimum electrode configurations suitable for different FC operations.

4.3.11 AI modeling of FC performance

As stressed by Ogaji et al. (2006), over the last few years, fuel cell technology has been increasing its share in the generation of stationary power. Numerous pilot projects are operating worldwide, continuously increasing the amount of operating hours either as stand-alone devices or as part of gas turbine combined cycles. An essential tool for adequate and dynamic analysis of such systems is a software model which enables the user to assess a large number of alternative options in the least possible time. On the other hand, the sphere of application of artificial-neural networks has widened covering such endeavors of life such as medicine, finance, and unsurprisingly engineering (diagnostics of faults in machines). Neural network models represent an important tool of Artificial Intelligence for fuel cell researchers in order to help them to elucidate the processes within the cells, by allowing optimization of materials, cells, stacks, and systems and support control systems. Artificial-neural networks are described as diagrammatic representation of a mathematical equation which receives values (inputs) and gives out results (outputs). Artificial-neural networks systems have the capacity to recognize and associate patterns, and because of their inherent design features, they can be applied to linear and nonlinear problem domains.

Ogaji et al. (2006) model the performance of fuel cells by using an artificial-neural network. The inputs to the network are variables critical to a fuel cell performance, while the outputs are some FC design variables affecting the network performance. Critical parameters for the cell include those describing the geometrical configuration as well as the operating conditions. For the neural network, discussed are effects of various network design parameters such as: network size, training algorithm, activation functions, and their effects on the effectiveness of the performance modeling. Results of the analysis as well as limitations of the approach are discussed.

4.3.12 Fuel cell applications

4.3.12.1 ANN simulator for SOFC performance

Arriagada et al. (2002) describe the development of a novel modeling tool for the evaluation of a SOFC performance. An artificial-neural network (ANN) is trained with a reduced amount of data generated by a validated cell model, and it is then capable of learning the generic functional relationship between inputs and outputs of the system. Once the network is trained, the ANN-driven simulator can predict different operational parameters of the SOFC (i.e., gas flows, operational voltages, current density, etc.), thus avoiding detailed description of the fuel cell processes. The highly parallel connectivity within the ANN further reduces the computational time. In a real case, the necessary data for training the ANN simulator would be extracted from experiments. This simulator could be suitable for different applications in the fuel cell field, such as, the construction of performance maps and operating point optimization and analysis. All these are performed with minimum time demand and good accuracy. This intelligent model together with the operational conditions may provide useful insight into SOFC operating characteristics and improved means of selecting operating conditions, reducing costs, and the need for extensive experiments.

The methanol concentrations, temperature, and current are considered as inputs, the cell voltage is taken as output, and the performance of a direct methanol fuel cell (DMFC) is modeled by adaptive-network-based fuzzy inference systems (ANFIS).

4.3.12.2 NN with SOFC control of power supply improvement

Jurado (2003) demonstrates the potential of a solid oxide fuel cell (SOFC) to perform functions other than the supply of real power to the grid. However, these additional functions require the use of an inverter. The flux-vector control is used very effectively for the control of this inverter, where the space-vector pulse width modulation (SVM) is implemented by neural networks (NNs). The results show the effect of the fuel cell behavior on the voltage at the sensitive load point. Interesting aspects are: FC modeling, improved performance of fuel cell, utility-connected inverter control, and neural network control schemes. The performance of the fuel cell is found to be excellent.

4.4 Neural networks for emission prediction of dust pollutants

Bator and Sieniutycz (2006) and Sieniutycz and Szwast (2018) review in detail the recent applications of artificial-neural network (ANN) for predicting suspended particulate concentrations in urban air, taking into account meteorological conditions. See detailed report in Section 4.6. Calculations are based on pollution measurements taken in Poland in the period 2001–02. PM10 emission and primary meteorological data, which were obtained from the Inspectorate for Environmental Protection (IEP) in Radom, were used to train and test the applied network. Two different methods of emission calculation can be applied. Firstly, artificial-neural network method based on multilayer perceptron with unidirectional information flow is used. Secondly, a hybrid model based on a modified Gaussian model of Pasquille's type and artificial-neural network with radial base function (RBF) is applied. Network architecture and transition function types are described. Statistical assessment of the results is described. In addition, hybrid model results are compared with emission calculations of dust pollution based on the Gaussian model, including various methods of calculation of pollution dispersion coefficients (see more in Section 4.6).

4.5 Hybrid and fuzzy systems

4.5.1 NN hybrid model of direct internal reforming SOFC

Chaichana et al. (2012) develop a mathematical model as an important tool for analysis and design of fuel cell stacks and systems. In general, a complete description of fuel cells requires an electrochemical model to predict their electrical characteristics, i.e., cell voltage and current density. However, obtaining the electrochemical model is quite a difficult and complicated task as it involves various operational, structural, and electrochemical reaction parameters. In their study, a neural network model is first proposed to predict the electrochemical characteristics of solid oxide fuel cell (SOFC). Various NN structures are trained based on the back-propagation feed-forward approach. The results show that the NN with optimal structure reliably provides a good estimation of fuel cell electrical characteristics. Then, a neural network hybrid model of a direct internal reforming SOFC, combining mass conservation equations with the NN model, is developed to determine the distributions of gaseous components in fuel and air channels of SOFC as well as the performance of the SOFC in terms of power density and fuel cell efficiency. The effects of various key parameters, e.g., temperature, pressure, steam to carbon ratio, degree of prereforming, and inlet fuel flow rate on the SOFC performance under steady-state and isothermal conditions, are also investigated. A combination of the first principle model and ANN provides a significant tool for an adequate prediction of SOFC performance and the reduction of computational time.

4.5.2 Hybrid NN model for PEM FC

Ou and Achenie (2005) discuss a neural network modeling approach for developing a quantitatively good model for proton exchange membrane (PEM) fuel cells. Various ANN approaches are tested. The back-propagation feed-forward networks and radial basis function networks ensure satisfactory performance with regard to cell voltage prediction. The effects of Pt loading on the performance of the PEM fuel cell are specifically studied. The results show that the ANN model is capable of simulating the effects for which there are currently no valid fundamental models available from the open literature.

Two novel hybrid neural network models (multiplicative and additive), each consisting of an ANN component and a physical component, are developed and compared with the full-blown ANN model. The results of the cell voltage predictions from hybrid models ensure the performance comparable to those from the ANN model. In addition, the hybrid models show performance gains over the physical model. The additive hybrid model ensures better accuracy than the multiplicative hybrid model.

4.5.3 ANN of the mechanical behavior of PEM FC

Fuel cell system model able to mimic and analyze vibration tests is sometimes demanded. However, in the literature this mechanical aspect of FC systems is usually neglected. Rouss and Charon (2008) propose neural network modeling for mechanical nonlinear behavior of a proton exchange membrane (PEM) fuel cell system. An experimental set is designed for this purpose: a fuel cell system in operation is subjected to random and swept-sine excitations on a vibrating platform in three directions. Its mechanical response is measured with three-dimensional accelerometers. Raw experimental data are exploited to create a multiinput and multioutput (MI-MO) model using a multilayer perceptron neural network combined with a time regression input vector. The model is trained and tested. Results from the analysis show good prediction accuracy. This approach is promising because it can be extended to more complex applications. In the future, the mechanical FC system controller will be implemented on a real-time system which provides an environment to analyze the performance and optimize mechanical parameters' design of the PEM fuel system and its auxiliaries.

4.5.4 ANN model of performance of a direct methanol FC

Wang et al. (2008) model the performance of a direct methanol fuel cell (DMFC) by adaptive-network-based fuzzy inference systems (ANFIS). Concentrations, temperature, and current are inputs, the cell voltage is an output. The artificial-neural network (ANN) and polynomial-based models are selected to be compared with the ANFIS in respect of quality and accuracy. Next, based on the obtained ANFIS model, they study characteristics of the DMFC. Methanol concentrations and temperatures are adjusted according to the load on the system. Within a restricted current range, methanol concentration does not affect stack voltage substantially.

4.6 Bator's example

Below, we review a valuable application of artificial-neural network (ANN) for predicting suspended particulate concentrations in urban air, taking into account meteorological conditions. Described calculations are based on pollution measurements taken in the city of Radom, Poland, in the period 2001–02. PM10 emission and primary meteorological data, which were obtained from IEP in Radom, were used to train and test the applied network. Two different methods of emission calculation can be applied. Firstly, artificial-neural network method based on multilayer perceptron with unidirectional information flow is used. Secondly, a hybrid model based on a modified Gaussian model of Pasquille's type and artificial-neural network with radial base function (RBF) is applied. Network architecture and transition function types are described. Statistical assessment of the results is described. In addition, the hybrid model results are compared with emission calculations of dust pollution based on the Gaussian model, including various methods of calculation of pollution dispersion coefficients. There are many interesting practical topics as: dust emission, reduction of air pollution, hybrid models of neural networks, and role of statistical tests.

4.6.1 Introduction

While it is commonly known that industrial energy systems often have deleterious impact on the environment (in particular, the quality of environmental air), the reverse effect is usually ignored. It is usually assumed that variations in the environmental state have little impact on the performance of an energy system. Nonetheless, energy and chemical systems can be found in which contamination of the local atmosphere may have a clearly negative impact on the system performance. For example, automobile catalysts' life will be shorter in a polluted atmosphere. As the oxygen in air is a common oxidizing medium in industrial burners, pollution of environmental air can cause improper combustion. In particular, local burning of flammable dust particles can enhance chemical instabilities and facilitate run-away of exothermic reactors. Fouling of heat exchanger pipes caused by dusty air flow may reduce the exchanger's thermal efficiency over a long-run performance period.

4.6.1.1 Significance of pollution reduction

The above examples show that problems of air pollution reduction may be considered not only in the most common context of the ecological policy, but also in the context of energy management. In the energy framework, the attitude focused on energy systems substantiates the perusal of the research into management of atmospheric quality on local and regional scales, the case of the following sections.

4.6.1.2 Atmospheric air

Atmospheric air is a main component of natural environment and air quality is determined by its purity. A large accumulation of people, companies, transportation, and municipal infrastructure in rapidly growing urban areas causes large concentration of anthropogenic sources of air pollutions emission, including dust pollutants (Meadows et al., 1995; Głowiak et al., 1985). Environmental conditions, and especially the condition of atmospheric air, are significant factors in estimating life quality in urban and industrial areas and main factors in estimating ecological policy progress. Atmospheric quality management problems on local and regional scales must be taken into consideration as a component of ecological policy (Borys, 1999; Noga, 1996). Administrative decisions concerning atmospheric air protection, and improvement of living conditions, can be properly undertaken, if time and spatial distribution of air pollution concentration are known (Bator and Kacperski, 2001a,b).

4.6.1.3 Dusts

Dusts, as distinct from other air pollutants, are heterogeneous media, and their harmfulness depends on their chemical and mineralogical constitution and on particle size. Fine dust PM10 (the respirable suspended fraction of dust with aerodynamic diameter of particles smaller than 10 µm) is significant for human organism. The dusts are accumulated in bottom section of respiratory system. This phenomenon is a main reason for pathological changes in the lungs, especially lung diseases (Ensor, 1998). So far, a great deal of chemical compounds contained in dust particles accumulated in the respiratory system have not been studied for their influence on the human organism. However, estimation and analysis of air polluted by dust, especially particles of PM10, are very important (Bator and Kacperski, 2001a,b).

4.6.1.4 Quantitative aspects of pollution emission

Quantity of pollution emission can be found in two ways: by direct measurement or through mathematical simulations.

Direct measurement of pollution emission

The first method requires a suitable monitoring network and is implemented by all local authorities, the responsibility for air pollution control resting with local units of Inspectorate for Environmental Protection (IEP). The established monitoring policy serves mainly to determine the actual concentration of pollutants with respect to their admissible concentration in the air. Yet, while this policy conforms with the current law of environmental protection, it is not sufficiently effective to warrant correct and nonambiguous administrative decisions which would ensure better standards of living. The current practice can be improved because a new way of monitoring atmosphere protection strategies has been advanced in the last few years. The new approach exploits several-years-monitoring data to predict pollution concentration in monitored regions by application of artificial-neural network (Skrzypski, 2002). The prediction results will indicate regions where the concerted effort of local authorities should be focused to eliminate pollutions.

Mathematical simulations of pollution emission

The second method of determining pollution emission consists in mathematical simulation of pollution propagation in atmospheric air. The simulation can be used to estimate the present condition of air pollution (diagnostic models) and to predict future emission concentrates of pollution (prediction models). Lots of mathematical models of air pollution propagation exist in literature (Madany and Bartchowska, 1995; Budziński, 1997; Juda-Rezler, 1998; Zannetti, 1990, 1992; Markiewicz, 1996; Straszko and Paulo, 1996).

4.6.1.5 Management of environmental protection

Artificial-neural networks, monitoring of air pollution in atmosphere and predicting emission concentrations, are a suitable tool in constructing a management system for environmental protection. Prediction by ANN takes into account essential characteristics of pollution, i.e., variation of pollution parameters of air flowing into town and variation of pollution emissions from unorganized sources of emission. However, in the current practice, the latter is omitted in a majority of applied models in some countries, including the Gaussian model, which is used to predict pollution emission from industrial premises situated locally.

4.6.2 Aims, scope, and assumptions

The study described below is both theoretical and experimental. It is principally based on updated Bator's past work (Bator, 2005; Bator and Sieniutycz, 2006). The primary purpose of this section is to develop a mathematical model for predicting emission concentration of dust pollutions (PM10) and to characterize basic virtues of this model.

The immission prediction method involves two approaches:

- hybrid model based on modified Gaussian model of Pasquille's type and artificial-neural network with radial base function (RBF),
- artificial-neural network—multilayer network with unidirectional flow of information (multilayer perceptron MLP).

An equation to determine a pollution concentration in a receptor point in Gaussian model of Pasquille's type contains two coefficients of pollution dispersion. There are a number of methods to calculate these coefficients. In this section, the following methods are used:

- Pasquille—Gifford's (Zannetti, 1990)
- Brookhaven's (Zannetti, 1990)
- Briss' (Reible, 1999; Briggs, 1983)
- Christian's (Zannetti, 1990; Paterkowski et al., 1997),
- Nowicki's (Directives of Polish Ministry of Environmental Protection, 1981; Directives for Calculation of Pollution Index of Atmospheric Air, 1983) (presently applied in Poland)
- Bator's (Bator, 2005; Bator and Sieniutycz, 2006).

The method applied in this volume is, in fact, the method based on Bator's works (Bator, 2005; Bator and Sieniutycz, 2006).

In the classical Gaussian model of Pasquille's type, meteorological conditions are taken into consideration in the form of a constant meteorological parameter which is determined experimentally for given meteorological station (Directives of Polish Ministry of Environmental Protection, 1981; Directives for Calculation of Pollution Index of Atmospheric

Air, 1983). In the hybrid model used in this study, the meteorological parameter is calculated by artificial-neural network with radial base function (RBF). The numerical value of this parameter is estimated on the basis of actual weather conditions in the considered area and for the analyzed time period. This way enables using the Gaussian model to calculate average concentrations for any period of time (period for which the meteorological parameter, m, was estimated). Application of classical Pasquille's model is restricted to determination of maximal and annual average concentrations.

4.6.3 Experimental data

Portable sampler MiniVol of AIRmetrics was used for the measurement of atmospheric air pollution by dust PM10. The measurement technique using the sampler is a modified standard method applied for PM10 dusts, described in Code of Federal Regulations (40 CFR 50).

To satisfy these criteria, sampler PM10 must be equipped with:

- system of tested inlet air, ensuring distinction of dust particles' size,
- flow rate controller, ensuring constant flowrate in specified limits,
- system measuring total air flow in the investigation period,
- work time controller, starting and stopping sampler.

All these requirements are compiled with portable sampler MiniVol Airmetrics. In fact, this sampler is a useful device to carry out studies of atmospheric air dustiness by PM10 dust.

Monitoring of air pollution by PM10 was carried out (Bator, 2005) for three measurement series.

> **SERIES I**—measurements were made every day from January 1, 2001 to December 31, 2002 at one point located near the building of the Faculty of Material Science and Footwear Technology of Radom Technical University, 27 Chrobrego Street in Radom (point P5 in Fig. 4.1),
> **SERIES II**—measurements were made for 30 days from September 19 to October 18, 2001 in 9 measurement points,
> **SERIES III**—measurements were made for 30 days from October 25 to November 25, 2001 in 9 measurement points.

Localization of all measurement points in Radom for the second and third series are shown in Fig. 4.1.

The samplers worked for 23 h daily. During 1-h breaks, the filter and the battery were replaced and the samples were inspected to ensure correct work of the devices. Measurement results are presented as graphs of time series for 1-day averages of PM10 dust concentrations (1st series—Figs. 4.2 and 4.3; 2nd series—Table 4.1; 3rd series—Table 4.2).

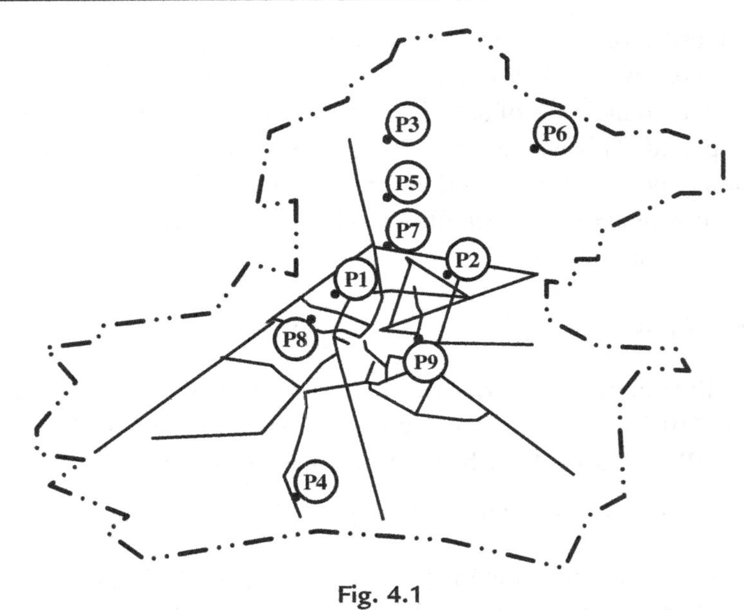

Fig. 4.1
Localization of measurement points for II and III series.

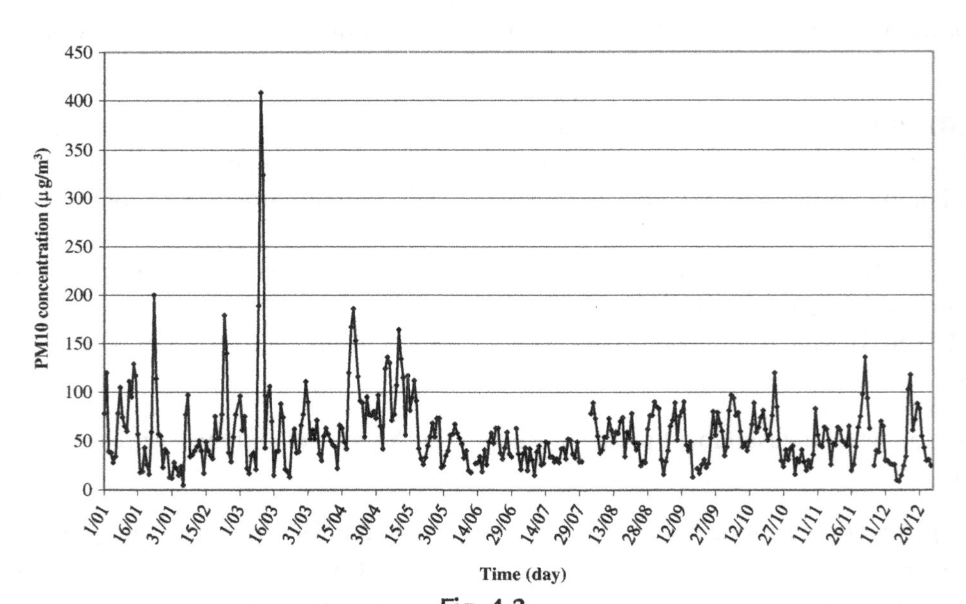

Fig. 4.2
Twenty-four-hour-mean concentration distribution of PM10 dust for 2001.

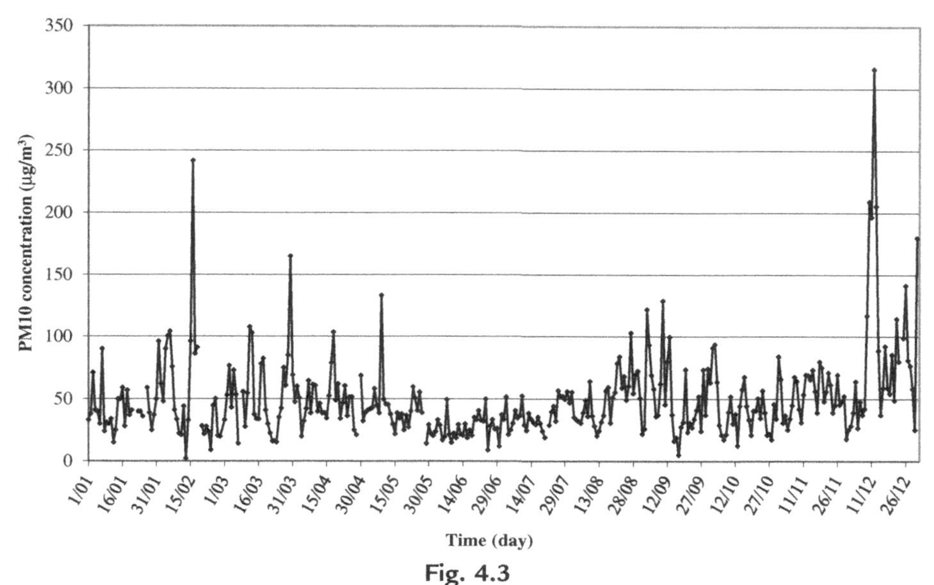

Fig. 4.3
Twenty-four-hour-mean concentration distribution of PM10 dust for 2002.

Primary meteorological data (temperature, pressure, and humidity of atmospheric air and direction and speed of wind) for investigation period were obtained from IEP in Radom (performing continuous monitoring of these air parameters).

4.6.4 Artificial-neural networks

Artificial-neural networks (ANN) are a system of parallel and dispersed information processing. They are capable of "network generalizing," that is, they can find correct solution for data that do not appear in the set of training data (data that is not presented during the network training).

Network structure for each new problem is created by suitable connections of neurons, forming many types of neural networks. Each type of network is closely connected with appropriate method of weights estimation, i.e., network training. Over 10 types of ANN accomplishing different tasks can be distinguished (Osowski, 1996).

For prediction of atmospheric pollution emission, multilayer networks MLP and networks with radial base function (RBF) are mainly applied.

4.6.5 Emission prediction with multilayer perceptron method

In a proposed method, multilayer perceptron (MPL) is applied. The ANN consists of four main layers: input layer, two hidden layers, and output one. Scheme of used networks is shown in Fig. 4.4.

Table 4.1 Twenty-four-mean concentration distribution of PM10 dust from 19.09.2001 to 18.10.2001.

SERIES II	P1	P2	P3	P4	P5	P6	P7	P8	P9
19-09	36	41	43	38	61	28	52	46	45
20-09	52	42	42	31	53	38	36	47	
21-09	59	47	44	50	71	47	49	35	56
22-09	31		22	27	26	39	32		39
23-09	38		31	65	32	23	50	53	43
24-09	43	31	43	38	56	32		44	42
25-09	36	37	38	27	26	31	24	29	38
26-09	51	52	36	47	52		69	50	47
27-09	49	59		38	57		58	50	34
28-09	43	68	54	26	51	46	67	53	70
29-09	81	72	58	40	68		59	105	88
30-09	59	51	36	44	42	81	50	55	58
1-10	54	70	64	67	24	54	67	78	68
2-10	54	42	40	42	49	11	42	40	42
3-10	56	41	54	41	52	59	53	52	
4-10	20	48	23	28	57	18	35	36	44
5-10	67	60		57	59	40	68	45	73
6-10	62	60	47	39	62	45		59	59
7-10	43	63	67	48	38	66	80	67	71
8-10	74	81	85	67	80	72	75	76	82
9-10	107	109	120	53	71	96	128	109	125
10-10	56	61	47	25	61	28	67	65	60
11-10	31	35	28	30	46	24	34	41	42
12-10	54	52	80	61	60	49	63	81	63
13-10	112	100	111	83	116	97	105	102	85
14-10	83	87	96	60	41	50	81		47
15-10	92	83	82	54	88	82	93		101
16-10		72	79	82	91	71	89	79	88
17-10	97	85	80	77	77	73	64	76	75
18-10	117	164	153	108	119		114	142	177

Input layer is composed of six neurons for following input parameters: wind direction and velocity, air temperature, atmospheric pressure, air humidity (meteorological conditions), and emission concentration of PM10 dust. A second hidden layer is connected to output layer consisting of one neuron. One output parameter constitutes predicted value of 24-h average concentration of PM10. For hidden and output layers, the bipolar sigmoidal function, described by Eq. (4.1), is used as transition function (Bator, 2003):

$$f_b(u_i) = tgh(\beta u_i) \tag{4.1}$$

The backward propagation method was used for network training. Network training was accomplished according to paradigm of control student. The prediction for the next day was made based on the results known for the three previous days.

Table 4.2 Twenty-four-mean concentration distribution of PM10 dust from 25.10.2001 to 25.11.2001.

SERIES III	P1	P2	P3	P4	P5	P6	P7	P8	P9
25-10	83		46	67	77	62	101	81	72
26-10	95	105	101	86	95	45	96	107	
27-10	79	51	36	71	45	70	95	78	73
28-10	62	69	58	55	58	59	62	62	39
29-10	29	45	62	21	40	23	42	30	38
30-10	41	48	37	39	40	40	39	31	45
31-10	21	26	23	23	20			25	31
03-11		64	55	58	70		75		
04-11		89	46	40		43	50	49	
05-11	36	45	30	25	31	29	24		40
06-11		60	51	45	33	54	51	54	55
07-11	31	35	18	25		26		58	37
08-11	21	26	24	16	21	15	23	22	23
09-11	30	30	20	17	23	21	29	27	59
10-11	70	61	55	48	49	36	54	67	69
11-11	39	53	45	32	45	86	48	47	45
12-11	60	69	50	70	37	55	69	55	72
13-11	64	52	49	60	44	65	61	117	68
14-11	90	87	76	60	86	63	64	101	83
15-11	37	49	28	36		30	38		42
16-11	32	22		26	27	21	24	32	31
17-11	45			42		29	53	47	
18-11	42	37		41	44	42	45	51	54
19-11	31	28		21		32		39	34
20-11	87	36		53	84	37	70		86
21-11	32	51		42	43	40	50	32	45
22-11	17	31		29	19	8		26	23
23-11	20	28		44	19	22	11	26	28
24-11		20		20	12	20	22		29
25-11	30	24			24		26		47

The data of PM10 emission introduced to the network input layer were used to train and test the network. These data were divided into two parts: the training part and the testing one.

4.6.6 Working parameters of the ANN model

The calculations described here were performed by R. Bator with the help of his own program Proglmis (Bator, 2003). The results were reported in his PhD thesis (Bator, 2005) and a paper (Bator and Sieniutycz, 2006).

To test the influence of training data number on network training, calculations were carried out for three different data sets. Sizes and ranges of the data sets are presented in Table 4.6.

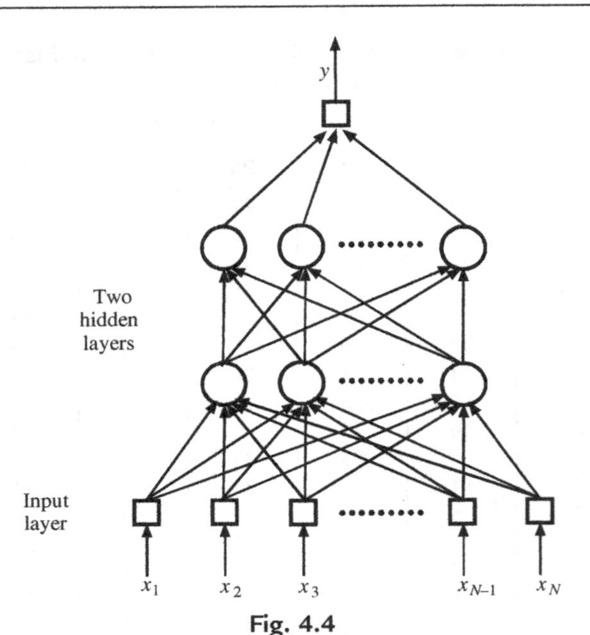

Fig. 4.4
Scheme of unidirectional, two-layer MLP artificial-neural network (Osowski, 1996; Hertz et al., 1993).

Input data were normalized to interval [−1, 1] with the help of a linear transformation, as described in the following equation:

$$z_i = \frac{x_i - x_{i(\min)}}{x_{i(\max)} - x_{i(\min)}} \tag{4.2}$$

where

i—parameter,
z—normalized value,
x—parameter value.

Data normalization was performed to equalize the effect of all parameters regardless their value. As shown elsewhere, the numerical values of the wind velocity can be even 1000 times less than atmospheric pressure (Kucharski et al., 2003; Wojtylak et al., 1997). Using the transition function described by Eq. (4.1), which assumes its values in the range from −1 to 1, was another reason to normalize the data.

Sizes of training and testing sets for a particular calculation series are presented in Tables 4.3 and 4.4.

Neuron number, in particular hidden layers, was determined experimentally. Using the program ProgImis2003 (Bator, 2003) calculation, series of PM10 concentrations were carried

Table 4.3 Size of input data for MLP network.

Input parameter	Size of data set			Data ranges	
	Z1	Z2	Z3	min	max
PM10 concentration (μg/m^3)	721	540	270	10	180
Air temperature (°C)	719			−15.2	27.4
Air humidity (%)	722			44.6.5	98
Atmospheric pressure (hPa)	722			972	1024
Direction of wind (deg)	720			0	359
Velocity of wind (m/s)	719			0	8.9

Table 4.4 Samples number of training and testing data sets.

Data set	Training data set		Testing data set	
	Samples number	%	Samples number	%
Z1	637	88.3	84	11.7
Z2	480	88.9	60	11.1
Z3	230	85.2	40	14.8

out. The number of neurons in the hidden layers was variable. The deviation of testing data set was assumed as choice criterion. The minimum deviation was obtained for the network with seven neurons in each hidden layer. The calculated results are shown in Fig. 4.5.

Too many neurons in the hidden layers cause the effect called "learning by heart." The effect consists in training deviation decreasing all the time, whereas generalization deviation after approaching of minimum begins a rapid increase. In the case of too little number of neurons in the hidden layers, training and generalization deviation is stable after short time of learning.

A very significant parameter of the ANN work is training time defined by the number of training epochs. Each epoch consists of all calculations performed by the network expressed as a response to the output neuron. Too long training causes the same effect of "learning by heart" as too many neurons in the hidden layers as well an effect of learning dwindles. Therefore, the time of training was individually determined for each set of input data. Maximal time of the network training amounts to about 500 epochs for network architecture assumed in this study.

The network training consists in modification of weights to minimize a performance index with many local minima. The effect of training is dependent on the starting point. To find the global minimum, the training process was repeated a 100 times for various starting values of weights. A calculation with minimal deviation was accepted as a final result of all 100 repetitions.

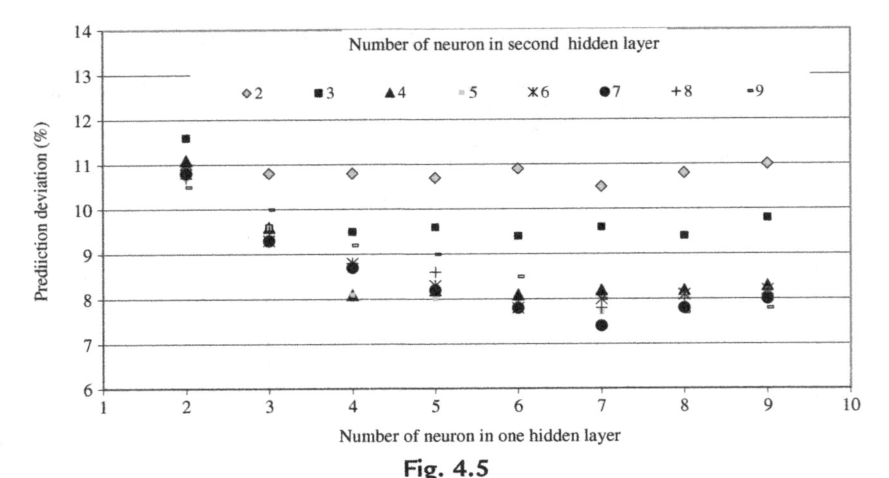

Fig. 4.5
Determination of neurons number in hidden layers.

4.6.7 Prediction results

4.6.7.1 Statistical estimates

Statistical estimation of the prediction was performed comparing calculated values with measurement values; for that purpose the mean-square deviations δ_{sr}, Pearson's correlation coefficient R^2, and average difference between the measured value and the calculated one $\overline{\Delta S}$ [8, 26]. The results of the estimation are presented in Table 4.5.

4.6.7.2 Comparison of the measured and calculated values

The graphical comparison of the measured values with the calculated ones is shown in Figs. 4.6–4.11.

4.6.8 Emission prediction by hybrid model

The hybrid model used in this study is the Gaussian model of Pasquille's type adjoint with the model of artificial-neural network with radial base function (RBF), Fig. 4.12. The RBF neural network is applied to calculate a meteorological parameter using standard meteorological data (velocity and direction of wind, air humidity, temperature, and atmospheric pressure). In this depiction, the neural network is part and parcel of the hybrid model.

Table 4.5 Statistical estimate of ANN learning and prediction efficiency.

Data set	Learning series			Testing series		
	R^2	$\delta_{śr}$ (%)	$\overline{\Delta S}$	R^2	$\delta_{śr}$ (%)	$\overline{\Delta S}$
Z1	0.828	5.8	10.2	0.715	6.4	11.5
Z2	0.764	6.7	14.1	0.653	7.5	14.6
Z3	0.753	7.1	17.1	0.561	9.8	18.7

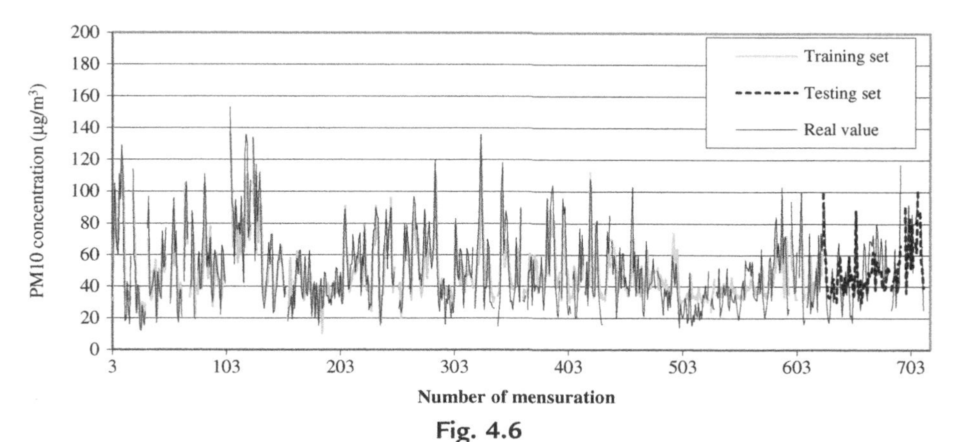

Fig. 4.6
Comparison of series of PM10 concentrations for data with 721 samples.

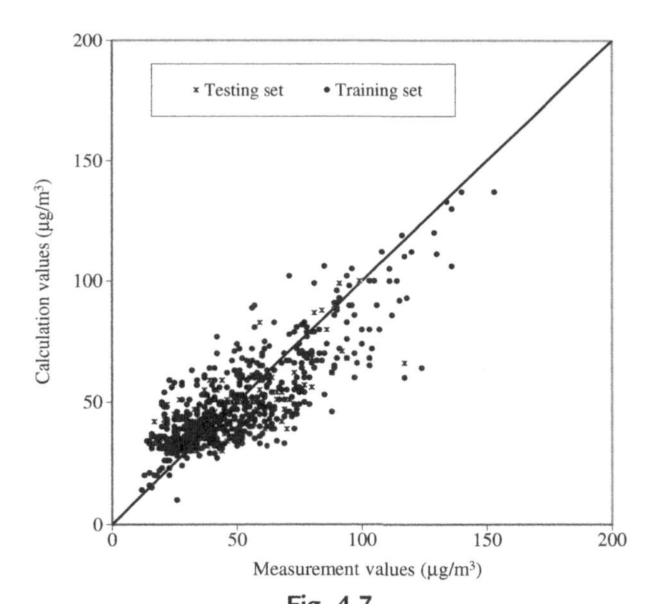

Fig. 4.7
Comparison of measured and calculated by ANN values of PM10 concentrations; data set with 721 samples.

4.6.9 Work parameters of RBF network

The following parameters serve as the input data of the RBF network:

1. meteorological conditions
   - direction of wind,
   - velocity of wind,
   - air temperature,

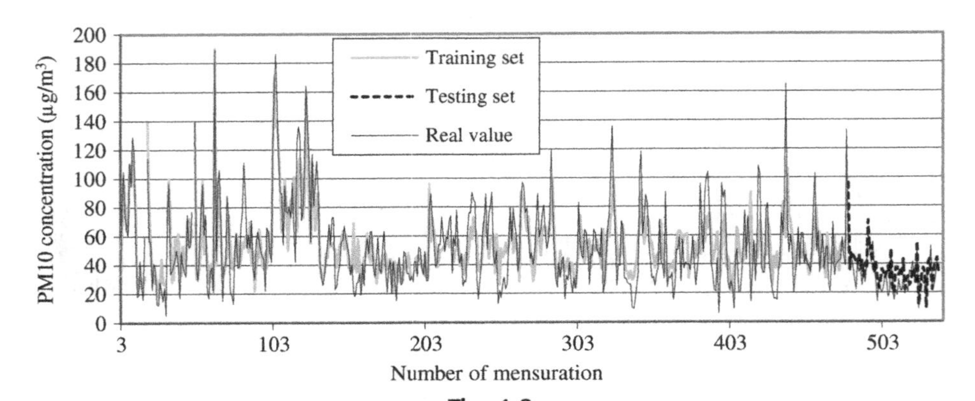

Fig. 4.8
Comparison of series of PM10 concentrations for data with 540 samples.

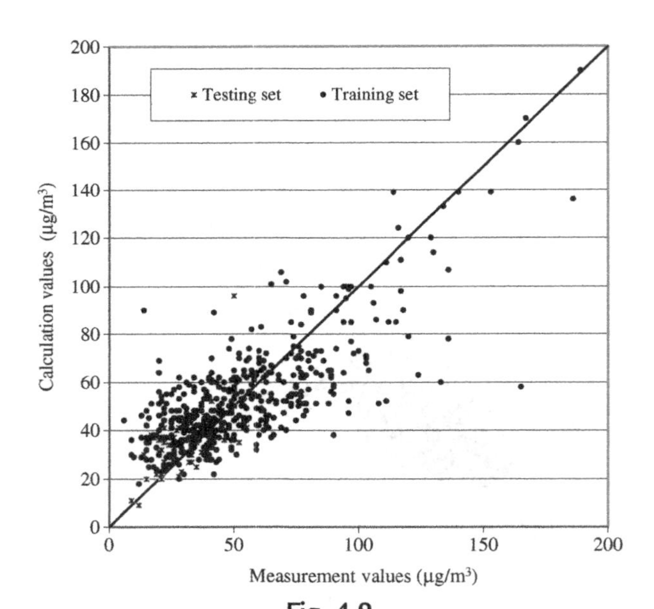

Fig. 4.9
Comparison of measured and calculated by ANN values of PM10 concentrations; data set with 540 samples.

- atmospheric pressure,
- humidity of atmospheric air.

2. emission concentration of PM10 dust

The meteorological parameter, m, is obtained on the output from network. Then, this parameter is used in the classical Pasquille's model. To the network training and testing, back propagation is applied. Radial network architecture has a structure analogical to the structure of multilayer

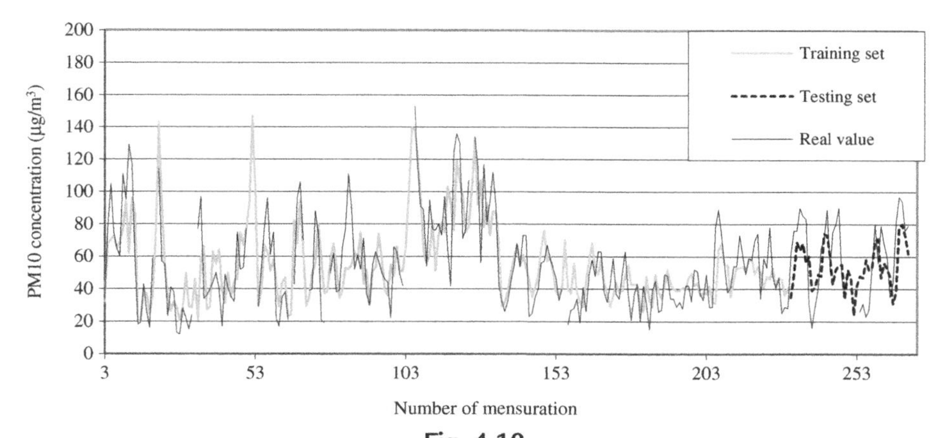

Fig. 4.10
Comparison of series of PM10 concentrations for data with 270 samples.

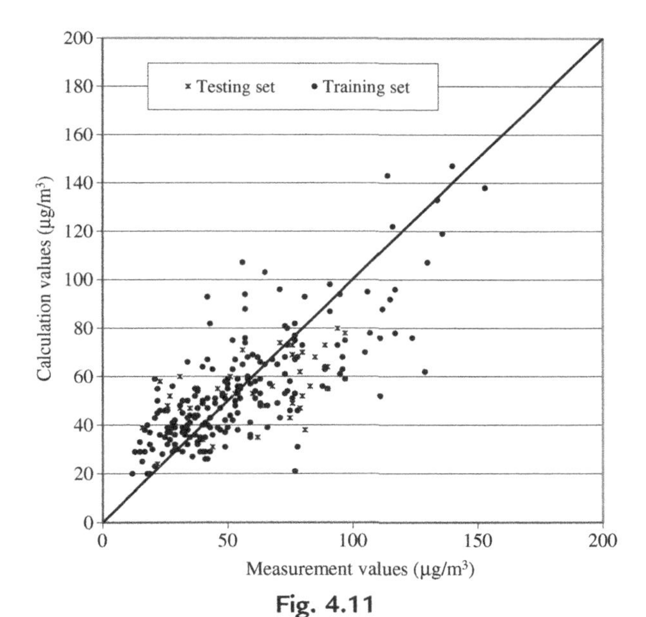

Fig. 4.11
Comparison of measured and calculated by ANN values of PM10 concentrations, data set with 270 samples.

sigmoid network with one hidden layer consisting of neurons with radial base function (Fig. 4.13). The Gaussian base function is used. The network is configured for six input neurons, seven neuron in the hidden layer, and six outputs. The number of outputs is dictated by the number of classes of atmosphere stability. The sizes of training and testing sets are, respectively, 305 (83.6% of all samples), 60 (16.4%) samples for annual calculation period, and 23 (76.7%) and 7 (23.3%) samples for 30 days one.

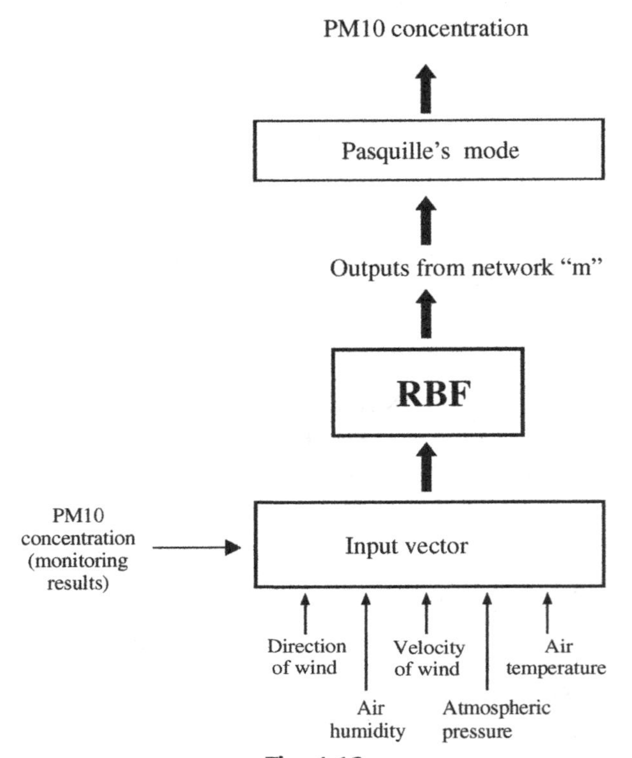

Fig. 4.12
Scheme of hybrid model: Pasquille-RBF model.

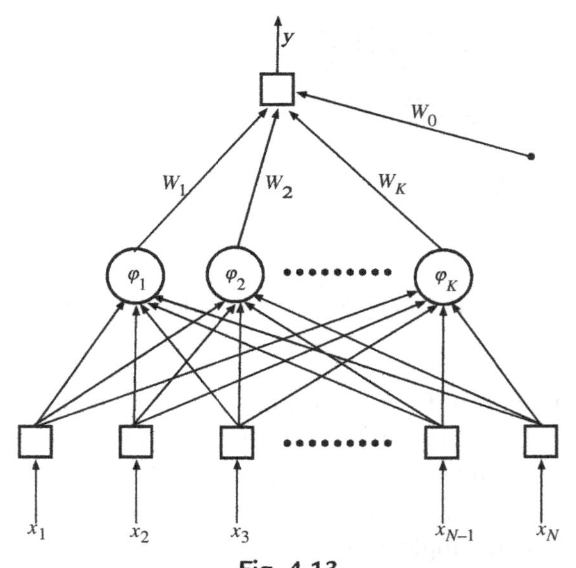

Fig. 4.13
General architecture of RBF network.

The number of inputs is dependent on the number of the introduced parameters; the number of outputs depends on analytical needs. However, number of base functions, identified with neurons in hidden layers, is one of fundamental data required for correct solution of the approximation problem. An insufficient number of hidden neurons render satisfactory reductions of deviation impossible, while too many neurons cause an increase of generalization deviation in the testing set—the so-called effect of "learning by heart" occurs. In this study, the number of base functions is found by the trial and error method by searching for the number of neurons, which ensures the best fitting results of the hybrid model to experimental data.

4.6.10 Prediction results

The concentration of PM10 dust in the measurement points in Radom was calculated form the emission and meteorological data received for an analyzed period. These data were obtained by using the program OPERAT2000, based on the Gaussian model of Pasquille's type. The model enables a suitable choice of a calculation method for diffusion coefficients. The computations for years 2001 and 2002 were performed for all tested calculation methods of diffusion coefficients. For the measurement SERIES II and III (30-day series), maximum concentrations were calculated using all methods, whereas average concentrations were only computed by the method of Pasquille-RBF (the hybrid model) because the remaining methods exclude calculations for averaging times other than for a year.

To compare real concentrations with those calculated by particular methods, the chi-square test was used. Results of the test are shown in Table 4.6. The graphs of scattering for calculation results were made (Figs. 4.14–4.17).

Table 4.6 Results of chi-square test.

	Diffusion coefficients calculations method:					
Series	**Nowicki**	**Christian**	**Brookhaven**	**Briggs M**	**Giford**	**Pasquill RBF**
			Average concentration			
2001	0.058	0.944	0.919	0.888	0.930	0.971
2002	0.935	0.982	0.972	0.940	0.968	0.985
Series II	0.673	0.793	0.731	0.685	0.732	0.927
Series III	0.937	0.962	0.946	0.946	0.959	0.989
			Maximal concentration			
2001	0.059	0.732	0.058	0.089	0.041	0.579
2002	0.125	0.124	0.245	0.147	0.111	0.620
Series II	0.012	0.329	0.145	0.025	0.036	0.884
Series III	0.115	0.223	0.152	0.135	0.182	0.597

4.6.11 Summary and conclusions

The calculations show that the best results are obtained by using a ANN-Pasquille's model and Christian's model, but the second one is only a little worse. The application of Nowicki's diffusion coefficients (one of the present routines to calculate the propagation of air pollution from emitters of economic units) gives results with the maximum deviation.

Now, easy conditions for concentration of population and industry in a large industrial conurbation cause a considerable deterioration of air quality. Thus, reduction of air pollution is one of the main factors for better living conditions in urban areas. Analyses and models applied in diagnosis and prediction of air-sanitary conditions constitute an essential component in management of environmental quality.

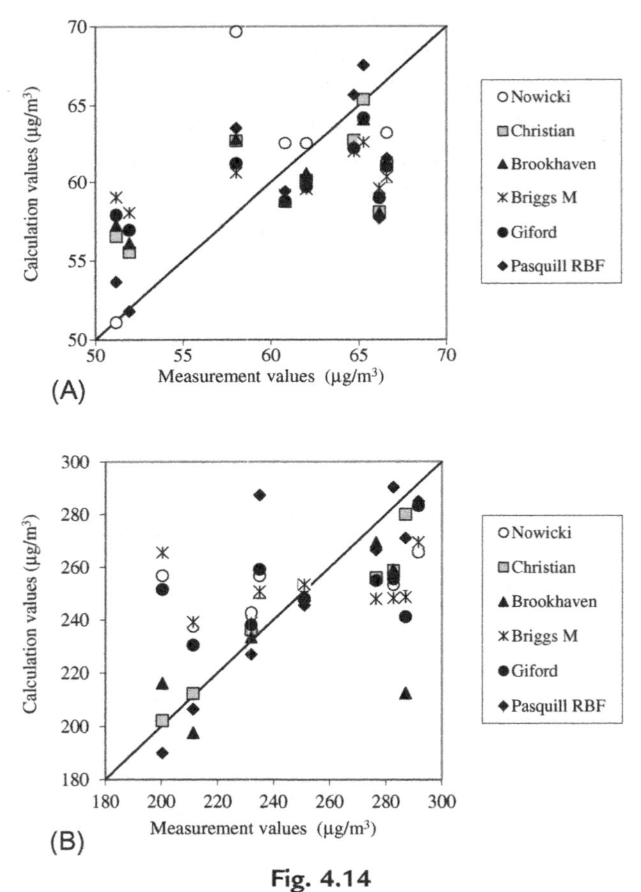

Fig. 4.14
Comparison of prediction results with real values (2001—SERIES I) (A) average concentration and (B) maximal concentration.

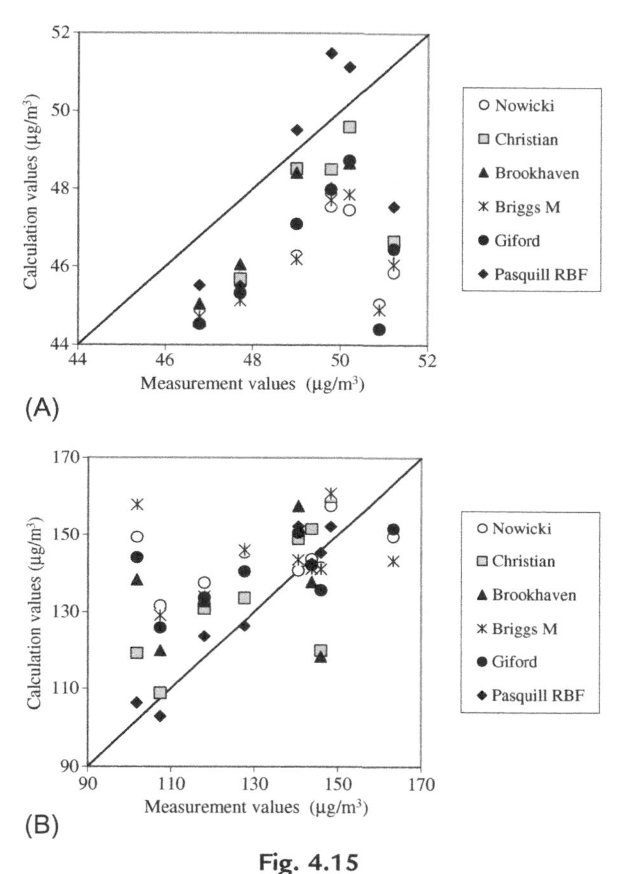

Fig. 4.15
Comparison of prediction results with real values (2002—SERIES I) (A) average concentration and (B) maximal concentration.

The most significant conclusions and observations derived from performed studies, calculations, and theoretical analyses are the following:

- Concentrations of dust PM10 modeled by using the classical Pasquille's model differ considerably, in accordance with the applied method of calculation of diffusion coefficients. The assumption of essential influence of meteorological parameters on the propagation process was confirmed. The results conforming to reality the best were obtained for coefficients calculated by Christian's method ($\chi^2 = 0.920$ for average concentrations, $\chi^2 = 0.352$ for maximal concentrations). Nowicki's method, now applied relatively seldom, appears the least efficient ($\chi^2 = 0.651$ for average concentrations, $\chi^2 = 0.078$ for maximal concentrations).
- The results obtained from hybrid model were the most conforming to real data ($\chi^2 = 0.927$ – minimal value, $\chi^2 = 0.968$ – average value for average concentrations and $\chi^2 = 0.579$ – minimal value, $\chi^2 = 0.670$ average value for maximal concentrations).

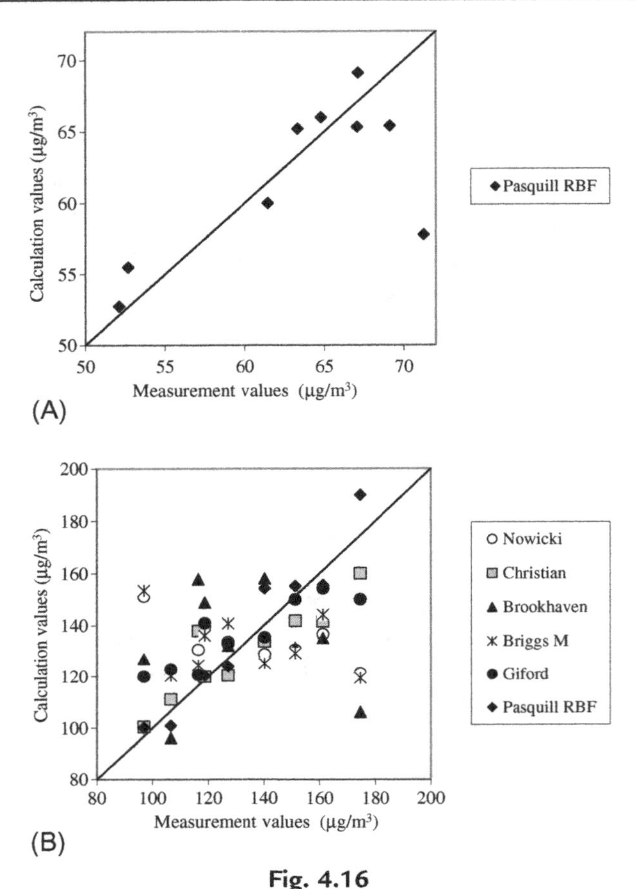

Fig. 4.16
Comparison of prediction results with real values (SERIES II) (A) average concentration and (B) maximal concentration.

- The neural model used in prediction of dust PM10 concentrations consists of input layer (six neurons), two hidden layers (each seven neurons), and output layer (one neuron). This network architecture gave results with a minimal deviation.
- Networks with excessively large quantity of neurons in the hidden layers learned fast and effectively, but their aptitude to prediction was low and became lower in the course of time of learning.
- The sizes of training data sets decided about the prediction effects. Reduction of data set to 30% caused the decrease of prediction by about 50%.
- The prediction performed for the next day, based on the data of three previous days, yields results where the mean-square error didn't exceed 6.4% and the average difference between measured and calculated concentrations was lower than 11.5 $\mu g/m^3$.
- Artificial-neural networks are a convenient and effective tool in management of environment. They are especially useful in cases with input data discontinuity.

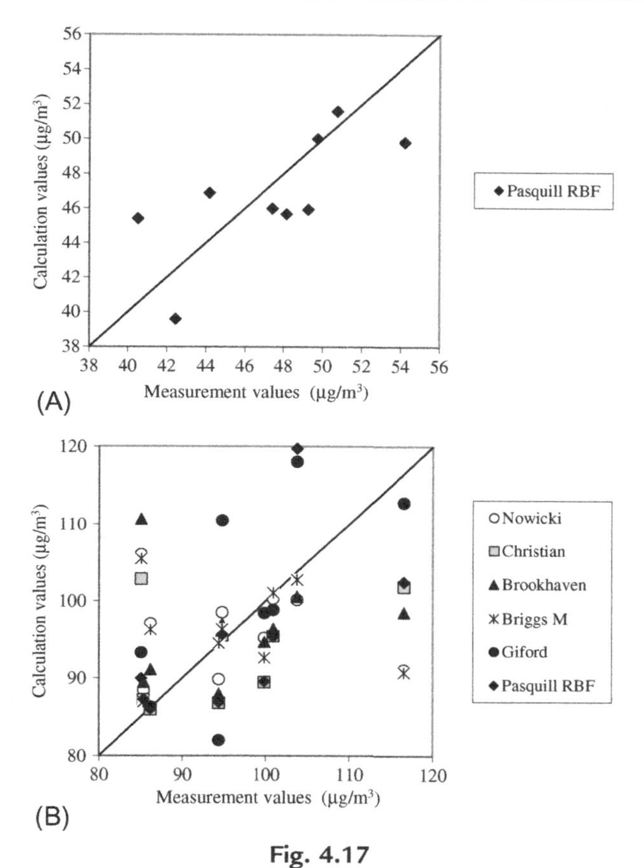

Fig. 4.17
Comparison of prediction results with real values (SERIES III) (A) average concentration and (B) maximal concentration.

- The program ProgImis2003 (Bator, 2003) constructed for the purpose of the predictive investigations can successively be applied in researches of emission prediction by ANN. It should also be used by local authorities in order to control environmental pollution.
- Use of the data of predicted environmental pollution in optimization should determine optimal control minimizing the environmental pollution under prescribed conditions.

4.7 Summarizing remarks

In this chapter, we have presented a collection of brief reviews of selected papers which apply artificial-neural networks. Various approaches to ANN fuel cell modeling have been considered. Some methods apply the multilayer perceptron with unidirectional information flow. Hybrid models involving artificial-neural network with radial base function (RBF) are mentioned without considering the network architecture and transition functions. Significant difficulties with obtaining reliable models are particularly troubling in case of Molten Carbonate Fuel Cells (MCFCs). The complexity of many FC processes causes complications in their mathematical description; therefore, it seems reasonable to expect an increasing number of ANN FC experiments in the near future. An expanded report on extensive application of RBF neural networks for emission prediction of dust pollutants was provided.

References

Arriagada, J., Olausson, P., Selimovic, A., 2002. Artificial neural network simulator for SOFC performance prediction. J. Power Sources 112 (1), 54–60.

Bator, R., 2003. Computer Program *Progimis*. Faculty of Materials Science and Technology (WMiTO), Pulaski Technical University, Department of Environmental Protection, Division of Process and Environmental Engineering, Poland.

Bator, R., 2005. Analysis and Modelling of Fields of Pollutants Immission in Atmospheric Air (Ph.D. thesis). Faculty of Chemical and Process Engineering at the Warsaw University of Technology.

Bator, R., Kacperski, W.T., 2001a. Monitoring and computer simulation in environmental planning and development (Monitoring i symulacja komputerowa w strategii kształtowania środowiska). In: Polish National Conference "Environmental Planning and Development - Natural, Technical and Social-Economic Conditions" (Konferencja ogólnopolska "Kształtowanie środowiska – uwarunkowania przyrodnicze, techniczne i społeczno – ekonomiczne"), Olsztyn, Poland, September 12–14, 2001.

Bator, R., Kacperski, W.T., 2001b. Potential for dust pollution control in urban air (Możliwości kształtowania zapylenia w powietrzu miejskim). Zesz. Probl. Post. Nauk Rol. 476, 149–155.

Bator, R., Sieniutycz, S., 2006. Application of artificial neural network for emission prediction of dust pollutants. Int. J. Energy Res. 30 (13), 1023–1036.

Borys, T., 1999. Indicators of eco-development (Wskaźniki Ekorozwoju). Ekon. Śr. 22, 12–25. Białystok, Poland.

Briggs, G.A., 1983. Diffusion estimation for small emissions. Air. Res. Atmos. Turbul. Diffus. 2, 2–24.

Budziński, K., 1997. Analytical model of trajectory for engineering applications in atmospheric air pollution control (Analityczny model trajektorii do zastosowań inżynierskich w ochronie powietrza atmosferycznego). In: Reports of Warsaw University of Technology, Environmental Engineering. Prace Naukowe Politechniki Warszawskiej, Inżynieria Środowiska vol. 23. Warszawa, Poland.

Caponetto, R.C., Fortunà, L., Rizzo, A., 2005. Neural network modelling of fuel cell systems for vehicles. In: Proc. of 10th IEEE ETFA Conf. Emerg. Tech. Fact. Automation, 19–22 Sept, 2005, pp. 6–192. https://doi.org/10.1109/ETFA.2005.1612519 ieeexplore.ieee.org.

Chaichana, K., Patcharavorachot, Y., Chutichai, B., Saebea, D., Assabumrungrat, S., Arpornwichanop, P., 2012. Neural network hybrid model of a direct internal reforming solid oxide fuel cell. Int. J. Hydrogen Energy 37 (3), 2498–2508. https://doi.org/10.1016/j.ijhydene.2011.10.05.

Directives for Calculation of Pollution Index of Atmospheric Air (Wytyczne obliczania stanu zanieczyszczenia powietrza atmosferycznego), 1983. Ministerstwo Administracji, Gospodarki Terenowej i Ochrony Środowiska, Departament Ochrony Środowiska. Warszawa, Poland.

Directives of Polish Ministry of Environmental Protection, for assessment of admissible levels of substances in air (Rozporządzenie Ministra Ochrony Środowiska w sprawie oceny poziomów substancji w powietrzu), 1981. Dz. U.1, Warszawa, Poland.

El-Sharkh, M.Y., Rahman, A., Alam, M.S., 2004. Neural networks-based control of active and reactive power of a stand-alone PEM fuel cell power plant. J. Power Sources 135 (1–2), 88–94.

Ensor, D.S., 1998. Hypotenuse. Eviron. Prot. Issue 35(1).

Entchev, E., Yang, L., 2007. Application of adaptive neuro-fuzzy inference system techniques and artificial neural networks to predict solid oxide fuel cell performance in residential microgeneration installation. J. Power Sources 170 (1), 122–129.

Głowiak, B., Kempa, E., Winnicki, T., 1985. Fundamentals of Environmental Management (Podstawy ochrony środowiska) PWN, Warszawa.
Hattia, M., Tioursi, M., 2009. Dynamic neural network controller model of PEM fuel cell system. Int. J. Hydrog. Energy 34 (11), 5015–5021.
Hertz, J., Krogh, A., Palmer, R.G., 1993. Introduction to Theory of Neural Computations (Wstęp do teorii obliczeń neuronowych). WNT, Warszawa.
Jemeï, S., Hissel, D., Péra, M.C., Kauffmann, J.M., 2003. On-board fuel cell power supply modeling on the basis of neural network methodology. J. Power Sources 124 (2), 479–486.
Juda-Rezler, K., 1998. Air pollution modelling. In: The Encyclopedia of Environmental Control Technology. vol. 2. Gulf Publishing Company, Houston, USA.
Jurado, F., 2003. Power supply quality improvement with a SOFC plant by neural-network-based control. J. Power Sources 117 (1–2), 75–83.
Kamiński, W., Tomczak, E., 1999. An integrated neural model for drying and thermal degradation of selected products. Dry. Technol. 17 (7–8), 1291–1301. https://doi.org/10.1080/07373939908917615.
Khajeh-Hosseini-Dalasm, N., Ahadian, S., Fushinobu, K., Okazaki, K., Kawazoe, Y., 2011. Prediction and analysis of the cathode catalyst layer performance of proton exchange membrane fuel cells using artificial neural network and statistical methods. J. Power Sources 196 (8), 3750–3756.
Kucharski, M., Kamiński, W., Petera, J., 2003. Application of neural networks for prediction of carbon monooxide concentrations near roads with large traffic (Zastosowanie sztucznych sieci neuronowych do przewidywania stężeń tlenku węgla w obrębie dróg o dużym natężeniu ruchu). Inżynieria i Aparatura Chemiczna 4, 12–25.
Kumbur, E.C., Sharp, K.V., Mench, M.,.M., 2008. A design tool for predicting the capillary transport characteristics of fuel cell diffusion media using an artificial neural network. J. Power Sources 176 (1), 191–199.
Lobato, J., Cañizares, P., Rodrigo, M.A., Linares, J.J., Piuleac, C.-G., Curteanu, S., 2009. The neural networks based modeling of a polybenzimidazole-based polymer electrolyte membrane fuel cell: effect of temperature. J. Power Sources 192 (1), 190–194.
Madany, A., Bartchowska, M., 1995. Review of Polish models for atmospheric pollution propagation (Przegląd polskich modeli rozprzestrzeniania się zanieczyszczeń atmosferycznych). In: Reports of Warsaw University of Technology, Environmental Engineering.Prace Naukowe Politechniki Warszawskiej, Inżynieria Środowiska Issue 19, Warszawa, Poland.
Mammar, K., Chaker, A., 2012. A neural network-based modeling of PEM fuel cell and controller synthesis of a stand-alone system for residential application. Int. J. Comput. Sci. Issues 1694-0814. 9 (6). No 1.
Markiewicz, M., 1996. Review of models for pollutant propagation in atmospheric air (Przegląd modeli rozprzestrzeniania się zanieczyszczeń w powietrzu atmosferycznym). In: Reports of Warsaw University of Technology, Environmental Engineering.Prace Naukowe Politechniki Warszawskiej, Inżynieria Środowiska Issue 21, Warszawa.
Meadows, D.H., Meadows, D.L., Randers, J., 1995. Crossing Boundaries. Global Disaster or Safe Future (Przekroczenie Granic. Globalne Załamanie czy Bezpieczna Przyszłość). WNT, Warszawa.
Milewski, J., Świrski, K., 2009. Modelling the SOFC behaviours by artificial neural network. Int. J. Hydrog. Energy 34 (13), 5546–5554.6.
Noga, M., 1996. Measurement of social affluence taking into account the state and quality of human natural environment (Pomiar dobrobytu społecznego uwzględniający stan i jakość środowiska przyrodniczego człowieka). Ekon. Śr. 19, 12–25. Białystok, Poland.
Ogaji, S., Singh, R., Pilidis, P., Diacakis, M., 2006. Modelling fuel cell performance using artificial intelligence. J. Power Sources 154 (1), 192–197.
Osowski, S., 1996. Neural Networks - An Algorithmic Approach (Sieci Neuronowe w Ujęciu Algorytmicznym). WNT, Warszawa.
Ou, S., Achenie, L.E.K., 2005. A hybrid neural network model for PEM fuel cells. J. Power Sources 140 (2), 319–330.
Palancar, M.C., Aragon, J.M., Castellanos, J.A., 2001. Neural network model for fluidized bed dryers. Dry. Technol. 19 (6), 1023–1044. https://doi.org/10.1081/DRT-100104803.

Paterkowski, W., Wolnowska, A., Straszko, J., 1997. Assessment of selected models for pollution propagation in atmospheric air (Ocena wybranych modeli rozprzestrzeniania zanieczyszczeń w powietrzu atmosferycznym). Ochr. Powiet. Probl. Odpad. 2, 12–24.

Reible, D.D., 1999. Fundamentals of Environmental Engineering. Springer Verlag, Heidelberg.

Rouss, V., Charon, W., 2008. Multi-input and multi-output neural method of the mechanical nonlinear behaviour of a PEM fuel cell system. J. Power Sources 175, 1–17.

Saengrung, A., Abtahi, A., Zilouchian, A., 2007. Neural network model for a commercial PEM fuel cel system. J. Power Sources 172 (2), 749–759.

Sieniutycz, S., Szwast, Z., 2018. Optimizing Thermal, Chemical, and Environmental Systems. Elsevier, Amsterdam/ Oxford, UK.

Skrzypski, J., 2002. Analysis and Modeling of Emission Pollution Fields in Large Cities (Analiza i modelowanie pól imisji zanieczyszczeń w dużych miastach). Polish Academy of Science, Łódź.

Sousa, T., 2007. Thermodynamics as a Substantive and Formal Theory for the Analysis of Economic and Biological Systems. Ph.D. thesis carried out at the Department of Theoretical Life Sciences, Vrije Universiteit Amsterdam, The Netherlands and at the Environment and Energy Section, Instituto Superior T´ecnico, Lisbon, Portugal.

Sousa Jr., R., Colmati, F., Gonzalez, E.R., 2006. Modeling techniques applied to the study of gas diffusion electrodes and proton exchange membrane biochemical fuel cells. J. Power Sources 161 (1), 183–190.

Straszko, J., Paulo, L.A., 1996. Diffusional models for pollutants propagation in the atmosphere (Modele dyfuzyjne procesu propagacji zanieczyszczeń w atmosferze). Inż. Chem. Procesowa 17, 12–25.

Świrski, K., Milewski, J., 2009. Optimization in electric power installation by neural networks. Przegl. Elektrotech. 85 (8), 155–157.

Van Hemmen, J.L., Kühn, R., 1991. Collective phenomena in neural networks. In: Models of Neural Networks. Springer, Berlin, pp. 1–105 (Chapter 1).

Wang, R., Qi, L., Xie, X., Ding, Q., Li, C., Ma, C.M., 2008. Modeling of a 5-cell direct methanol fuel cell using adaptive-network-based fuzzy inference systems. J. Power Sources 185 (2), 1201–1208.

Wojtylak, M., Ośródka, K., Ośródka, L., 1997. Implementation and investigation of pollutant emission forecast based on arificial neural networks (Wdrożenie i badanie prognozy imisji zanieczyszczeń opartej na metodzie sztucznych sieci neuronowych). Wiadomości IMGW XX (XLI), 12–26.

Wu, X.-J., Zhu, X.-J., Cao, G.-Y., Tu, H.-Y., 2007. Modeling a SOFC stack based on GA-RBF neural networks identification. J. Power Sources 167 (1), 145–150.

Zannetti, P.A., 1990. Air Pollution Modelling. Van Nostrad Reinhold, New York.

Zannetti, P., 1992. Particle Modeling and Its Application for Simulation Air Pollution Phenomena. Environmental Modeling Computation Mechanics Publication, Elsevier, Applied Science.

Further reading

Straszko, J., Dziubakiewicz, D., 2000. Pollution dynamics of atmospheric air (Dynamika skażania powietrza atmosferycznego). Ochr. Powiet. Probl. Odpad. 35, 1–12.

CHAPTER 5

Systems design: Modeling, analysis, synthesis, and optimization

5.1 Introducing components of system design

The system design involves modeling, analysis, synthesis, and optimization. Some of these parts can be performed interchangeably, yet there are parts which necessarily must be done at the end of the procedure. One of these parts is system optimization, which, especially because of the systems complexity, should never be accomplished too early. In particular, the procedures running before the systems optimizing should involve issues such as: system modeling, system analysis, and system synthesis.

The scheme in Fig. 5.1 shows elements of the system. Systems design is the process of defining architecture, modules, interfaces, and data for a system to satisfy specified requirements. Systems design could be understood as the application of systems theory to product development. There is some overlap with the disciplines of systems analysis, systems architecture, and systems engineering.

If the broader topic of product development blends the perspective of marketing, design, and manufacturing into a single approach to product development, then design is the act of taking the marketing information and creating the design of the product to be manufactured. Systems design is, therefore, the process of defining and developing systems to satisfy specified requirements of the user.

Until the 1990s, systems design had a crucial and respected role in the data processing industry. In the 1990s, standardization of hardware and software resulted in the ability to build modular systems. The increasing importance of software running on generic platforms enhanced the discipline of software engineering. Architectural design turned out to be important.

The architectural design of a system emphasizes system architecture which involves both system structure and system behavior. The architectural design is associated with the logical design of the system which pertains to an abstract representation of the data for flows, inputs, and outputs of the system. The procedure is often conducted via modeling by using an overabstract and (sometimes graphical) model of the actual system. When dealing with

Complexity and Complex Thermo-Economic Systems. https://doi.org/10.1016/B978-0-12-818594-0.00005-2

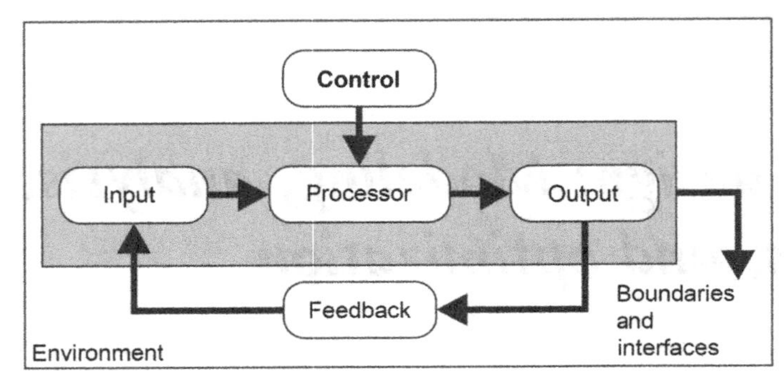

Fig. 5.1
A diagram showing the elements of a system.

complex systems, these designs are mandatory. Logical design also includes preparation of entity-relationship diagrams (ER diagrams). The next step is physical design.

The physical design relates to the actual input and output processes of the system. This is explained in terms of how data is embedded into a system, how it is verified/authenticated, how it is processed, and how it is displayed. In physical design, the following requirements about the system are decided.

1. Input requirement,
2. Output requirements,
3. Storage requirements,
4. Processing requirements,
5. System control and backup or recovery.

In other words, the physical portion of system design can generally be broken down into three subtasks:

a. User interface design
b. Data design
c. Process design

User interface design is concerned with how users add information to the system and with how the system presents information back to them. *Data design* is concerned with how the data is represented and stored within the system. Consequently, *process design* is concerned with how data moves through the system, and with how and where it is validated, secured, and/or transformed as it flows into, through, and out of the system. At the end of the system design phase, documentation describing the three subtasks is produced and made available for use in the next phase.

In this context, physical design does not refer to the tangible physical design of an information system. To use an analogy, a personal computer's physical design involves input via a keyboard, processing within the CPU, and output via a monitor, printer, etc. It would not concern the actual layout of the tangible hardware, which for a PC would be a monitor, CPU, motherboard, hard drive, modems, video/graphics cards, USB slots, etc. Physical design involves a detailed characteristic of a user, a product database structure processor, and a control processor. The personal specification is developed for the proposed system.

Related disciplines are:

- Benchmarking—an effort to evaluate how current systems perform
- Computer programming and debugging in the software world, or the design in the consumer, enterprise, or commercial world, to lead to the final systems components.
- Model design—designers will produce one or more "models" of what they think a system should finally look like. The ideas from the analysis section will then either be used or discarded. A document will be produced with a description of the system, but with nothing specific (yet without mention of any specific brands)
- Requirements analysis—analyzes the needs of the end-users or customers
- System architecture—creates a blueprint for the design with the necessary structure and behavior specifications for the hardware, software, people, and data resources. In many cases, multiple architectures are evaluated before one is selected.
- System testing—evaluates the system's actual functionality in relation to expected or intended functionality, including all integration aspects.

Alternative design methodologies are:

Rapid application development (RAD)
Rapid application development (RAD) is a methodology in which a system designer produces prototypes for an end-user. The end-user reviews the prototype and offers feedback on its suitability. This process is repeated until the end-user is satisfied with the final system.
Joint application design (JAD)
Joint application design (JAD) is a methodology which evolved from RAD, in which a system designer consults with a group consisting of the following parties:

- Executive sponsor
- System designer
- Managers of the system

JAD involves a number of stages, in which the group collectively develops an agreed pattern for the design and implementation of the system.

5.2 Microscopic and macroscopic bases for kinetic equations

A discussion of basic properties of kinetic equations and the resulting theory of transport and rate phenomena can be found in Section 1.7.2 of Sieniutycz (2016). The research considered therein pertains to rate processes in both homogeneous and heterogeneous systems, with possible electric effects. For the reader's convenience, we recapitulate here a part of relevant information.

The typical ingredients of kinetic theories are distribution functions (functions fi ($\mathbf{r}$, $\mathbf{c}_i$, t) for each component of a multicomponent mixture, $i = 1, 2, \ldots n$) and governing kinetic equation (equations) which describe time evolution of the distribution function fi in the phase space, where $\mathbf{r}$ is the radius vector and $\mathbf{c}_i$ is the molecular velocity of ith species at time t.

The use of kinetic equations is usually preceded by summarizing macroscopic properties which should result from any physically acceptable kinetic equation. It has to be shown that an H-theorem holds, a macroscopic expression for the entropy source follows, and that the resulting kinetic equations lead to the Onsager reciprocity relations in the linear case. In fact, these ingredients reflect the fact that irreversibility is already built into the fundamental kinetic equation whether this involves the Boltzmann integro-differential equation or its extensions.

For the standard derivation and interpretation of the Boltzmann equation, based on the assumption of molecular chaos, the reader is referred to many textbooks on kinetic theory and nonequilibrium statistical mechanics (Chapman and Cowling, 1973; Huang, 1967; Keizer, 1987).

For a one-component system, the Boltzmann equation can be written in the form

$$\frac{\partial f}{\partial t} + \mathrm{c}\nabla_r f + \mathbf{F}\nabla_c f = \int \sigma_T g[f'f'_1 - ff_1]d\mathrm{c}_1^3 \tag{5.1}$$

In the above formula, simplified notation is introduced:

$$f(\mathbf{r}, \mathbf{c}', t) = f', \ f(\mathbf{r}, \mathbf{c}_1', t) = f_1', \quad \text{and } \sigma_T$$

for a linear operator corresponding to the differential scattering cross section $\sigma(\Omega, g)$, where Ω is the solid angle and g the constant magnitude of a relative velocity. Gradient operators involve derivatives with respect to position and velocity (Keizer, 1987).

The Boltzmann equation is extremely difficult to solve in its general form (Eq. 5.1). Therefore, a relaxation time approximation is frequently applied. Since the collision term of Eq. (5.1) tends toward the thermal equilibrium, the relaxation time approximation usually has the form

$$\frac{\partial f}{\partial t} + \mathbf{c}\nabla_r f + \mathbf{F}\nabla_c f = -\frac{f - f^{\mathrm{eq}}}{\tau} \tag{5.2}$$

This approximation, which contains the relaxation time τ, leads frequently to reasonable estimates of transport coefficients (Warner, 1966). When investigating nonequilibrium phenomena, time evolution of the distribution function must be predicted by solving a kinetic equation subject to given initial conditions. This is a mathematically involved problem (Enskog, 1917; Chapman and Cowling, 1973; Cercignani, 1975). Yet, some rigorous properties of any solution to the Boltzmann equation follow from the fact that, in each molecular collision, some dynamical quantities are conserved. As the result, the well-known macroscopic conservation laws follow for mass, energy, and momentum (Huang, 1967; Kreuzer, 1981).

Summing up, conservation laws appear in the kinetic theory as the result of collisional invariants for mass, momentum, and kinetic energy of molecules. When they are combined with equations describing irreversible fluxes of mass, heat, and momentum, so-called equations of change are obtained which describe the hydrodynamic fields of velocity, temperature, and concentrations. The formalism is very useful when describing a variety of reacting and nonreacting systems on the hydrodynamic level (Hirschfelder et al., 1954; Bird et al., 1960).

The kinetic theory leads to definitions of temperature, pressure (scalar and tensor), internal energy, heat flux density, diffusion fluxes, entropy flux, and entropy source in terms of definite integrals of the distribution function with respect to molecular velocities.

The inequality $\sigma > 0$, which expresses the fact that the entropy source strength must be positive in any irreversible process, constitutes, within the framework of the kinetic theory, a statement of the second law. This is known as Boltzmann's H-theorem.

One of the most important benefits of kinetic theories is that they provide us with expressions for transport coefficients. In the limit of an infinite time and, whenever boundary conditions are compatible with equilibrium, the entropy source vanishes identically, detailed balance follows, and logarithms of the distribution functions become summational invariants. They must, therefore, be equal to a linear combination of the microscopic quantities m_i, $m_i\mathbf{c}_i$, and $(1/2)m_i\mathbf{c}_i$. This observation allows to recover the Maxwell equilibrium distribution, the definition of hydrostatic pressure, and the ideal gas law. It also shows that the equilibrium value of the kinetic entropy density coincides with the thermodynamic one. It follows that the diffusion fluxes and the nondiagonal elements of the pressure tensor must be nonzero only if the entropy source σ does not vanish. In particular, the result obtained confirms that the diffusional part of the entropy flux vanishes at the equilibrium.

Importantly, at disequilibrium, the classical phenomenological form of expressions for the irreversible entropy flux and entropy source (involving products of fluxes and forces) can be obtained from a, usually approximate, solution of the Boltzmann kinetic equation. This corresponds to the linear phenomenological laws and can be shown by the Enskog iteration method (Enskog, 1917; de Groot and Mazur, 1984). More adequately, if one calculates fluxes in hydrodynamic equations by retaining in the distribution functions *fi* only the first order perturbations $\Phi^{(1)}$, then it follows that the diffusion fluxes $\mathbf{j}_i$, the heat flux $\mathbf{q}$, and the off-diagonal elements of the pressure tensor $\mathbf{P}$ are linear functions of the gradients of the macroscopic functions μ_i, T, and $\mathbf{u}$.

Onsager's symmetry results from the kinetic theory in a straightforward manner under the essential condition of microscopic reversibility (de Groot and Mazur, 1984). In fact, the first approximation of Enskog corresponds to the linear laws of the phenomenological theory. On the other hand, if terms quadratic in $\Phi^{(1)}$ or linear in $\Phi^{(2)}$ are retained in the entropy density expansion, $\rho s = (\rho s)^0 + (\rho s)^1 + (\rho s)^2$ ....., this entropy will be a function of the macroscopic gradients in the Enskog formalism (Enskog, 1917). Therefore, terms nonlinear in ρs correspond to nonequilibrium entropies.

The Enskog method only enables us to calculate deviations from the local Maxwellian distribution which are related to the spatial nonuniformities in the system. However, the Maxwellian distribution is not the most general solution of the Boltzmann equation. Higher-order hydrodynamic theories (Burnett, 1935) and/or analyzes of short-time effects require more detailed approaches. It can be shown that homogeneous (spatially uniform) perturbations have a relaxation time, τ, of the order of a few collision times. Only for times larger than τ can a true solution of the Boltzmann equation be approximated by the Enskog solution.

5.2.1 Basic structure of disequilibrium thermodynamic theory

It is well-known that the classical macroscopic theory of equilibrium systems is just thermostatics, while the word "thermodynamics" would be the proper choice for a word that expresses jointly the dynamic and thermal aspects of nonequilibrium behavior. However, this word is commonly used for what is, in fact, the thermostatics. Accordingly, nonequilibrium theories have received the expanded name "thermodynamics of nonequilibrium processes" or simpler "nonequilibrium thermodynamics."

In absence of the electric and magnetic fields, thermostatics is imbedded in the familiar state space of the "classical" variables S, V, N_i. Thermostatics can rigorously be derived from statistical mechanics (Warner, 1966; Huang, 1967; Lewis, 1967; Landau and Lifshitz, 1975; Lavenda, 1985; Keizer, 1987) by using the classical ensemble concept (Landau and Lifshitz, 1975) or by the method of information theory (Jaynes, 1957, 1962; Schlögl, 1980; Jaworski, 1981; Grandy, 1988; Månsson and Lindgren, 1990). Transfer of routine methods of statistical mechanics to disequilibrium to describe diverse nonequilibrium phenomena (Kestin, 1979; de Groot and Mazur, 1984; Sieniutycz, 2001) is possible, while frequently is not easy (Bedeaux, 1986, 1992; Kreuzer, 1981; Lavenda, 1985; Keizer, 1987; Grandy, 1988).

Two fundamental laws of thermostatics are the law of conservation of energy and the law expressing the positiveness of the entropy production. The latter constitutes, in fact, a bridge between thermostatics and thermodynamics. It is perceivable that whenever entropy production rate σ is finite, an irreversible behavior is involved, sometimes associated with a certain dynamics. By postulating that the entropy production σ can never be negative, the second law comprises both irreversible transients and dissipative steady states, for positive σ, and equilibrium steady states (equilibrium statics), for vanishing σ.

Doubts still persist about the appropriateness of parameter spaces for describing irreversible processes. Vilar and Rubi (2001) describe difficulties and ambiguities associated with the absence of the local thermal equilibrium. They analyze explicitly inertial effects in diffusion and show how the main ideas of an improved description can be applied to other situations.

Kinetic theory (Enskog, 1917; Burnett, 1935; Grad, 1958; Lifshitz and Pitajevsky, 1984) makes it possible to determine the limitations on the domain of validity of the local equilibrium hypothesis (Prigogine, 1949). Experiments show that the hypothesis applies frequently quite far from equilibrium in the case of such transport phenomena as heat and mass transfer. However, when chemical reactions, viscoelastic media, rarefied gases, shock waves, and high frequency phenomena are in question, the hypothesis is inapplicable. To describe such systems, it is necessary to include so-called hidden or internal variables (Kestin, 1979; Bampi and Morro, 1984). Simple local disequilibrium effects appear in the macroscopic description of diffusion where the diffusion velocities contribute to the kinetic energy (de Groot and Mazur, 1984). The kinetic energy of diffusion does not vanish in the barycentric frame. In effect, one can modify the flux-force relationship (classical Fick's law) by an additional force of inertial type. This force involves the time derivative of diffusion velocity and accompanies the classical gradient of chemical potential in the phenomenological equation of diffusion (de Groot and Mazur, 1984; Sieniutycz, 1983).

Let us discuss several shortcomings of the classical theory of nonequilibrium thermodynamics. Using the Eulerian (field) description of a continuum, we begin by formulating the conservation laws. For the case of a one-component fluid, they can be written in the form:

$$\partial\rho/\partial t + \nabla\cdot\mathbf{J} = 0 \tag{5.3}$$

$$\partial(\rho\mathbf{u})/\partial t + \nabla\cdot(\rho\mathbf{u}\mathbf{u} + \mathbf{1}P + \mathbf{\Pi}) = \rho\mathbf{F} \tag{5.4}$$

$$\frac{\partial\rho e}{\partial t} + \nabla\cdot(\rho\mathbf{u}e) = -\nabla\cdot\mathbf{q} - P\nabla\cdot\mathbf{u} - \mathbf{\Pi}:\nabla\mathbf{u} \tag{5.5}$$

describing, respectively, the conservation of mass and the balances of the momentum and the internal energy. Here ρ is the mass density, $\mathbf{v}$ the barycentric velocity, P the pressure, $\mathbf{\Pi}$ the viscous shear tensor, $\mathbf{q}$ the heat flux vector, and e the specific internal energy. $\mathbf{F}$ is an external force acting per unit mass of the fluid. Extracting from these equations, the substantial time derivatives of ρ, $\mathbf{u}$, and e and substituting them into the Gibbs equation written in the form

$$\rho\frac{ds}{dt} = \rho T^{-1}\frac{de}{dt} - \rho^{-1}PT^{-1}\frac{d\rho}{dt} \tag{5.6}$$

lead to the well-known expression for the entropy balance

$$\rho\frac{ds}{dt}=-\nabla\cdot\left(\frac{\mathbf{q}}{T}\right)-\mathbf{q}\cdot\nabla T^{-1}-T^{-1}\mathbf{\Pi}:\nabla\mathbf{u} \tag{5.7}$$

When this equation is combined with the mass continuity equation

$$\frac{\partial\rho}{\partial t}+\nabla\cdot(\rho\mathbf{u})=0 \tag{5.8}$$

we obtain the entropy balance in the form containing on the left-hand side of the four-dimensional (space-time) divergence of a quantity which is identified with the four-dimensional entropy flux

$$\frac{\partial\rho s}{\partial t}+\nabla\cdot\left(\rho s\mathbf{u}+\frac{\mathbf{q}}{T}\right)=-\mathbf{q}\cdot\nabla T^{-1}-T^{-1}\mathbf{\Pi}:\nabla\mathbf{u} \tag{5.9}$$

whereas the right-hand side contains two source terms, the first related to the conductive heat transfer, and the second to the diffusive momentum transfer. Having these source terms, we can postulate forms of phenomenological relationships, linking thermodynamic fluxes and forces and obeying the second law (nonnegative entropy source σ)

$$\sigma=\mathbf{q}\cdot\nabla T^{-1}-T^{-1}\mathbf{\Pi}:\nabla\mathbf{u}\geq 0 \tag{5.10}$$

In the present considerations of this inequality and phenomenological relationships, we shall omit criticisms associated with methodology leading to inequality (5.10). These issues are reviewed by Lebon and Mathieu (1983). We shall also ignore dissipative phenomena associated with the second term in Eq. (5.10), which is related to momentum diffusion and effects of first and second viscosities. For a comprehensive analysis of these effects, the reader is directed to books on nonequilibrium thermodynamics (Fitts, 1962; de Groot and Mazur, 1984). We restrict to isotropic systems, where thermal and viscous terms in Eq. (5.10) are independent. This will allow us to analyze solely consequences of the thermal term in Eq. (5.10). The nonnegativeness of the entropy production leads to a simple equation of heat transfer (sometimes called "phenomenological equation")

$$\mathbf{q}=L_q\nabla T^{-1}\cong-\lambda\nabla T \tag{5.11}$$

where

$$\lambda\cong L_qT^{-2} \tag{5.12}$$

is the well-known coefficient of thermal conductivity. The result is the well-known Fourier's law of heat conduction. Following this reasoning, much more involved phenomenological equations can be obtained which work with $\boldsymbol{L}$, as a symmetric matrix rather than scalar; this matrix satisfying the well-known Onsager's symmetries, $L_{ik}=L_{ki}$. As an example of advanced development along this line, Rowley and Horne (1978) and Rowley (1992) predicted values of heat and mass transport coefficients in mixtures and designed experiments measuring such values. Their approach constitutes a combined experimental and theoretical examination of diffusion, thermal conductivity, and heat of transport in nonelectrolyte liquid mixtures. Associated conclusions of methodological nature contributed to prediction of lasting role for molecular dynamic simulations.

5.3 System analysis

System design involves modeling, analysis, synthesis, and optimization. System analysis is a procedure or approach that serves to determine the system's performance for a given (known) structure of this system. An example may be a typical student project with a given input data which should be made for a defined system structure. The resulting calculation data characterize system outputs.

The Merriam-Webster dictionary defines system analysis as "the process of studying a procedure or business in order to identify its goals and purposes and create systems and procedures that will achieve them in an efficient way" (Webster, 1989). Another viewpoint sees system analysis as a problem-solving technique which breaks down a system into its component pieces for the purpose of studying how well those component parts work and interact to accomplish their purpose. The field of system analysis relates closely to requirements analysis or to operations research. It is also "an explicit formal inquiry carried out to help decision makers identify a better course of action and make a better decision than they might otherwise have made."

Analysis, as stems from Greek, is defined as "the procedure by which we break down an intellectual or substantial whole into parts," while synthesis means "the procedure by which we combine separate elements or components in order to form a coherent whole." System analysis researchers apply methodology to the systems involved, forming an overall picture.

System analysis is used in every situation where something is developed. Analysis can also involve a series of components which perform organic functions together, such as system engineering. System engineering is an interdisciplinary field of engineering which focuses on how complex engineering projects should be designed and managed. Practitioners of system analysis are often called up to dissect systems that have grown haphazardly to determine the current components of the system. This was shown during the year 2000 re-engineering effort when business and manufacturing processes were examined as part of the automation upgrades. Employment utilizing system analysis include system analyst, business analyst, manufacturing engineer, system architect, enterprise architect, software architect, etc.

While practitioners of system analysis can be called upon to create new systems, they often modify, expand, or document existing systems (processes, procedures, and methods).

Researchers and practitioners rely on system analysis. Activity system analysis has been applied to assess various research and practice studies including business management, educational reform, educational technology, etc.

5.4 System synthesis

Synthesis can be defined as:

- composition or combination of parts or elements to form a whole
- production of a substance by the union of chemical elements, groups, or simpler compounds or by the degradation of a complex compound
- combining of often diverse conceptions into a coherent whole

Synthesis is a combination of parts, or elements, to form a more complete view or system. The attitude is that the coherent whole that follows is considered to show the truth more completely than would a mere collection of parts. The term synthesis also refers in the dialectical philosophy of the 19th-century German philosopher G.W.F. Hegel and includes synthesis to the higher stage of truth which combines the truth of a thesis and an antithesis. Jean-Paul Sartre's philosophy underscores an existential type of synthesis. In Being and Nothingness, consciousness always tries to become being, to achieve a synthesis, as it were, between no-thing and some-thing.

The terms analysis and synthesis stem from Greek, meaning "to take apart" and "to put together," respectively. These terms are used in many disciplines, from mathematics and logic to economics and psychology, to substantiate similar search procedures.

In systems theory, synthesis is an approach serving to determine the system structure for a given (known) performance of the system. A related problem involves finding a structure most appropriate to a given performance. This is a very important problem which may be linked to a huge diversity of structures modeled by black-boxes (Mynarski, 1979). Reduction of the number of these structures can be achieved by the well-known method of maximum entropy which was promoted by Jaynes (1957, 1962) and Ingarden and Urbanik (1962). Applying maximum entropy approaches, systematic derivations of all findings of classical thermodynamics are available (Tribus, 1970).

5.5 Macro-, meso-, and micro-approaches

We should keep in mind that any unbiased analysis always leads to a unique solution, whereas synthesis does not ensure one-to-one solutions. This property of synthesis is consistent with the observation that the same performance (behavior) can occasionally exhibit systems of quite different structures.

Before making the decision to investigate the more specific details of paradigms and theories, let's look broadly at three possible levels of inquiry on which scientific investigations might be based. These three levels of inquiry correspond to: a microlevel approach, a mesolevel approach and a macrolevel approach.

At the microlevel, researchers examine the smallest levels of interaction; even in some cases, just "the self" alone. Microlevel analyses might include one-on-one interactions between couples or friends. Or perhaps a researcher might be interested in how a person's perception of self is influenced by his or her social context. In each of these cases, the level of inquiry is micro. When researchers investigate groups, their inquiry is at the mesolevel. Researchers who investigate mesolevel might study how norms of workplace behavior vary across professions or how children's sporting clubs are organized, to cite two examples. Researchers at the macrolevel examine large-scale patterns such as structures and institutions.

5.6 Functional vs structural features

David C. Brown (Artificial Intelligence in Design Group of Computer Science Dep., Worcester, Massachusetts, United States) stresses, in an unpublished work, the need to examine the definition of the terms "feature," and "functional feature" in particular. The goal is to shed some light on the reason for the profusion of types of features which have been discussed in the literature, to revisit the general definition of the term and to attempt a definition that uses concepts from artificial intelligence. By separating structure, behavior and function, and by defining function, obtained are alternative interpretations of "functional feature."

Working in the areas of intelligent information systems, databases, and knowledge base systems, Guan and Bell (1998) review the state of art for a rough set theory, a quite new mathematical tool for use in computer applications in circumstances which are characterized by vagueness and uncertainty. Their technique called "rough analysis" can be applied in artificial intelligence and cognitive sciences as a tool for dealing with vagueness and uncertainty of facts and in classification. The objective of their work is to enhance the application of their technique by designing a series of algorithms to implement the technique in a knowledge representation context. Although the technique has been shown to be successful in dealing with the vagueness of many practical applications, there are still several theoretical and practical problems to be solved. It is the set of issues researchers address, in the context of handling and analyzing large data sets during the knowledge representation process. The researchers find it important to seek efficient computational methods for the theory. In their rough set theory, a table called an information system or a database relation is used as a special kind of formal language to represent knowledge syntactically. Semantically, knowledge is defined as classification of information systems. The use of rough analysis does not involve the details of rough set theory directly, but it uses the same classification techniques. With this premise, the authors discuss how to apply the rough analysis technique to databases and how to assess computational methods for the rough analysis of databases.

The practical motivation and significance of Guan's and Bell's (1998) work is to use a tool for computer applications in which reasoning and learning are based on (often large) collections of data stored in computers, and possibly managed by database management systems. Their examples show how evidence can be obtained for reasoning purposes by examining properties of data as expressed in certain kinds of integrity constraints. The rough analysis can, thus, be considered as a possible approach for deriving evidence from data to see which hypothesis is best supported by these data. A related problem is to find which set of attributes or features (perhaps minimal) can be used to distinguish individual skills from one to another. The solution to this "classical" problem in database theory can provide essential support in underpinning the reasoning and learning applications encountered in artificial intelligence. Their discovery of "keys" can also provide insights into the structure of data which are not easy to obtain by alternative means.

5.7 Identification of complex systems

5.7.1 System identification

System identification deals with building mathematical models of dynamical systems based on observed data. It has been decades since system identification algorithms were developed as powerful techniques to analyze system structures and approximate real systems behavior in specific operational ranges. Since dynamical systems are abundant in our lives, techniques of system identification have many applications, such as, modeling, prediction, signal detection, fault diagnosis, etc. Available identification algorithms are not bullet-proof methodologies which can be used without any interaction from the user. A few reasons for this include:

- An appropriate model structure must be found. This can be a difficult problem, in particular if the dynamics of the system is nonlinear.
- There are certainly no "perfect" data in real life. The fact that the recorded data are disturbed by noise must be taken into consideration.
- The process may vary with time, which can cause problems if an attempt is made to describe it with a time-invariant model.
- It may be difficult or impossible to measure some variables/signals which are of central importance for the model. The essential ingredients of a system identification problem are: model selection, system analysis, experimental design, criteria of best fit, algorithm development, and model validation. Below we discuss these topics in detail.

5.7.2 Model importance

Models of systems are of fundamental importance in virtually all disciplines. In engineering, models can be useful for system analysis, control, prediction, and simulation. Advanced techniques for the controller design, optimization, supervision, fault detection, and diagnosis are also based on system models. Since the quality of the model determines the quality of the final solution, modeling is essential in developing the whole system. When we intend to identify a system, we need some preknowledge of how its components relate to each other. Generally speaking, we shall call such an assumed relationship among observed signals the model of the system. Obviously, models may be phrased in varying degrees of mathematical formula. The intended use will determine the degree of sophistication that is required to make the model purposeful.

5.7.3 Linear and nonlinear models

In fact, there is an infinite collection of mathematical models. In your research, you will have to restrict yourselves to a definite class of models (e.g., discrete-time state space models). This might seem like a highly restricted class; however, many industrial processes can be described very accurately by this type of models. Moreover, the number of control system design tools based on these models is almost without bound. For these reasons, this model class is a very interesting one. Nonlinear systems have more complicated structures than the linear systems. System identification on nonlinear systems which have a known structure but unknown parameters is referred to semiphysical modeling. Engineers will typically use system identification techniques to build their models. Two different philosophies may guide the choice of parameterized model sets, which are:

- Black-box model structures. The goal is to choose model structure from a model set which is flexible to accommodate a variety of systems, without looking into their internal structures.
- Gray-box model structures. The goal is to incorporate physical insight into the model set, with a certain number of adjustable parameters actually unknown about the system. Compared to models obtained from physical insight, black-box models are relatively easy to obtain and use and, even more importantly, these models are simple enough to make model-based control system design tractable from a mathematical and practical point of view. Yet, black-box models have a limited validity, limited working range, and in some cases, have no direct physical meaning. In control system design problems, it is essential to develop a mathematical model of the system to be controlled.
  Therefore, it is always wise first to try and utilize physical insight to characterize possible structures which will be reflected by the model. In contrast to models based on mathematical modeling, the models obtained solely on system identification have the following properties:
- They have limited validity (they are valid for a certain working point, a certain type of input, a certain process, etc.).
- They provide little physical insight, since in most cases the parameters of the model have no direct physical meaning. The parameters are used only as tools to give a reasonable description of the system's overall behavior.
- They are relatively easy to construct and use.

5.7.4 Procedure for nonlinear system identification

Modeling and identification of nonlinear systems is a challenging task because nonlinear processes do not share many common properties. A major goal of research in this field is to extend possessed capabilities of modeling to describe a wide class of nonlinear systems.

This section specifies the major steps that have to be performed for a successful system identification. The input/output data are recorded during a specifically designed identification experiment, where the user may determine which signals to measure and when to measure them and may also choose the input signals. If the process under consideration cannot be actively excited, a training data set still has to be designed by selecting a data set from the gathered measurements that start model inputs/outputs, model architecture, dynamics representation, model order, and model structure.

In this book, we are often led to consider problems of identification of large-scale interconnected systems. Accurate models of complex nets are needed especially for optimal control in process engineering, power production, and transportation systems. The specifics lie in the fact that individual elements cannot be disconnected and excited by arbitrary inputs for identification purposes. Moreover, structural interactions cause correlations between interaction signals. In particular, any output random disturbances can be transferred into other inputs. This leads to cross-correlation problems, very difficult from the effective modeling point of view. First attempts in 1980s were limited to static linear blocks, and in practice, the results were preferentially collected for linear dynamic systems working in steady state. Yet, we are now able to formulate the approach for components which are both dynamic and nonlinear. All blocks are represented by two-channel (Hammerstein) systems, used, e.g., in modeling of heating processes. The least squares estimate may be applied to identify unknown parameters of a system. The parameters of particular elements are obtained in a singular value decomposition procedure; as a whole, such an algorithm represents a simple simulation example.

5.8 Hierarchical structures and hierarchy of controls

A hierarchical control system (HCS) is a form of control system in which a set of devices and governing software is arranged in a hierarchical tree. When the links in the tree are implemented by a computer network, then the hierarchical control system has also the form of a networked control system.

A human-built system with complex behavior is often organized as a hierarchy. For example, a command hierarchy has among its notable features the organizational chart of superiors, subordinates, and lines of organizational communication. Hierarchical control systems are organized similarly to divide the decision-making responsibility.

Each element of the hierarchy is a linked node in the tree. Commands, tasks, and goals to be achieved flow down the tree from superior nodes to subordinate nodes, whereas sensations and command results flow up the tree from subordinate to superior nodes. Nodes may also exchange messages with their siblings. The two distinguishing features of a hierarchical control system are related to its layers.

- Each higher layer of the tree operates with a longer interval of planning and execution time than its immediately lower layer.
- The lower layers have local tasks, goals, and sensations, and their activities are planned and coordinated by higher layers which do not generally override their decisions. The layers form a *hybrid intelligent system* in which the lowest, reactive layers are subsymbolic. The higher layers, having relaxed time constraints, are capable of reasoning from an abstract world model and performing planning. A *hierarchical task network* is a good fit for planning in a hierarchical control system.

Besides artificial systems, an animal's control systems are proposed to be organized as a hierarchy. In *perceptual control theory* which postulates that an organism's behavior is a means of controlling its perceptions, the organism's control systems are suggested to be organized in a hierarchical pattern as their perceptions are constructed so. Control system structure is illustrated in Fig. 5.2.

The diagram in Fig. 5.2 is a general hierarchical model which shows functional manufacturing levels using computerized control of an industrial control system.

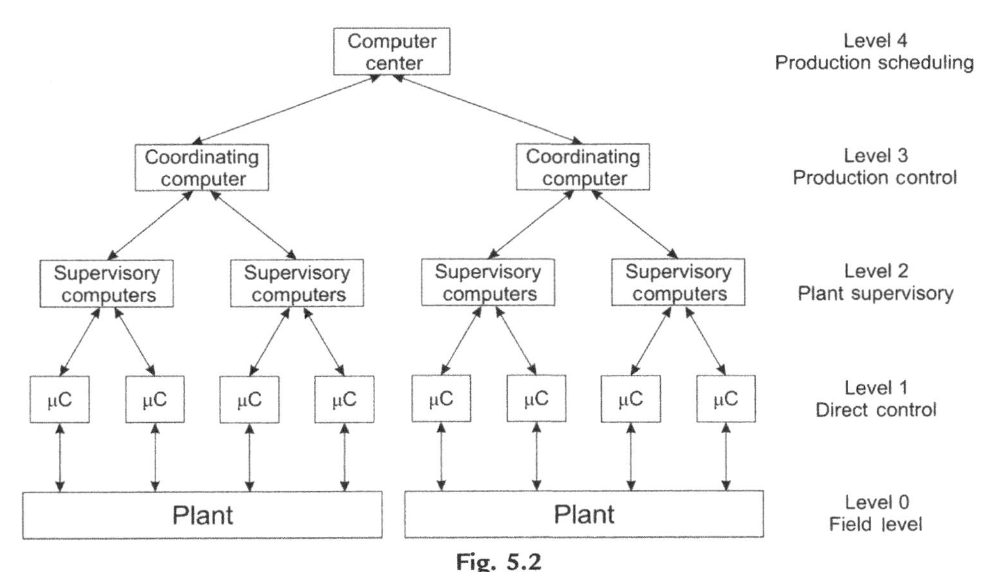

Fig. 5.2
Functional levels of manufacturing control operation.

Referring to the diagram:

- Level 0 contains the field devices such as flow and temperature sensors, and final control elements, such as control valves
- Level 1 contains the industrialized input/output (I/O) modules, and their associated distributed electronic processors
- Level 2 contains the supervisory computers, which collate information from processor nodes on the system and provide the operator control screens
- Level 3 is the production control level, which does not directly control the process, but is concerned with monitoring production and monitoring targets
- Level 4 is the production scheduling level

5.8.1 Applications: Manufacturing, robotics, and vehicles

Among the robotic paradigms is the hierarchical paradigm in which a robot operates in a top-down fashion, heavy on planning, especially motion planning. Computer-aided production engineering has been a research focus at NIST since the 1980s. Its Automated Manufacturing Research Facility was used to develop a five-layer production control model. In the early 1990s, American Defense Advanced Research Projects Agency (DARPA) sponsored research to develop distributed (i.e., networked) intelligent control systems for applications such as military command and control systems. NIST built on earlier research to develop its Real-Time Control System (RCS) and Real-time Control System Software which is a generic hierarchical control system which has been used to operate a manufacturing cell, a robot, and an automated vehicle.

In November 2007, DARPA held the Urban Challenge. The winning entry, Tartan Racing employed a hierarchical control system, with layered mission planning, motion planning, behavior generation, perception, world modeling, and mechatronics.

5.9 Communication systems

Basic structure of communication systems is illustrated in Fig. 5.3.

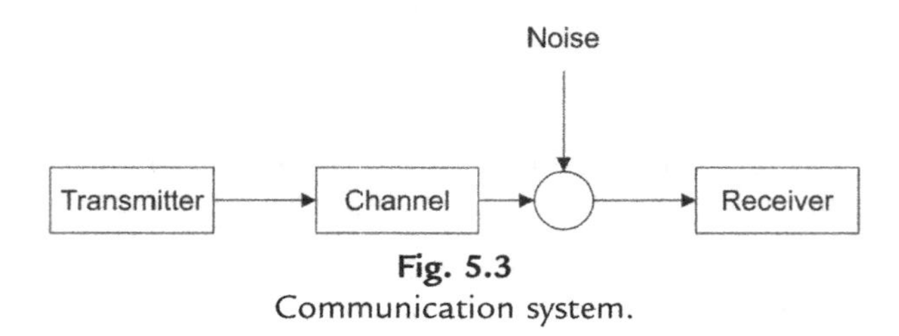

Fig. 5.3
Communication system.

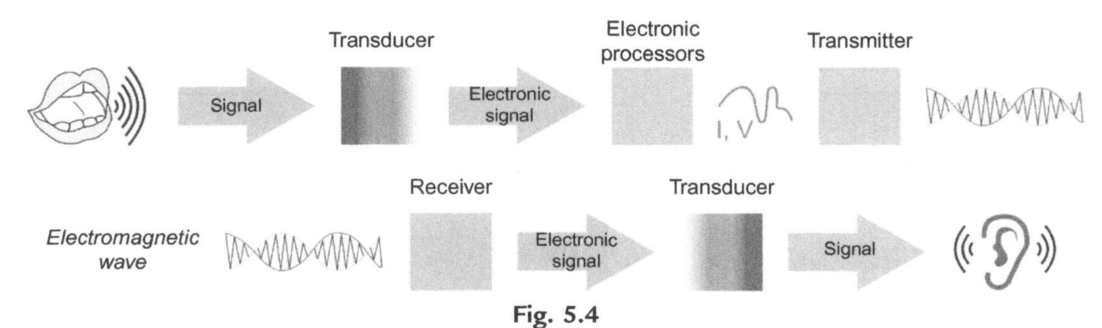

Fig. 5.4
An electronic communication system using electronic signals.

In his book, Mynarski (1979) presents a number of more complicated structures of communication systems. He also specifies a number of detrimental factors influencing negatively the quality of informational output. While in the matter-energy relation determined are thermodynamic properties of the system, associated with transformation of heat into work, in the matter-information relation determined are structural properties of the system associated with the location and linkages of the separate parts of the system. On the other hand, in the relation energy-information determined are communication properties of the systems associated with information transfer.

Fig. 5.4 schematizes an electronic communication system applying electronic signals. In fact, in information systems, control properties are determined only in a three-dimensional space: matter-energy-information (Mynarski, 1979). Information is the most important component as it ensures order and organization in every system; its effective transfer is especially important in large systems. The close link between information and energy ensures good contacting in the system, which means the satisfaction of the susceptibility condition for not only real control, but also for the proper degree of cohesion, or the so-called coherence of the system (Chapter 9). The mutual relation of matter, energy, and information is the basis for contemporary human existence and the development of civilization.

Categories of information related to managerial levels and the decisions of managers are distinguished in Fig. 5.5. Compare, for example, the difference between strategic and operational operations, and corresponding personal privileges.

5.10 Adaptive controls

Adaptive control is the control method used by a controller which must adapt to a controlled system with parameters which vary, or are initially uncertain. For example, as an aircraft flies, its mass will slowly decrease as a result of fuel consumption; a control law is needed that adapts itself to such changing conditions. Adaptive control is different from *robust control* in that it does not need a priori information about the bounds on these uncertain or time-varying

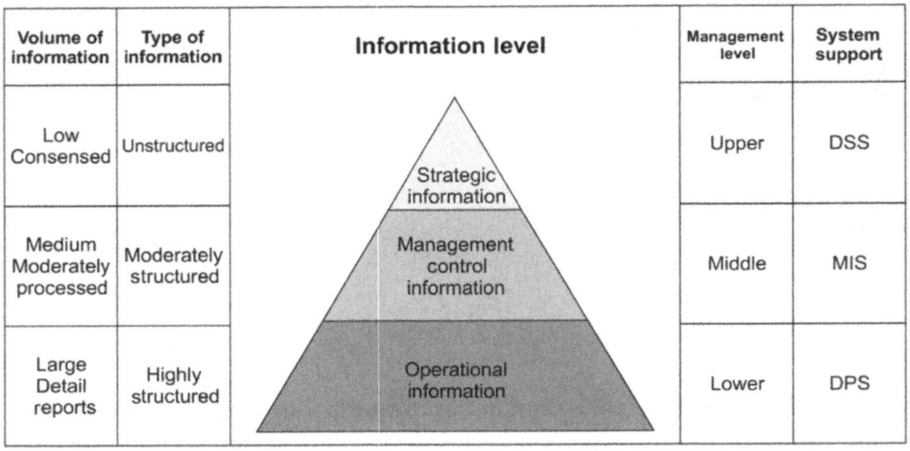

Volume of information	Type of information	Information level	Management level	System support
Low Consensed	Unstructured	Strategic information	Upper	DSS
Medium Moderately processed	Moderately structured	Management control information	Middle	MIS
Large Detail reports	Highly structured	Operational information	Lower	DPS

Fig. 5.5
Categories of information related to managerial levels and the decisions of managers.

parameters; robust control guarantees that if the changes are within given bounds the control law need not be changed, while adaptive control is concerned with control law changing itself.

Parameter estimation. The foundation of adaptive control is *parameter estimation*, which is a branch of *system identification*. Common methods of estimation include *recursive least squares* and *gradient descent*. Both these methods provide update laws which are used to modify estimates in real-time (i.e., as the system operates). *Lyapunov stability* property is used to derive these update laws and show convergence criteria (typically persistent excitation; relaxation of this condition are studied in Concurrent Learning adaptive control). *Projection (mathematics)* and normalization are commonly used to improve the robustness of estimation algorithms.

5.10.1 Classification of adaptive control techniques

In general, one should distinguish between:

1. Feedforward adaptive control
2. Feedback adaptive control

as well as between

1. Direct methods
2. Indirect methods
3. Hybrid methods

Direct methods are ones wherein the estimated parameters are those directly used in the adaptive controller. In contrast, indirect methods are those in which the estimated parameters are used to calculate required controller parameters. Hybrid methods rely on both estimation of parameters and direct modification of the control law (Figs. 5.6 and 5.7).

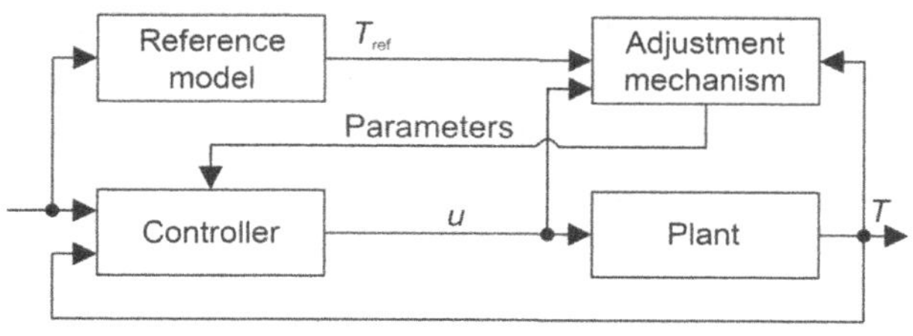

Fig. 5.6
Model reference adaptive control (MRAC).

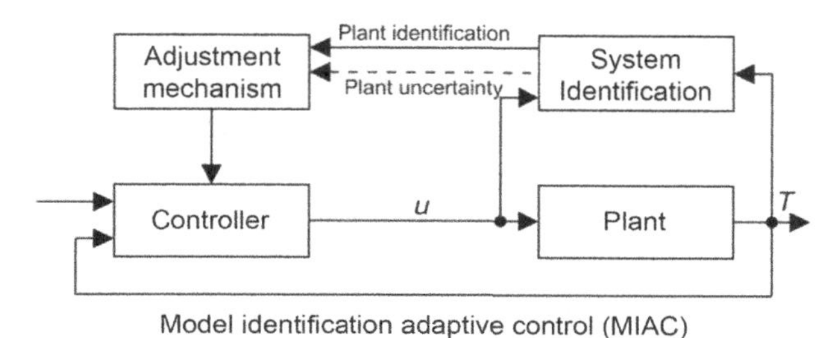

Fig. 5.7
Model identification adaptive control (MIAC).

Example Various adaptive control problems arise when studying decision-making in systems involving fuzzy sets; Butt and Akram (2016) describe a novel intuitionistic fuzzy rule-based decision-making system based on intuitionistic fuzzy sets with a process scheduler of a batch operating system. The intuitionistic algorithm of Butt and Akram (2016) imposes the nice value and burst time of all available processes in the ready queue, intuitionistically fuzzifies the input values, triggers appropriate rules of their intuitionistic fuzzy inference engine, and finally calculates the dynamic priority (dp) of all processes in the ready queue. Once the dp of a process is calculated, the ready queue is sorted in decreasing order of dp of every process. The process with maximum dp value is sent to the central processing unit for execution. Finally, the authors show nice working of their algorithm on two different data sets and give comparisons with some standard nonpreemptive process schedulers.

5.11 Optimization of complex systems

Mathematically, optimization is seeking of the best solution under given conditions (constraints). Such a general definition of optimization includes most conscious activity pursued by mankind during its long history. Its mathematical pursuit also has a long history, but several major strides have been made during the last three decades.

All rational human activity is characterized by a continuous search for progress and development. The tendency of searching to obtain the best solution under definite circumstances is called optimization—in the broad sense of this word. So-understood optimization has been the property of the rational human activity for years. However, in recent years, the need for methods which would lead to improvement of the quality of industrial and practical processes has grown stronger, leading to a fast development of a group of mathematical methods of optimum seeking, which are now called collectively methods of optimization. Clearly, what brought about a rapid development of these methods was the progress in computer science which made numerical solutions of many practical problems possible. In mathematical terms, optimization is seeking for the best solution within imposed constraints.

Process engineering is an important source of inspiration in optimization. Most technological processes are characterized by flexibility in the choice of some parameters and/or functions; by changing these quantities, it is possible to correct process performance and development. In other words, decisions need to be made which make it possible to control an actually running process. There are also decisions that need to be made in designing a new process or new equipment. Thanks to these decisions (controls), some goals can be reached. For example, it may be possible to achieve a sufficiently high concentration of a valuable product at the end of a tubular rector at minimum costs, or in another problem, to assure both a relatively low decrease of fuel value and a maximum amount of work delivered from an engine. How to accomplish such particular tasks is the problem of control in which some constraints are represented by transformations of the system's state and others by boundary conditions of the system. If this problem can be solved, then usually a number of solutions may be found to satisfy process constraints. Therefore, it is possible to go further and require that a defined objective function (process performance index) should be reached in the best way possible, for example: in a shortest time, with a least expenditure of valuable energy, with minimum costs, etc. An optimization problem emerges, related to an optimal choice of the process decisions.

In testing a process, it is necessary to quantify the related knowledge in mathematical terms; this leads to a mathematical model of optimization which formulates the problem in the language of functions, functionals, equations, inequalities, etc. The mathematical model should be strongly connected with reality, as it emerges from it and finds its application in it. However, the mathematical model deals often with very abstract meanings; thus, the finding of an optimal solution requires the knowledge of advanced methods. We shall present here only selected methods, suited to the content of this book.

In technology, practically every problem of design, control, and planning can be approached through an analysis leading to the determination of the least (minimum) or the greatest (maximum) value of some particular quantity, of physical, technological, or economical nature which is called the optimization criterion, performance index, objective function, or profit function. The choice of decisions (also physical, technological, or economical nature), which can vary in a definite range, affects the optimization criterion, the criterion being a measure of the effectiveness of the decisions. The task of an optimization is to find decisions to assure the minimum or maximum value of the optimization criterion.

The existence of decisions as quantities whose values are not prescribed, but rather chosen freely or within certain limits, makes optimization possible. Optimization—understood as an activity leading to the achievement of the best result under given conditions was always an inevitable part of human activity—received a solid scientific basis only when its meanings and methods became mathematically described. Thanks to the recent computational techniques and use of high-speed computers, optimization research has gained economic ground, and the range of the solved problems increased enormously. Apart from the use of digital computers, many optimization problems have been solved by using analogue or hybrid computers.

In this book, we assume that each optimization problem can be represented by a suitable mathematical model. Clearly, the mathematical model can simulate the behavior of a real system in a more exact or in a less exact way. Whenever good agreement is observed between the real system behavior and its model, the optimization results can be used to improve the performance of the system. However, cases may exist when the process data are not reliable enough and oversimplifications may occur in model constructing; in these cases, the results of optimization cannot be accepted without criticism. Clearly, in case of large data inaccuracy or model invalidity, any optimization results will not be reliable. However, the models and the data which are now used for optimization are, in fact, the same as those used in design and process control. In many cases, these models are well-established, so the related optimization is desirable.

It is not seldom that the technical implementation of a correct optimization solution may prove difficult. In these cases, optimization results can still be useful to expose extremal or limiting possibilities of the system from the viewpoint of accepted optimization criterion. For example, an obtained limit can be represented by upper (lower) bound on the amount of the electrical or mechanical energy delivered from (supplied to) the system. Real system characteristics, which lie below (above) the limits predicted by the optimal solution, can sometimes be taken into account by considering suboptimal solutions. The latter may be easier to accomplish than the optimal solution.

The mathematical model of optimization is the system of all the equations and inequalities which characterize the considered process, including the optimization criterion. The model makes it possible to determine how the optimization criterion changes with the variations in the decisions. In principle, mathematical models can be obtained in two ways. With the knowledge of physical laws, the so-called analytical models are formulated. By performing identification of the system, experimental models are determined, often based on the regression analysis. Sometimes they are represented by polynomial equations linking outputs and inputs of the system.

In design, analytical models are usually used because only they make possible a wide extrapolation of data necessary when the process scale is changed. In analytical models, the number of unknown coefficients to be determined is usually much lower than in empirical ones. However, when controlling existing processes, empirical models are still, quite frequently, applied.

If an optimization is associated with planning and doing experiments and its partial purpose is finding the data which help to determine optimal decisions, we are dealing with the *experimental optimization*. If a mathematical description is used, which takes into account the process, its environment, and a control action, we are dealing with the *analytical optimization*. A part of this book deals with the analytical optimization. The models we use in most of its chapters are deterministic ones. Yet, since some results in the fields of energy limits and environmental care may be linked with processes, uncertainty, and simulated annealing criteria (Nulton and Salamon, 1988; Harland and Salamon, 1988; Andresen and Gordon, 1994), the final part of this chapter discusses several basic techniques of stochastic optimization.

At the working state of a technological process, the problem of adaptation of the mathematical model plays an important role. Adaptation should always be made whenever variations are observed in uncontrolled variables of the system, which normally should remain constant. For slowly varying changes, adaptation of the continuous type is possible. Otherwise, fast varying changes require periodic adaptation, for the averaged values of changes which are regarded as a noise. Optimization *on line* is carried out simultaneously with the adaptation of the model; in this optimization, a control action is accomplished directly by the computer. Yet, a computer-aided control involves optimization *off line*.

Optimization criterion (performance index) is the important quantity that appears in the mathematical model in the form of a function or a functional. Usually, this is an explicit and analytical form. The choice of the optimization criterion in an industrial process must be a subject of very accurate analysis, which involves, most often, both technological and economic terms. This is because the definition of the criterion has the important effect on the problem solution and expected improvements.

Along with the performance index, there appear in the mathematical model some equality and inequality equations (algebraic, differential, integral, etc.) which characterize *constraints* imposed on the process. Both constraints and performance index can contain *decision variables* or *controls* that are adjustable variables which an engineer or a researcher can change.

Some other variables can also appear both in the constraining equations and in performance index which are called *uncontrolled variables*. These variables are determined by certain external factors independent of the observers and industrial leaders (composition of a raw material, for example). They cannot be controlled, but they can often be measured. For optimization, control variables are most essential, as they are those which characterize external action performed on the process and are of utmost importance on the optimization criterion. Important is also the sensitivity analysis of the optimization criterion with respect to the decisions, because their number determines to a large extent the difficulty of the optimization problem. Leaving out decisions which affect the optimization criterion in an insignificant way should be treated as a natural procedure which contributes to the increased transparency of the results and often facilitates problem solving.

Imposing constraints on decision variables is typical for all practical problems of optimization. There are, for example, constraints on consumption of some resources, process output, product purity, concentrations of contaminants, etc. Constraints can also be formulated for thermodynamic parameters of the process in order to specify allowable ranges of temperatures and pressures, intervals of catalyst activities, reaction selectivities, etc. Important are constraints which assure reliability and safety of the equipment, for example those imposed on reagent concentrations in combustible mixtures (preventing the explosion), on gas and fluid flows in absorbers (prevention of flood), on the superficial velocity of fluidization (prevention of the blowing out of the material), etc. Constraints may also be imposed on construction parameters, e.g., there are constraints imposed on the sizes of apparatuses in closed residential areas or constraints on lengths of pipes in the heat exchangers, which arise from the standardization. Important, although requiring some experience, is leaving out from the optimization formulation these constrains which have a negligible effect on the optimal solution. This enables one to use easier techniques for the problem solving, yet still maintains exactness of the optimization result.

Variety of optimization problems is strongly connected with the variety of constraints (algebraic, differential, integral—each may appear in equality or inequality form). In this chapter, only examples of simplest algebraic constraints are considered; more involved ones may be found in further chapters of the book.

Sets of algebraic constraints characterize often a system at its steady state, hence the name "static optimization."

The mathematical nature of the optimization problem lends itself naturally to a division between finite and infinite dimensional problems. In the first, called the mathematical programming problem, the optimization criterion (performance index) is the *function* of the decisions (and possibly also other additional variables like constants, parameters, and so forth, called collectively the uncontrolled variables). In the second, called the dynamic optimization problem or the variational problem, the optimization criterion is the *functional* of the decisions (as well as state variables and constant parameters). In the first case, one is searching for an optimum in the form of an n-tuple *of numbers*. In the second case, the optimizer's task is to determine the *functions* describing the dynamic behavior of the optimal control as a function of time or as a function of any time-like independent variable characterizing the evolution of the process (length, holdup time, etc.). These functions constitute the optimal control which characterizes the best external action on the system in time. The optimal control is obtained as the basic ingredient of the optimization solution simultaneously with another vector function, which characterizes the best time behavior of the process state, called the optimal trajectory. The stable asymptotic solution of the optimal control problem (if it exists) is associated with a steady-state situation when the process time tends to infinity and the external controls are time-independent. These asymptotic solutions are often solutions to a corresponding static problem. However, this fact is seldom exploited directly, that is, we seldom seek solutions to the static problems as the special (steady-state) solutions of dynamic problems. The static optimization problems (mathematical programming problems) have their own methods which are simpler than those of the dynamic problems. Hence, tractable static problems may involve many more components of the control vector than their dynamic counterparts.

The complexity of the optimization problems is as a rule connected with the complexity and variety of the constraints. Constraining equations and inequalities may be algebraic, difference, differential, integral, and integro-differential. For mathematical programming problems (static optimization problems), constraints are algebraic. For variational problems (dynamic optimization), differential equations are the natural constraints, although any other type of constraint, specified previously, can additionally appear in more involved variational problems. This may result in an extremely complex structure for dynamic optimization problems. This is not to say that static problems of mathematical programming cannot be complex and difficult to solve. These difficulties are due to the nonlinearities that may appear in the constraining equations and/or the performance index.

Instead of listing some main facts in the field history, we only stress here several of most important recent milestones for its development. These are: Dantzig's methods for linear programming (Dantzig, 1968; Llewellyn, 1963), Karush-Kuhn-Tucker conditions for nonlinear programming (Kuhn and Tucker, 1951; Zangwill, 1969), Bellman's dynamic programming (Bellman, 1957, 1961, 1967; Bellman and Kalaba, 1965; Aris, 1964), and Pontryagin's maximum principle (Pontryagin et al., 1962; Fan and Wang, 1964; Lee and Marcus, 1967; Leitman, 1981; Sieniutycz, 1978; Findeisen et al., 1980, Sieniutycz and Jeżowski, 2018). These methods are complemented by the classical methods of differential calculus, Lagrange multipliers, and the calculus of variations (functional optima; Gelfand and Fomin, 1963). Together, they make a powerful collection of optimization tools which can solve many difficult practical problems (Beveridge and Schechter, 1970). Of newer optimization techniques, methods of averaged optimization and sliding regimes in optimized problems (Tsirlin, 1974, 1997; Berry et al., 2000) received the status of effective tools in applied finite-time thermodynamics. For these developments, progress in computational techniques has also been essential.

As optimization is a very vast field, studied in detail in many sources, we shall not describe here the above methods and techniques. Rather we shall refer the reader to the above books and sources and finalize this section by limiting ourselves to the most often considered topics. They are listed below.

5.11.1 Topics in the optimization theory and applications

- Query optimization (QO)
  QO is a function of many relational database management systems. The query optimizer attempts to determine the most efficient way to formulate a problem.
- Engineering optimization (EO)
  EO deals, among others, with work or power yield processes, involves practical optimization and risk management tools. It is difficult when dealing with large or complex projects.
- Genetic algorithms (GA)
  GA are algorithms or computational optimization procedures which mimic a simplistic mechanism of biological evolution.
- Search optimization (SO)
  SO advocates integration of subsymbolic machine learning techniques into search heuristics for solving complex optimization problems.
- Semidefinite programming (SDP)
  SDP is a subfield of convex optimization concerned with the optimization of a linear objective function.
- Complex systems optimization (CSO)
  CSO involves optimizing cohesive conglomerations of interrelated and interdependent objects (subsystems). Applied in interdisciplinary study of systems.
- Multiobjective programming
  Multiobjective programming (also known as multiobjective optimization) involves vector optimization, multicriterial and multiattribute optimization, or Pareto problems.
- Virtual engineering optimization
  Involves fluid dynamics (CFD), finite elements analysis (FEA), and optimization of complicated virtual systems.
- Price optimization
  Uses data which may describe operating costs, capital costs, inventories, and historic prices and sales.
- Feedback system optimization. An optimization approach that necessitates expansion of the apparent space of a system. The term bipolar feedback has been coined in bio-systems where positive and negative feedback systems can interact.
- Process engineering or Process systems engineering
  scheduling process networks, multiperiod planning and optimization, data reconciliation, real-time optimization, flexibility measures, fault diagnosis.
- Mathematical programming
  Optimization of nonlinear problems subject to equality and inequality constraints. Includes nonlinear optimization methods widely applied in conformational analysis and optimization techniques used in computational systems biology.

- Stochastic optimization
  Stochastic optimization (SO) methods are optimization methods that generate and use random variables. For stochastic problems, random variables appear.
- Optimizing compiler
  overhead of compiler optimization: Any extra work which takes time; a whole-program optimization is the time consumption for large programs. The often-complex interaction.
- Trajectory optimization
  Applies to the trajectory improvement in the maximum-principle-based calculations. Recently, trajectory optimization has also been used in a wide variety of industrial process and robotics applications.
- Scalability and scalable systems
  It is a highly significant issue in electronics systems, databases, routers, and networking. A system whose performance improves after adding hardware.
- Underdetermined system
  In general, a system of linear equations which has an infinite number of solutions, if any.
- Dynamical systems theory
  Dynamical systems theory is an area of mathematics used to describe the behavior of the complex dynamical systems, usually by employing differential equations.
- Mathematical optimization
  generalization of optimization theory and formulations constituting a large area of applied mathematics (see mathematical programming).
- Search engine optimization (SEO)
  SEO differs from local search engine optimization in that the latter is focused on optimizing a business' online presence.
- Particle swarm optimization (PSO)
  PSO in the problem optimized means that the method does not require that the optimization problem is differentiable as is required in classic optimization methods.
- Optimization as disambiguation
  Mathematical optimization is the theory and computation of extrema or stationary points of functions.
- Test functions for optimization
  Deal first with single-objective optimization cases. Next, test functions deal with their respective Pareto fronts for multiobjective optimization problems.
- Convex optimization
  Convex optimization is a subfield of mathematical optimization which studies the problem of minimizing convex functions over convex sets.
- Optimization problem
  In mathematics and computer science, an optimization problem is the problem of finding the best solution from all feasible solutions.

- Bayesian optimization
  Bayesian optimization is a sequential design strategy for global optimization of black-box functions that doesn't require derivatives.
- Stochastic optimization
  Stochastic optimization (SO) methods are optimization methods that generate and use random variables. For stochastic problems, the random variables appear.
- Multiobjective optimization
  Multiobjective optimization is also known as multiobjective programming, vector optimization, multicriteria optimization, multiattribute optimization.
- Ant colony optimization algorithms
  This approach involves optimization algorithms modeled on the actions of an ant colony.
- Topology optimization (TO)
  TO is different from shape optimization and sizing optimization in the sense that the design can attain any shape.
- Constrained optimization
  In mathematical optimization, constrained optimization (in some contexts called constraint optimization) is the process of optimizing an objective function under constraints.
- Program optimization
  In computer science, program optimization or software optimization is the process of modifying a software system to make some aspect of it work more efficiently.
- Discrete optimization
  Discrete optimization is a branch of optimization in applied mathematics and computer science. As opposed to continuous optimization, some or all parts of the mathematical model are given in the form of difference equations, sometimes called "state transformations." Other part of the model may have the form of arbitrary algebraic constraints. The optimizer's task is to find optimal values of discrete states, controls, and sometimes, some global characteristics, e.g., the sequence describing the changes of an optimal performance function.
- Effective optimization (EO)
  Effective optimization is a function of many relational database management systems. A qualified optimizer attempts to determine and apply the most efficient way to execute the problem.
- Engineering optimization
  Engineering optimization is the subject which uses optimization techniques to achieve design goals in engineering. It is sometimes referred to as design.
- Combinatorial optimization
  Combinatorial optimization is a discrete optimization which consists of the integer programming.

5.11.2 Topics in computational problems of optimization

The development of a computer-based information system includes a system analysis phase. This helps produce the data model, a precursor to creating or enhancing a database. There are numerous and different approaches to system analysis.

When a computer-based information system is developed, system analysis (according to a Waterfall model) would constitute the following steps:

- The development of a feasibility study: determining whether a project is economically, socially, technologically, and organizationally feasible
- Fact-finding measures, designed to ascertain the requirements of the system's end-users (typically involving interviews, questionnaires, or visual observations of work on the existing system)
- Gauging how the end-users would operate the system (in terms of general experience in using computer hardware or software), what the system would be used for, and so on.
- Another view outlines a phase approach to the process. This approach breaks system analysis into five phases:
- Scope definition: Clearly defined objectives and requirements necessary to meet a project's requirements as defined by its stakeholders
- Problem analysis: the process of understanding problems and needs and arriving at solutions that meet them
- Requirements analysis: determining the conditions that need to be met
- Logical design: looking at the logical relationship among the objects
- Decision analysis: making a final decision

These cases are elements of widely used system analysis modeling tools for identifying and expressing functional requirements of a system. Each case represents a business scenario or event for which the system must provide a defined response. These cases evolve from an object-oriented analysis.

5.12 Final remarks

In this chapter, we described basic components of systems design which involve modeling, analysis, synthesis, and optimization. Also, we briefly characterized kinetic equations for systems modeling. We pointed out essential role of communication systems in the system theory and briefly characterized their structure and properties. A comparison has been made between macro-, meso-, and micro-approaches as well as a distinction was depicted between functional and structural features of systems. Moreover, we stressed the importance of identification methods in systems investigation and characterized basic methods of constrained optimization (usually hierarchical and with adaptive controls). Example of Butt and Akram (2016) was outlined in which adaptive control problems arise when studying decision-making in the systems involving fuzzy sets.

References

Andresen, B., Gordon, J., 1994. Constant thermodynamic speed for minimizing entropy production in thermodynamic processes and simulated annealing. Phys. Rev. E 50, 4346–4351.

Aris, R., 1964. Discrete Dynamic Programming. Blaisdell, New York.

Bampi, F., Morro, A., 1984. Nonequilibrium thermodynamics: a hidden variable approach. In: Cassas Vazquez, J., Jou, D., Lebon, G. (Eds.), Lecture Notes in Physics. In: vol. 199. Springer, Berlin, pp. 221–232.

Bedeaux, D., 1986. Nonequilibrium thermodynamics and statistical physics. In: Prigogine, I., Rice, S.A. (Eds.), Advances in Chemical Physics. In: vol. 64. Wiley, New York, p. 67.

Bedeaux, D., 1992. Nonequilibrium thermodynamics of surfaces. Flow, diffusion and transport processes. In: Advances in Thermodynamics Series. vol. 6. Taylor and Francis, New York, pp. 430–459.

Bellman, R.E., 1957. Dynamic Programming. Princeton University Press, Princeton.

Bellman, R.E., 1961. *Adaptive Control Processes*: A Guided Tour. Princeton University Press, Princeton, NJ.

Bellman, R.E., 1967. Introduction to Mathematical Theory of Control Processes. Academic Press, New York.

Bellman, R.E., Kalaba, R., 1965. Dynamic Programming and Modern Control Theory. Academic Press, New York.

Berry, R.S., Kazakov, V.A., Sieniutycz, S., Szwast Z., Tsirlin, A.M., 2000. Thermodynamic Optimization of Finite Time Processes. Wiley, Chichester. ISBN: 0-471-96752-1.

Beveridge, G.S., Schechter, S., 1970. Optimization: Theory and Practice. McGraw-Hill, New York.

Bird, R.B., Steward, W.E., Lightfoot, E.N., 1960. Transport Phenomena. Wiley, New York.

Burnett, D., 1935. The distribution of the molecular velocities and the mean motion in a nonuniform gas. Proc. Lond. Math. Soc. 40, 382–435.

Butt, M.A., Akram, M., 2016. A new intuitionistic fuzzy rule-based decision-making system for an operating system process scheduler. Springerplus 5, 1547. https://doi.org/10.1186/s40064-016-3216-z.

Cercignani, C., 1975. Theory and Application of the Boltzmann Equation. Scottish Academic Press, Edinburgh.

Chapman, S., Cowling, T.G., 1973. The Mathematical Theory of Non-Uniform Gases, third ed. Cambridge University Press, Cambridge, UK.

Dantzig, G.B., 1968. Linear Programming and Extensions. Princeton University Press, Princeton, NJ.

de Groot, S.R., Mazur, P., 1984. Nonequilibrium Thermodynamics. Dover, New York.

Enskog, D., 1917. Kinetic Theory of Processes in Moderately Dense Gases (Dissertation).Upsala University, Upsala.

Fan, L.T., Wang, C.S., 1964. The Discrete Maximum Principle: A Study of Multistage System Optimization. Wiley, New York.

Findeisen, W., Szymanowski, J., Wierzbicki, A., 1980. Theory and Computational Methods of Optimization. Państwowe Wydawnictwa Naukowe, Warszawa (in Polish).

Fitts, D.D., 1962. Nonequilibrium Thermodynamics. McGraw-Hill, New York.

Gelfand, J.M., Fomin, S.W., 1963. Calculus of Variations. Prentice-Hall, Englewood Cliffs, NJ.

Grad, H., 1958. Principles of the theory of gases. In: Flugge, S. (Ed.), Handbook der Physik, vol. 12. Springer, Berlin.

Grandy, W.T., 1988. Foundations of Statistical Mechanics. Equilibrium Phenomena/Nonequilibrium Phenomena, vol. I/vol. II. Kluwer Academics, Amsterdam.

Guan, J.W., Bell, D.A., 1998. Rough computational methods for information systems. Artif. Intell. 105, 77–103.

Harland, J.R., Salamon, P., 1988. Simulated annealing: a review of the thermodynamic approach. Nucl. Phys. B (Proc. Suppl.) 5A, 109.

Hirschfelder, J.O., Curtiss, C.F., Bird, R.B., 1954. Molecular Theory of Gases and Liquids. Wiley, New York.

Huang, K., 1967. Statistical Mechanics. Wiley, New York.

Ingarden, R.S., Urbanik, K., 1962. Quantum informational thermodynamics. Acta Phys. Pol. 21, 281–304.

Jaworski, W., 1981. Information thermodynamics with the second order temperatures for the simplest classical systems. Acta Phys. Pol. A 60, 645–659.

Jaynes, E.T., 1957. Information theory and statistical mechanics. Phys. Rev. 106, 620–630.

Jaynes, E.T., 1962. Information theory and statistical mechanics. In: Ford, W.K. (Ed.), Brandeis Lectures in Theoretical Physics, vol. 3. W.A. Benjamin, New York, p. 181.

Keizer, J., 1987. Statistical Thermodynamics of Nonequilibrium Processes. Springer, New York.

Kestin, J., 1979. In: Domingos, et al., (Ed.), Foundations of Non-Equilibrium Thermodynamics. Macmillan Press, London.

Kreuzer, H.J., 1981. Nonequilibrium Thermodynamics and Its Statistical Foundations. Clarendon Press, Oxford.

Kuhn, H.W., Tucker, A.W., 1951. Nonlinear programming. In: Neyman, J. (Ed.), Proceedings of the Second Berkeley Symposium on Mathematical Statistics and Probability. University of California Press, Berkeley, pp. 481–493.

Landau, L., Lifshitz, E., 1975. Statistical Physics. Pergamon, Oxford.

Lavenda, B.H., 1985. Nonequilibrium Statistical Thermodynamics. Wiley, Chichester, UK.

Lebon, G., Mathieu, P., 1983. Comparison of the diverse theories of nonequilibrium thermodynamics. Int. Chem. Eng. 23, 651–662.

Lee, E.B., Marcus, L., 1967. Foundations of Optimal Control Theory. Wiley, New York.

Leitman, G., 1981. The Calculus of Variations and Optimal Control. Plenum Press, New York.

Lewis, R.M., 1967. A unifying principle in statistical mechanics. J. Math. Phys. 8, 1448–1459.

Lifshitz, E., Pitajevsky, L., 1984. Physical Kinetics. Pergamon, London.

Llewellyn, R.W., 1963. Linear Programming. Holt, Rinehart & Winston, New York.

Månsson, B.Å.G., Lindgren, K., 1990. Thermodynamics, information and structure. In: Nonequilibrium Theory and Extremum Principles. Advances in Thermodynamics Series, vol. 3. Taylor and Francis, New York, pp. 95–128.

Mynarski, S., 1979. Elementy Teorii Systemów i Cybernetyki (Elements of Systems Theory and Cybernetics). Państwowe Wydawnictwo Naukowe, Warszawa (in Polish).

Nulton, J., Salamon, P., 1988. Statistical mechanics of combinatorial optimization. Phys. Rev. A 37, 1351.

Pontryagin, L.S., Boltyanski, V.G., Gamkrelidze, R.V., Mishchenko, E.F., 1962. The Mathematical Theory of Optimal Processes. Interscience Publishers, New York.

Prigogine, I., 1949. On the domain of validity of the local equilibrium hypothesis. Physica 15, 1942.

Rowley, R.L., Horne, F.H., 1978. The Dufour effect: experimental confirmation of the Onsager heat-mass reciprocity relation for a binary liquid mixture. J. Chem. Phys. 68, 325–326.

Rowley, R.L., 1992. Application of nonequilibrium thermodynamics to heat and mass transport properties: measurement and prediction in nonelectrolyte liquid mixtures. In: Flow, Diffusion and Transport Processes. Advances in Thermodynamics Series, vol. 6. Taylor and Francis, New York, pp. 82–109.

Schlögl, F., 1980. Stochastic measures in nonequilibrium thermodynamics. Phys. Rep. 62, 267–280 (and literature therein).

Sieniutycz, S., 1978. Optimization in Process Engineering. Wydawnictwa Naukowo-Techniczne, Warszawa (in Polish).

Sieniutycz, S., 1983. The inertial relaxation terms and the variational principles of least action type for nonstationary energy and mass diffusion. Int. J. Heat Mass Transf. 26, 55–63.

Sieniutycz, S., 2001. Thermodynamic limits for work-assisted and solar-assisted drying operations. Archiv. Thermodyn. 22 (3–4), 17–36.

Sieniutycz, S., 2016. Thermodynamic Approaches in Engineering Systems. Elsevier, Amsterdam, Oxford/Cambridge, MA.

Sieniutycz, S., Jeżowski, J., 2018. Energy Optimization in Process Systems and Fuel Cells, third ed. Elsevier, Oxford, Amsterdam. pp. xi–xvii and 1–791.

Tribus, M., 1970. Thermostatics and Thermodynamics. Energia, Moscow (in Russian).

Tsirlin, A.M., 1974. Averaged optimization and sliding regimes in optimal control problems. Izv. AN SSSR Ser Tech. Cybern. 2, 143–151.

Tsirlin, A.M., 1997. Methods of Averaged Optimization and Their Applications. Nauka-Fizmatlit (in Russian).

Vilar, J.M.G., Rubi, J.M., 2001. Thermodynamics beyond local equilibrium. Proc. Natl. Acad. Sci. U. S. A. 98, 11081–11084.
Warner, G.H., 1966. Statistical Physics. Wiley, New York.
Webster, N., 1989. Webster's 1989 Encyclopedic Unabridged Dictionary of the English Language. Portland House, a Division of Dilithium Press, Ltd., Distributed by Crown Publishers, Inc., New York, NY.
Zangwill, W.I., 1969. Nonlinear Programming: A Unified Approach. Prentice-Hall, New York.

Further reading

Levine, R.D., Tribus, M. (Eds.), 1979. The Maximum Entropy Principle. MIT Press, Cambridge, MA.

CHAPTER 6

System analysis in energy engineering and ecology

6.1 Introducing physical systems, basic notions, and contributors

This section sketches the applications of nonequilibrium thermodynamics to physically complex systems such as solutions of macromolecules, magnetic hysteresis bodies, viscoelastic fluids, polarizable media, and so forth. These systems require some extra variables (so-called internal variables) to be introduced into the fundamental equation of Gibbs. Irreversible thermodynamic systems understood as EIT systems are also analyzed. In this approach, the dissipative fluxes are the independent variables treated on an equal footing with the classical variables of thermostatics. The significance of systems with Gibbs equations which incorporate additional internal variables is studied. Such systems include viscoelastic and viscoplastic materials and fluids with high-diffusive fluxes of heat, mass, and momentum. Rational and extended thermodynamic theories are appropriate tools for the analysis of these systems. The relative merits of these approaches are discussed and corresponding research papers are reviewed.

6.1.1 Classical and quasiclassical complex systems

Traditionally, we use the adjective "complex" to refer to systems such as solutions of macromolecules, magnetic hysteresis bodies, viscoelastic fluids, polarizable media, and so forth, all of which exhibit involved thermodynamic behavior and for which the role of internal variables is essential in thermodynamic descriptions (Sieniutycz, 2016, TAES).

The idea of an internal state variable, that is, a variable not controllable through external conditions, originated from the analysis of some models in rheology (Lemaitre and Chaboche, 1988; Maugin, 1987, 1990a) and electromagnetic bodies (Meixner, 1961; Kluitenberg, 1966, 1977, 1981). The goal was a theoretical framework for studying relaxation processes such as a shock wave passing through an electrically polarizable substance. However, relaxation phenomena are not the only ones to which the concept of internal variables is useful. Theories of plasticity and fracture and description of dissipative effects in electro-deformable solids can also make good use of this concept (Maugin, 1990a,b). Furthermore, some nontrivial analogies exist between plastic behavior and magnetic hysteresis; hence, the

Complexity and Complex Thermo-Economic Systems. https://doi.org/10.1016/B978-0-12-818594-0.00006-4

developments in the plasticity theory have implications for magnetic hysteresis systems. The idea of internal variable has been successfully applied to analyze: relaxation in strained solids (Kestin and Rice, 1970), dielectric and magnetic relaxation (Kluitenberg, 1977, 1981; Meixner, 1961), spin relaxation in ferromagnets (Maugin, 1975), magnetoelasticity (Maugin, 1979), thermoplastic fracture effects (Maugin, 1990a,b), electromechanical and other couplings (Maugin, 1990a), and magnetically nonsaturated fluids (Maugin and Drouot, 1983a). As contrasted with previous approaches, magnetic hysteresis (Maugin, 1987; Maugin and Sabir, 1990) has been cast in the framework of nonequilibrium thermodynamics describing this phenomenon exactly in terms of the constitutive equations, internal variables, and dissipation function. An interesting analogy between solid mechanics, viscoplastic Bingham fluids, and magnetism has been shown (Maugin, 1990a). A clear distinction has been made between relaxation processes and hysteresis effects. Thermodynamic framework incorporating internal variables is summarized in reviews by Coleman and Gurtin (1964) and Bampi and Morro (1984).

It is essential that hysteresis processes can be described by using the concept of the internal variables of state, provided one takes into account some residual fields at the vanishing load. However, we should be careful to make a distinction between the relaxational recovery of equilibrium and hysteresis processes. In relaxation, time scale is an essential issue; in hysteresis, the relatively slow response exhibited by a ferromagnetic sample is practically rate-independent. Similarly, the low-temperature plastic effects are practically independent of the rate of strain. A detailed comparison of these effects is available (Maugin, 1992). For the theory of internal variables, see also Kestin (1979), Kestin and Rice (1970), Kestin and Bataille (1980), and Muschik (1981, 1986, 1990a,b, 2007, 2008).

Maugin (1992) treats magnetic hysteresis as a thermodynamic process. As contrasted with many previous formal approaches, hysteresis is cast in the framework of nonequilibrium thermodynamics. He describes the process at the phenomenological level, in terms of the constitutive equations, internal variables of state, and the dissipation function which is homogeneous of degree one. The model accounts for the cumulative effect of the residual magnetization and describes magnetic hardening in a natural way. Both local and global stability criteria are obtained for hysteresis loops. The model, which is incremental, provides an operational way to construct hysteresis loops from a "virgin state" by alternate loads with increasing maximum amplitude. The interaction between the magnetic properties and stress or temperature variations is taken into account. An interesting analogy between solid mechanics (viscoplastic Bingham fluid) and magnetism is shown, while a clear distinction is made between relaxation processes and hysteresis-like effects.

The continuum modeling of polarizable macromolecules in solutions is another example showing the role of internal variables (Maugin and Drouot, 1983a, 1992; Maugin, 1987). Their use allows one to construct unknown constitutive equations and evolution equations describing mechanical, electrical, and chemical behavior. Also, the effects that do not contribute to dissipation can be singled out from the general formalism. Flow-induced and electrically induced phase transitions can be studied as well as diffusion, migration, and mechanochemical effects. In addition, Maugin and Drouot (1983b) treat thermomagnetic behavior of magnetically nonsaturated fluids, whereas Maugin and Sabir (1990) determine nondestructive testing in mechanical and magnetic hardening of ferromagnetic bodies.

Rheology of viscoelastic media (Leonov, 1976; Lebon et al., 1988), isotropic-nematic transitions in polymer solutions (Sluckin, 1981; Giesekus, 1986), and stress-induced diffusion (Tirrell and Malone, 1977) are other applications of internal variables. A number of reviews describe the requisite nonequilibrium thermodynamic framework (Bampi and Morro, 1984; Morro, 1985; Maugin, 1987). Bampi and Morro (1984) developed a hidden or internal variable approach to nonequilibrium thermodynamics. An application of the internal variable concept using rational thermodynamics to describe viscoelasticity is also available.

Morro (1992) imbeds the thermodynamics of linear viscoelasticity into the framework of rational thermodynamics and extremum principles. First, a general scheme for thermodynamics of simple materials is outlined. This scheme is applicable to materials with internal variables and those with fading memory. Morro's rigorous definitions are given for states, processes, cycles, and thermodynamic laws. His thermodynamic approach is then applied to viscoelastic solids and fluids. The approach is essentially entropy-free in the sense that the dissipative nature of the theory is assured by the negative definiteness of the half-range Fourier sine transform of fluids. The approach is essentially entropy-free in the sense that the dissipative nature of the theory is assured by the negative definiteness of the half-range Fourier sine transform of a (Boltzmann) relaxation kernel. The fact that such negative definiteness yields all conditions derived so far by various procedures is a new result. In addition, such a condition has been proved to be sufficient for the validity of the second law. Necessary and sufficient conditions for the validity of the second law are derived for fluids. Evolution equations are analyzed.

The Navier-Stokes model of a viscous fluid follows as a limiting case when some relaxation functions are equal to delta-functions. The extremum principles are shown to be closely related to thermodynamic restrictions on the relaxation functions. The treatment is quite involved from the mathematical viewpoint, but the reader gets to reexamine many basic features of the rational theory in a brief and precise way. In this aspect, the reader is referred to the paper by Altenberger and Dahler (1992) which discusses applications of convolution integrals to the entropy-less statistical-mechanical theory of multicomponent fluids.

Historically, the starting point for extended thermodynamic theories was the paradox of infinite propagation rate, encountered in the classical phenomenological equations describing the transport of mass, energy, and momentum. Attempts to overcome the paradox resulted in equations of change in terms of (damped) wave equations rather than classical parabolic equations. This has been demonstrated in a number of articles. The topics ranged from theoretical developments (based on phenomenological thermodynamics or nonequilibrium statistical mechanics), to applications in fluid mechanics, chemical engineering, and aerosol science. It has been recognized that taking into account local nonequilibrium and/or inertial effects (resulting from finite masses of the diffusing particles), the constitutive equations become non-Fourier, non-Fick, and non-Newtonian. A tutorial review of the historical development leading to elimination of the paradox is available (Sieniutycz, 1992).

Various simple forms of wave equations for coupled heat and mass transfer may arise. Simple (flux-containing) extensions of the second differential of entropy and excess entropy production can be constructed. These criteria allow to prove, by the second method of Liapounoff, the stability of equations describing coupled wave transfer of heat and mass. Dissipative variational formulations can be found, leading to approximate solutions by a direct variational method.

Applications of the hyperbolic equations to the description of short-time effects and high-frequency behavior can be studied. We shall not enter here into analytical details of diverse treatments, as they are commonly accessible in the literature.

6.1.2 Classical understanding of a system

System is a functional integrity accomplishing a certain goal. In thermal and chemical engineering, the term is usually referred to existing plants in which individual units are interconnected by suitable links. A system is usually constructed by the design of various units (subsystems) and the subsequent integration of these units. The links between the units are generally nonlinear; hence, the properties of the system are not determined by additive combination of system outputs. Also, a performance criterion of the system is essentially different than a performance index of an individual unit.

The idea of system approach is as old as European civilization. It has generally been accepted that the Aristotle statement...*the whole is more than sum of its parts*...is a first attempt to define the system problem (Klir, 1972; Ziębik, 1996). Yet, in past research the mechanistic approach prevailed, in which a problem is divided into parts analyzed separately. Because of the approach application so many inventions have been made that, for many many years, no changes of the approach seemed to be necessary. Only a short time before the Second World War, the system approach was discovered anew by Ludvig van Bertallanffy, philosopher and biologist, who stated that investigations of the parts must not only be supplemented by recognition of the role of the whole, but that there exists a separate discipline called theory of systems dealing with investigations of the whole (Bertallanffy, 1973).

When the system structure is known, we can ask about its performance; this problem is known as "system analysis." Inversely, one can know the system's performance and ask about its structure; this problem is known as "system synthesis." While analysis leads usually to a unique solution, synthesis does not generally provide a unique solution. In fact, the same performance can be shown by systems of various structures. Therefore, finding a structure that is most relevant to a given performance is a very important problem in the system's theory. The method of maximum entropy applied, e.g., in thermodynamics, physics, and urban modeling (Martyushev and Seleznev, 2006), is an example of the synthesis-type approach which leads to most relevant structures of systems described in probability spaces.

In engineering, two basic approaches to investigations can be distinguished—process analysis and system analysis. Results of the process analysis are often applied as input data for the system analysis. Ziębik's (1996) paper, discussed below, and other works of this author (Ziębik, 1986, 1990a,b, 1991a,b, 1995; Ziębik and Presz, 1993; Ziębik et al., 1994) deal with the methods of system analysis in the energy management of industrial plants. Ziębik's (1996) paper reviews the construction principles for mathematical models of industrial energy management (linear and nonlinear) and their applications and provides an optimizing mathematical model for the preliminary design of industrial energy management. It is interesting because, to date, process analyses have been predominant.

The scope of the chapter is not only to review and classify main methods and results obtained in the field, but also to consider common objections caused by misunderstanding of these methods and results. In fact, the second part of the chapter contains a critical comparison of various methods applied for systems of energy generation, such as second law analyses, entropy generation minimization, approaches coming from ecology, and finite time thermodynamics. Systems considered are those with transfer or rate processes which occur in a finite time and in equipment of finite dimension. These processes include heat and separation operations and are found in heat and mass exchangers, thermal networks, energy convertors, energy recovery units, storage systems, chemical reactors, and chemical plants.

6.2 System energy analyzes

In the investigations developed to date in engineering sciences, two approaches can be distinguished—process analysis and system analysis. In the process analysis, the assessment of energy impact is constrained to the interior of the balance shield of the considered unit; this is how the direct energy consumption is determined. Yet it should be realized that no real energy unit can operate as a separate system, but as a component of a global energy system. In fact, the production and consumption of energy carriers always take place within the network of interconnected thermal units. Therefore, both the direct consumption of energy carriers in the investigated unit and the indirect consumption occurring in other units must be analyzed. For this purpose, the system approach should be applied (Bertallanffy, 1973; Klir, 1972; Leontieff, 1951; Mielentilew, 1982). The applications of the system approach to investigate problems of industrial thermal engineering started in the sixties of our century. As their beginning, the elaboration of a mathematical model for material and energy balance of an industrial plant exemplified by ironworks may be considered (Szargut and Ziębik, 1972).

6.3 Mathematical modeling of industrial energy management

To proceed in a systematic way, a total system is often decomposed into subsystems. In an industrial plant, technological subsystem (consisting of technological processes) and the energy subsystem (energy management) are distinguished. The production of the energy branches is intended to cover the needs of the technological branches and partially also the plant's own consumption. The complexity of connections between energy management and technological branches, as well as the interdependences between the energy branches, results in the total energy management of an industrial plant which is being more than the sum of energy of processes considered separately. Some of these relations are of feedback character. Therefore, all balance equations of energy carriers should be investigated as a whole. Thus, the energy management of an industrial plant is a system defined as a set of energy equipment and engines, as well as the inner relations between them and the external relation between energy management and the environment, the aim of which is the production, conversion, transmission, and distribution of energy carriers consumed in industrial plants. Because of these relations the energy management, treated as a complex, has attributes which its parts (the particular energy branches considered separately) do not possess.

As a simple example of an energy system, a combined heat-and-power generating plant (Fig. 6.1) can be considered. Another form of presenting such an energy system, besides a schematic diagram, is the binary input-output matrix (Table 6.1). Same relations situated under the main diagonal have a feedback character. The existence of feedback relations is responsible for the fact that the partial balances of energy carriers lead to an agreement of the balance by means of subsequent approximations. If, for example, the production of electric energy is increased, the production of high-pressure steam grows, too. But this increased production of high-pressure steam leads to the next increase of electric energy production due to the own consumption of

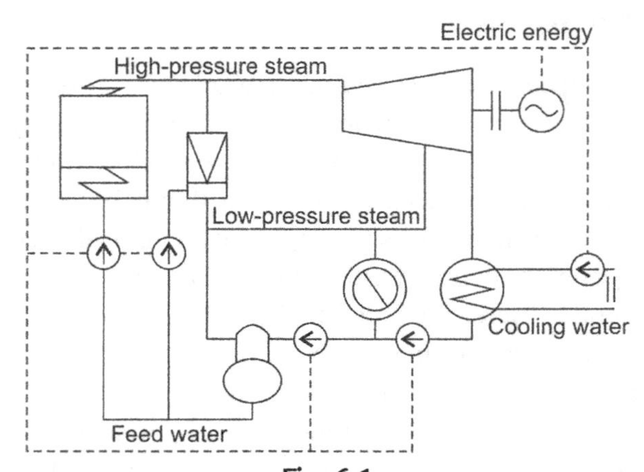

Fig. 6.1
Heat and power generating plant—interbranch flows. *From Ziębik, A., 1996. System analysis in thermal engineering. Arch. Thermodyn. 17, 81–97, with permission.*

Table 6.1 Input-output binary matrix.

	Electric energy	Feed Water	High-pressure steam	Low-pressure steam	Cooling water
Electric energy	1	1	1	1	1
Feed water	0	0	1	1	1
High-pressure steam	1	0	0	1	0
Low-pressure steam	0	1	0	0	0
Cooling water	1	0	0	0	0

From Ziębik, A., 1996. System analysis in thermal engineering. Arch. Thermodyn. 17, 81–97, with permission.

electric energy in the boiler house. And again, the demand for high-pressure steam will grow, causing again an increase of the demand for electric energy in the boiler house and so on, due to the existence of the feedback relation between the turbine and the boilers. Therefore, a mathematical model of the balance of energy systems of industrial plants has been prepared.

This model is a development of the *input-output analysis* (Ziębik, 1996), applied to the energy management of industrial plants. The productive branch of an industrial plant is a technological and energy process, producing a given major product, as well as optional by-products. If there is more than one source of energy carrier produced as the major product, the production must be divided into its basic part and peak part (e.g., the steam extraction nozzle of the turbine and the steam tram the pressure-reducing valve). If a given energy carrier is the major product in one branch and a by-product in another, it should be considered to be a whole in the balance equations of this major product (e.g., steam tram the waste-heat boiler). In another case, an energy carrier produced as a by-product can be treated as individual fuel (e.g., blast-furnace gag or coke-oven gas). In some cases, the own production of energy carriers must be supplemented by external supplies (e.g., electric energy). Some energy carriers are only provided from outside (mainly fuels). Sometimes, a part of the production of energy carriers is sold to external consumers (e.g., heat and hot water). The possibility of accumulating energy carriers (e.g., steam, hot water, fuel, and technical gases) has been taken into account.

The set of balance equations of energy carriers is presented in Table 6.2, in which the particular symbols denote:

G_i, P_i—peak and basic part of the main production of energy carrier,
U_{ij}, $\overline{U}_{ik}$—by-production of energy carriers in energy and technological branch, respectively,
D_i—supply of the energy carrier,
Z_{ij}, Z_{ik}—consumption of the energy carrier in energy and technological branch, respectively,
Y_i—consumption of the energy carrier for the general needs of an industrial plant,

Table 6.2 Balance of energy carriers.

Input part					
	Main production		By-production		
Energy carrier	Peak part	Basic part	Energy subsystem	Technol. subsystem	Ext. supply
i	G_i	P_i	$\sum_{j=1}^{n}[U_{ij}(G_j)+U_{ij}(P_j)]$	$\sum_{j=1}^{p}[\overline{U}_{ik}(\overline{G}_k)$	D_i
Output part					
	Interbranch flows				
Energy carrier	Energy subsystem	Technol. subsystem	General needs	Storage	Sale
i	$\sum_{j=1}^{n}[Z_{ij}(G_j)+Z_{ij}(P_j)]$	$\sum_{j=1}^{p}[\overline{Z}_{ik}(\overline{G}_k)$	Y_i	V_i	K_i

Ziębik, A., 1996. System analysis in thermal engineering. Arch. Thermodyn. 17, 81–97, with permission.

V_i—increase of the energy carrier in the energy storage system,
K_i—sale of the energy carrier,
$i,j=1,2, \ldots, n$—number of the energy branch,
$k=1,2, \ldots, p$—number of the technological branch.

This model can be considered to be a simulation model, but it can also be used in optimizations. In the simulation model, values of the following variables can be calculated: peak part of the production of energy carriers—G_i, some of the external supplies or sales of energy carriers unknown a priori—D_i or K_i, increase of energy carriers in the energy storage system—V_i. This model may also be considered as a linear or nonlinear mathematical model.

6.4 Linear model of the energy balance for an industrial plant

6.4.1 Model equations

The general principle of balance equations is the assumption of linearity for the relations between consumption, by-production, and main production. It means that the dependences $U_{ij}(G_j)$, $U_{ij}(P_j)$, $\overline{U}_{ik}(\overline{G}_k)$, $Z_{ij}(G_j)$, $Z_{ij}(P_j)$, $Z_{ik}(\overline{G}_k)$ occurring in Table 6.2 have a linear form. This assumption limits the application of this model to cases in which the coefficients of the consumption and by-production can be assumed constant in the considered period of time. The increase of the energy carrier in the energy storage system V_i can be neglected in this case. The set of balance equations of energy carriers in matrix notation looks as follows:

$$\mathbf{G}+\mathbf{F}_G\mathbf{G}+\mathbf{D}=\mathbf{A}_G\mathbf{G}+\mathbf{K}+\mathbf{T} \tag{6.1}$$

$$\mathbf{T}=\left(\overline{\mathbf{A}}-\overline{\mathbf{F}}\right)\overline{\mathbf{G}}+\mathbf{Y}-\left(\mathbf{E}-\mathbf{A}_p+\mathbf{F}_P\right)\mathbf{P} \tag{6.2}$$

where

- $\mathbf{G}$, $\mathbf{P}$—vectors of the peak and basic part of the production of energy carriers,
- $\overline{\mathbf{G}}$—vector of the production of technological branches,
- $\mathbf{F}_G$, $\mathbf{F}_P$, $\overline{\mathbf{F}}$—matrices of coefficients of energy carriers by-production,
- $\mathbf{A}_G$, $\mathbf{A}_p$, $\overline{\mathbf{A}}$—matrices of coefficients of energy carriers consumption,
- $\mathbf{D}$—vector of external supplies of energy carriers,
- $\mathbf{K}$—vector of sale of energy carriers,
- $\mathbf{E}$—unit matrix.

The vector **T** contains quantities known a priori. After solving Eq. (6.1), usually the element s of the vectors **G** or **D**, sometimes **K**, are obtained (the number of equations equals the number of unknown values). This is a simulation model. If the vectors **G**, **P**, and **D** are calculated simultaneously (surplus of unknown values), this problem is solved by means of linear programming.

6.4.2 Simulation of a long-term balance of the energy system of an industrial plant

Due to the complexity of thermal processes in the energy subsystem, a change in the production of one branch affects the production processes in all other branches because of interbranch dependences. Particularly, due to a change of production of the technological subsystem, a new variant of the energy balance must be prepared. In this case Eq. (6.1), after transformation, takes the following form:

$$\mathbf{G} = (\mathbf{E} - \mathbf{A}_G + \mathbf{F}_G)^{-1}(\mathbf{T} + \mathbf{K} - \mathbf{D}) \tag{6.3}$$

The elements of the inverse matrix take into account the direct as well as indirect relations between energy processes. The entries of the inverse matrix can be called coefficients of cumulative energy consumption concerning the energy management of the considered industrial plant.

Many variants of forecasting of the energy balance plan may be calculated by making use of the matrix equations (6.3). In particular, the influence of changes of the selected element in the production of the technological subsystem (vector **T**) on the energy balance can be investigated.

6.4.3 Analysis of exergy balances

All values in this model can be expressed by exergy units (Szargut, 1983; Szargut and Ziębik, 1972; Ziębik, 1995). The losses of exergy may be calculated by means of a linear mathematical model of the energy balance (Table 6.2). The amount of supplied exergy is obtained by summing up the element s of the *j*th column for the energy carriers passing to the energy subsystem. The useful effects of operation of the respective energy branch are: the exergy of the main product and the exergy of by-products. The difference between the supplied exergy and the useful effects expresses the losses of exergy. The vector $\delta\mathbf{B}$ expressing the exergy losses of all energy branches takes the following form:

$$\delta\mathbf{B} = \left[(\mathbf{A}_G^T - \mathbf{F}_G^T) - \mathbf{E})\mathbf{G}^D + (\mathbf{A}_P^T - \mathbf{F}_P^T - \mathbf{E})\mathbf{P}^D + \mathbf{Y}^T\right]\mathbf{b}_e \tag{6.4}$$

where $\mathbf{b}_e$ denotes the vector of specific exergy of energy carriers; the upper index D denotes the operation of creating a diagonal matrix from a vector.

6.4.4 System analysis for the rationalization of the energy management of industrial plants

The algorithm of the evaluation of system effects is based on Eq. (6.3). In the system method, the effects of rationalization of energy management are calculated at the boundary of the balance shield of an industrial plant. In this way, the system of interior relations between energy processes has been taken into account.

The rationalization of industrial energy management affects, first of all, directly the coefficients of consumption and by-production of energy carriers (element s of the "matrices" $\mathbf{A}_G$, $\mathbf{A}_P$, $\overline{\mathbf{A}}$, $\mathbf{F}_G$, $\mathbf{F}_P$, $\overline{\mathbf{F}}$). New values of these coefficients can be determined by means of the process analysis (thermodynamic analysis).

If the rationalization has taken place in the kth technological branch of an industrial plant, the change in the consumption of energy carriers can be calculated from the equation:

$$\Delta\overline{\mathbf{Z}}_k = \left[(\overline{\mathbf{A}} - \overline{\mathbf{F}})_k'' - (\overline{\mathbf{A}} - \overline{\mathbf{F}})_k'\right]\overline{G}_k \tag{6.5}$$

where

$\Delta\overline{\mathbf{Z}}_k$—vector of changes of energy consumption in kth technological branch due to energy rationalization in this branch,

$(\overline{\mathbf{A}} - \overline{\mathbf{F}})_k''$—column vector k belongs to matrix $(\overline{\mathbf{A}} - \overline{\mathbf{F}})$ after rationalization,

$(\overline{\mathbf{A}} - \overline{\mathbf{F}})_k'$—the same as above but before rationalization

$\overline{G}_k$—amount of production of the kth technological branch.

Substituting Eq. (6.5) into Eq. (6.3) we obtain:

$$\Delta\mathbf{G}_k = (\mathbf{E} - \mathbf{A}_G + \mathbf{F}_G)^{-1}\left[(\overline{\mathbf{A}} - \overline{\mathbf{F}})_k'' - (\overline{\mathbf{A}} - \overline{\mathbf{F}})_k'\right]\overline{G}_k \tag{6.6}$$

where $\Delta\mathbf{G}_k$ denotes a change of the vector $\mathbf{G}$ due to rationalization in the technological branch k.

In Eq. (6.6) the interior relations between energy processes are taken into account by means of the inverse matrix $(\mathbf{E} - \mathbf{A}_G + \mathbf{F}_G)^{-1}$. In this way, the direct and indirect connections between energy carriers have been taken into account. The application of the inverse input-output matrix eliminates the laborious work of successive approximations in investigations of the influence of the rationalization of the energy and technological process upon the industrial energy system as a whole.

As an example, the system analysis of evaporative cooling in a heating furnace has been considered. The change of traditional water cooling without evaporation to evaporative cooling brings direct energy benefits due to the decrease of industrial water consumption and the production of steam. The increase of energy consumption for the preparation of soft water is compensated with a surplus by the useful effects of evaporative cooling. Table 6.3 contains values of coefficients of direct consumption and by-production of energy carriers before and after the installation of evaporative cooling.

The by-production of steam in the installation of evaporative cooling of a heating furnace substitutes the main production of medium-pressure steam in the heat-and-power generating plant. Due to mutual connections existing in the energy subsystem of an industrial plant (among other connections of feedback character), the by-production of medium-pressure steam influences the whole energy balance of the combined heat-and-power generating plant. Fig. 6.2 presents the relations between pressurized industrial water and other energy branches.

The example of calculations and denotations in Fig. 6.2 complies with the data basis of the examples of calculations contained in Ziębik (1990a,b). The numbers correspond to the following energy carriers: *l*—low-pressure steam; m—medium-pressure steam; *h*—high-pressure steam; *d*—demineralized water; c—compressed air; *p*—pressurized industrial water; *n*—nonpressurized industrial water; *e*—electric energy. The first index of the coefficients of specific consumption of energy a_{ij} denotes the consumed energy carrier, the second one concerns the consuming in energy branch.

Table 6.4 shows Ziębik's results of system analysis obtained by means of a linear mathematical model of the energy balance of an industrial plant. Direct changes due to the rationalization of the cooling system of the heating furnace have been underlined. The other changes presented in Table 6.4 result from the interdependences existing in the energy system of

Table 6.3 Coefficients of consumption and by-production of energy carries before and after the installation of evaporative cooling.

Coefficients of consumption and by-production of energy carries	Before rationalization	After rationalization
Coefficient of industrial water consumption ($\Delta T_w = 10\,K$), Mg/Mg r.p.	0	0.098
Coefficients of soft water consumption, Mg/Mg r.p.	0	0.098
Coefficient of by-production of medium-pressure steam, Mg/Mg r.p.	0	0.0833

Ziębik, A., 1996. System analysis in thermal engineering. Arch. Thermodyn. 17, 81–97, with permission.

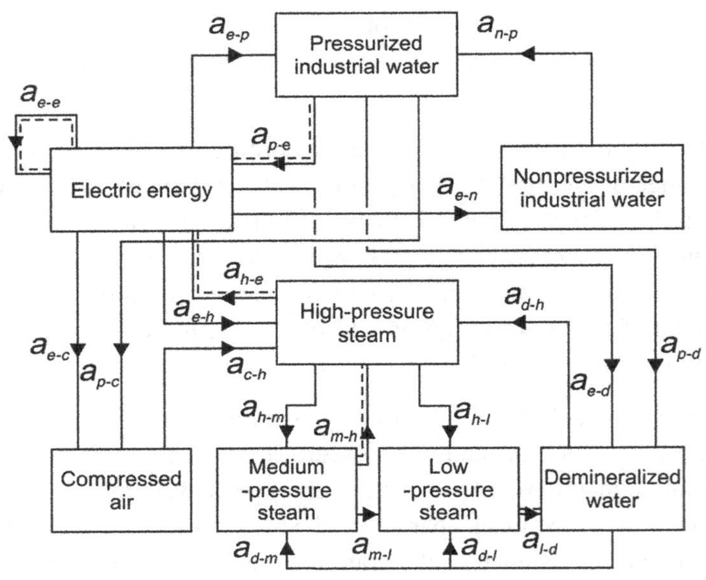

Fig. 6.2

Schematic diagram of direct and indirect connections between pressurized industrial water and other energy branches. *From Zię bik, A., 1996. System analysis in thermal engineering. Arch. Thermodyn. 17, 81–97, with permission.*

Table 6.4 Results or the system analysis or evaporative cooling.

Energy carrier	Changes of major production, by-production, and external supplies due to evaporative cooling			
	Unit	**Major production**	**By-production**	**External supply**
Soft water	kg/Mg r.p.	+86.3	+13.2	–
Demineralized water	kg/Mg r.p.	−41.8	−22.7	–
Low-pressure steam	kg/Mg r.p.	−9.8	0	–
Medium-pressure steam	kg/Mg r.p.	−98.9	+86.3	–
High-pressure steam	kg/Mg r.p.	−114.4	–	–
Compressed air	kmol/Mg r.p.	−0.021	–	–
Industrial water	Mg/Mg r.p.	−7.1	–	–
Electric energy	kWh/Mg r.p	−6.4	0	0
Power coal	MJ/Mg r.p.	–	–	−380.0
Natural gas	MJ/Mg r.p.	–	–	−4.0

Zię bik, A., 1996. System analysis in thermal engineering. Arch. Thermodyn. 17, 81–97, with permission.

the considered industrial plant. Because of some dependences of feedback character, accurate results can be obtained only by means of a computer-aided mathematical model of energy management. The final result of the system analysis is a decrease of exterior supplies of fundamental fuels (mainly power coal).

6.5 Nonlinear model of short-term balance

In order to achieve an efficient control of the energy management of an industrial plant, it is necessary to set up the energy balance for a shift and for 24 h. In this case, the assumption about the linearity of relations between production and consumption is a far-fetched simplification. In the balance equations of the mathematical model of the 24 h balance, the energy characteristics of the particular engines and a complex of engines are applied. These energy characteristics are mostly nonlinear or piece-wise linear functions. The mathematical model is, therefore, a nonlinear one.

The following assumptions have been made in the nonlinear mathematical model:

- the balances of energy carriers are set up for time intervals of 1 h; the energy balances for a shift and 24 h are assembled by means of 1 h balances,
- the timetables of work and repair idle-time for energy and energy-technological equipment are known; the plan of repair based on a long-term plan of energy balance is determined; this results from the connection of the model of long-term energy balance with that of short-term balance,
- from forecasts of hour-diagrams of the demand for energy carriers for a technological subsystem, the general needs of the plant and external consumers are known; the hour-diagrams show the average demands for energy carriers at particular hours of the considered shift or 24 h,
- the characteristics of engines or a complex of these are given; these may be nonlinear piece-wise linear functions and sometimes linear dependences,
- the dependences of the consumption of energy carriers on the parameters of energy-technological processes are taken into account (e.g., the influence of the blast parameters and the injection of auxiliary fuels on the energy characteristics of the blast-furnace)
- the storage volume of energy carriers (gasholders, steam-storage cells, and hot-water accumulators) has been taken into account,
- short-time fluctuations between the production and consumption of energy carriers existing in time intervals of 1 h are covered by the ability to accumulate the heat and gag distribution network.

This model can be considered a simulation model, but it can also be used for optimization purposes.

For the simulation model, the energy characteristics of the complexes of engines are assumed to be known. Also, the structure of the fuel-feeding system of an industrial plant and a part of the supplementary external supply are known.

The main aims of the model can be listed as follows:

- forecast of the energy balance of an industrial plant for a work-shift and 24 h for the purpose of production control,
- hour by hour correction of the forecast of the energy balance,
- preparation of the energy balance of an industrial plant in case of failure.

The mathematical simulation model presents a set of balance equations of energy carriers for time intervals of 1 h as described in Table 6.2.

For the optimization model, it has been assumed that the amount of production of a technological subsystem is known. The objective function is expressed by the following formula:

$$K_e = \sum_{i=1}^{n} \left[\kappa_{Gi}\dot{G}_i + \kappa_{Pi}\dot{P}_i + \kappa_{Di}\dot{D}_i + (\kappa_{Dsi} - \kappa_{Di})\dot{D}_{is}\right] + K_T \to \min \tag{6.7}$$

where K_e is the variable operating cost of energy management, κ_{Gi} is the variable operating unit cost of peak energy equipment (without the costs of external energy carriers), κ_{Pi} is the variable operating unit cost of basic energy equipment (without the costs of external energy carriers), κ_{Di} is the unit cost of the basic part of the external supply of an energy carrier, κ_{Dsi} is the unit cost of the peak part of the external supply of an energy carrier, D_{si} is the peak part of the external supply of an energy carrier, K_T are the losses in the technological system due to the deficiency of energy carriers.

The balance equation (Table 6.2) is the global constraint. The production capacity and limits of external supplies are local inequality constraints. This optimization problem is solved by means of the decomposition of the global optimization problem.

6.6 Mathematical model of optimization

6.6.1 Aims

The aim of preliminary design is to choose an optimal variant of the structure of a designed industrial energy system from among numerous possible variants. This results from the variety of elements (engines and energy equipment) constituting the industrial energy system, the different variants of thermodynamic parameters characterizing these elements, as well as many possible combinations of mutual connections between the energy carriers themselves. Therefore, the application of the mathematical model of energy management balance presented in Section 6.3 is very expedient.

Based on a brief foredesign, the vectors $\overline{\mathbf{AF}}$ and $\overline{\mathbf{FG}}$ concerning the consumption and by-production of energy carriers in the technological subsystem and the vector of energy consumption for general needs—$\mathbf{Y}$, as well as the vector of sale—$\mathbf{K}$, are given. The matrices of the coefficients of consumption and by-production—$\mathbf{A}_G$, $\mathbf{A}_P$, $\mathbf{F}_G$, and $\mathbf{F}_P$ are formulated in the preliminary design. The vectors of the basic and peak part of main production—$\mathbf{P}$ and $\mathbf{G}$, as well as the vector of supplied energy carriers—$\mathbf{D}$, are calculated by means of optimization. The algorithm determining the industrial energy systems structure in the preliminary design contains solutions of the following partial problems:

(a) generation of a set of variants of the energy systems based on the scenario of the energy management of a given industrial plant,
(b) determination of the structure of the binary input-output matrix and its structural analysis,
(c) determination of time-distribution functions of the demand for energy carriers,
(d) determination of the elements of the input-output matrix,
(e) determination of the optimal power rating and capacity of engines and other energy equipment.

The problems (c), (d), and (e) are closely interconnected. In order to solve these partial problems, the decomposition method of the global optimization problem must be applied.

6.6.2 Choice of structure of the input-output matrix and its structural analysis

First of all, a scenario of energy management has to be formulated based on which a general list of energy carriers is set up, bringing a set of information concerning the engines and energy equipment used to produce the major product enumerated therein. To each energy carrier, being a major product, corresponds a project. Connected with this is a set of designs comprising all technically possible and economically justified methods of production of a given energy carrier. The set of designs is created making use of information about engines and energy equipment. Each design is described by its binary input-output matrix characterizing the structure of the consumption and by-production of energy carriers. From designs, a set of all possible variants of structure of industrial energy systems is formed, choosing one design from each project. For each variant, matrices $\mathbf{A}_G^b$, $\mathbf{A}_P^b$, $\mathbf{F}_G^b$, and $\mathbf{F}_P^b$ are created by means of Boolean algebra. Next, the input-output matrix $(\mathbf{A}_b+\mathbf{F}_b)$ characterizing the interdependences in an industrial energy system is set up:

$$\mathbf{A}^b + \mathbf{F}^b = \mathbf{A}_P^b + \mathbf{A}_G^b + \mathbf{F}_P^b + \mathbf{F}_G^b \tag{6.8}$$

where b denotes the binary matrix. This matrix is subjected to structural analysis in order to obtain a structure close to a triangular matrix, which is more adequate for further calculations. First, the energy carriers are divided into three groups: input-type (supplies from outside), center-type (energy carriers produced and consumed inside the energy systems), and output-type (for outside consumers). Next, inside the center group of energy carriers, strongly coherent subsystems are to be distinguished. Relations of the feedback type exist only among energy carriers belonging to strongly coherent subsystems. The separation of such subsystems transforms the *center* matrix to a matrix with diagonally arranged blocks (Popyrin, 1978).

6.6.3 Objective function and constraints

For each considered variant of the designed industrial energy systems, the power rating and capacity of engines and energy equipment, as well as the nominal amount of external supplies, are determined by optimization. The objective function takes the following form:

$$\begin{aligned} K_R = (\boldsymbol{\rho}_P + \boldsymbol{\beta}_P)\mathbf{I}_P + (\boldsymbol{\rho}_P + \boldsymbol{\beta}_G)\mathbf{I}_G + (\boldsymbol{\rho}_D + \boldsymbol{\beta}_D)\mathbf{I}_D \\ + \boldsymbol{\alpha}_P\dot{\mathbf{P}}_n + \boldsymbol{\alpha}_G\dot{\mathbf{G}}_n + \mathbf{a}_D\dot{\mathbf{D}}_n + \boldsymbol{\kappa}_P\mathbf{P} + \boldsymbol{\kappa}_G\mathbf{G} + \boldsymbol{\kappa}_D\mathbf{D} + \mathbf{K}_T \rightarrow \min \end{aligned} \tag{6.9}$$

where

K_R—annual costs of the industrial energy system,
$\boldsymbol{\rho}$—row vector of the discount rates,
$\boldsymbol{\beta}$—row vector of the rates of constant costs of repairs and maintenance,
$\mathbf{I}$—column vector of capital expenditure,
$\boldsymbol{\alpha}$—row vector of the rates of prime costs,
$\dot{\mathbf{P}}_n, \dot{\mathbf{G}}_n, \dot{\mathbf{D}}_n$—power ratings or nominal capacity and supply,
$\boldsymbol{\kappa}$—row vector of the rates of operating costs,
$\mathbf{P}$, $\mathbf{G}$, $\mathbf{D}$—column vector of production and supply,
K_T—annual costs of losses in the technological subsystem due to a deficiency of energy carriers.

Eq. (6.1) is the global constraint. Local inequality constraints result from the maximum demands for particular energy carriers and from the limitations of external supplies and capital expenditures:

$$\dot{P}_{ni} + \dot{G}_{ni} + \dot{D}_{ni} \leq \dot{\Omega}_{imax} \tag{6.10}$$

$$\dot{D}_{ni} \leq \dot{v}_{i_{imax}} \tag{6.11}$$

$$D_i \leq v_i \tag{6.12}$$

$$I_{Pi} + I_{Gi} + I_{Di} \leq I_i, \tag{6.13}$$

where

$\dot{\Omega}_{imax}$—maximum demand for energy carriers,
$\dot{v}_{i_{imax}}$—maximum flux of external supply,
v_i—annual limit of external supply,
I_i—capital expenditure.

In order to solve the optimization problem with the cost function of Eq. (6.9), the time-distribution function of the demand for energy carriers and the values of element s of the input-output matrix must be known. The technical coefficients of the consumption and by-production of energy carriers (that means element s of the input-output matrix) depend on the power rating and load of the engines and energy equipment. But in order to get the time-distribution function, we must know the technical coefficients. Therefore, the setting up of time-distribution functions and the calculation of technical coefficients, as well as the determination of the power rating of engines and energy equipment, are related. In order to solve these problems, the global optimization must be decomposed.

6.6.4 Decomposition of the global optimization problem

Lagrange's method of decomposing the global optimization problem has been applied. Based on Eqs. (6.9), (6.1) and neglecting the term in Eq. (6.1) without influencing the results of optimization, the Lagrangian function takes the form:

$$L = \mathbf{K}_R + \boldsymbol{\lambda}\left[(\mathbf{A}_p - \mathbf{F}_P - \mathbf{E})\mathbf{P} + (\mathbf{A}_G - \mathbf{F}_G - \mathbf{E})\mathbf{G} - \mathbf{D}\right] \rightarrow \min \tag{6.14}$$

where $\boldsymbol{\lambda}$ denotes the row vector of Lagrange's multipliers.

In order to determine Lagrange's multipliers, the procedure of coordination must be known. It has been proved that the matrix method of calculating the unit costs of energy carriers can be used as a coordination procedure.

Lagrange's decomposition method leads to an iterative procedure (Fig. 6.3). In successive iteration, Lagrange's multipliers (unit costs of energy carriers) are determined by means of the matrix method on a higher level (complex of energy management). Next, optimization problems (optimal rower rating, capacity of engines, and energy equipment) are solved on lower level (particular energy carriers). Problems of the optimization of the particular energy carriers are solved according to their sequence in the upper triangular input-output matrix. In strongly coherent subsystems (i.e., subsystems with feedback), the inner iterative loops are solved. The determination of the optimum values of all decision variables in successive iterations is followed by a return to the level of coordination. Then a corrected balance of energy carriers is set up and the corrected values of the unit costs of energy carriers are calculated by means of the matrix method. Next, the corrected vector of the unit costs of energy carriers is applied on the level of optimization of the particular energy carriers (Fig. 6.3).
In conclusion, process analysis in thermal engineering is not a sufficient tool for the evaluation of the energy effects of operations aiming at an improvement of thermal processes. (Ziębik, 1996). As energy carriers are produced and consumed in the network of interconnected thermal processes, the thermodynamic rationalization realized in one thermal process influences the energy balances of other processes. Industrial thermal engineering constitutes one whole, displaying properties which the respective thermal processes, treated separately, do not possess.

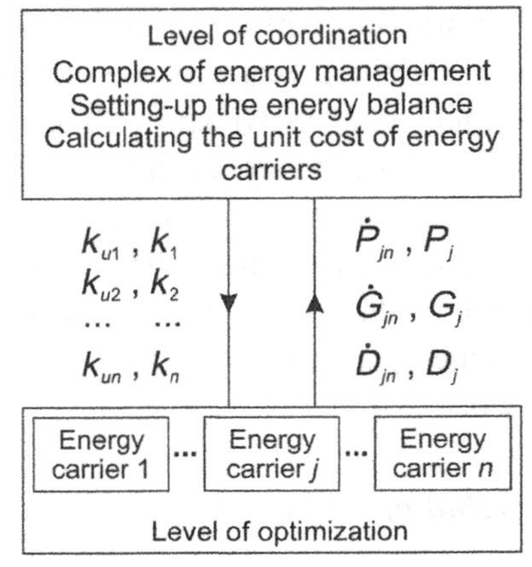

Fig. 6.3

Schematic diagram of the decomposition method; k, k_u—average unit costs of energy carrier and unit cost of by-production of energy carriers, $\dot{P}_{jn}$, $\dot{G}_{jn}$, $\dot{D}_{jn}$—power rating and nominal supply, P_j, G_j, D—annual production and supply. Evaluation of the energy effects of operations aiming at an improvement of thermal processes. *From Ziębik, A., 1996. System analysis in thermal engineering. Arch. Thermodyn. 17, 81–97, with permission.*

Therefore, besides process analysis, also system analysis ought to be applied in investigations concerning heat engineering.

Summing up, process analysis in thermal engineering is not a sufficient tool for the evaluation of the energy effects of operations aiming at an improvement of thermal processes. (Ziębik, 1996). System analysis in industrial thermal engineering is based on a mathematical model of the energy balance. The linear input-output model serves to analyze the heat management in the case of long-term balances. The nonlinear model is applied in short-term balances. Both models can be used as simulative or optimizing ones. For the purpose of preliminary designing of the heat management, the optimization model of selecting the structure of heat engineering is used.

6.7 Remarks on diverse methodologies and link with ecological criteria

In this book, we investigate "nonclassical" thermodynamic limits which include kinetic effects. In particular, we are interested in "dynamic" limits for systems evolving in time. The limits usually depend on operational constraints established under the condition that, in any circumstances, the key process will run with a required mean intensity, yet yielding a desired product. The "nonclassicity" requirement usually yields bounds that may be orders of magnitude higher than those classical ones known from the textbooks.

Consider, for example, heat consumption in a distillation column and evaluate a realistic limit that corresponds with the lower limit of heat associated with the use of the theoretical plates instead of real plates of certain efficiency (lower than unity). This bound is usually two to three times larger than another (lower) bound, the heat consumed at minimum reflux conditions (in the case with an infinite number of theoretical stages). Next, the heat at minimum reflux is usually several times larger than that the reversible evaporation heat. In effect, a design engineer must expect that heat consumption should be at least of the order of magnitude higher than the evaporation heat. In complex separation processes as those with cycles, losses, and nonlinearities, such evaluations are nontrivial. Complex optimization techniques must be used to get (possibly dynamic) limits for various operations. The union of applied thermodynamics and optimal control theory that derives these limits has found applications in design of solar engines, solar cells, semiconductor devices, photosynthesis engines, and other sophisticated devices, e.g., a book of De Vos (1992).

Already, early papers on FTT stressed that its dynamical models can be optimized with respect to their external controls subject to arbitrary in principle optimization criteria (Andresen et al., 1977; Andresen, 1983). Thus, not only power or entropy source, but also economic criteria, such as profit or the sum of investment and operational costs, can be used in FTT. Examples of such optimization, which, of course, should not be mixed with those leading to energy limits, are available (Berry et al., 2000). While it is true that the economic costs are influenced by irreversible properties of the system, the irreversibility does not follow from economic costs since no one-to-one relation does exist between both quantities. No prior recourse to economics to consider the irreversibility impact on the investment and operational costs is mandatory to get energy limits; physical considerations are sufficient as for other physiochemical properties. In particular, all irreversible extensions of exergy can be obtained without any prior considering of process economies. This substantiates the relevance of energy limits and irreversible exergies obtained in a number of papers (Andresen et al., 1977, 1983; Andresen, 1983; Berry et al., 2000; Bejan, 1982; Tsirlin and Kazakov, 2002a,b; Bejan and Errera, 1998; Sieniutycz, 1999, 2001).

Endoreversible modeling (imperfect reservoirs and perfect power generators) (Rubin, 1979a,b; Rubin and Andresen, 1979) is an anathema to adversaries of FTT or EGM, but the magic word in the language of their followers. With the endoreversibility assumption, classical thermodynamic equations for reversible engines and chillers may easily be combined with irreversible expressions describing energy and mass transports in imperfect reservoirs (those containing lossy elements in form of various "conductances," "resistances," "boundary layers," "penetration depths," etc.). This results in relatively simple synthesizing models that often are susceptible to analytic treatments (Rubin, 1979a,b; Rubin and Andresen, 1979; Hoffmann et al., 1997; Sieniutycz and Kubiak, 2002). The reason for this simplicity is the linearity of mathematical equations describing lossy parts of the system. When the linearity does not hold, e.g., in the case of systems with radiative transport, then the description of the overall system is more complicated, yet it can still be obtained in a workable form (de Vos, 1992; Sieniutycz and Berry, 2000; Sieniutycz and de Vos, 2000; Sieniutycz, 2000). When internal irreversibilities are included, simplicity of endoreversible description is generally lost unless a simple but rough description of the internal entropy generation, based on the so-called internal factor, Φ, is accepted (Salah El-Din, 2001a,b; Hoffmann et al., 1997; Sieniutycz and Szwast, 2003). Yet, models treated in classical thermodynamics of real thermal machines often give up external irreversibilities rather than internal ones (Ng et al., 1997, 1998; Gordon and Ng, 2000; Haywood, 1975). Thus, all sort of irreversibilities can be treated in FTT and EGM approaches, although often at the expense of replacing analytical solutions by numerical ones.

Recent works on second law analyses often deal with complex thermal systems composed of many objects and links; for their optimization, the reader is referred to a review (Tsatsaronis et al., 1993). Advanced applications of thermodynamic limits, which include separation processes, may be regarded as prolongation of earlier ideas developed for second law analyses of thermal systems (Gaggioli, 1980). Other works using second law analyses include ecological applications of exergy; they are of interest here in view of their link with the theory of energy limits. A basic notion therein, supposedly of value in thermal technology, is the so-called cumulative exergy cost (CEC) defined as total consumption of exergy of natural resources necessary to yield (the unit of) a final product (Szargut, 1986). Also introduced is the notion of cumulative exergy loss, as the difference between the cumulative exergy cost and exergy of the considered product. In ecological research, various analogues and other criteria are introduced. There are general criteria (Odum, 1971; Ulanowicz, 1997; Jørgensen, 1988, 1997, 2000, 2001; Kay and Schneider, 1992) and special ones, related to thermal technology (Angulo-Brown, 1991; Yan, 1993).

The ecological cost (Szargut, 1986) measures the cumulative consumption of exergy of unrestorable resources burdening a definite product. Resulting technical indicators are used to forecast changes in demand for heat agents caused by changes in production level and technology of product yield and in costs of heat agents. This provides information about diverse exergy-consuming technologies, and this is also suggested as a way to compare technologies. Also, the so-called pro-ecological tax is proposed as the penalty for negative effects of action causing exhaust of natural resources and contamination of natural environment (Szargut, 2001a,b). All these applications involve nonequilibrium systems in which the sole use of the classical exergy is insufficient without including the associated notion of minimal (residual) dissipation of this exergy. This is, in fact, the realm into which we are driven with many analyses which lead to nonequilibrium applications of the exergy thermodynamics. They emerge since engineering processes must be limited by some irreversible processes allowing a minimum entropy production rather than by purely reversible processes. Limits following from reversible processes are most often too far from reality to be most useful.

However, the method of cumulative exergy costs has its own imperfections and difficulties. Its definition of the sequential process is vague. Total consumption of exergy of natural resources, necessary to yield a product which defines the cumulative exergy cost, is, in fact, a dynamical notion burdened by sorts, locations, and dates of various technologies, all affecting process efficiencies, semiproducts, controls, etc., and thus, in effect, influencing the cost definition. One way to improve this definition would be to include statistical measures of the process and its exergy consumption. Yet, no statistical procedure leading to an averaged sequence process that would add rigor to the definition of cumulative exergy costs was defined in the original work (Szargut, 1986, 2001a,b, 2002a,b). Moreover, as its definition shows, the cumulative exergy cost (CEC) is merely the exergy change of the whole (sequential) production process.

Last but not the least, the CEC method underestimates the significance of kinetic terms in treating practical processes. Thus, despite some suggestions implying its economic ramification (Szargut, 1986, 2001a,b, 2002a,b), the cumulative exergy cost is not a perfect measure of economic costs of consumed unrestorable resources. In fact, this quantity is a rough representative of thermodynamic limit (work limit) of the same sort which is applied in the methods of EGM and FTT when they deal with limits of sequential operations. According to the Gouy-Stodola law (Gouy, 1889; Stodola, 1905; Bejan, 1982), the exergy change represented by the CEC is proportional to the associated entropy production; this quantity should be minimized, yet, no precise definition of the minimization constraints and operational quantities is given in theory of cumulative exergy costs.

While applying the CEC method to replace the minimization of the entropy production in both methods, EGM (Bejan, 1996a,b) and FTT (Berry et al., 2000), its originators seem unaware that the origin of their approach is, in fact, similar. To define exergy costs as unique and usable quantities, effects of external controls and disturbances should be eliminated from their definition. This requires either statistical averaging or extremizing cumulative costs with respect to controls and disturbances. But, unlike in FTT and EGM, the method of CEC does not attribute to its merit any potential functions and does not use techniques securing a unique result. Thus, in the current definitions of the cumulative exergy cost and ecological cost, mathematical structures of cost functions remain largely unknown. In fact, cumulative costs are not functions, but rather functionals of controls, disturbances, and path coordinates. To ensure potential properties in optimal costs, their definition should imply a method that would eliminate effect of controls; its absence makes the definition inexact. Without optimization, unique (potential) properties of costs are lost.

A solution to the above difficulties lies in *finite time thermodynamics* (Berry et al., 2000) where extremal, potential cost functions are generated via optimization of costs with respect to controls (but not states). Mathematical structures of these costs as continuous and discrete functionals are recognized (Sieniutycz, 1973, 1978, 1997a,b,c; Rubin and Andresen, 1979; Ondrechen et al., 1981, 1983; Andresen, 1983; Andresen et al., 1977, 1983, 1984; Berry et al., 2000). They are often called finite time potentials.

The problem of finite time potentials originated from the knowledge of the fact that principal functions of extremal solutions to variational problems are potentials depending on process time and state is nearly as old as analytical mechanics. Functions of this sort were first obtained in physics as extremum actions. Yet, for an arbitrary variational problem, where an analytical solution does not exist and only numerical solutions are possible, an effective numerical method of finding potential functions was established only in 1957, with the advent of Bellman's method of dynamic programming (Aris, 1964; Bellman, 1961; Bellman and Dreyfus, 1967).

Classical thermodynamic potentials, which are time-independent quantities, are special cases of such generalized (time-dependent) potentials in the case of identically vanishing Hamiltonians. In particular, the purpose of some FTT analyses is the exposition of how the classical thermodynamic potential, entropy $S(x)$, emerges for an irreversible deterministic dynamics (Berry et al., 2000). Furthermore, these analyses can also demonstrate the emergence of "nonclassical thermodynamic potentials" from the same general approach. One of nonclassical potentials is finite time availability (Andresen et al., 1977; Andresen, 1983; Berry et al., 2000; Tsirlin and Kazakov, 2002a,b).

The basic problem of extracting the most exergy from a continuous hot stream was formulated in the reversible limit by Bejan (1982). While its endoreversible generalization was posed soon after (Ondrechen et al., 1981; Andresen et al., 1983), all details were not completed until a decade and half later (Bejan and Errera, 1998; Sieniutycz, 1999, 2001). Sieniutycz's (1999) paper deals explicitly with cascades as genuine discrete systems with a small number of stages described by difference equations, and his 2001 work discusses the continuous limit and abandoning endoreversibility.

Criticizing endoreversible limits and general methodologies of FTT and EGM Szargut (2001b) mixes the problem of energy limits (a physical problem) with a different problem of economic optimum for the system (an economical problem). Whereas a general analysis of the system's limiting possibilities (Sieniutycz, 2001) is focused on the physical problem, much of Szargut's discussion (2001b) is centered on the economic one. This confusion is, perhaps, motivated by the observation that minimum energy data are sometimes used in evaluation of economic optimum. A well-known example in the theory of Linde operation shows that the real work supplied to the compressor at economically optimal conditions may be a dozen times larger than the reversible (exergy) limit for the production of condensed air (Bosniakovic, 1965; Ciborowski, 1976). It is just this large difference between economically optimal work and reversible work limit (lower bound) that makes irreversible exergies useful. They provide higher lower limits for work consumption (lower upper limits for work production) than the corresponding reversible results, and thus, they are closer to real work.

The hierarchy of energy limits is an important issue (Andresen et al., 1984; Sieniutycz, 2001). It begins with perfect (reversible) limits for reversible models and stretches by more and more exact models of imperfect processes, thus including more and more detailed contributions to the entropy production. The limits of zero rank are reversible ones; they stem from the classical exergy; they are the weakest and thus the worst. The limits of first rank are endoreversible limits; by assumption they correspond with the situation when the device that consumes (generates) work is perfect, as in the case of Carnot machine, whereas the reservoirs can be imperfect. An endoreversible model is more general than the classical Carnot model, where no dissipation is possible in either of two reservoirs. The endoreversible modeling yields a locus of limiting irreversible states with minimum dissipation in bulks of reservoirs. Still the work generators or consumers remain ideal. When a work generator becomes imperfect, limits of higher rank are defined (Berry et al., 2000; Sieniutycz and Kubiak, 2002; Sieniutycz and Szwast, 2003). The concept of such enhanced limits is motivated by the fact that only imperfect machines are encountered in the thermal technology.

Entropy source minimization has no relevance to an economic optimum of a product yield, where economic cost criteria are attributed to a valuable final product (Tsatsaronis et al., 1993; El-Sayed, 1999; Clark, 1986). On the other hand, these economic criteria have little in common with physical limits on energy consumption or production. However, economic optimization of an energy network may require different methods than currently used in thermodynamics.

Network or system thermodynamics (Peusner, 1986) can help in economic optimization of energy systems, such as, e.g., the system shown in Fig. 6.4. Thermodynamics, especially as the subject of the thermal networks theory, is currently being extended to combine technical and economic approaches; the extended discipline is referred to as thermoeconomics (Section 6.9); it links the principles of economics with thermodynamic analyses of energy or resources (El-Sayed, 1999). Extra variables related to equipment and capital costs are introduced simultaneously along with purely thermodynamic variables. By including still more variables related to environmental effects, ecological impact is treated.

Clark (1986) has stressed that economic cost analysis introduces a new and different set of variables which must be considered simultaneously with thermodynamic ones to obtain the best technically acceptable design. He has developed evaluation methods which bridge the interface between thermodynamic and economic considerations for the purpose of optimum design. An optimum configuration will vary depending on such economic factors as equipment and capital costs, sources of capital, period of investment, cost of displaced fuel, resale value, etc. Other constraints follow from ecology when one takes into account not only the exergy consumption, but also the exergy losses resulting from the deleterious impact of wastes on human activity and health, crops, forestry, natural resources, etc.

Energy and resource analysis has been defined as a particular set of procedures for evaluating the total energy (resource) requirements for the supply of a service or product (Berry, 1989). The quantification of economic analyses in technological systems is described in the literature on engineering economy. Thermodynamics has nonetheless helped correlate and predict selling prices of chemicals. The old, incorrect concept of prices based on enthalpy analysis has been

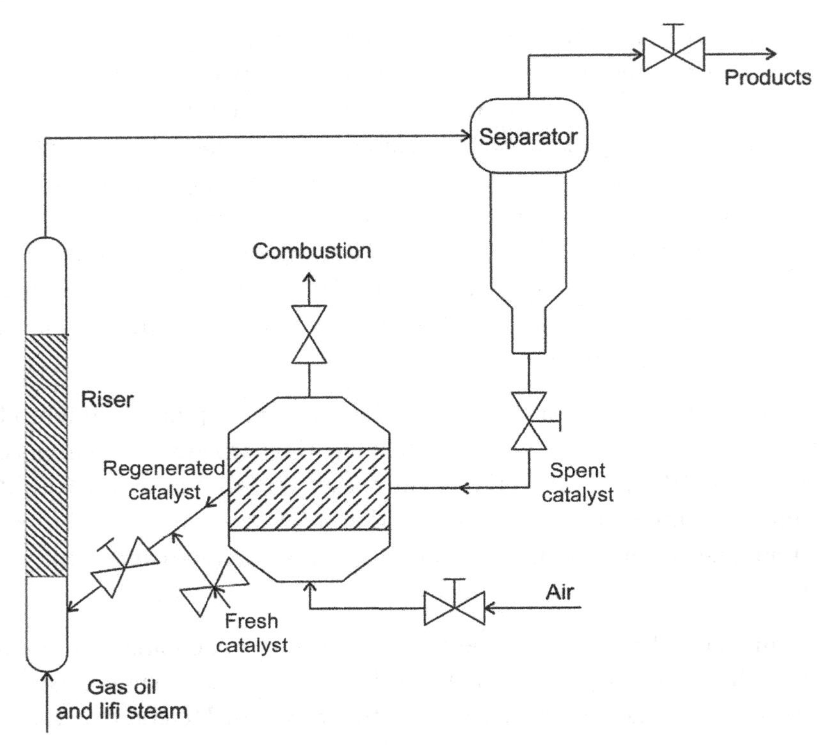

Fig. 6.4
A typical structure of the unit of fluidized catalytic cracking.

replaced by that of prices based on analyses of free energy and exergy (Szargut and Petela, 1965). Exergy optimization, economic optimization, and ecological criteria have been compared and considerable differences in various optima have been shown (Clark, 1986; El-Sayed, 1999).

Consideration of the irreversibility's impact on investment and operational costs is not only possible in the FTT method, but it is more effective in FTT than in the classical thermal analyses due to the optionality of the optimization criterion in FTT (Andresen et al., 1984). In fact, FTT can effectively treat both categories of problems: those leading to limits and those evaluating economic decisions. Opinions that only endoreversible models can be treated in FTT are disproved by current analyses of real thermal machines with internal entropy generation, describing either imperfect engine cycles (Chen et al., 1996; Cheng et al., 1999; Salah El-Din, 2001a) or imperfect refrigeration cycles (El Haj Assad, 1999; Chen et al., 2000; Salah El-Din, 2001b). Experiments in the realm of chillers confirm essential role of thermodynamics in describing and organizing diagnostic procedures for practical devices (Edera and Kojima, 2002; Chua et al., 1996; Ng et al., 1997, 1998). *Internal* entropy production is a basic parameter of analyses contained in recent books (Berry et al., 2000; Gordon and Ng, 2000). Alternative method of cumulative exergy costs (Szargut, 1986, 2001a,b, 2002a,b) does not ensure objective energy limits (Sieniutycz and Kubiak, 2002).

The definition of the limits for energy generation or consumption might be the subject of the attitude that *limits* should be attributed to systems with extreme efficiencies of energy convertors. Since the Carnot efficiency is the upper limit of all thermal generators, a proposal was made (Sieniutycz, 1998) that, as the perfect cycle, the Carnot cycle is a limiting abstract system that works with superconducting circulating fluid, to eliminate the internal entropy production and to secure the closed loop of the cycle without cooling the circulating fluid. Moreover, following Landau and Lifshitz (1974), it was assumed that the power generating device (machine) is both perfect and very small. In the case of a CNCA system, this assumption avoids the quantitative effect of the energy generator properties on those of the two-reservoir system. The assumption leads to the notion of a small Carnot machine, infinitesimal in relation to reservoirs. In cascade systems, each such Carnot machine is a mechanical energy generator that only transforms the small amount of internal energy of two-reservoir (disequilibrium) system into the mechanical energy, but otherwise is negligible and does not change the limiting work. Circulating Carnot fluid is, in this case, a medium that does not change after a cycle, as a sort of catalyst. In fact, it also substantiates the use of "endoreversible" models that admit arbitrary dissipative properties within fluids of upper and lower reservoirs (resistances or boundary layers), but otherwise involve small Carnot units as ideal energy generators. Such an attitude is assumed by the researcher who wants to define energy limits in two-reservoir system without considerations of a related three-body network composed of two reservoirs and a real thermal machine of a finite size.

The attitude has generated both followers and adversaries. The former claim that the system considered is actually the two-body (two-reservoir) system, whereas the (small) Carnot machine is just an addition enabling the conversion of a small amount of energy. The representatives of this idea (Landau and Lifshitz, 1974; Sieniutycz, 1998; Orlov and Rudenko, 1984; Szwast, 1994) claim is that, only in this case, the adiabatic availability or reversible work from the two-body system, $\Delta(E_1 + E_2)_{(S1+S2)}$, involves the properties of two bodies (both reservoirs). In fact, the associated exergy function does involve only the reservoir properties (Ondrechen et al., 1981). Yet, the adversaries (Gyftopoulos, 1999, 2002; Moran, 1998; Lior, 2002; Szargut, 1998, 2001b) want supposedly to see the CNCA system as a simple network (implying a three-body problem) in which there are two reservoirs and a thermal machine of a finite size. When the latter viewpoint is accepted, the size and other properties of the machine influence even reversible work, thus the universality of the results decreases. In conclusion, the introduction of thermal machines of finite sizes into the analyses, however valuable for real network descriptions, decreases simplicity and universality of the modeling and results describing the limiting work.

6.8 Control thermodynamics for explicitly dynamical systems

Engineering systems deal most often with far from equilibrium processes, i.e., those outside of the realm of linear irreversible thermodynamics. These processes must go to the desired extent of completion within a finite time and must produce at least a minimum amount of product. Furthermore, efficiency based solely on physical grounds is more often than not an insufficient criterion for the performance of a chemical engineering process. In view of resources scarcity, economic and ecological considerations also play important roles. Thus, classical thermodynamics can at most place lowest limits on chemical engineering processes and that one must turn to irreversible thermodynamics for more information.

When one (say, upper) reservoir is finite, a finite resource problem emerges in which power yield is necessarily a dynamical operation and the reservoir's temperature changes in time. Within this class, various dynamical energy systems have been analyzed to date; here we shall make only a few introductory remarks. Even for purely thermodynamic criteria, such as those associated with integrals of entropy ore exergy dissipated, a finite time constraint leads to the optimal performance function as a state function which is not classical from the viewpoint of equilibrium thermodynamics. In approaches involving optimal control (Salamon et al., 2001), optimization of a cost consumption (or an associated entropy production) automatically eliminates controls from optimal costs expressions, thus generating a potential (R or R_σ). A potential depends only on initial and final states, duration, and (in multistage processes) total number of stages. Names "control thermodynamics" and "optimization thermodynamics" introduced in Russian literature are particularly suitable for dynamical cases.

Suitable averaging procedures are proposed along with methods that use averaged criteria and models in optimization (Berry et al., 2000). It follows that an optimal sequence has a quasi-Hamiltonian structure that becomes Hamiltonian in the special cases of processes with optimal sizes of stages or in continuous limit (Sieniutycz, 1999). Thus, the well-known machineries of Pontryagin's maximum principle (Fan, 1966) and dynamic programming (Aris, 1964) can be included to generate functions of optimal cost. These theoretical achievements enter also the realm of economic criteria (Berry et al., 2000). In fact, problems in the method of cumulative exergy costs (Szargut, 1986) also belong to the group of dynamical operations as they involve sequences of operations and finite resources.

Control thermodynamics, which is the union of nonequilibrium thermodynamics and optimal control (Salamon et al., 2001), investigates the effect of constraints imposed on the process duration and its average rate on the optimal performance of some generic processes. This is accomplished through integral or sum expressions describing quantities such as internal entropy generation, total exergy input, system's exergy change, work, etc. Usually, the goal of thermodynamic analysis is:

(1) find paths minimizing driving exergy or internally produced entropy and set realistic bounds on consumption of energy and resources in thermal, separation, and chemical processes incorporating minimal irreversibility,
(2) determine optimal strategies or controls for such processes, and.
(3) attribute the bounds to an actual process, to verify its possible improvement. The bounds constructed on the basis of thermodynamic criteria, in particular exergy, are both relevant and useful. Note that these bounds are in general functions of state and duration rather than numbers. They generalize the well-known thermostatic bounds for finite rates and/or finite time. In this book, they define thermodynamic limits rather than the economical consumption of exergy or resources for various generic processes.

Optimization techniques play a central role in obtaining the majority of bounds in control thermodynamics. The methods of linear programming (Dantzig, 1968) and nonlinear programming (Zangwill, 1974) are as a rule insufficient in those situations where functional extrema are sought. Instead, the application of optimal control techniques is necessary (Pontryagin et al., 1962; Leitman, 1966; Sieniutycz, 1991). Control thermodynamics often retains the philosophy of model idealization known from reversible thermodynamics (the Carnot cycle), but uses somewhat more realistic models which have basic irreversibilities incorporated. Notion of thermodynamic metric and its consequence for establishing bounds as well as providing optimal paths deserves a mention (Hoffmann et al., 1989, and the literature therein). These authors show that when a part of the dissipated energy remains within the system, not all of the availability is necessarily lost. The bound defined by the thermodynamic length no longer limits the availability losses, but rather the so-called work deficiency, W_d (where usually $W_d > -\Delta A$), or the total loss of availability that would have resulted if all the available work were lost to the environment.

6.9 Interface of energy limits, structure design, thermoeconomics, and ecology

Complex energy systems, which exchange with their surroundings mass in the form of many cold and heat streams and may be chemically active, appear in almost every industry. The recognition that a common principle is the basis for the generation of the shape and structure of systems, whether living or inanimate, leads to the Constructal Theory, whose developments are systematized in his book (Bejan, 2000). Starting with the idea of competition between mechanical and thermal losses of exergy in flow systems, Bejan developed his *constructal theory of organization in Nature* (Bejan, 1997a,b,c, 2000) and has shown its applicability in many diverse fields including urban growth, economics, and physiological processes. This theory provides an understanding of how naturally organized systems emerge and evolve. In addition to engineering, the theory has exposed its inquiring and predictive potential in other areas, such as biology and medicine (allometric laws, structure of the respiratory and circulatory systems, bodily rhythms, organ, and tissue structures) and earth sciences (circulation of planetary fluids, structures of river basins, etc.). For example, the theory explains why we have a bronchial tree with 23 levels of bifurcation (Reis et al., 2004 and Chapter 10 of Sieniutycz and Jeżowski, 2013). It also shows promising developments in social sciences.

The constructal theory describes a deterministic principle for the generation of geometric form in natural systems. Shape and structure emerge from the endeavor for better performance in both engineering and Nature. The constructal theory applies the idea that the objective and constraints principle used in engineering is also the mechanism from which the geometry of natural flows emerges. The principle accounts for tree-shaped flows and other geometric forms found in engineering and Nature: round ducts, regularly spaced internal channels, and the proportionality between width and depth in rivers. Flow systems with geometric structure exhibit at least two flow regimes. One regime is slow (diffusion, walking), with high resistivity that fills the volumes of smallest finite scale, and one or more regimes are fast, with low resistivity (streams, channels, and streets). The balancing of the regions with different regimes assures that material and channels are distributed optimally. Despite the differences between the regions of high and low resistance, better global performance is assured when the distribution is relatively uniform. The system works best when its internal flow resistances are spread around, so that more and more of the internal points are stressed as much as the hardest working points. One good form leads to another, the forms being constantly improved with time. Bejan's book (2000) synthesizes a vast spectrum of publications (e.g., Bejan and Ledezma, 1998; Bejan, 1997a,b,c, and many others).

The contemporary formulation of the constructal principle is as follows: For a finite-size system to persist, it must evolve and organize in such a way that it provides easier access to the imposed currents that flow through it (Bejan, 2000). This formulation recognizes the natural tendency of imposed currents to construct paths of optimal access through constrained open system. In the Constructal theory, the process of construction and shape optimization proceeds stagewise from smaller to larger scales until a given volume is fully covered. High-conductivity paths emerge that form a tree, and the low-conductivity paths appear that fill the infinity of points of the given volume. Tree networks abound in nature in both animate and inanimate flow systems. They can be found in plants, roots, lungs, leaves, vascular tissues, river drainage basins, and dendritic clusters. In fluid trees, the smallest-scale volumetric flow is by slow diffusion, while the larger scale flow is organized into faster channeled streams forming a tree-like structure. Channels of lower conductivity are tributaries of a channel of higher conductivity. An optimal geometry composed of low- and high-conductivity flow regimes coalesce all the volume-to-point flows. Artificial constructs and natural structures, such as the internal computer arrangements or the arterial-vascular system in animals, exhibit the same cooperation between slow and fast transfer mechanisms, with the slow mode operating at the smallest scale. Tree networks are also observed in living communities; they can be explained in terms of minimization of the travel time between one point and a finite-size area of infinite destinations. The theory predicts the system's growth, in particular urban growth, from alleys to streets, avenues, and highways and drives multimodal transport systems that are set up to minimize travel time over long distances (Bejan, 2000; Bejan and Lorente, 2003, 2004). Systems with porous structures and fuel cells can be analyzed by the Constructal Theory (see, respectively, discussions in Chapters 9 and 10 of Sieuiutycz and Jeżowski, 2013). Some applications which determine energy and size limits and flow architectures in living systems are described in Section 10.13 therein.

Reis and Bejan, 2006 and Bejan (2006) examined the largest flow system on earth from the point of view of the constructal theory (Bejan, 1997a,b,c, 2000), which now constitutes the thermodynamics of nonequilibrium (flow) systems with configuration (Bejan and Lorente, 2004, 2005). They believe that there are two time arrows in physics. The old is the time arrow of the second law of thermodynamics, the arrow of irreversibility. The new is the time arrow of the constructal law, the arrow of how everything that flows acquires architecture or configuration. The constructal theory explains that existing configurations assure their survival by morphing in time toward easier flow configurations. Its time arrow unites physics with biology and engineering. A vision suggested (Bejan, 1997a,b,c; Reis and Bejan, 2006) involves the earth with its solar heat input, heat rejection, and wheels of atmospheric and oceanic circulation. Such an overall system constitutes a heat engine without shaft whose maximized (not ideal) mechanical power output cannot be delivered to an extraterrestrial system. Instead, the earth engine dissipates through air and water friction and other irreversibilities (e.g., heat leaks) all the mechanical power it produces. It does so by "spinning in its brake" the fastest that it can, and hence, the winds and the ocean currents, which proceed along easiest routes.

Because the flowing Earth is a constructal heat engine, its flow configuration has evolved in such a way that it is least imperfect it can be. It produces maximum power, which it then dissipates at maximum rate. The heat engines of engineering and biology (power plants, animal motors) have shafts, rods, legs, and wings that deliver the mechanical power to external entities that use the power (e.g., vehicles and animal bodies needing propulsion). Because the engines of engineering and biology are constructal, they morph in time toward flow configurations that make them the least imperfect that they can be. Therefore, they evolve toward producing maximum power, which means a time evolution toward minimum entropy generation rate (Reis and Bejan, 2006). Yet, the standard system design includes economics (Moran and Shapiro, 2003), i.e., an optimum is not associated with the maximum power. And, a critical analysis of constructal approach shows that it does not necessarily improve the flow performance if the internal branching of the flow is increased (Kuddusi and Egrican, 2008).

The standard system design involves principles of thermodynamics, hydrodynamics, material engineering, economics, and mechanical design. The term thermoeconomics may be referred to this complex field of application, although it is frequently applied in narrowed sense to methodologies combining economics and exergy theory to optimize the design and operation of energy systems.

Bejan (1982, 1987, 1996a,b, 1988) has synthesized the great deal of thermodynamic analyses of heat and mass transfer, fluid flow, and related design, on the basis of the minimization of entropy production σ. The minimum of the entropy produced, with terms containing temperature and pressure gradients, has been used in optimizations of fins and fin arrays of various geometries (Poulikakos and Bejan, 1982). Approximate optimization of the tradeoff between investment costs and exergy-based exploitation costs offers useful estimates, as summarized by various researchers (Bejan, 1982, 1987; Szargut and Petela, 1965; Szargut, 1970). On the other hand, approaches to "optimal design" that use the entropy (or exergy) generated as their optimization criteria make little sense from the standpoint of economic or thermoeconomic design.

The solution of a thermoeconomic problem is, in general, not equivalent to that of the corresponding thermodynamic problem. It does, however, reduce to thermodynamic optimizations in two special cases. The first case appears when the price of certain thermodynamic quantities such as the power produced becomes much larger than the prices of other participating quantities (Berry et al., 1978). This limit represents an energy theory of value, i.e., a value system in which one considers energy as the single valuable commodity (Berry et al., 1978). In the second case, the economic value of the exergy unit is the same for all forms of matter and energy taking part in the process. Then the thermodynamic problem of the minimum exergy loss is equivalent to that of the minimum of the economic costs. This case is, however, quite special since the prices of the exergy units generally differ (Szargut and Petela, 1965).

Nevertheless, a number of complex economic and ecological analyses have been born as generalizations of thermodynamic irreversibility analyses. While various performance criteria serve in various problems (this variety has already been admitted by control thermodynamics and extended versions of FTT), physical constraints are the product of thermodynamic analyses. Even if one replaces thermodynamic criteria by economic ones, the same optimum search method can often be used in both cases. Most of the methodological experience gained during a model formulation is preserved when passing from thermodynamic to economic optimization.

Systems with nontraditional energy sources have begun to become economically realistic. Examples of such sources are solar energy (photothermal and photovoltaic converters), wind energy, biomass energy of waterfalls, waves and tides, geothermal energy, and convective-hydrothermal resources (de Vos, 1992; Bodvarsson and Witherspoon, 1989; Neill and Jensen, 1976; Gustavson, 1979). While the economic cost of renewables approaches zero, limits on their exergy consumption and thermodynamic efficiencies are finite and should be evaluated because the results of such evaluations may influence the equipment dimension and hence its investment cost. Photothermal and photovoltaic conversions have been treated by Landsberg (1990) and Jarzębski (1990), and, in the framework of FTT, by De Vos (1992). The solar-driven convection known as winds has been modeled in terms of the FTT of heat engines by treating the Earth's atmosphere as the working fluid (de Vos, 1992; Gordon, 1990, and references therein). Upper and lower limits for the performance coefficient of solar absorption cooling cycles followed from the first and second laws (Mansoori and Patel, 1979). These limits depend not only on the environmental temperatures of the cycle components, but also on the thermodynamic properties of refrigerants, absorbents, and mixtures thereof. Comparative studies of refrigerant-absorbent combinations are now possible.

Attempts at systematizing process synthesis by the energy limits theory and second law analyses are known; an example is that of a methane liquefaction plant with the irreversibility of a unit as the selection criterion (King et al., 1972). The irreversibility of the plant decreases as a rule monotonically with each set of iterations during evolutionary synthesis; however, the final criteria are economic and operative (Gundersen and Naess, 1988). Thermodynamics is a valuable tool when dealing with new processes (Ranger et al., 1990; Le Goff et al., 1990a,b,c,d; Grant, 1979). Performance studies of single units can help in the design of systems.

Complex power systems are hybrid by nature. This is illustrated in Fig. 6.5 which depicts complex catalytic system with reactor, regenerator, and separator Szwast (1994), and in Fig. 6.6, which exemplifies a complex hybrid power system with turbine and fuel cell.

System approaches incorporate control and optimization of whole chemical or petrochemical industries and constitute a new trend in thermoeconomic analyses.

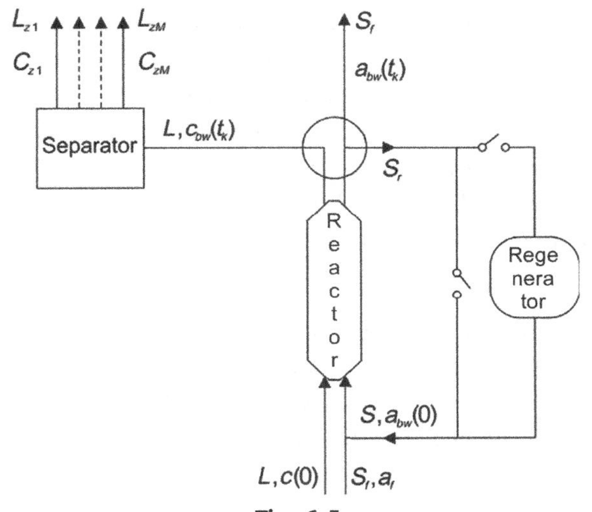

Fig. 6.5

An example of a complex catalytic system with reactor, regenerator, and separator (Szwast, 1994).

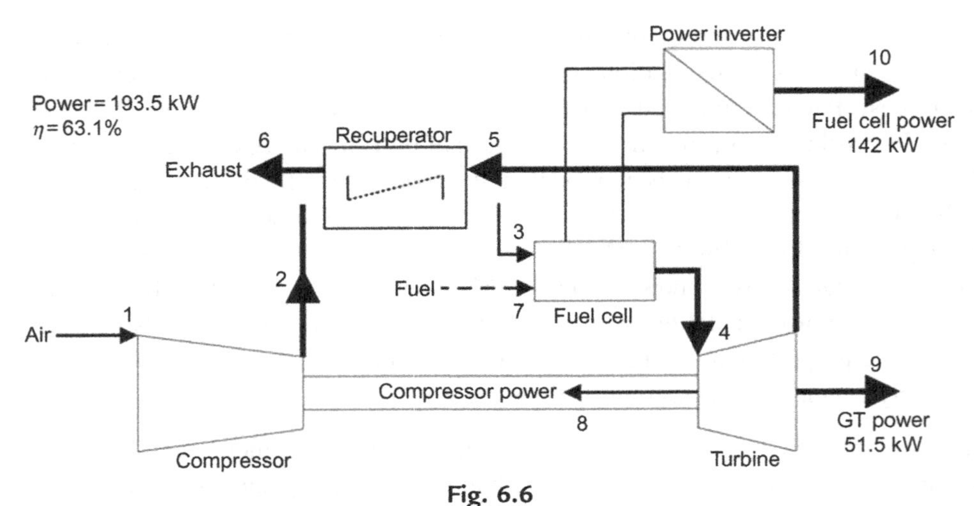

Fig. 6.6
An example of a complex hybrid power system with turbine and fuel cell.

6.10 Toward the thermoeconomics and integration of heat energy

For chemical plants, the irreversibility rate $I = T_0\sigma$ provides a means for assessing the thermodynamic performance of individual plant components (I_k) as well as that of the whole plant (I_{tot}). This enables one to compare the exergy losses of the different plant components and to search for their relative changes caused by a selected parameter ξ_i. These relative changes are in the form of the efficient of structural bounds, C_{sb} (Brodyanskii, 1973)

$$C_{sb} = \frac{\partial I_{tot}}{\partial \xi_i} \bigg/ \frac{\partial I_k}{\partial \xi_i} \tag{6.15}$$

C_{sb} is a thermodynamic indicator of the system structure which offers a basis for systematic study and improvement of system structures (Gundersen and Naess, 1988). Second law analyses help in establishing operating conditions of many industrial chemical processes such as coal gasification, coal combustion, ammonia synthesis, nitric acid production, production of methanol from naphtha and natural gas, and formaldehyde from methanol, and in reacting plasma systems (Denbigh, 1956; Bosniakovic, 1965; Riekert, 1974, 1979). Szargut et al. (1988) have indicated where a potential for improvement in fuel exists in chemical and metallurgical plants with large fossil fuel consumption.

As the theory of energy limits applies minimal values of the entropy production or exergy dissipation, the issue of exergy savings is related to economic optima and deserves consideration. At the interface of energy limits, theory and (thermo)economics is the research of Tsatsaronis, Valero, and associates and that of von Spakovsky and coworkers. The first group has developed a general theory of exergy saving and applied it in problems of exergetic cost and thermoeconomic cost, for the purpose of energy savings and thermoeconomics (Tsatsaronis,

1984; Valero et al., 1986a,b,c). A system for energetic/exergetic optimization of power plants (GAUDEAMO) was worked out (Valero et al., 1986d). Aspects of unification of thermoeconomic theories were considered (Valero et al., 1989). Thermoeconomics was applied in structural analysis and relative free energy function (Valero et al., 1992, 1993). Thermoeconomic analyses have shown their effectiveness in modeling gas turbine cogeneration systems and for calculating exergy in chemical processes (Lozano et al., 1993). Tsatsaronis, Lozano, and Valero have proposed the methodology of exergoeconomics and applied it to the exergetic cost theory (Tsatsaronis, 1984; Tsatsaronis and Winhold, 1984; Lozano and Valero, 1988, 1993). The exergetic cost theory was applied, among others, to a stream boiler in a thermal generating station (Lozano and Valero, 1987). Other applications of the exergetic cost theory are also known (Valero et al., 1994a,b; CGAM Problem). These treatments have contributed to practical and industrial applications of thermodynamic analyses in which the classical function of exergy and theory of exergetic cost are tools in thermoeconomic diagnoses of energy systems (Lozano et al., 1993, 1994).

The term exergoeconomics was coined by Tsatsaronis (1984), to indicate a particular combination of an exergy analysis with an economic analysis, in which combination the exergy costing principle is applied. For any other combination of a thermodynamic analysis with an economic analysis, the broader term thermoeconomics should be used. Traditional exergy efficiencies (Sciubba and Wall, 2007) could mislead because they use the concept of "exergy input" instead of the more appropriate concept of the "exergy of fuel". The general concepts of "fuel" and "product" (in conjunction with exergy analysis and exergoeconomics), "cost per unit of exergy of fuel and product," "cost of exergy destruction" as well as the general formulation of exergy balances and cost balances were developed in Tsatsaronis group (Tsatsaronis, 1984; Tsatsaronis and Winhold, 1984, 1985a,b; Lazzaretto and Tsatsaronis, 1999, 2006). Energy conversion plants have been treated by exergoeconomic analyses (Tsatsaronis and Winhold, 1984, 1985a; Tsatsaronis et al., 1986, 1991, 1993, 1994). Examples have shown that the exergy approach leads engineers to develop new concepts of energy conversion systems based on exergetic evaluations or to significantly change the system design (Tsatsaronis et al., 1991, 1992; Tsatsaronis, 1999). Valero, Lozano, and Munoz exploited some of these developments, in which the contributions of investment costs and operating and maintenance expenses are neglected in the cost balances, to propose the "exergetic cost theory" (Valero et al., 1986a,b,c,d; Lozano and Valero, 1993; Lozano et al., 1993). In this theory, the useful additions are (a) the division of the variables by the cost of the fuel to the total system, and (b) the matrix formulation. The approaches used by Valero and coworkers mainly apply to existing systems, whereas the approaches used by Tsatsaronis group focus on the design optimization of new systems. Significant contributions also came from the Padova group of researchers (Lazzaretto and Andreatta, 1995; Lazzaretto and Macor, 1995; Reini, 1994; Toffolo and Lazzaretto, 2003).

The field of thermoeconomics was benefited by discussions of aspects of economic activity from an energy perspective and applications of statistical-mechanical theory to economics

(Samuelson, 1947; Georgescu-Roegen, 1971; Gong et al., 1997; El-Sayed, 2003; Valero et al., 2006). The idea of linking thermodynamics and costing considerations was explored first by Lotka (1921), Keenan (1932), Benedict and Gyftopoulos (1980), based on Benedict's 1949 original work, and Gilbert (1956): the essential idea that emerged from their very general papers is that entropic considerations should somehow be accounted for in monetary cost calculations. Beckmann (1953), Henatsch (1957), and Szargut (1957) explicitly address the cost allocation between co-generated steam and power in energy systems.

In natural sciences, however, thermoeconomics is defined as the statistical physics of economic value (Georgescu-Roegen, 1971; Chen, 2005). According to Corning (Corning and Kline, 2000; Corning, 2002), thermoeconomics is based on the position that the role of energy in the human development should be defined and understood through the second law of thermodynamics, but in terms of such economic criteria as productivity, efficiency, and the costs and benefits of the various mechanisms for capturing and utilizing available energy to build biomass and do work. Thermoeconomists claim that human economic systems can be modeled as thermodynamic systems attempt to develop theoretical economic analogs of the first and second laws of thermodynamics (Burley and Foster, 1994). With this attitude, the exergy, i.e., the measure of the useful work energy of a system, is the most important measure of value. In thermodynamics, thermal systems exchange heat, work, and or mass with their surroundings; in this direction, relations between the energy associated with the production, distribution, and consumption of goods and services can be determined.

As the aim of many economic activities is to achieve a certain structure, thermoeconomics attempts to apply nonequilibrium thermodynamics, in which structures are dissipative in form, and information theory, in which information entropy is a central entity, to the modeling of economic activities in which the flows of energy and materials act to create scarce resources (Sieniutycz and Salamon, 1990; Baumgärtner, 2004). Thermoeconomic activity is regarded as a dissipative structure sustained by transforming and exchanging resources, goods, and services. In engineering and industrial design, thermoeconomics applies to methodologies combining exergy and economics for optimizing the design and operation of thermal systems, in particular power generation units. Cost estimates accompany the design studies (Moran and Shapiro, 2003; Szargut et al., 1988).

An isomorphism between problems in thermodynamics and economics (Baumgärtner, 2004) makes the laws of thermodynamics inspiring for economists. Boulding (1966), Ayres and Kneese (1969), and Georgescu-Roegen (1971) apply thermodynamics when analyzing economy-environment interactions and formally imbed the economy in its biogeophysical context. In a first step, they formulate the Materials Balance Principle, based on the thermodynamic law of mass conservation (Boulding, 1966; Ayres and Kneese, 1969; Ayres, 1978; Kneese et al., 1972), thus concluding that all resource inputs that enter a production eventually become waste. Up to now, this is an accepted theorem of resource, environmental, and ecological economics. Simultaneously, Georgescu-Roegen (1971) develops an extensive critique of economics based on the laws of thermodynamics, and in particular the Entropy Law, which he considers to be "the most economical of all physical laws" (Georgescu-Roegen, 1971, p. 280). This initiates a lively discussion whether the thermodynamic laws are relevant to economics (Burness et al., 1980; Daly, 1992; Kåberger and Månsson, 2001; Khalil, 1990; Lozada, 1991, 1995; Norgaard, 1986; Townsend, 1992; Williamson, 1993; Young, 1991).

While Georgescu-Roegen has drawn a principally correct picture of the irreversible transformations of energy and matter in economies, his analysis suffers from some flaws (Ayres, 1999; Baumgärtner, 2004; Lozada, 1991, 1995). But as Georgescu-Roegen's research and many later studies show (Baumgärtner et al., 1996), the Entropy Law, if properly applied, yields insights into the irreversible nature of economy-environment interactions that are not available otherwise. Both the First and the Second Laws of Thermodynamics, therefore, need to be combined in the study of how natural resources are extracted, used in production, and give rise to emissions and waste, thus leading to integrated models of ecological-economic systems (Odum, 1971; Ayres and Martinás, 1995; Ayres, 1998, 1999; Ayres et al., 1998, 2003; Baumgärtner, 2000; Faber, 1985; Faber et al., 1995; Perrings, 1987; Ruth, 1993, 1999).

The interconnection of the exergy concept with "environmental issues" (taken in their broad sense) is of importance in the system context. Exergy per se is not a measure of environmental impact (Sciubba and Ulgiati, 2005; Sciubba and Wall, 2007), but, in essence, at the end of the life cycle of a device, plant and product, the exergy balance of the extraction-transformation-production-distribution-use-disposal cycle shows how many primary exergy resources have actually been consumed. In fact, the use of the entropy concept in ecology and ecological economics has been the subject of many works (Baumgärtner et al., 1996; Faber et al., 1996; Valero, 1998). Corning and Kline (2000) discuss what they call the thermoeconomics of living systems—a cybernetic and economic approach to analyzing the role of available energy in biological evolution—and relate this paradigm to a distinction they draw between various statistical or structural definitions of information and what they call "control information." In addition, with the "global warming" in mind, some papers discuss exergy analysis of a "CO_2 zero emission" high efficiency plants (Calabro et al., 2004).

All chemical processes consume unrestorable natural resources; the quicker civilization develops, the sooner these are exhausted. Ecological components have been added to modify the formulations of energy limits theory and traditional problems of thermal technology (Angulo-Brown, 1991; Yan, 1993). Exergy is used as a measure of the quality of resources (Szargut, 1986, 1987, 1989, 1990; Szargut and Morris, 1987; Szargut and Majza, 1989; Szargut et al., 1988; Månsson, 1990). It is important to calculate the rate at which industrial processes consume exergy resources. The cumulative consumption of exergy from unrestorable natural resources appearing in the chain of processes leading from natural raw materials to product expresses the ecological cost of the product (Szargut, 1978, 1986, 1990; Szargut and Majza, 1989). Exergetic ecological costs are applied to the optimization of production processes from the viewpoint of minimization of the consumption of natural resources.

Indices of ecological costs determining the extent to which technological processes exhaust natural resources have been summarized (Szargut, 1989); related analyses of chemical processes are available in a book (Szargut et al., 1988). One example treats the ecological second law analysis of heat delivery from a complex heat-power station; the minimization of the consumption of unrestorable natural materials is achieved by using exergy (Szargut, 1990). Cumulative exergy cost seems to be suitable for industrial chemistry; in fact, it is a minimal work, as every exergy change is. However, in view of its nonuniqueness (dependence on accepted technology) and absence of statistical averaging in its definition, it is not an exact counterpart of an energy limit that must be a state-dependent physical quantity. As stated earlier, the method of cumulative exergy costs has serious imperfections. These are: (a) vague sequential process; (b) date- and location-affected exergy consumption of resources defining cumulative cost, and (c) undefined mathematical structure of costs. Moreover, cumulative costs are not functions, but rather functionals of controls and state coordinates and, as such, they depend not only on process paths, but also on the external controls and disturbances. Only optimal costs generated by optimization acquire properties of functions (potentials).

Starting with some older works (Beckmann, 1953; Szargut, 1957, 1970; Sieniutycz, 1973; Szwast, 1990), the possibility of establishing a univocal and direct correlation between monetary price and physical value is debated, associated with various aspects of thermoeconomic analysis and methods of exergy accounting (Szargut, 1986, 1987, 1989, 1990; Tsatsaronis, 1987, 1993, 2007; Valero, 1999, 2003; Sciubba, 2000, 2001a,b; Möller et al., 2006; Yantovski, 2007a,b). Discussion of advanced zero emissions plant (AZEP) is included (Möller et al., 2006). Also, Szargut introduces concepts of cumulative exergy consumption (CEC: Szargut, 1987) and thermo-ecological cost (TEC: Szargut, 2005) and postulates the application of exergy for the determination of the pro-ecological tax replacing the actual personal taxes (Szargut, 2002b). See a discussion of Szargut's TEC definition and criticism of his exergy cost of human work-based nonrenewable natural resources (Yantovski, 2007b). As, in Yantovski's opinion, the involvement of monetary prices is not compatible with exergy analysis; economic analysis, which is crucial for decision making, should be done separately after the thermodynamic one. Yet, the data of cumulative exergy will be valid for years, as they are sensitive to technology changes only. Also, according to Sciubba and Wall (2007), the method of cumulative exergy consumption (CEC) provides a very clear picture of the "resources" incorporated in the production of goods. The method can be extended to include immaterial services, and like in extended exergy accounting (EEA: Sciubba, 2000, 2001a,b), it can account for labor and capital as well, thus paving the way to the calculation of an "exergy cost" of commodities measured in kJ/unit (common for CEC and EEA) instead of €/kJ (as in thermoeconomics). A problem for CEC is its neglect of labor and in general of all immaterial production factors. Finally, there are topological limitations for systems treated by the CEC method, which are, in principle, sequential systems (Sieniutycz, 2003).

Engineering of sustainable energy and, in particular, sustainable energy design can be promoted in the complex thermoeconomic context (Gevorkian, 2007). In the energy conversion field, there is today practically no process or cycle analysis that does not include exergy considerations (Sciubba and Wall, 2007). Monographs on identification, modeling, and simulation of energy technology systems by methods of systems theory involve the graphs theory and matrix description of topological structures (Radwański et al., 1993; Portacha, 2002). The latter source gives also many examples of topological structures of electric power stations and discusses costs of heat and power generation. Analyses of vapor power systems and case studies of exergy accounting of power plants are available (Bejan et al., 1996; Moran and Shapiro, 2003). Various cost-effective methods are used to improve energy resource utilization. One method is cogeneration which sequentially yields power and heat (or steam). Its aim is to accomplish the work and heat fluxes in an integrated system of the total expenditure less than that attributed for the sum of the process components. Two other methods are power recovery (by inserting a turbine into a pressurized gas or a liquid stream to capture a residual work) and waste-heat recovery (to capture a residual exergy of heat, as in the case of exhaust gases of combustion engines). System-related optimization criteria include thermodynamic, economic, ecological, reliability, and other terms. Multilevel optimization is the characteristic feature of models describing energy systems and industrial plants (Ziębik, 1990a). Considerations on renewable energy sources, exploitation of waste energy contained in biomass, biofuels, and recycled wastes accompany discussion of various aspects of pro-ecological policy (Lewandowski, 2001).

Heat pumps may be components of optimum operation of complex thermal plants with cogeneration (Dentice d'Accadia et al., 2000). A plant may contain many gas-fueled engines with heat recovery. Each engine drives simultaneously an electric generator and the compressor of a heat pump. The problem consists in selecting the operation mode with lowest cost subject to process constraints. In another problem, a method for determining the optimized dimensions of a ground source heat pump system (GSHPS) heat exchanger is given (Marzbanrad et al., 2007). Optimum length and diameter for the heat exchanger is found for different mass flows by using the exergy form of second law. The controls minimize entropy generation and result in increased efficiency of the heat pump.

The book by Hirs (2000) contains in its Part 3 a number of examples on process integration, discussed in the context of: process systems, process units, processes with combined heat and power, and optimal design. Intelligent functional approach (Frangopoulos, 1983, 1990) is postulated to be a method for analysis and optimal synthesis-design-operation of complex systems including those with cogeneration (Dentice d'Accadia et al., 2000). Reviews of exergoeconomic methodologies are available (Tsatsaronis, 1987, 1993, 2007;

Lozano et al., 1993; El-Sayed, 2003, 2007; Sciubba and Wall, 2007). A considerable amount of research deals with various aspects of thermoeconomic monitoring and diagnostics (Toffolo and Lazaretto, 2004; Valero et al., 2004; Zaleta-Aguilar et al., 2004; Verda and Zaleta, 2004; Sciubba, 2004) and the methodologies to generate fingerprints for malfunctioning devices (El-Sayed, 2007; Reini and Taccani, 2004; Toffolo and Lazaretto, 2004; Correas, 2004).

While there are still some unsolved issues as to the inclusion of environmental considerations into thermoeconomics, it is rather clear that thermoeconomic approaches will be more and more extensively adopted in the assessment of industrial processes and production cycles (Sciubba and Wall, 2007). Thermoeconomic mode of system synthesis is most likely their "next frontier": where system's topological structure is sought for a required performance, and an optimization problem involves decisions not only related to process variables, but also to the system's configuration. The objective functions can be formulated in the context of the methods of thermoeconomics, cumulative exergy content, and extended exergy accounting, and thermoeconomics seems to be the one that is more likely to be used in the applications in the near future (Sciubba and Wall, 2007). The problem can be solved by mixed-integer linear programming (MILP) techniques (Kelahan and Gaddy, 1977; Salcedo, 1992; Floudas, 1995), genetic algorithms (Goldberg, 1989; Androulakis and Venkatasubramanian, 1991; Michalewicz, 1996; Garrard and Fraga, 1998; Upreti and Deb, 1997; Tayal and Fu, 1999; Yu et al., 2000; Toffolo and Lazzaretto, 2002; Ravagnani et al., 2005; Selba et al., 2006; Shopova and Vaklieva-Bancheva, 2006), and artificial intelligence techniques (Melli et al., 1992; Sciubba and Melli, 1998; Munos and Moore, 1999).

A need is postulated for the analysis of energy systems not only on the bases of their operational life, but also on the basis of considerations for the manufacturing and recycling phases (Curti et al., 1993). Environmental and resource scarcity factors are taken into consideration. The main idea is to minimize the cumulated exergetic losses of operation, manufacture and recycling, distributing the exergy of manufacture (the so-called *gray exergy*), and recycling over the life time of the system. Additional levels of analysis consist of introducing pollution and resource scarcity factors and using these to artificially penalize the above-mentioned exergy terms. In this approach, the first level of analysis (exergy analysis with gray exergy) is applied to domestic heat pumps operating with different evaporators. The pinch analysis (Linnhoff, 1979 and Chapter 12 of Sieniutycz and Jeżowski, 2013), extended to include exergy factors, falls within the framework of a global multidisciplinary analysis which considers economic, energetic, exergetic, and environmental factors in the light of sustainable development (Staine and Favrat, 1995). One of the essential themes of such a vast framework is the system's *life cycle analysis* that involves the manufacture, the exploitation, and the recycling of components. This is more than in the original pinch method, which focuses primarily on economic and heat transfer issues. The extension of the pinch analysis to include exergy factors involves a global exergy balance and irreversibility considerations due to heat transfer, dissipation, and the manufacture of components. The thermodynamic optimization of heat exchangers based on an optimal distribution of exergy losses is performed, and the gray exergy associated with the manufacture of shell and tube heat exchangers is calculated. Inclusion of electrical energy balance is also possible. Such a balance is particularly useful when introducing heat pumps or power units. The extended composite curves offer a graphic representation of all the exergy losses by using a Carnot factor-heat rate diagram and an electric power—Carnot factor diagram. With such diagrams, the choice of the optimal pinch value (T_{min}) is determined for the minimum total exergy loss of the process. A methodology for the heat integration of industrial batch processes based on Pinch Analysis principles is also proposed (Krummenacher and Favrat, 2000). This methodology resorts to intermediate heat storage to prevent adverse effects on the operating flexibility. A minimum number of storage units and their range of feasible operation follows in terms of the amount of heat recovery. Heuristic rules emerge, related to minimum cost solutions. Heat-pump-surrounding (HPS) systems are modeled as gray systems, where the gray model serves to assess the system exergy efficiency (Ao et al., 2007). Gray system theory and its applications are summarized in Luo et al. (2001) and Liu and Lin (2005).

As approaches to complex system design based on second law analyses have been developed, emphasis has changed from exergy (energy) minimal to cost optimal units and networks (Gundersen and Naess, 1988; Tsatsaronis et al., 1993; Zięby, 1996; Valero et al., 1994a,b; Frangopoulos et al., 2002; Sciubba, 2001a,b). While this reorientation does not constitute a real problem in FTT, where performance criteria can be arbitrary, analyses restricted to the classical availability and exergy criteria have more limited chance of proving their full usefulness in such extended schemes.

Apparently, the optimal design point of a heat exchanger network can be calculated only by taking into proper account entropic losses, i.e., exergy destruction (Sciubba and Wall, 2007). However, as stressed by Tsatsaronis (2007), such an approach has a number of severe flaws and no optimal design can be obtained by it for the following reasons:

(a) An optimal design point can, in general, be obtained by considering the economic and the environmental performance together with the thermodynamic one, but not just the thermodynamic performance.
(b) As the economic values of the exergy unit differ for various forms of exergy, the thermodynamic optimum is also questionable because the exergy destruction caused by friction is covered by mechanical (or electric) energy, whereas the exergy destruction caused by heat transfer is usually covered by the exergy of fossil fuels (or renewable energy). These exergy forms have not only different costs per unit of exergy, but also different thermodynamic values, if we refer them to the same form of primary energy.

(c) The optimal design point of a heat exchanger cannot be obtained in isolation, because such a design depends on the relative position of the heat exchanger within the overall energy conversion system being studied.

In fact, contemporary design of heat exchanger networks involves real-life objectives that contain both a quantitative part (cost of equipment, energy, and resources) and a qualitative part (safety, operability, controllability, flexibility, etc.). The industrial problem is very complex and involves a combinatorial approach to the match between hot and cold streams, flow configuration choice, equality and inequality constraints on the temperature-dependent properties, materials, pressure drop limitations, and so on. Many of these issues are discussed in the second part (Chapters 11–20) of the book by Sieniutycz and Jeżowski (2013). All these lead to the trend toward complex thermoeconomics and require reduction or abandonment of thermodynamic concepts. Optimizing of energy systems works at various levels: synthesis (configuration), design (component characteristics), and operation (Frangopoulos et al., 2002), with possibly including human work (Sciubba, 2001a,b). As one goes from operation optimization to design and synthesis, the problem is difficult from methodological and computational viewpoints.

When physical criteria of performance are abandoned, methodologies developed for the analysis, improvement, and optimization of energy systems must take into account not only energy consumption and financial resources expended, but the scarcity of all resources as well as any possible thermal and/or chemical pollution and degradation of the environment. The current trend is to combine all these aspects into a global methodology (Frangopoulos and von Spakovsky, 1993; Von Spakovsky and Frangopoulos, 1993). Its essence is to mathematically combine in a single model second law analysis with the economics, subject process constraints including pollution, resource availability, and the so-called gray exergy associated with the manufacture and recycling of the capital equipment used by an energy system. The benefit of such global approach is a systematic search for the optimum system configuration, performance, and cost (von Spakovsky, 1994). Industry has the capability of finding such optima for individual components, taking into account solely thermodynamic and economic considerations. On a system level, however, such a capability usually does not presently exist in industry, since the component optimums found on individual basis do not necessarily and more than likely do not correspond to an overall system optimum (Von Spakovsky, 1993, 1994; Von Spakovsky and Evans, 1990). An approach to exergy costing in exergoeconomics improves the fairness of the costing by taking a closer look at the cost formation and the monetary value of processes (Tsatsaronis et al., 1993).

6.11 Final remarks

In various editions of the book by Sieniutycz and Jeżowski (2009, 2013, 2018), we apply system approaches to develop techniques of heat integration, a collection of approaches designed to reduce cost of utilities and environmental impacts. The basic idea of heat integration is to apply heat energy of hot streams to heat up cold streams; this causes cooling of hot streams and heating of cold streams along with decreased consumption of external utilities. An example is the use of hot bottom product from distillation to increase the temperature of feed stream directed to distillation tower and to achieve a significant reduction of an external heat. To proceed in systematic way, total system is decomposed into three subsystems: (i) basic process subsystem (BPS) composed of processes and apparatuses typical for the industry we are working in, (ii) utility subsystem (US) that produces heating and cooling utilities and shaft work, and, (iii), heat exchanger network (HEN), that is a network where process streams from BPS and utilities exchange heat. Interconnections among subsystems, main features of heat integration as well as related issues, are discussed elsewhere (Sieniutycz and Jeżowski, 2009, 2013, 2018).

We shall terminate this chapter with a brief discussion of some applications of thermodynamics in complex engineering of power yield and exergy-valued biofuels.

Berry (1989) presented a plausible review of methods involving energy, resting on understanding of its role in the contemporary world. Berthiaume et al. (2001) described exergetic evaluation of the renewability of a biofuel in the context of a method to quantify the renewability of ethanol produced from corn. An ideal CO_2-glucose-ethanol cycle is considered to show that exergy can be potentially produced through harnessing of natural thermochemical cycles. Then exergy accounting is used to evaluate the departure from ideal behavior caused by nonrenewable resource consumption through the concept of restoration work. This procedure leads the authors to propose a renewability indicator. Different cycles and processes involved in ethanol production from corn are described.

Biomass has great potential as a clean, renewable feedstock for producing modern energy carriers. Piekarczyk et al. (2013) have performed a thermodynamic evaluation of biomass-to-biofuels production systems. Biomass is a renewable feedstock for producing modern energy carriers. However, its usage is accompanied by possible drawbacks, mainly due to limitation of land and water and competition with food production. The analysis concerns the so-called second-generation biofuels, like Fischer-Tropsch fuels or Substitute Natural Gas, which are produced either from wood or from waste biomass. The most promising conversion case is the one which involves production of syngas from biomass gasification, followed by synthesis of biofuels. The thermodynamic efficiency of biofuels production is analyzed and compared using both direct exergy analysis and thermo-ecological cost.

A paper by Li et al. (2004) presents the results from biomass gasification tests in a pilot-scale (6.5-m tall $\times$ 0.1-m diameter) air-blown circulating fluidized bed gasifier and compares them with model predictions. The operating temperature was maintained in the range of 700–850°C, while the sawdust feed rate varied from 16 to 45 kg/h. Temperature, air ratio, suspension density, fly ash reinjection, and steam injection were found to influence the composition and heating value of the product gas. Tar yield from the biomass gasification decreases exponentially with increasing operating temperature for the range studied.

A nonstoichiometric equilibrium model based on direct minimization of Gibbs free energy is developed to predict the performance of the gasifier. Experimental evidence indicated that the pilot gasifier deviated from chemical equilibrium due to kinetic limitations. A phenomenological model adapted from the pure equilibrium model, incorporating experimental results regarding unconverted carbon and methane to account for nonequilibrium factors, predicts product gas compositions, heating value, and cold-gas efficiency in good agreement with the experimental data.

Ptasinski et al. (2007) focus on exergetic evaluation of biomass gasification, where the synthesis gas may subsequently be used for the production of electricity, fuels, and chemicals. The gasifier is one of the least-efficient unit operations in the whole biomass-to-energy chain and an efficiency analysis of the gasifier alone can substantially contribute to the efficiency improvement of this chain. Ptasinski et al. (2007) compare different types of biofuels for their gasification efficiency and benchmark this against gasification of coal. Exergy-based efficiencies, defined as the ratio of chemical and physical exergy of the synthesis gas to chemical exergy of a biofuel, are proposed to quantify the value of the gasification process. Biofuels considered by Ptasinski et al. (2007) include various types of wood, vegetable oil, sludge, and manure. Exergetic efficiencies are evaluated for an idealized equilibrium gasifier with ignored heat losses. Gasification efficiencies are evaluated at the carbon-boundary point, where exactly enough air is added to avoid carbon formation and achieve complete gasification. The cold-gas efficiency of biofuels is comparable to that of coal.

Further on, Ptasinski (2008) reviews the thermodynamic efficiency of biomass gasification and biofuels conversion. Biomass has great potential as a clean renewable feedstock for producing biofuels such as Fischer-Tropsch biodiesel, methanol, and hydrogen. The use of biomass is accompanied by possible ecological drawbacks, however, such as limitation of land or water and competition with food production. For biomass-based systems, a key challenge is thus to develop efficient conversion technologies which can also compete with fossil fuels. The development of efficient technologies for biomass gasification and synthesis of biofuels requires correct use of thermodynamics. Energy systems are analyzed traditionally by energetic analysis based on the first law of thermodynamics. This type of analysis estimates only magnitudes of mass and energy flows and does not provide information on how quality of the energy and material streams degrades through the process.

Ptasinski (2008) uses the exergy analysis to analyze the biomass gasification and conversion of biomass to biofuels. He reviews thermodynamic efficiencies of biomass gasification for air-blown as well as steam-blown gasifiers. Finally, he evaluates overall technological chains for biomass-biofuels, including methanol, Fischer-Tropsch hydrocarbons, and hydrogen. The efficiency of biofuels production is compared with that of fossil fuels.

Vitasari et al. (2011) present an exergy analysis of biomass-to-synthetic natural gas (SNG) production by indirect gasification of various biomass feedstock, including virgin (woody) biomass as well as waste biomass (municipal solid waste and sludge). In indirect gasification, heat needed for endothermic gasification reactions is produced by burning char in a separate combustion section of the gasifier and subsequently the heat is transferred to the gasification section. The advantages of indirect gasification are no syngas dilution with nitrogen and no external heat source required. The production process involves several process units, including biomass gasification, syngas cooler, cleaning and compression, methanation reactors, and SNG conditioning. The process was simulated with a computer model using the flow-sheeting program Aspen Plus. The exergy analysis was performed for various operating conditions such as gasifier pressure, methanation pressure, and temperature.

Papadikis et al. (2009) study fluid-particle interaction and the impact of shrinkage on pyrolysis of biomass inside a 150 g/h fluidized bed reactor was modeled. Juraščík et al. (2010) present the results of exergy analysis for a biomass-to-synthetic natural gas (SNG) conversion process. The main process units of biomass-to-SNG conversion technology are gasifier, gas cleaning, synthesis gas compression, CH_4 synthesis, and final SNG conditioning. The study was based on wood gasification, analyzed for different gasification conditions. The results show that the largest exergy losses take place in the biomass gasifier, CH_4 synthesis part, and CO_2 capture unit.

Szewczyk (2002, 2006) derived balances of microbial growth and the related applications in biological fuel cells. Correspondingly, Logan (2008) described mechanism of electric current generation in microbial fuel cells. Szewczyk (2006) introduced two categories of BFCs: EFCs and microbial (MFCs). The former, due to the immobilization of biocatalysts, offer high current densities and possibilities of the device miniaturization. The latter, because of the presence of living organisms, use complex substrates and can do prolonged work. The catalytic steam reforming of bio-based alcohols, mainly bioethanol, is a new technology based on the environmental compatibility of hydrogen energy when compared with other feedstocks. There are numerous BFC units working effectively at the present time (Sieniutycz and Jeżowski, 2013).

Optimization of an energy network may require different methods from those currently used in thermodynamics. Network of the system thermodynamics (Peusner, 1986, 1990) can help in economic optimization of energy networks. Thermodynamics, especially as the subject of the thermal networks theory, is currently extended to combine technical and economic approaches into thermoeconomics (see, e.g., chapter: Thermodynamics and Optimization of Practical Processes in Sieniutycz (2016) (TAES)). Ruth (1993, 1999) considers the integration of economics, ecology, and thermodynamics and formulates physical principles in the environmental economic analysis. Kåberger and Månsson (2001) contribute to the discussion on the relation between thermodynamics and economic theory. With respect to thermodynamic constraints on the economy, there are two diametrically opposite positions in this discussion. One claims that constraints are insignificant ("of no immediate practical importance for modeling"), as in the intermediate run they do not limit economic activity and need not be incorporated in the economic theory. The other holds that thermodynamics tells us that there are practical limits to materials recycling, which already puts bounds on the economy and, therefore, must be included in the economic models.

Using the thermodynamic concept of entropy, Kåberger and Månsson (2001) show that there are fundamental problems with both positions. Even in the long run, entropy production associated with material dissipation need not be a limiting factor for economic development. Abundant energy resources from solar radiation may be used to recover dissipated elements (Kåberger and Månsson, 2001). With a simple, quantitative analysis, it may be shown that the rate of entropy production caused by human economic activities is very small compared to the continuous natural entropy production in the atmosphere and on the Earth's surface (Aoki, 1995, 2001; Mady and Oliveira, 2013). It is possible to replace a part of the natural entropy production with societal entropy production by making use of solar energy. Otherwise, society consumes resources available for coming generations. However, future generations need not have fewer resources available to them than the present generation. Using solar energy as a driving resource, human industrial activities could be transformed into a sustainable system, where more abundant elements are industrially used and recycled. An economic theory, fit to guide industrial society in that development, must not disregard thermodynamics, nor must it overstate the consequences of the laws of thermodynamics (Kåberger and Månsson, 2001).

References

Altenberger, A., Dahler, J., 1992. Statistical mechanical theory of diffusion and heat conduction in multicomponent systems. In: Flow, Diffusion and Transport Processes. Advances in Thermodynamics Series, vol. 6. Taylor and Francis, New York, pp. 58–81.

Andresen, B., 1983. Finite-Time Thermodynamics. University of Copenhagen, Englewood Cliffs.

Andresen, B., Salamon, P., Berry, R.S., 1977. Thermodynamics in finite time: extremals for imperfect engines. J. Chem. Phys. 66, 1571.

Andresen, B., Rubin, M.H., Berry, R.S., 1983. Availability for finite time processes. General theory and model. J. Phys. Chem. 87, 2704–2713.

Andresen, B., Salamon, P., Berry, R.S., 1984. Thermodynamics in finite time. Phys. Today (September issue), 62–70.

Androulakis, I.P., Venkatasubramanian, V., 1991. A genetic algorithmic frame-work for process design and optimization. Comput. Chem. Eng. 15 (4), 217–228.

Angulo-Brown, F., 1991. An ecological optimization-criterion for finite-time heat engines. J. Appl. Phys. 69, 7465–7469.

Ao, Y., Chen, J., Duan, M., Shen, S., 2007. Gray modeling of heat-pump-surrounding system exergy efficiency. In: Proceedings of 2007 IEEE International Conference on Grey Systems and Intelligent Services, Nanjing, China, November 18–20, 2007.

Aoki, I., 1995. Entropy production in living systems: from organisms to ecosystems. Thermochim. Acta 250, 359–370.

Aoki, I., 2001. Entropy and exergy principles in living systems. In: Jorgensen, S.E. (Ed.), Thermodynamics and Ecological Modelling. Lewis (Publishers of CRC Press), Boca Raton, pp. 165–190.

Aris, R., 1964. Discrete Dynamic Programming. Blaisdell, New York.
Ayres, R.U., 1978. Resources Environment and Economics—Applications of the Materials/Energy Balance Principle. Wiley, New York.
Ayres, R.U., 1998. Eco-thermodynamics: economics and the second law. Ecol. Econ. 26, 189–209.
Ayres, R.U., 1999. The second law, the fourth law, recycling, and limits to growth. Ecol. Econ. 29, 473–483.
Ayres, R.U., Kneese, A.V., 1969. Production, consumption, and externalities. Am. Econ. Rev. 59, 282–297.
Ayres, R.U., Martinás, K., 1995. Waste potential entropy: the ultimate ecotoxic? Econ. Appl. 48, 95–120.
Ayres, R.U., Ayres, L.W., Martinás, K., 1998. Exergy, waste accounting, and life-cycle analysis. Energy 23, 355–363.
Ayres, R.U., Ayres, L.W., Warr, B., 2003. Exergy, power and work in the US economy 1900–1998. Energy Int. J. 28, 219–273.
Bampi, F., Morro, A., 1984. Nonequilibrium thermodynamics: a hidden variable approach. In: Cassas Vazquez, J., Jou, D., Lebon, G. (Eds.), Lecture Notes in Physics. vol. 199. Springer, Berlin, pp. 221–232.
Baumgärtner, S., 2000. Ambivalent Joint Production and the Natural Environment. An Economic and Thermodynamic Analysis. Physica Verlag, Heidelberg.
Baumgärtner, S., 2004. Thermodynamic models. In: Proops, J., Safonov, P. (Eds.), Modelling in Ecological Economics. Edward Elgar, Cheltenham, pp. 102–129.
Baumgärtner, S., Faber, M., Proops, R., 1996. The use of the entropy concept in ecological economics. In: Faber, M., Manstetten, R., Proops, J. (Eds.), Ecological Economics—Concepts and Methods. Edward Elgar, Cheltenham, pp. 115–135.
Beckmann, H., 1953. Die Verteilung der Selbstkosten in Industrie und Heizkraftwerken auf Strom und Heizdampf. BWK 5, 37–44.
Bejan, A., 1982. Entropy Generation Through Heat and Fluid Flow. Wiley, New York.
Bejan, A., 1987. The thermodynamic design of heat and mass transfer processes and devices. Heat Fluid Flow 8, 258–276.
Bejan, A., 1988. Advanced Engineering Thermodynamics. Wiley Interscience, New York.
Bejan, A., 1996a. Entropy Generation Minimization: The Method of Thermodynamic Optimization of Finite-Size Systems and Finite-Time Processes. CRC Press, Boca Raton.
Bejan, A., 1996b. Entropy generation minimization: the new thermodynamics of finitesize devices and finite time processes. J. Appl. Phys. 79, 1191–1218.
Bejan, A., 1997a. Theory of organization in nature: pulsating physiological processes. Int. J. Heat Mass Transf. 40, 2097–2104.
Bejan, A., 1997b. Constructal theory: from thermodynamic and geometric optimization to predicting shape in nature. Energy Convers. Manag. 39, 1705–1718.
Bejan, A., 1997c. Constructal theory network of conducting paths for cooling a heat generating volume. Int. J. Heat Mass Transf. 40, 799–816.
Bejan, A., 2000. Shape and Structure From Engineering to Nature. Cambridge University Press, Cambridge, pp. 234–245.
Bejan, A., 2006. Advanced Engineering Thermodynamics, third ed. Wiley, Hoboken.
Bejan, A., Errera, M.R., 1998. Maximum power from a hot stream. Int. J. Heat Mass Transf. 41, 2025–2036.
Bejan, A., Ledezma, G.A., 1998. Streets tree networks and urban growth: optimal geometry for quickest access between a finite-size volume and one point. Physica A 255, 211–217.
Bejan, A., Lorente, S., 2003. In: Rosa, R.N., Reis, A.H., Miguel, A.F. (Eds.), Bejan's Constructal Theory of Shape and Structure. Evora Geophysics Center, Evora, Portugal (Chapter 4).
Bejan, A., Lorente, S., 2004. The constructal law and the thermodynamics of flow systems with configuration. Int. J. Heat Mass Transf. 47, 3203–3214.
Bejan, A., Lorente, S., 2005. La Loi Constructale. L'Harmattan, Paris.
Bejan, A., Tsatsaronis, G., Moran, M., 1996. Thermal Design and Optimization. Wiley, New York.
Bellman, R.E., 1961. Adaptive Control Processes: A Guided Tour. Princeton University Press, Princeton, pp. 1–35.
Bellman, R., Dreyfus, S.E., 1967. Dynamic Programming Applications. Państwowe Wydawnictwa Ekonomiczne, Warsaw (in Polish).

Benedict, M., Gyftopoulos, E., 1980. Economic selection of the components of an air separation process. In: Gaggioli, R.A. (Ed.), Thermo-dynamics: Second Law Analysis. In: ACS Symposium Series, vol. 122. American Chemical Society, Washington, DC.
Berry, R.S., 1989. Energy future: time horizons and instability. Environment 31, 42.
Berry, R.S., Salamon, P., Heal, G., 1978. On the relation between thermodynamic and economic optima. Resour. Energy 1, 125–127.
Berry, R.S., Kazakov, V.A., Sieniutycz, S., Szwast, Z., Tsirlin, A.M., 2000. Thermodynamic Optimization of Finite Time Processes. Wiley, Chichester, p. 117.
Bertallanffy, L., 1973. General System Theory. Foundations, Development, Applications, revised ed. George Braziler, New York.
Berthiaume, R., Bouchard, Ch., Rosen, M.A., 2001. Exergetic evaluation of the renewability of a biofuel. Exergy Int. J. 1 (4), 256–268.
Bodvarsson, G.S., Witherspoon, P.A., 1989. Geothermal reservoir engineering, part 1. Geotherm. Sci. Technol. 1, 1–2.
Bosniakovic, F., 1965. Technische Thermodynamik. I und II. Theodor Steinkopff, Dresden.
Boulding, K.E., 1966. The economics of the coming spaceship Earth. In: Jarrett, H. (Ed.), Environmental Quality in a Growing Economy. Johns Hopkins University Press, Baltimore, pp. 3–14.
Brodyanskii, V.M., 1973. Exergy Method of Thermodynamic Analysis. Energia, Moscow (in Russian).
Burley, P., Foster, J., 1994. Economics and Thermodynamics—New Perspectives on Economic Analysis. Kluwer Academic Publishers, Dordrecht.
Burness, H.S., Cummings, R.G., Morris, G., Paik, I., 1980. Thermodynamic and economic concepts as related to resource-use policies. Land Econ. 56, 1–9.
Calabro, A., Girardi, G., Fiorini, P., Sciubba, E., 2004. Exergy analysis of a "CO_2 zero emission" high efficiency plant. In: Proceedings of ECOS'04, Guanajuato, Mexico, July 2004.
Chen, J., 2005. The Physical Foundation of Economics—An Analytical Thermodynamic Theory. World Scientific, Singapore.
Chen, J., Wu, C., Kiang, R.J., 1996. Maximum specific power output of an irreversible radiant heat engine. Energy Convers. Manag. 37, 17–22.
Chen, J., Yan, Z., Chen, L., Andresen, B., 2000. On the performance of irreversible combined heat pump cycles. In: Hirs, G.G. (Ed.), ECOS 2000 Proceedings. University Twente, The Netherlands, pp. 269–277.
Cheng, C.-Y., Lee, S.-H., Chen, C.-K., 1999. Power optimizations of an irreversible Brayton heat engine with finite heat reservoirs. In: Wu, C., Chen, L., Chen, J. (Eds.), Advances in Recent Finite Time Thermodynamics. Nova Science, New York, pp. 425–448.
Chua, H.T., Ng, K.C., Gordon, J.M., 1996. Experimental study of the fundamental properties of reciprocating chillers and its relation to thermodynamic modelling and chiller design. Int. J. Heat Mass Transf. 39, 2195–2204.
Ciborowski, J., 1976. Basic Chemical Engineering. Wydawnictwa Naukowo Techniczne, Warsaw (in Polish).
Clark, J.A., 1986. Thermodynamic optimization. Interface with economic analysis. J. Non-Equilib. Thermodyn. 10, 85–122.
Coleman, B.D., Gurtin, M., 1964. Thermodynamics with internal variables. J. Chem. Phys. 47, 597–613.
Corning, P.A., 2002. Thermoeconomics—beyond the second law. J. Bioecon. 4, 57–88. Available at:www.complexsystems.org.
Corning, P.A., Kline, S.J., 2000. Thermodynamics, information and life revisited, Part II. Thermoeconomics and control information. Syst. Res. Behav. Sci. 15, 453–482.
Correas, L., 2004. On the thermoeconomic approach to the diagnosis of energy system malfunctions suitability to real time monitoring. Int. J. Thermodyn. 7, 85–94 (the special issue on thermoeconomic diagnosis).
Curti, V., Favrat, D., von Spakovsky, M., 1993. Contribution to a life cycle exergy analysis of heat pumps with various evaporator sources. In: LENI-CONF-1993-027, Lausanne.
Daly, H.E., 1992. Is the entropy law relevant to the economics of natural resources? Yes, of course it is!. J. Environ. Econ. Manage. 23, 91–95.

Dantzig, G.B., 1968. Linear Programming and Extensions. Princeton University Press, Princeton.
de Vos, A., 1992. Endoreversible Thermodynamics of Solar Energy Conversion. Clarendon Press, Oxford, pp. 29–51.
Denbigh, K.G., 1956. The second law efficiency of chemical processes. Chem. Eng. Sci. 6, 1–9.
Dentice d'Accadia, M., Sasso, M., Sibilio, S., 2000. Optimum operation of a thermal plant with cogeneration and heat pumps. In: Hirs, G.G. (Ed.), ECOS 2000 Proceedings. Part 3: Process Integration. University Twente, The Netherlands, pp. 1411–1424.
Edera, M., Kojima, H., 2002. Development of a new gas absorption chiller heater advanced utilization of waste heat from gas-driven co-generation systems for air conditioning. Energy Convers. Manag. 43, 1493–1501.
El Haj Assad, M., 1999. Performance characteristics of an irreversible refrigerator. In: Wu, C., Chen, L., Chen, J. (Eds.), Advance in Recent Finite Time Thermodynamics. Nova Science, New York, pp. 181–188.
El-Sayed, Y.M., 1999. Thermodynamics and thermoeconomics. Int. J. Appl. Thermodyn. 2, 5–18.
El-Sayed, Y.M., 2003. The Thermoeconomics of Energy Conversions. Pergamon, Oxford, p. 4.
El-Sayed, Y., 2007. Fingerprinting the malfunction of devices. Int. J. Thermodyn. 10, 79–85.
Faber, M., 1985. A biophysical approach to the economy: entropy, environment and resources. In: van Gool, W., Bruggink, J.J.C. (Eds.), Energy and Time in the Economic and Physical Sciences. North Holland, Amsterdam, pp. 315–335.
Faber, M., Jöst, F., Manstetten, R., 1995. Limits and perspectives on the concept of sustainable development. Econ. Appl. 48, 233–251.
Faber, M., Manstetten, R., Proops, J., 1996. Ecological Economics—Concepts and Methods. Edward Elgar, Cheltenham.
Fan, L.T., 1966. The Continuous Maximum Principle: A Study of Complex System Optimization. Wiley, New York.
Floudas, C.A., 1995. Nonlinear and Mixed-Integer Optimization. Fundamentals and Applications. Oxford University Press, New York.
Frangopoulos, C.A., 1983. Thermodynamic functional analysis: a method for optimal design or improvement of complex thermal systems (Ph.D. thesis). Georgia State University, Atlanta, USA.
Frangopoulos, C.A., 1990. Optimization of synthesis-design-operation of a cogeneration system by the Intelligent Functional Approach. In: Proceedings of Florence World Energy Research Symposium, Florence, Italy. pp. 597–609 and 805–815.
Frangopoulos, C.A., Von Spakovsky, M.R., 1993. A global environomic approach for energy systems analysis and optimization – I. In: Proceedings of the International Conference Energy Systems and Ecology ENSEC' 93, Cracow, Poland, July 5–9, 1993, pp. 123–132.
Frangopoulos, C.A., von Spakovsky, M.R., Sciubba, E., 2002. A brief review of methods for the design and synthesis optimisation of energy systems. In: Proceedings of ECOS 2002, vol. I, Berlin, Germany, July 3–5, 2002, pp. 306–316.
Gaggioli, R.A. (Ed.), 1980. Thermodynamics: Second Law Analysis. American Chemical Society, Washington, DC.
Garrard, A., Fraga, E.S., 1998. Mass exchange network synthesis using geneticalgorithms. Comput. Chem. Eng. 22 (12), 1837–1850.
Georgescu-Roegen, N., 1971. The Entropy Law and the Economic Process. Harvard University Press, ISBN: 1-58348-600-3. Harvard (copyright by the President and Fellows of Harvard College).
Geworkian, P., 2007. Sustainable Energy Systems Engineering. The Complete Green Building Design Structure. McGraw-Hill, New York.
Giesekus, H., 1986. Constitutive models of polymer fluids: toward a unified approach. In: Kroner, E., Kirchgassner, K. (Eds.), Trends in Applications of Pure Mathematics to Mechanics. Springer, Berlin, pp. 331–348.
Gilbert, R., 1956. La recherche des economiques de l'energie par l'analyse entropique. Genie Chem 75 (4), 89–94.
Goldberg, D.E., 1989. Genetic Algorithm in Search, Optimization and Machine Learning. Adison-Wesley, Reading, MA.
Gong, M., Wall, G., et al., 1997. On exergetics, economics and optimization of technical processes to meet environmental conditions. In: Cai, R. (Ed.), Presented at TAIES'97. Thermodynamic Analysis and

Improvement of Energy Systems, Beijing, China, June 10–13. Beijing World, Warsaw, pp. 403–413. ISBN: 7-5062-3264-Z, http://www.exergy.se/ftp/execopt.pdf.
Gordon, J.M., 1990. Nonequilibrium thermodynamics for solar energy applications. In: Sieniutycz, S., Salamon, P. (Eds.), Finite-Time Thermodynamics and Thermoeconomics. In: Advances in Thermodynamics, vol. 4. Taylor and Francis, New York, pp. 95–120.
Gordon, J.M., Ng, K.C., 2000. Cool Thermodynamics. Cambridge International Science Publishing, Cambridge, UK.
Gouy, G., 1889. Sur l'Energie utilisable. J. Phys. 8, 501.
Grant, C.D., 1979. Energy Conservation in the Chemical and Process Industries. George Goldwin, London.
Gundersen, T., Naess, L., 1988. The synthesis of cost optimal heat exchanger networks. Comput. Chem. Eng. 12 (6), 503–530.
Gustavson, M., 1979. Limits to wind power utilization. Science 204, 13.
Gyftopoulos, E., 1999. Infinite time (reversible) versus finite time (irreversible) thermodynamics: a misconceived distinction. Energy Int. J. 24, 1035–1039.
Gyftopoulos, E., 2002. On the Courzon-Ahlborn efficiency and its lack of connection to power-producing processes. Energy Convers. Manag. 42, 609–615.
Haywood, R.W., 1975. Analysis of Engineering Cycles. Pergamon, Oxford.
Henatsch, A., 1957. Thermodynamische-energie wirtschaftliche Bewertung von Warmeaustauscher. Wiss. Zeitsch. der Techn. Hochs. f. Verkehrswes. Dresden 14, 57–67.
Hirs, G.G. (Ed.), 2000. ECOS 2000 Proceedings. Part 3: Process Integration. University Twente, The Netherlands.
Hoffmann, K.H., Andresen, B., Salamon, P., 1989. Measures of dissipation. Phys. Rev. A 39, 3618–3621.
Hoffmann, K.-H., Burzler, J.M., Schubert, M., 1997. Endoreversible thermodynamics. J. Non-Equilib. Thermodyn. 22, 311–355.
Jarzebski, Z.M., 1990. Solar Energy Photovoltaic Conversion. PWN, Warsaw.
Jørgensen, S.E., 1988. Use of models as an experimental tool to show that structural changes are accompanied by increased exergy. Ecol. Model. 41, 117–126.
Jørgensen, S.E., 1997. Integration of Ecosystem Theories: A Pattern, second ed. Kluwer Academic Publishers, Dordrecht.
Jørgensen, S.E., 2000. Application of exergy and specific exergy as ecological indicators of coastal areas. Aquat. Ecosyst. Health Manag. 3 (3), 419–430.
Jørgensen, S.E. (Ed.), 2001. Thermodynamics and Ecological Modelling. Lewis Publishers of CRC Press, Boca Raton.
Juraščík, M., Sues, A., Ptasinski, K.J., 2010. Exergy analysis of synthetic natural gas production method from biomass. Energy 35 (2), 880–888.
Kåberger, T., Månsson, B., 2001. Entropy and economic processes—physics perspectives. Ecol. Econ. 36, 165–179.
Kay, J., Schneider, E.D., 1992. Thermodynamics and measures of ecological integrity. In: Proceedings of Ecological Indicators. Elsevier, Amsterdam, pp. 159–182.
Keenan, J.H., 1932. A steam chart for second law analysis. Mech. Eng. ASME 54, 195–204.
Kelahan, R.C., Gaddy, J.J., 1977. Synthesis of heat exchange networks by mixed integer optimization. AIChE J. 23 (6), 423–435.
Kestin, J., 1979. In: Domingos, et al., (Ed.), Foundations of Non-Equilibrium Thermodynamics. Macmillan Press, London.
Kestin, J., Bataille, J., 1980. Thermodynamics of solids. In: Continuum Models of Discrete Systems. University of Waterloo Press, Ontario.
Kestin, J., Rice, J.M., 1970. Paradoxes in the application of thermodynamics to strained solids. In: Stuart, F.B., Gal Or, B., Brainard, A.J. (Eds.), A Critical Review of Thermodynamics. Mono Book Corp, Baltimore, pp. 275–298.
Khalil, E.L., 1990. Entropy law and exhaustion of natural resources: Is Nicholas Georgescu–Roegen's paradigm defensible? Ecol. Econ. 2, 163–178.
King, C.J., Gantz, D.W., Barnes, F.J., 1972. Systematic evolutionary system synthesis. Ind. Eng. Chem. Process. Des. Dev. 11, 271.

Klir, G.J., 1972. Trends in General Systems Theory. John Wiley, New York.
Kluitenberg, G.A., 1966. Application of the theory of irreversible processes to continuum mechanics. In: Donnelly, R., Herman, R., Prigogine, I. (Eds.), Nonequilibrium Thermodynamics, Variational Techniques, Stability. University Press, Chicago.
Kluitenberg, G.A., 1977. On dielectric and magnetic relaxation and vectorial internal degrees of freedom in thermodynamics. Physica 87a, 541–563.
Kluitenberg, G.A., 1981. On vectorial internal variables and dielectric and magnetic relaxation phenomena. Physica A 109 (1–2), 91–122.
Kneese, A.V., Ayres, R.U., d'Arge, R.C., 1972. Economics and the Environment: A Materials Balance Approach. Resources for the Future, Washington.
Krummenacher, P., Favrat, D., 2000. Indirect and mixed direct–indirect heat integration of batch processes based on Pinch analysis. In: Hirs, G.G. (Ed.), ECOS 2000 Proceedings. Part 3: Process Integration. University Twente, The Netherlands, pp. 1411–1424.
Kuddusi, L., Egrican, N., 2008. A critical review of constructal theory. Energy Convers. Manage. 49, 1283–1294.
Landau, L.D., Lifshitz, E., 1974. Statistical Physics. Pergamon, Oxford.
Landsberg, P., 1990. Statistics and thermodynamics of energy conversion from radiation. In: Nonequilibrium Theory and Extremum principles.Advances in Thermodynamics Series, vol. 3. pp. 482–536.
Lazzaretto, A., Andreatta, R., 1995. Algebraic formulation of a process-based exergycosting method. In: Krane, R.J. (Ed.), Symposium on Thermodynamics and the Design, Analysis, and Improvement of Energy Systems, AES.vol. 35. ASME, New York, pp. 395–403.
Lazzaretto, A., Macor, A., 1995. Direct calculation of average and marginal costs from the productive structure of an energy system. J. Energy Resour. Technol. 117, 171–178.
Lazzaretto, A., Tsatsaronis, G., 1999. On the calculation of efficiencies and costs in thermal systems. In: ASME, AES. vol. 39, pp. 421–430.
Lazzaretto, A., Tsatsaronis, G., 2006. SPECO: a systematic and general methodology for calculating efficiencies and costs in thermal systems. Energy Int. J. 31, 1257–1289.
Lebon, G., Perez Garcia, C., Casas-Vazquez, J., 1988. On the thermodynamic foundations of viscoelasticity. J. Chem. Phys. 88, 5068–5075.
Le Goff, P., Matsuda, H., Rivero, R., 1990a. Advances in chemical plants and heat transformers. In: The 3rd IEA Heat Pump Conference, Tokyo.
Le Goff, P., Rivero, R., de Oliveira, S., Cachot, T., 1990b. Application of the enthalpy–Carnot factor diagram to the exergy analysis of distillation process. In: Symposium on Thermodynamics and the Design Analysis and Improvement of Energy Systems, ASME Winter Annual Meeting, Dallas, TX, November 25–30.
Le Goff, P., Schwarzer, B., Rivero, R., de Oliveira, S., Liu, B., Aoufoussi, Z., 1990c. Exergy effectiveness and exergo-economic efficiency of sorption heat pumps and heat transformers. In: 3rd International Workshop on Research Activities on Advanced Heat Pumps, Graz, September 24–26.
Le Goff, P., de Oliveira, S., Matsuda, H., Ranger, P.M., Rivero, R., Liu, B., 1990d. Heat transformers for upgrading waste heat from industrial processes. In: Proceedings of the FLOWER'S 90 Conference, Firenze, May.
Leitman, G., 1966. An Introduction to Optimal Control. McGraw-Hill, New York.
Lemaitre, J., Chaboche, J.L., 1988. Mechanics of Solid Materials. Cambridge University Press, Cambridge, UK.
Leonov, A.I., 1976. Nonequilibrium thermodynamics and rheology of viscoelastic polymer media. Rheol. Acta 21, 683–691.
Leontief, W.W., 1951. The Structure of American Economy 1919–1939. Oxford University Press, New York.
Lewandowski, W.M., 2001. Proecological Sources of Renewable Energy. Wydawnictwa Naukowo Techniczne, Warsaw (in Polish).
Li, X.T., Grace, J.R., Lim, C.J., Watkinson, A.P., Chen, H.P., Kim, J.R., 2004. Biomass gasification in a circulating fluidized bed. Biomass Bioenerg 26 (2), 171–193.
Linnhoff, B., 1979. Thermodynamic Analysis in the Design of Process Networks (Ph.D. dissertation). University of Leeds, UK.
Lior, N., 2002. Thoughts about future power generation systems and the role of exergy analysis in their development. Energy Convers. Manag. 43, 1187–1198.

Liu, S., Lin, Y., 2005. Grey Information: Theory and Practical Applications. (Advanced Information and Knowledge Processing). Springer Verlag, Berlin.
Logan, B.E., 2008. Microbial fuel cells. Wiley, New York.
Lotka, A.J., 1921. Note on economic conversion factors of energy. Am. J. Proc. Natl. Acad. Sci. 7, 192–197.
Lozada, G.A., 1991. A defense of Nicholas Georgescu-Roegens's paradigm. Ecol. Econ. 3, 157–160.
Lozada, G.A., 1995. Georgescu–Roegens's defence of classical thermodynamics revisited. Ecol. Econ. 14, 31–44.
Lozano, M.A., Valero, A., 1987. Application of the exergetic cost theory to a stream boiler in a thermal generating station. In: Moran, M.J., Stecco, S.S., Reistad, G.M. (Eds.), Analysis and Design of Advanced Energy Systems: Applications, AES, ASME Book No. G0377B. In: vol. 3–2. ASME, New York, pp. 41–51.
Lozano, M.A., Valero, A., 1988. Methodology for calculating exergy in chemical process. In: Wepfer, A.J., Tsatsaronis, G., Bajura, R.A. (Eds.), Thermodynamic Analysis of Chemically Reactive Systems, ASME Book No. G00449. ASME, New York, pp. 77–86.
Lozano, M.A., Valero, A., 1993. Theory of exergetic cost. Energy 18, 939–960.
Lozano, M.A., Valero, A., Serra, L., 1993. Theory of exergetic cost and thermoeconomic optimization. In: Szargut, J., Kolenda, Z., Tsatsaronis, G., Ziębik, A. (Eds.), Proceedings of the International Conference Energy Systems and Ecology ENSEC'93, Cracow, Poland, July 5–9, pp. 339–350.
Lozano, M.A., Bartolome, J.L., Valero, A., Reini, M., 1994. Thermoeconomic diagnosis of energy systems. In: FLOWERS'94, Proceedings of the Florence World Energy Research Symposium, pp. 149–156.
Luo, Y., Zhang, L., Li, M., 2001. Application of the Gray System Theory in the Mechanical Engineering. China National Defense Science and Technology University Publisher, Changsha.
Mady, C.E.K., Oliveira Jr., S., 2013. Human body exergy metabolism. Int. J. Thermodyn. 16 (2), 73–80.
Mansoori, G.A., Patel, V., 1979. Thermodynamic basis for the choice of working fluids for solar absorption cooling systems. Sol. Energy 22, 483–491.
Månsson, B.Å., 1990. Thermodynamics and economics. In: Sieniutycz, S., Salamon, P. (Eds.), Finite-Time Thermodynamics and Thermoeconomics. In: Advances in Thermodynamics, vol. 4. Taylor and Francis, New York, pp. 153–174.
Martyushev, L.M., Seleznev, V.D., 2006. Maximum entropy production principle in physics, chemistry and biology. Phys. Rep. 426, 1–45.
Marzbanrad, J., Sharifzadegan, A., Kahrobaeian, A., 2007. Thermodynamic optimization of GSHPS heat exchangers. Int. J. Thermodyn. 10, 107–112.
Maugin, G.A., 1975. On the spin relaxation in deformable ferromagnets. Physica 81a, 454–458.
Maugin, G.A., 1979. Vectorial internal variables in magnetoelasticity. J. Mecanique 18, 541–563.
Maugin, G.A., 1987. Thermodynamique a variables internes et applications. In: Seminar: Thermodynamics of Irreversible Processes. Institut Francais du Petrole, Rueil-Malmaison, France.
Maugin, G.A., 1990a. Thermomechanics of Plasticity and Fracture. Cambridge University Press, Cambridge, UK.
Maugin, G.A., 1990b. Nonlinear dissipative effects in electrodeformable solids. In: Maugin, G.A., Collet, B., Drouout, R., Pouget, J. (Eds.), Nonlinear Mechanical Couplings. Manchester University Press, Manchester, UK (Chapter 6).
Maugin, G.A., 1992. Thermodynamic of hysteresis. In: Extended Thermodynamic Systems. Advances in Thermodynamics Series, vol. 7. Taylor and Francis, New York, pp. 25–52.
Maugin, G.A., Drouot, R., 1983a. Internal variables and the thermodynamics of macromolecules solutions. Int. J. Eng. Sci. 21, 705–724.
Maugin, G.A., Drouot, R., 1983b. Thermomagnetic behavior of magnetically nonsaturated fluids. J. Magn. Magn. Mater. 39, 7–10.
Maugin, G.A., Drouot, R., 1992. Nonequilibrium thermodynamics of solutions of macromolecules. In: Advances in Thermodynamics Series, vol. 7. Taylor and Francis, New York, pp. 53–75.
Maugin, G.A., Sabir, M., 1990. Mechanical and magnetic hardening of ferromagnetic bodies: influence of residual stresses and applications to nondestructive testing. Int. J. Plast. 6, 573–589.
Meixner, J., 1961. Der Drehimpulssatz in der Thermodynamik der irreversiblen Prozesse. Zeit. Phys. 16, 144–165.
Melli, R., Sciubba, E., Paoletti, B., 1992. Second-Law based synthesis and optimization of thermal systems: a third-generation A.I. code. In: Proceedings of ECOS'92.

Michalewicz, Z., 1996. Genetic Algorithms + Data Structures = Evolution Programs, third ed. Springer-Verlag, New York.
Mielentilew, L.A., 1982. Optimization of the Development and Control of Large Scale Energy Systems. High School, Moscow (in Russian).
Möller, B.F., Torisson, T., Assadi, M., 2006. AZEP gas turbine combined cycle power plants—thermo-economic analysis. Int. J. Thermodyn. 9, 21–28.
Moran, M.J., 1998. On second-law analysis and the failed promise of finite-time thermodynamics. Energy 6, 517–519.
Moran, M.J., Shapiro, H.N., 2003. Fundamentals of Engineering Thermodynamics. Wiley, New York.
Morro, A., 1985. Relaxation phenomena via hidden variable thermodynamics. In: Grioli, G. (Ed.), Thermodynamics and Constitutive Equations. Springer, Berlin.
Morro, A., 1992. Thermodynamics and extremum principles in viscoelasticity. In: Extended Thermodynamic Systems. Advances in Thermodynamics Series, vol. 7. Taylor and Francis, New York, pp. 76–106.
Munos, R., Moore, A., 1999. Variable resolution discretization for high accuracy solutions of optimal control problems. In: International Joint Conference on Artificial Intelligence. http://www.cmap.polytechnique.fr/~munos/papers/ijcai99.ps.gz.
Muschik, W., 1981. Thermodynamical theories, survey and comparison. ZAMM 61, T213.
Muschik, W., 1986. Thermodynamical theories, survey, and comparison. J. Appl. Sci. 4, 189.
Muschik, W., 1990a. Aspects of Non-equilibrium Thermodynamics: Six Lectures on Fundamentals and Methods. World Scientific, Singapore.
Muschik, W., 1990b. Internal variables in the non-equilibrium thermodynamics. J. Non-Equilib. Thermodyn. 15, 127–137.
Muschik, W., 2007. Why so many "schools" of thermodynamics? Forsch. Ingenieurwes. 71, 149–161.
Muschik, W., 2008. Survey of some branches of thermodynamics. J. Non-Equilib. Thermodyn. 33, 165–198.
Neill, D.T., Jensen, W.P., 1976. Geothermal powered heat pumps to produce process heat. In: Proceedings of the 11th Intersociety Energy Conversion Engineering Conference, USA.
Ng, K.C., Chua, H.T., Ong, W., Lee, S.S., Gordon, J.M., 1997. Diagnostics and optimization of reciprocating chillers: theory and experiment. Appl. Therm. Eng. 17, 263–276. Erratum ibid., 17 (1997), 601–602.
Ng, K.C., Chua, H.T., Tu, K., Gordon, J.M., Kashiwagi, T., Akisawa, A., Saha, B.B., 1998. The role of internal dissipation and process average temperature in chiller performance and diagnostics. J. Appl. Phys. 83, 1831–1836.
Norgaard, R.B., 1986. Thermodynamic and economic concepts as related to resource-use policies: synthesis. Land Econ. 62, 325–328.
Odum, H.T., 1971. Environment, Power and Society. Wiley, New York.
Ondrechen, M.J., Andresen, B., Mozurkiewicz, M., Berry, R.S., 1981. Maximum work from a finite reservoir by sequential Carnot cycles. Am. J. Phys. 49, 681–685.
Ondrechen, M.J., Rubin, M.H., Band, Y.B., 1983. The generalized Carnot cycle: a working fluid operating in finite time between finite heat sources and sinks. J. Chem. Phys. 78, 4721–4727.
Orlov, V.N., Rudenko, A.V., 1984. Optimal control in the problems involving limiting possibilities of irreversible thermodynamic processes—a review. Autom. Telemekh. 5, 7–42.
Papadikis, K., Gub, S., Bridgwater, A.V., 2009. CFD modeling of the fast pyrolysis of biomass in fluidised bed reactors: modelling the impact of biomass shrinkage. Chem. Eng. J. 149, 417–427.
Perrings, C., 1987. Economy and Environment: A Theoretical Essay on the Interdependence of Economic and Environmental Systems. Cambridge University Press, Cambridge.
Peusner, L., 1986. Studies in Network Thermodynamics. Elsevier, Amsterdam.
Peusner, L., 1990. Lumped parameter networks: a topological graph approach to thermodynamics. Nonequilibrium Theory and Extremum Principles. Advances in Thermodynamics Series, vol. 3. Taylor and Francis, New York, pp. 39–94.
Piekarczyk, W., Czarnowska, L., Ptasiński, K., Stanek, W., 2013. Thermodynamic evaluation of biomass-to biofuels production systems. Energy 62, 95–104.

Pontryagin, L.S., Boltyanski, V.A., Gamkrelidze, R.V., Mischenko, E.F., 1962. The Mathematical Theory of the Optimal Processes. Wiley, New York.

Popyrin, L.S., 1978. Mathematical Modelling and Optimization of Thermoenergy Facilities. Energy, Moscow (in Russian).

Portacha, J., 2002. Energy Research in Heat Systems of Power Stations. Technological University Publications, ITC PW, Warsaw (in Polish).

Poulikakos, D., Bejan, A., 1982. Fin geometry for minimum entropy generation in forced convections. Trans. ASME, J. Heat Transf. 104, 616–623.

Ptasinski, K.J., 2008. Thermodynamic efficiency of biomass gasification and biofuels conversion (Review). Biofuels Bioprod. Biorefin. 2 (3), 239–253.

Ptasinski, K.J., Prins, M.J., Pierik, A., 2007. Exergetic evaluation of biomass gasification. Energy 32 (4), 568–574.

Radwański, E., Skowroński, P., Twarowski, A., 1993. Modeling Problems of Energy Technology Systems. Technological University Publications, ITC PW, Warsaw (in Polish).

Ranger, P.-M., Matsuda, H., Le Goff, P., 1990. Modelling of a new type of absorption heat pump combining rectification and reverse rectification. J. Chem. Eng. Japan 23, 530.

Ravagnani, M.A.S.S., Silva, A.P., Arroyo, A.A., Constantino, A.A., 2005. Heat exchanger network synthesis and optimisation using genetic algorithms. Appl. Therm. Eng. 25, 1003–1017.

Reini, M., 1994. Analisi e sviluppo dei metodi termoeconomici per lo studio degli impianti di conversione dell'energia (Ph.D. thesis). Dipartimento di Ingegneria Meccanica, University of Padova.

Reini, M., Taccani, R., 2004. On the thermoeconomic approach to the diagnosis of energy system malfunctions. The role of the fuel impact formula. Int. J. Thermodyn. 7, 61–72 (the special issue on thermoeconomic diagnosis).

Reis, A.H., Bejan, A., 2006. Constructal theory of global circulation and climate. Int. J. Heat Mass Transf. 49, 1857–1875.

Reis, A.H., Miguel, A.F., Aydin, M., 2004. Constructal theory of flow architectures of the lungs. Med. Phys. 31, 1135–1140.

Riekert, L., 1974. The efficiency of energy-utilization in chemical processes. Chem. Eng. Sci. 29, 1613.

Riekert, L., 1979. Large chemical plants. In: Froment, G.F. (Ed.), Proceedings 4th International Symposium, Antwerp, October. pp. 35–64.

Rubin, M.H., 1979a. Optimal configuration of a class of irreversible heat engines. I. Phys. Rev. A 19 (1979), 1272–1276.

Rubin, M.H., 1979b. Optimal configuration of a class of irreversible heat engines II. Phys. Rev. A 19 (1979), 1277–1289.

Rubin, M.H., Andresen, B., 1979. Optimal staging of endoreversible heat engines. J. Appl. Phys. 53, 1–7.

Ruth, M., 1993. Integrating Economics, Ecology and Thermodynamics. Kluwer, Dordrecht.

Ruth, M., 1999. Physical principles in environmental economic analysis. In: van den Bergh, J.C.J.M. (Ed.), Handbook of Environmental and Resource Economics. Edward Elgar, Cheltenham, pp. 855–866.

Salah El-Din, M.M., 2001a. Second law analysis of irreversible heat engines with variable temperature heat reservoirs. Energy Convers. Manag. 42, 189–200.

Salah El-Din, M.M., 2001b. Performance analysis of heat pumps and refrigerators with variable reservoir temperatures. Energy Convers. Manag. 42, 201–216.

Salamon, P., Nulton, J.D., Siragusa, G., Andersen, T.R., Limon, A., 2001. Principles of control thermodynamics. Energy 26, 307–319.

Salcedo, R.L., 1992. Solving nonconvex nonlinear programming and mixed-integer nonlinear programming problems with adaptive random search. Ind. Eng. Chem. Res. 31, 262–273.

Samuelson, P.A., 1947. Foundations of Economic Analysis. Harvard University Press, Cambridge, MA.

Sciubba, E., 2000. On the possibility of establishing a univocal and direct correlation between monetary price and physical value: the concept of Extended Exergy Accounting. In: Proceedings of the 2nd International Workshop on Advanced Energy Studies, Porto Venere, Italy.

Sciubba, E., 2001a. Extended energy accounting: a proposal for a new measure of value, a talk at the Gordon research conference. In: Modern Developments in Thermodynamics, 11–16 March, Ventura, California.
Sciubba, E., 2001b. Beyond thermoeconomics? The concept of Extended Exergy Accounting and its application to the analysis and design of thermal systems. Exergy Int. J. 1, 68–84.
Sciubba, E., 2004. Hybrid semi-quantitative monitoring and diagnostics of energy conversion processes. Int. J. Thermodyn. 7, 95–106 (the special issue on thermoeconomic diagnosis).
Sciubba, E., Melli, R., 1998. Artificial Intelligence in Thermal Systems Design: Concepts and Applications. Nova Science-Pergamon Press, New York.
Sciubba, E., Ulgiati, S., 2005. Energy and exergy analyses: complementary methods or irreducible ideological options? Energy. 30(10).
Sciubba, E., Wall, G., 2007. A brief commented history of exergy from the beginnings to 2004. Int. J. Thermodyn. 10, 1–26.
Selba, R., Kizilkan, Ő., Reppich, M., 2006. A new design approach for shell and tube heat exchangers using genetic algorithms from economic point of view. Chem. Eng. Process. 45, 268–275.
Shopova, E.G., Vaklieva-Bancheva, N.G., 2006. BASIC—a genetic algorithm for engineering problems solution. Comput. Chem. Eng. 20, 1293–1309.
Sieniutycz, S., 1973. The thermodynamic approach to fluidized drying and moistening optimization. AIChE J. 19, 277–285.
Sieniutycz, S., 1978. Optimization in Process Engineering, first ed. Wydawnictwa Naukowo Techniczne, Warsaw.
Sieniutycz, S., 1991. Optimization in Process Engineering, second ed. Wydawnictwa Naukowo Techniczne, Warsaw, pp. 151–194.
Sieniutycz, S., 1992. Wave equations of heat and mass transfer. In: Flow, Diffusion and Transport Processes. Advances in Thermodynamics Series, vol. 6. Taylor and Francis, New York, pp. 146–167.
Sieniutycz, S., 1997a. Hamilton–Jacobi–Bellman theory of dissipative thermal availability. Phys. Rev. 56, 5051–5064.
Sieniutycz, S., 1997b. Irreversible Carnot problem of maximum work in a finite time via Hamilton–Jacobi–Bellman Theory. J. Non-Equilib. Thermodyn. 22, 260–284.
Sieniutycz, S., 1997c. Carnot problem of maximum work from a finite resource interacting with environment in a finite time. Physica A 264, 234–263.
Sieniutycz, S., 1998. Generalized thermodynamics of maximum work in finite time. Open Syst. Inf. Dyn. 5, 369–390.
Sieniutycz, S., 1999. Optimal control for multistage endoreversible engines with heat and mass transfer. In: Bejan, A., Mamut, E. (Eds.), Thermodynamics and Optimization of Complex Energy Systems. Kluwer, Dordrecht, pp. 363–384.
Sieniutycz, S., 2000. Hamilton–Jacobi–Bellman framework for optimal control in multistage energy systems. Phys. Rep. 326, 165–285.
Sieniutycz, S., 2001. Thermodynamic optimization for work-assisted and solar-assisted heat and mass transfer operations. Arch. Thermodyn. 22, 17–36.
Sieniutycz, S., 2003. Thermodynamic limits on production or consumption of mechanical energy in practical and industrial systems. Progr. Energy Combust. Sci. 29, 193–246.
Sieniutycz, S., 2016. Thermodynamic Approaches in Engineering Systems. Elsevier, Amsterdam, Oxford/ Cambridge, MA.
Sieniutycz, S., Berry, R.S., 2000. Discrete Hamiltonian analysis of endoreversible thermal cascades. In: Sieniutycz, S., de Vos, A. (Eds.), Thermodynamics of Energy Conversion and Transport. Springer, New York, pp. 143–172 (Chapter 6).
Sieniutycz, S., de Vos, A., 2000. Thermodynamics of Energy Conversion and Transport. Springer, New York, pp. 143–172 (Chapter 6).
Sieniutycz, S., Jeżowski, J., 2009. Optimal decisions for chemical and electrochemical reactors. In: Energy Optimization in Process Systems, first ed. Elsevier, Oxford, pp. 321–366 (Chapter 9).

Sieniutycz, S., Jeżowski, J., 2013. Energy Optimization in Process Systems and Fuel Cells. Elsevier, Oxford.

Sieniutycz, S., Jeżowski, J., 2018. Energy Optimization in Process Systems and Fuel Cells. Elsevier, Oxford.

Sieniutycz, S., Kubiak, M., 2002. Dynamical energy limits in traditional and work driven operations I. Heat–mechanical systems. Int. J. Heat Mass Transf. 21, 2115–2129.

Sieniutycz, S., Salamon, P. (Eds.), 1990. Finite Time Thermodynamics and Thermoeconomics. In: Advances in Thermodynamics, vol. 4. Taylor & Francis, New York.

Sieniutycz, S., Szwast, Z., 2003. Work limits in imperfect sequential systems with heat and fluid flow. J. Nonequilib. Thermodyn. 28, 85–114.

Sluckin, T.J., 1981. Influence of flow on the isotropic-nematic transition in polymer solutions: a thermodynamic approach. Macromolecules 14, 1676–1680.

Staine, F., Favrat, D., 1995. Intégration énergétique de procédés industriels par la méthode du pincement étendue aux facteurs exergétiques. In: LENI-REPORT-1995-022, Lausanne.

Stodola, A., 1905. Steam Turbines (With an Appendix on Gas Turbines and the Future of Heat Engines) (Loewenstein, L.C. (Trans.)). Van Nostrand, New York.

Szargut, J., 1957. Towards a rational evaluation of steam prices. Gospod. Cieplna 5, 104–106 (in Polish).

Szargut, J., 1970. Application of exergy in establishing of generalized technicaleconomical interdependences. Arch. Bud. Maszyn 17, 105–108.

Szargut, J., 1978. Minimization of the consumption of natural resources. Bull. Acad. Pol. Tech. 26, 41–45.

Szargut, J., 1983. Analiza Termodynamiczna i Ekonomiczna w Energetyce Przemysłowej (Thermodynamic and Economic Analysis of Industrial Energetics). Wydawnictwa Naukowo Techniczne, Warsaw (in Polish).

Szargut, J., 1986. Application of exergy for the calculation of ecological cost. Bull. Polish Acad. Sci. Tech. Sci. 34 (1986), 475–480.

Szargut, J., 1987. Analysis of cumulative exergy consumption. Int. J. Energy Res. 11, 541–547.

Szargut, J., 1989. Indices of the cumulative consumption of energy and exergy. Zeszyty Nauk. Polit. Śląskiej: Ser. Energetyka 106, 39–55.

Szargut, J., 1990. Analysis of cumulative exergy consumption and cumulative exergy losses. In: Sieniutycz, S., Salamon, P. (Eds.), Finite-Time Thermodynamics and Thermoeconomics. In: Advances in Thermodynamics, vol. 4. Taylor and Francis, New York, pp. 278–302.

Szargut, J., 1998. Problems of thermodynamic optimization. Arch. Thermodyn. 19, 85–94.

Szargut, J., 2001a. Exergy analysis in thermal technology. In: Bilicki, Z., Mikielewicz, J., Sieniutycz, S. (Eds.), Ekspertyza KTiSp PAN: Współczesne Kierunki w Termodynamice. Wydawnictwa Politechniki Gdańskiej.

Szargut, J., 2001b. Letter to the Editor. Arch. Thermodyn. 22, 36.

Szargut, J., 2002a. Minimization of the depletion of non-renewable resources by means of the optimization of the design parameters. In: Proceedings of ECOS 2002, vol. I, Berlin, Germany, July 3–5, pp. 326–333.

Szargut, J., 2002b. Application of exergy for the determination of the proecological tax replacing the actual personal taxes. Energy 27, 379–389.

Szargut, J., 2005. Exergy Method: Technical and Ecological Applications. WIT Press, Southampton.

Szargut, J., Majza, E., 1989. Thermodynamic imperfection and exergy losses at production of pig iron and steel. Arch. Metall. 34, 197–216.

Szargut, J., Morris, D.R., 1987. Cumulative exergy consumption and cumulative degree of perfection of chemical processes. Int. J. Energy Res. 11, 245–261.

Szargut, J., Petela, R., 1965. Exergy. Wydawnictwa Naukowo Techniczne, Warsaw.

Szargut, J., Ziębik, A., 1972. Linear mathematical model of the material and energy balance of ironworks. Arch. Energ. 2, 1–18 (in Polish).

Szargut, J., Morris, D.R., Steward, F., 1988. Exergy Analysis of Thermal Chemical and Metallurgical Processes. Hemisphere, New York.

Szewczyk, K., 2002. Balances of microbial growth (in Polish). Biotechnology 57, 15–32.

Szewczyk, K., 2006. Biological fuel cells—Biologiczne ogniwa paliwowe (in Polish). Na Pograniczu Chemii i Biologii 15, 179–208.

Szwast, Z., 1990. Exergy optimization in a class of drying systems with granular solids. In: Sieniutycz, S., Salamon, P. (Eds.), Finite-Time Thermodynamics and Thermoeconomics. In: Advances in Thermodynamics, vol. 4. Taylor and Francis, New York, pp. 209–248.

Szwast, Z., 1994. Discrete optimal control thermodynamic processes with a constant Hamiltonian. Period. Polytech. Phys. Nucl. Sci. Budapest Tech. Univ. 2, 85–109.

Tayal, M.C., Fu, Y., 1999. Optimal design of heat exchangers: a genetic algorithm framework. Ind. Eng. Chem. Res. 38, 456–467.

Tirrell, M., Malone, F., 1977. Stress-induced diffusion of macromolecules. J. Polym. Sci. 15, 1569–1583.

Toffolo, A., Lazaretto, A., 2004. On the thermoeconomic approach to the diagnosis of energy system malfunctions indicators to diagnose malfunctions: application of a new indicator for the location of causes. Int. J. Thermodyn. 7, 41–49 (the special issue on thermoeconomic diagnosis).

Toffolo, A., Lazzaretto, A., 2002. Evolutionary algorithms for multi-objective energetic and economic optimization in thermal system design. Energy 27, 549–567.

Toffolo, A., Lazzaretto, A., 2003. A new thermoeconomic method for the location of causes of malfunctions in energy systems. In: Proceedings of IMECE 2003, ASME International Mechanical Engineering Congress and R&D Exposition, Washington, DC, USA, November 15–21, 2003. file 2003-42 689.

Townsend, K.N., 1992. Is the entropy law relevant to the economics of natural resource scarcity? J. Environ. Econ. Manage. 23, 96–100.

Tsatsaronis, G., 1984. Combination of exergetic and economic analysis in energy conversion processes. In: Energy Economics and Management in Industry, Proceedings of the European Congress, Algarve, Portugal, April 2–5. vol. 1. Pergamon Press, Oxford, England, pp. 151–157.

Tsatsaronis, G., 1987. A review of exergoeconomic methodologies. In: Moran, M., Sciubba, E. (Eds.), Second Law Analysis of Thermal System. ASME, New York, pp. 81–97.

Tsatsaronis, G., 1993. Thermodynamic analysis and optimization of energy systems. Prog. Energy Combust. Sci. 19, 227–257.

Tsatsaronis, G., 1999. Strengths and limitations of exergy analysis. In: Bejan, A., Mamut, E. (Eds.), Thermodynamic Optimization of Complex Energy Systems. Kluwer Academic Publishers, Dordrecht, pp. 93–100.

Tsatsaronis, G., 2007. Comments on the paper 'A brief commented history of exergy from the beginnings to 2004' by E Sciubba and G. Wall. Int. J. Thermodyn. 10, 187–190.

Tsatsaronis, G., Winhold, M., 1984. Thermoeconomic Analysis of Power Plants, EPRI AP-3651, RP 2029-8. Final Report. Electric Power Research Institute, Palo Alto, CA. August.

Tsatsaronis, G., Winhold, M., 1985a. Exergoeconomic analysis and evaluation of energy conversion plants. Part I. A new general methodology. Energy Int. J. 10, 69–80.

Tsatsaronis, G., Winhold, M., 1985b. Exergoeconomic analysis and evaluation of energy conversion plants. Part II. Analysis of a coal-fired steam power plant. Energy Int. J. 10, 81–94.

Tsatsaronis, G., Winhold, M., Stojanoff, C.G., 1986. Thermoeconomic Analysis of a Gasification-Combined-Cycle Power Plant. EPRI AP-4734, RP 2029-8, Final Report. Electric Power Research Institute, Palo Alto, CA. August.

Tsatsaronis, G., Lin, L., Pisa, J., Tawfik, T., 1991. Thermoeconomic design optimization of a KRW-based IGCC power plant. In: Final Report Submitted to Southern Company Services and the U.S. Department of Energy, DE-FC21-89MC26019. Center for Electric Power, Tennessee eTechnological University. November.

Tsatsaronis, G., Pisa, J., Lin, L., Tawfik, T., Sears, R.E., Gallaspy, D.T., 1992. Thermoeconomics in search of cost-effective solutions in IGCC Power Plants (invited paper). In: Coal Energy and the Environment, Proceedings of the Ninth Annual International Pittsburgh Coal Conference, Pittsburgh, Pennsylvania, October 12–16, pp. 369–377.

Tsatsaronis, G., Lin, L., Pisa, J., 1993. Exergy costing in exergoeconomics. J. Energy Resour. Technol. 115, 9–16.

Tsatsaronis, G., Tawfik, T., Lin, L., 1994. Exergetic comparison of two KRW-based IGCC power plants. J. Eng. Gas Turbines Power 116, 291–299.

Tsirlin, A.M., Kazakov, V., 2002a. Finite-time thermodynamics: limiting possibilities of irreversible separation processes. J. Phys. Chem. 106 (R.S. Bery's issue), 10926–10936.

Tsirlin, A.M., Kazakov, V., 2002b. Realizability areas for thermodynamic systems with given productivity. J. Non-Equilib. Thermodyn. 27, 91–103.

Ulanowicz, R.E., 1997. Ecology, the Ascendent Perspective. Columbia University Press, New York.

Upreti, S.R., Deb, K., 1997. Optimal design of an ammonia synthesis reactor using genetic algorithms. Comput. Chem. Eng. 1 (21), 87–92.

Valero, A., 1998. Thermoeconomics as a conceptual basis for energy ecological analysis. In: Proceedings of the 1st International Workshop on Advanced Energy Studies, Porto Venere, Italy.
Valero, A., 1999. Qualifying irreversibilities through second law: exergy accounting. In: Gordon Research Conference: Modern Developments in Thermodynamics 18–23, April. Il Ciocco, Tuscany, Italy.
Valero, A., 2003. Thermodynamic process of cost formation. In: Encyclopedia of Life-Support Sciences. EOLSS, London.
Valero, A., Lozano, M.A., Munoz, M., 1986a. A general theory of exergy saving: I. On the exergetic cost. In: Gaggioli, R.A. (Ed.), Computer-Aided Engineering and Energy Systems, vol. 3: Second Law Analysis and Modelling. AES vol. 2–3: ASME Book no. H0341C. ASME, New York, pp. 1–8.
Valero, A., Munoz, M., Lozano, M.A., 1986b. A general theory of exergy saving: II. On the thermoeconomic cost. In: Gaggioli, R.A. (Ed.), Computer-Aided Engineering and Energy Systems, vol. 3: Second Law Analysis and Modelling. AES vol. 2–3: ASME Book no. H0341C. ASME, New York, pp. 9–15.
Valero, A., Munoz, M., Lozano, M.A., 1986c. A general theory of exergy saving: III. Energy saving and thermoeconomics. In: Gaggioli, R.A. (Ed.), Computer-Aided Engineering and Energy Systems, vol. 3: Second Law Analysis and Modelling. AES vol. 2–3: ASME Book no H0341C. ASME, New York, pp. 17–21.
Valero, A., Lozano, M.A., Alconchel, J.A., Munoz, M., Torres, C., 1986d. GAUDEAMO: a system for energetic/exergetic optimization of coal power plants. In: Gaggioli, R.A. (Ed.), Computer-Aided Engineering and Energy Systems, vol. 3: Second Law Analysis and Modelling. AES vol. 2–3: ASME Book no H0341A. ASME, New York, pp. 43–49.
Valero, A., Torres, C., Lozano, M.A., 1989. On the unification of thermoeconomic theories. In: Boehm, R.F., El-Sayed, Y.M. (Eds.), Simulation of Thermal Energy Systems, HTD vol. 124. ASME Book no H00527. ASME, New York, pp. 63–74.
Valero, A., Torres, C., Serra, L., 1992. A general theory of thermoeconomics: I. Structural analysis; II. The relative free energy function. In: Valero, A., Tsatsaronis, G. (Eds.), Proceedings of the International Symposium on Efficiency, Costs, Optimization and Simulation of Energy Systems ECOS' 92, Zaragoza, Spain, June 15–18, pp. 137–154.
Valero, A., Serra, L., Lozano, M.A., 1993. Structural theory of thermoeconomics. In: Richter, R. (Ed.), Proceedings of the Symposium on Thermodynamics and the Design, Analysis and Improvement of Energy Systems, November 28–December 3, New Orleans, USA. ASME Book no H00527. ASME, New York, pp. 1–10.
Valero, A., Lozano, M.A., Serra, L., Tsatsaronis, G., Pisa, J., Frangopoulos, C., Von Spakovsky, M.R., 1994a. CGAM problem: definition and conventional solution. Energy 19, 279–286.
Valero, A., Lozano, M.A., Torres, L.C., 1994b. Application of the exergetic cost theory to the CGAM problem. Energy 19, 365–381.
Valero, A., Correas, L., Lazaretto, A., Rangel, V., Reini, M., Taccani, R., Toffolo, A., Verda, V., Zaleta, A., 2004. Thermoeconomic philosophy applied to the operating analysis and diagnosis of energy utility systems. Int. J. Thermodyn. 7, 33–39 (the special issue on thermoeconomic diagnosis).
Valero, A., Serra, L., Uche, J., 2006. Fundamentals of exergy cost accounting and thermoeconomics. Part I: theory. J. Energy Resour. Technol. 128, 1–8.
Verda, V., Zaleta, A., 2004. Thermoeconomic philosophy applied to the operating analysis and diagnosis of energy utility systems. Int. J. Thermodyn. 7, 33–39 (the special issue on thermoeconomic diagnosis).
Vitasari, C.R., Jurascik, M., Ptasinski, K.J., 2011. Exergy analysis of biomass-to-synthetic natural gas (SNG) process via indirect gasification of various biomass feedstock. Energy 36 (6), 3825–3837.
Von Spakovsky, M.R., 1993. Aspects of the thermoeconomic modelling of energy systems with cogeneration. Rev. Gen. Therm. Fr. 383, 594–605.
Von Spakovsky, M.R., 1994. Application of engineering functional analysis to the analysis and optimization of the CGAM problem. Energy 19, 343–364.
Von Spakovsky, M.R., Evans, R.B., 1990. The design and performance optimization of thermal systems. J. Energy Resour. Technol. Trans. ASME 112.
Von Spakovsky, M.R., Frangopoulos, C.A., 1993. A global environomic approach for energy systems analysis and optimization – II. In: Proceedings of the International Conference Energy Systems and Ecology ENSEC' 93, Cracow, Poland, July 5–9, pp. 133–144.
Williamson, A.G., 1993. The second law of thermodynamics and the economic process. Ecol. Econ. 7, 69–71.

Yan, Z., 1993. Comment on "Ecological optimization criterion for finite-time heat engines." J. Appl. Phys. 73, 3583.
Yantovski, Y., 2007a. Comments on the paper 'A Brief Commented History of Exergy from the Beginnings to 2004' by E. Sciubba and G. Wall. Int. J. Thermodyn. 10, 193–194.
Yantovski, Y., 2007b. Review of the book: J. Szargut's book: exergy method: technical and ecological applications, WIT Press, Southampton, Boston, 2005. Int. J. Thermodyn. 10, 93–95.
Young, J.T., 1991. Is the entropy law relevant to the economics of natural resource scarcity? J. Environ. Econ. Manage. 21, 169–179.
Yu, H., Fang, H., Yao, P., Yuan, Y., 2000. A combined genetic/simulated annealing algorithm for large scale system energy integration. Comput. Chem. Eng. 24, 2023–2035.
Zaleta-Aguilar, A., Gallegos-Munoz, A., Rangel-Hernandez, V., Capilla, A.V., 2004. A reconciliation method based on a module simulator an approach to the diagnosis of energy system malfunctions. Int. J. Thermodyn. 7, 51–60 (the special issue on thermoeconomic diagnosis).
Zangwill, W.J., 1974. Nonlinear Programming. Państwowe Wydawnictwa Naukowe, Warsaw.
Ziębik, A., 1986. Mathematical Model of an Energy Balance for the Foredesign of the Energy Management of Industrial Plants. Problemy Projektowe: No 3 (in Polish).
Ziębik, A., 1990a. Mathematical Modelling of Energy Management Systems in Industrial Plants. Ossolineum, Wrocław.
Ziębik, A., 1990b. Energy Systems. Technical University of Silesia, Gliwice (in Polish).
Ziębik, A., 1991a. Mathematical model of industrial energy systems and its applications. In: Proceedings of the International Conference Analysis of Thermal and Energy Systems, Athens, pp. 745–756.
Ziębik, A., 1991b. Examples of Calculation Concerning Energy Systems. Technical University of Silesia, Gliwice (in Polish).
Ziębik, A., 1995. Applications of second law analysis in industrial energy and technological systems. In: Proceedings of the Conference on Second Law Analysis of Energy Systems: Towards the 21 st Century, Romepp. 349–359.
Ziębik, A., 1996. System analysis in thermal engineering. Arch. Thermodyn. 17, 81–97.
Ziębik, A., Presz, K., 1993. System method of the choice of the energy management structure of an industrial plant. Arch. Energ. 172 (in Polish).
Ziębik, A., Gwóźdź, J., Presz, K., 1994. Matrix method of calculating the unit costs of energy carriers as a coordination procedure in the optimization of industrial energy systems. In: Proceedings of the Second Biennial European Joint Conference on Engineering Systems Design and Analysis ASMB, London, pp. 19–26.

Further reading

Valero, A., Serra, L., Uche, J., 2005a. Fundamentals of exergy cost accounting and thermoeconomics. Part I: theory. J. Energy Resour. Technol. 128 (1), 1–8.
Valero, A., Serra, L., Uche, J., 2005b. Fundamentals of exergy cost accounting and thermoeconomics Part II: applications. J. Energy Resour. Technol. 128 (1), 9–15.

CHAPTER 7

Optimizing recovery of energy and resources: Theory and selected applications

7.1 Introduction

This chapter describes selected theoretical and numerical tools which are applied in the trajectory optimization of dynamic production problems, with special attention paid to nonlinear and complex systems. Numerical methods often provide the only way to solve the problem because in many cases governing Hamilton-Jacobi-Bellman (HJB) equations not only cannot be solved analytically, but may do not admit classical solutions. In the latter case, the viscosity solutions remain the only option (Crandall and Lions, 1983). Systems with nonlinear kinetics, e.g., radiation, are particularly difficult; therefore, discrete counterparts of continuous HJB equations and numerical approaches are useful. Discrete algorithms of dynamic programming (DP), which lead to work limits and generalized availabilities, are especially effective and helpful.

We consider convergence of discrete algorithms to viscosity solutions of HJB equations, solutions by discrete approximations, and the role of Lagrange multiplier λ associated with duration constraint. In the analytical discrete schemes, the Legendre transformation is a significant tool leading to original work function. We also describe numerical algorithms of dynamic programming and consider dimensionality reduction in these algorithms. Finally, we outline some applications of the method for other systems, in particular chemical energy systems.

Analytical and computational aspects of energy limits refer to fluids which are restricted in their amount or flow, and as such, play role of resources. In the practical processes of engineering and technology, the resource is a useful, valuable substance of a limited amount or flow. The value of the resource can be quantified thermodynamically by specifying its exergy or a maximum work which can be delivered when the resource is downgraded to the equilibrium with the environment. Reversible relaxation of the resource is associated with the classical exergy. When some dissipative phenomena are allowed, generalized exergies are obtained. These exergies incorporate both a limited availability and a minimum work supply for the resource's production. In the classical exergy, only the first property is taken into account (Sieniutycz, 2003a).

Complexity and Complex Thermo-Economic Systems. https://doi.org/10.1016/B978-0-12-818594-0.00007-6

Knowledge of a power integral is required when calculating the exergy or exergy-like functions. A generalized exergy follows as the extremum of this power integral subject to the exergy initial conditions. In thermal systems, this integral involves the product of thermal efficiency and the differential of exchanged energy. Various process models lead to diverse formulas for thermal efficiencies, which show how efficiency of a practical system deviates from the Carnot efficiency. In fact, the generalized exergies quantify somehow these deviations. Formally, an exergy follows from the principal function of variational problem of extremum work (maximum of work delivered or minimum of work supplied) subject to the boundary condition of equilibrium with the environment. Variational solution to the work extremum problem contains an exergy-type function as its principal component. Other components are the optimal trajectory and optimal control. In thermal systems, the trajectory is characterized by the temperature of the resource fluid $T(t)$, whereas the control variable may be efficiency η or Carnot temperature $T'(t)$. The latter quantity, defined as $T' \equiv T_2T_{1'}/T_{2'}$ (e.g., Sieniutycz, 2003b, 2007b), is particularly suitable in describing of driving forces in energy systems. Whenever $T'(t)$ differs from $T(t)$, the resource relaxes to the environment with a finite rate determined by the efficiency deviation from the Carnot efficiency. Only when $T'(t)=T(t)$, the efficiency is Carnot, but this corresponds with an infinitely slow relaxation rate of the resource to the thermodynamic equilibrium.

In Chapter 6 of a book (Sieniutycz and Jeżowski, 2018), we have presented analytical aspects of Hamilton-Jacobi-Bellman (HJB) and Hamilton-Jacobi (HJ) equations for nonlinear thermal systems with power generation. Solutions to these equations have been obtained for systems with linear kinetics. Rate-dependent exergies, generalizing classical exergies for systems with finite durations, have been evaluated. Processes bounded by these generalized exergies involve various imperfect phenomena as, e.g., heat conduction or nonideal compression and expansion. In process, modes departing from the equilibrium generalized exergies are larger than in their inversions approaching the equilibrium. Corresponding bounds for the mechanical energy yield or consumption are stronger than those defined by the classical exergy (enhanced bounds).

In order to obtain generalized exergies (or corresponding functions describing energy limits), one has to solve appropriate HJB equations. The problem is not only that most of optimal solutions cannot be obtained in the form of explicit analytical formulae. In fact, the HJB equations may not admit classical solutions at all, and one has to work with viscosity solutions (Section 7.3). This difficulty especially refers to systems with nonlinear kinetics (e.g., radiation systems). To overcome the difficulty, discrete counterparts of continuous equations are solved in numerical approaches to HJB equations. Especially, a few forms of discrete dynamic programming algorithms are efficient to solve continuous HJB equations.

Discrete Hamiltonians H^n are generally not constants of discrete autonomous paths. The motivation behind this chapter is to show that, even if the constancy of a H^n is violated (e.g., by complex discretization), the Lagrange multiplier λ of the duration constraint still exhibits the constancy property. Therefore, it is the constant λ that is capable of decreasing the discrete problem dimensionality, and thus, to simplify the numerical optimization procedure. In fact, in the presented examples, the Lagrange multiplier λ takes over the role of the energy integral, and this is what makes the problem interesting.

Section 7.2 of the present chapter shows an example of a continuous model of power production from the black radiation and difference equations following from the model's discretization. Section 7.3 discusses the constancy preservation conditions for discrete Hamiltonians and convergence conditions of discrete numerical schemes to viscosity solutions of HJB equations, whereas Section 7.4 presents a dynamic programming equation for power yield from radiation. Section 7.5 elucidates solving method by discrete approximations and introduces time adjoint as a Lagrange multiplier. Mean and local intensities of discrete processes are discussed in Section 7.6. Section 7.7 shows the significance of Legendre transform in recovering original work functions. Section 7.8 describes numerical procedures using dynamic programming, whereas Section 7.9 discusses dimensionality reduction in numerical DP algorithms. Section 7.10 presents most essential conclusions.

7.2 A discrete model for a nonlinear problem of maximum power from radiation

In the previous chapter, we considered HJB equations for continuous, active (work producing) systems working with finite rates. Here, we shall focus on discrete schemes solving these continuous equations, either analytically, via discrete approximations, or numerically, with the help of a computer.

Let us first recall a representative problem of minimum work consumed in a system subject to constraints imposed on dynamics and duration. Consider a dynamical radiation system, characterized by highly nonlinear kinetics (Sieniutycz and Kuran, 2005, 2006). For a *symmetric model* of power yield from radiation (both reservoirs consist of radiation), a suitable power integral is (Sieniutycz and Kuran, 2006)

$$\dot{W} = \int_{t^i}^{t^f} \dot{G}_c(T)\left(1 - \frac{\Phi T^e}{T'}\right)\beta\frac{T^a - T'^a}{\left(\Phi'(T'/T^e)^{a-1} + 1\right)T^{a-1}}dt \tag{7.1}$$

In the physical space, power exponent $a=4$ for radiation and $a=1$ for a linear resource. The integrand of Eq. (7.1) represents intensity of generalized profit, f_0. In the engine mode, integral (7.1) has to be maximized subject to the dynamical constraint ("state equation")

$$\frac{dT}{dt} = -\beta\frac{T^a - T'^a}{\left(\Phi'(T'/T^e)^{a-1} + 1\right)T^{a-1}} \tag{7.2}$$

As it follows from the general theory (Chapter 2 in Sieniutycz and Jeżowski, 2018), the extremum conditions for the optimization problem involving Eqs. (7.1), (7.2) are contained in the HJB equation

$$-\frac{\partial V}{\partial t}+\max_{T'(t)}\left\{\left(\dot{G}_c(T)\left(1-\Phi\frac{T^e}{T'}\right)+\frac{\partial V}{\partial T}\right)\beta\frac{T^a-T'^a}{\left(\Phi'(T'/T^e)^{a-1}+1\right)T^{a-1}}\right\}=0 \tag{7.3}$$

where $V=\max \dot{W}$. Since it is impossible to solve this equation analytically, except for the case when $a=1$, we shall describe a way for numerical solving based on the Bellman's method of dynamic programming (DP).

Considering computer needs, we introduce a related discrete scheme

$$\dot{W}^N=\sum_{k=1}^{N}\dot{G}_c(T^k)\cdot\left(1-\frac{\Phi T^e}{T'^k}\right)\beta\frac{T^{ka}-T'^{ka}}{\left(\Phi'\left(T'^k/T^e\right)^{a-1}+1\right)T^{ka-1}}\theta^k \tag{7.4}$$

$$T^k-T^{k-1}=\theta^k\beta\frac{T'^{ka}-T^{ka}}{\left(\Phi'\left(T'^k/T^e\right)^{a-1}+1\right)T^{ka-1}} \tag{7.5}$$

$$\tau^k-\tau^{k-1}=\theta^k \tag{7.6}$$

We search for a maximum of the sum (7.4), subject to discrete constraints (7.5) and (7.6). Our first task is to define conditions when the numerical schemes of dynamic programming for the set (7.4)–(7.6) converge to solutions of the Hamilton-Jacobi-Bellman equation (7.3). The following section analyses this problem in general terms (i.e., for arbitrary profit functions and constraints).

7.3 Nonconstant Hamiltonians and convergence of discrete DP algorithms to viscosity solutions of HJB equations

Conditions determining when *discrete* optimization schemes converge to solutions of Hamilton-Jacobi-Bellman equations (HJB equations) are quite involved. Moreover, in the present approach, they have to be linked with systematic studies of the problem of the Hamiltonian constancy preservation (Jordan and Polak, 1964a,b; Sieniutycz, 1973; Szwast, 1979, 1988; Marsden, 1992; McLachlan, 1993; Quispel, 1995; Berry et al., 2000; Sieniutycz, 2006b). For discrete autonomous models, the Hamiltonian constancy means that $H^n=H^{n-1}$; for nonautonomous ones, the constancy refers to the enlarged Hamiltonian involving the time adjoint. The problem is quite old and started with traditional mathematics yet before viscosity solutions of HJB equations were discovered and worked out. The theory of viscosity solutions (Crandall and Lions, 1983; Crandall et al., 1992) is, nowadays, the main theory concerning HJ and HJB equations. Its basic premise is that, in general, value functions (optimal performance functions) are nonsmooth and hence solutions are sought in the viscosity sense. The computation typically amounts to a suitable discretization and the use of discrete dynamic programming. The literature on this subject is wide, e.g., books by Barles (1994) and Bardi and Capuzzo-Dolcetta (1997). Nowadays, the theory for a single Hamilton-Jacobi equation is quite complete and ensures existence and uniqueness results for nondifferentiable and even discontinuous solutions. In parallel with the strong theoretical activity, many approximation techniques have been proposed to date, so that the numerical analysis of Hamilton-Jacobi equations has also experienced a fast growth (Barles and Souganidis, 1991; Falcone and Makridakis, 2001; Falcone and Ferretti, 2006). This progress has also an impact on the rectification and improvement of the optimization algorithms of the discrete maximum type for finite-difference control systems (Mordukhovich, 1988, 1995; Mordukhovich and Shvartsman, 2004). Moreover, since the work of Ishii (1985) on discontinuous Hamiltonians, it was known that HJ and HBJ equations with discontinuous ingredient may arise in some continuous systems, e.g., propagations of fronts in inhomogeneous media, geometric optics, in presence of layers and shape-from-shading problems (Coclite and Risebro, 2003, 2005, 2007). Hamilton-Jacobi equations are now applied whose coefficients have some kind of singularity and the Hamiltonian depends discontinuously on both the spatial and temporal location. The main results are the existence and well-posedness of a viscosity solution to the Cauchy problem (Tourin, 1992; Ishii and Ramaswamy, 1995). These works show that if the Hamiltonian H satisfies some structural conditions, essentially amounting to discontinuous jumps "in one direction," there exists a unique solution to the initial value problem (Capuzzo-Dolcetta and Perthame, 1996; Stromberg, 2003). In particular, for so-called shape-from-shading applications, Ostrov (1999, 2000) shows that a sequence of approximations defined by smoothing the coefficients converges to the unique viscosity solution defined as the solution of an auxiliary control problem. The technique of using the associated control problem (for convex Hamiltonians) is also exploited (Dal Maso and Frankowska, 2002; Karlsen et al., 2003) when studying some Hamilton-Jacobi equations with a discontinuous Hamiltonian. Connection with the conservation laws is also shown, and a quite general theory for conservation laws with discontinuous fluxes is established (Jin and Wen, 2005), with an entropy condition obtained for nonlinear parabolic equations with discontinuous coefficients.

Yet, whenever H is the total energy of an autonomous system, it has often been argued that its constancy must be preserved for physical reasons. In a series of work, Jin and his coworkers have considered Hamiltonian-preserving schemes for the Liouville equation with discontinuous potentials (Jin and Wen, 2005, 2006a,b; Jin and Liao, 2006). Their designing principle is to build the behavior of waves at the interface—either cross over with a changed velocity according to a constant Hamiltonian, or be reflected with a negative velocity (or momentum)—into the numerical flux. These schemes are the Hamiltonian-preserving schemes. In general, they build into the numerical flux the wave scattering information at the interface and use the Hamiltonian-preserving principle to couple the wave numbers at both sides of the interface. The constancy of H and the canonical structure of some discrete optimization algorithms (Jordan and Polak, 1964a,b; Szwast, 1988; Sieniutycz, 2006a,b) also conform to present tendencies in physics in constructing numerical integration schemes for ordinary differential equations (ODEs) in such a way that a qualitative property of the solution of the ODE is exactly preserved. For example, there are Poisson-structure-preserving integration schemes (symplectic integrators), symmetries and related invariants, etc. The optimal choice of free time intervals θ, necessary for the constancy of a discrete H, may be compared with the group of special-purpose integration methods collectively called the structure-preserving integrators (also called mechanical or geometric integrators). In these methods, local discretizing structure may be established without explicit recourse to an optimization criterion, although it has to preserve (exactly) a number of important properties known for OD equations (Marsden, 1992; McLachlan, 1993; Quispel, 1995; Quispel and Dyt, 1998; McLachlan and Quispel, 1998; McLachlan et al., 1999).

The preference of differential schemes with continuous H follows also from the fact that the "presently classical" theory for viscosity solutions of Hamilton-Jacobi equations (Crandall and Lions, 1983; Crandall et al., 1992) does not cover the case where the Hamiltonian H is discontinuous. This is because the straightforward method of comparing sub- and supersolutions does not work if H is discontinuous in x or t. Therefore, the H-preserving models contain the very core that may be pursued nowadays in the modern context. This core refers to the energy conservation in discontinuous systems, the property that is important when H has the meaning of physical energy (or steady energy flux) of an autonomous system, where the energy must be constant of an optimal path. This problem also appears in the context of variable resolution discretizations for high-accuracy solutions of optimal control problems (Munos and Moore, 2001; Falcone and Ferretti, 1994; Fialho and Georgiu, 1999). The latter of these works develops a variational approach involving the theory of viscosity solutions of Hamilton-Jacobi equations (Crandall and Lions, 1983) to establish the convergence of the Euler approximation schemes via discrete dynamic programming. Fialho and Georgiu (1999) also provide an algorithm to compute upper bounds for value functions and give the proof of convergence of approximation schemes for the control problem. Their case studies assess the robustness of a feedback system and the quality of trajectory tracking in the presence of uncertainty.

The attitude of Hamiltonian preservation has generated both followers and adversaries. The former claim that, in differential systems, H must be constant in an autonomous case because of physical symmetry requirements and resulting conservation laws, the latter admit violations of H preservation as the mathematical possibility supported by analyses of solutions. Sometimes, even the followers of the Hamiltonian preservation admit systems with discontinuous Hamiltonians. For example, in a paper by Jin et al. (2007), Hamiltonian is piecewise smooth with finitely many discontinuities. The problem arises when classical particles move through media which contain barriers or interfaces at which the Hamiltonian is discontinuous. It also arises in the Lagrangian description of geometrical optics, or in the propagation of the linear high frequency waves through interfaces. In fact, a vast research material has been collected to date showing that the constancy of Hamiltonians for both continuous and discrete models may not be preserved.

One may ask the question: how the physics allows for and manages the violation of the Hamiltonian constancy? The answer, stemming from the present reasoning and some earlier paper (Sieniutycz, 2006a,b), is simple: the Hamiltonian describes the system energy only when the HJB and HJ equations exist, whenever they do not (as in systems with discontinuous solutions to the value function), the role of the physical energy is taken over by another quantity. This quantity is the Lagrange multiplier of the process duration constraint, the time adjoint which is constant in both discrete and continuous autonomous systems. In other words, Hamiltonians represent the energy only in limited situations. Hamiltonians of discrete processes (even Pontryagin's of energy-like structure) are particularly susceptible for the violation of the Hamiltonian constancy. An arbitrary discretization mode of an original differential model violates the constancy preservation of the Hamiltonian (Sieniutycz, 2006a, b), and only discrete models linear with respect to the time interval $\Delta t^n \equiv \theta^n$ and without local constraints imposed on θ^n can admit constant discrete Hamiltonians. Importantly, when these conditions are not satisfied, the Lagrange multiplier λ of the duration constraint still exhibits the constancy property. Therefore, the motivation behind this chapter is to show how to implement this Lagrange multiplier λ to decrease the problem dimensionality and to simplify the numerical procedure of optimization.

To outline discrete structures and extremum conditions that may lead to either constant or nonconstant Hamiltonians, we consider a family of optimization models obtained by discretization of original continuous ones. In this case, one has to determine necessary optimality conditions of a general discrete process governed by, say, a work criterion W^N

$$W^N = \sum_{n=1}^{N} f_0(\mathbf{x}^n, t^n, \mathbf{u}^n, \theta^n)\theta^n \tag{7.7}$$

subject to constraints resulting from difference equations

$$x_i^n - x_i^{n-1} = f_i(\mathbf{x}^n, t^n, \mathbf{u}^n, \theta^n)\theta^n \equiv f_i(\tilde{\mathbf{x}}^n, \mathbf{u}^n, \theta^n)\theta^n \tag{7.8}$$

The scalar f_0 is the rate of the profit generation. Superscripts refer to stages and subscripts to coordinates. The integer n ($n = 1...N$) is usually called discrete time, the entity that should be distinguished from continuous time t. The latter is usually the physical time (t is the chronological time in unsteady-state operations and holdup or residence time in steady cascade operations). Both n and t are monotonously increasing. The s-dimensional vector $\mathbf{x} = (x_1, ..x_s)$ is the state vector, and the r-dimensional vector $\mathbf{u} = (u_1...u_r)$ is the control vector, where $\mathbf{x}^n \in E^s$,

$\mathbf{u}^n \in E^r$, and rate functions f_0^n and f_i^n are continuously differentiable always in $\mathbf{x}$ and θ, but not always in $\mathbf{u}$. The rate change of state coordinate x_i in time t is ith component of s-dimensional vector of rates, $\mathbf{f}$. The change of time t through the stage n defined as $\theta^n = t^n - t^{n-1}$ is called the time interval.

Various discretization schemes for constraining differential equations as Eq. (7.2) lead to discrete models in which Eq. (7.8) may be either linear or nonlinear in time interval θ^n. While θ^n is a control-type quantity, it is excluded from the coordinates of vector $\mathbf{u}$, i.e., it is treated separately in the optimization model. The $s+1$-dimensional vector $\widetilde{\mathbf{x}} = (\mathbf{x}, t)$ is also used which is the enlarged state vector describing the space-time. Usually, one assumes that a control sequence $\{\mathbf{u}^n\}$ and the corresponding trajectory $\{\mathbf{x}^n, t^n\}$ are admissible, i.e., that they satisfy the control constraint $u^n \in U^n$ and the state-space constraint $\widetilde{x}^n \in \widetilde{X}^n$.

For work production problems criterion, (7.7) is maximized, for work consumption a minimum of (7.1) is sought. In optimization problems with constrained duration $t^N - t^0$ (the so-called fixed-horizon problem), discrete model must explicitly include an equation defining time interval θ^n, either as the increment of a monotonously increasing state coordinate satisfying an equation $x_{s+1}^n - x_{s+1}^{n-1} = \theta^n$ or as the increment of "usual" time

$$t^n - t^{n-1} = \theta^n \tag{7.6}$$

The monotonic increase of the time-like coordinate, implying nonnegative θ at each stage n, is crucial for many properties of model (7.7)–(7.9). Throughout this chapter, models with free (unconstrained) intervals θ^n are considered as only those that are able to achieve their own continuous limit.

Two classes of discrete models, linear and nonlinear in free θ^n, should be distinguished when considering convergence of their optimality conditions to viscosity solutions of continuous Hamilton-Jacobi-Bellman (HJB) equations. In the first class, HJB equations follow straightforwardly from the optimality conditions. In the second class, a condition of weak nonlinearity of the discrete rates with respect to θ^n (discussed below) must be satisfied.

In the second class, the Hamiltonian of an autonomous system is not constant (as in the first class), but satisfies the condition

$$H^{n-1} + \theta^n \left(\partial H^{n-1} / \partial \theta^n\right) = H_1^{n-1} = \lambda \tag{7.9}$$

where H_1^{n-1} is the Hamiltonian of the first class. Below we outline the optimization scheme that substantiates the above condition and its further use in the dimensionality reduction of the dynamic programming algorithms (Bellman, 1961; Aris, 1964).

In each case, some locally sufficient optimality criteria are valid in discrete systems ("CBS criteria": Rund, 1966; Boltyanski, 1969; Sieniutycz, 1991). For a value function V^n (maximum work function in the radiation example)

$$V^n \equiv \max \sum_k^n f_0^k(\mathbf{x}^k, t^k, \mathbf{u}^k, \theta^k)\theta^k \tag{7.10}$$

a general optimality criterion at stage n that uses the θ-dependent rates has the form

$$0 = \max_{\mathbf{u}^n, \theta^n, \mathbf{x}^n, t^n} \{f_0^n(\mathbf{x}^n, t^n, \mathbf{u}^n, \theta^n)\theta^n - \left(V^n(\mathbf{x}^n, t^n) - V^{n-1}((\mathbf{x}^n - \mathbf{f}^n(\mathbf{x}^n, t^n, \mathbf{u}^n, \theta^n)\theta^n, t^n - \theta^n)\right\} \tag{7.11}$$

In what follows we shall also use functions of costs type, local costs $l_0 = -f_0$, and integral optimal costs $R = -V$.

For a special mode of discretization with θ-free rates $\mathbf{f}$ criterion, Eq. (7.11) refers to the "standard" or "canonical" model associated with all participating equations linear in θ^n. It may then be shown that these models are characterized by a constant Hamiltonian which satisfies a discrete *HJB* equation whose discrete value function converges under the reasonable assumption of the absence of local constraints on θ^n to the unique viscosity solution which is the value function of the original problem. Optimality criterion (7.11) can be applied to derive a set of practical (Hamiltonian-based) optimality conditions including those with respect to $\mathbf{x}^n$ and time t^n.

Using Eq. (7.11), one can pass to an algorithm of a discrete maximum principle and the related canonical equations. This issue is only briefly discussed here; for our purposes a dynamic programming equation (see below) is most essential as it provides direct information about the extremum performance function. Importantly, for applications in power systems, Bellman's equation of dynamic programming follows in its "forward" form from criterion (7.11).

Below, we present a set of special optimality conditions for the process described by Eqs. (7.7)–(7.11) and characterized by a differentiable V. Convexity of rate functions and constraining sets is assumed. From these conditions, those for subsystems linear in θ^n easily follow for θ-independent rates and the Hamiltonian.

Optimizing time intervals θ^n and controls $\mathbf{u}^n$ of Eq. (7.11) in the interior of the admissible control set leads, respectively, to two extremum conditions:

$$\begin{aligned} &-\partial V^{n-1}/\partial t^{n-1} + f_0^n(\mathbf{x}^n, t^n, \theta^n, \mathbf{u}^n) - (\partial V^{n-1}/\partial \mathbf{x}^{n-1}) \cdot \mathbf{f}^n(\mathbf{x}^n, t^n, \mathbf{u}^n, \theta^n) \\ &+\theta^n \partial\left\{f_0^n(\mathbf{x}^n, t^n, \theta^n, \mathbf{u}^n) - (\partial V^{n-1}/\partial \mathbf{x}^{n-1}) \cdot \mathbf{f}^n(\mathbf{x}^n, t^n, \mathbf{u}^n, \theta^n)\right\}/\partial \theta^n = 0 \end{aligned} \tag{7.12}$$

and

$$\partial\left\{f_0^n(\mathbf{x}^n, t^n, \theta^n, \mathbf{u}^n) - \left(\partial V^{n-1}/\partial \mathbf{x}^{n-1}\right) \cdot \mathbf{f}^n(\mathbf{x}^n, t^n, \mathbf{u}^n, \theta^n)\right\}/\partial \mathbf{u}^n = 0 \tag{7.13}$$

For convex functions and constraining set $u^n \in U^n$, stationarity condition (7.13) allows to find optimal control $\mathbf{u}^n$ from the maximum condition for following Hamiltonian expression

$$\mathbf{u}^n = \arg\max_{\mathbf{u}^n} \left\{ -\partial V^{n-1}/\partial t^{n-1} + f_0^n(\mathbf{x}^n, t^n, \theta^n, \mathbf{u}^n) - \left(\partial V^{n-1}/\partial \mathbf{x}^{n-1}\right) \cdot \mathbf{f^n}(\mathbf{x}^n, t^n, \theta^n, \mathbf{u}^n) \right\} \quad (7.14)$$

Discrete HJB equations It follows from Eqs. (7.12), (7.13) that a discrete *HJB* equation for the optimum profit function

$$-\partial V^{n-1}/\partial t^{n-1} + \max_{\mathbf{u}^n} \left\{ f_0^n(\mathbf{x}^n, t^n, \mathbf{u}^n) - \left(\partial V^{n-1}/\partial \mathbf{x}^{n-1}\right) \cdot \mathbf{f^n}(\mathbf{x}^n, t^n, \mathbf{u}^n) \right\} = 0 \quad (7.15)$$

or, an equivalent equation, written in terms of the cost functions $R = -V$ and $l_0 = -f_0$,

$$\partial R^{n-1}/\partial t^{n-1} + \max_{\mathbf{u}^n} \left\{ -l_0^n(\mathbf{x}^n, t^n, \mathbf{u}^n) + \left(\partial R^{n-1}/\partial \mathbf{x}^{n-1}\right) \cdot \mathbf{f^n}(\mathbf{x}^n, t^n, \mathbf{u}^n) \right\} = 0 \quad (7.16)$$

can be obtained for convex models linear in θ^n [this is why variable θ^n is absent in rates of Eqs. (7.15), (7.16)]. However, Eq. (7.12) proves that the classical Hamilton-Jacobi-Bellman structure is not attained when the discrete model is nonlinear in time intervals θ^n. Eq. (7.14) describes, in fact, a maximum principle with respect to $\mathbf{u}^n$ for a "Hamiltonian," i.e., the expression in braces of this equation. The principle is here written for the so-called enlarged Hamiltonian which includes the partial time derivative of V.
The above findings are summarized by the following lemma.

Lemma (maximum principle) *Let control constraint $u^n \in U^n$ and state-space constraint $\tilde{x}^n \in \tilde{X}^n$ are convex sets. Then the necessary condition for the maximum of criterion W^N with respect to the control sequence $\{\mathbf{u}^n\}$ is the maximum for the Hamiltonian expression*

$$H^{n-1} \equiv -l_0^n(\mathbf{x}^n, t^n, \mathbf{u}^n) + \left(\partial R^{n-1}/\partial \mathbf{x}^{n-1}\right) \cdot \mathbf{f^n}(\mathbf{x}^n, t^n, \mathbf{u}^n) \quad (7.17)$$

which appears in the discrete Hamilton-Jacobi-Bellman equations of the optimization problem, Eq. (7.15) *or* (7.16). *The proof is immediate consequence of Eqs.* (7.12), (7.13), *and Eq.* (7.14) (*the latter follows from Eqs.* (7.11), (7.13) *under convexity assumptions*). *Note that while the maximum of H^{n-1} holds in the state-space $\mathbf{x}^n$, the lemma leads directly to the maximum principle in the phase space, where phase-space Hamiltonian is defined as follows*

$$H^{n-1}\left(\mathbf{x}^n, z^{n-1}, t^n, \mathbf{u}^n, \theta^n\right) \equiv -l_0^n(\mathbf{x}^n, t^n, \mathbf{u}^n, \theta^n) + \sum_{i=1}^{s} z_i^{n-1} f_i^n(\mathbf{x}^n, t^n, \mathbf{u}^n, \theta^n) \quad (7.18)$$

The phase-space Hamiltonian (*of energy type*) *is the sum of profit intensity $f_0^n = -l_0^n$ and the scalar product of state adjoints*

$$z_k^{n-1} \equiv \partial R^{n-1}/\partial x_k^{n-1} \quad (k = 1, 2, \ldots s) \quad (7.19)$$

and corresponding rates f_k^n. In the continuous limit and for problems of analytical mechanics, state adjoints are identical with the generalized momenta, yet the extremum Hamiltonian does not represent the system's energy when rates f_i^n contain explicitly time intervals θ^n. This is the price of arbitrary discretization.

Discrete Hamilton-Jacobi equations. When optimal controls θ^n and $\mathbf{u}^n$ are evaluated from Eqs. (7.12), (7.13) and next substituted into Eq. (7.12), the resulting equation constitutes a discrete counterpart of the Hamilton-Jacobi equation of continuous processes.

For models linear with respect to θ^n, one obtains

$$\partial R^{n-1}/\partial t^{n-1} + H^{n-1}\left(x_1^n, \ldots x_s^n, t^n, \partial R^{n-1}/\partial x_1^{n-1} \ldots \partial R^{n-1}/\partial x_s^{n-1}\right) = 0 \tag{7.20}$$

whereas for those nonlinear in θ^n Hamiltonian H^{n-1} is replaced by partial derivative $h_\theta \equiv \partial(\theta^n H^{n-1})/\partial\theta^n$

$$\partial R^{n-1}/\partial t^{n-1} + h_\theta{}^{n-1}\left(x_1^n, \ldots x_s^n, t^n, \partial R^{n-1}/\partial x_1^{n-1} \ldots \partial R^{n-1}/\partial x_s^{n-1}\right) = 0 \tag{7.21}$$

In both cases, optimal controls are expressed in terms of state coordinates and state adjoints. However strange, the replacement of H^{n-1} by the derivative $h_\theta \equiv \partial(\theta^n H^{n-1})/\partial\theta^n$ in the second case leads, in fact, to a valid structure which is explained in the enlarged space-time in terms of (vanishing) partial derivative of an enlarged Hamiltonian function with respect to free control, θ^n (Sieniutycz, 2006b). Note that both equations are nonlinear in terms of $\partial R^{n-1}/\partial \mathbf{x}^{n-1}$.

In the general case of nonsmooth solutions, the results discussed above lead to convergence conditions of discrete computational schemes to viscosity solutions of continuous Hamilton-Jacobi equations of physical processes. To formulate these convergence conditions for discrete models whose rates contain explicitly intervals θ, a notion of the weak dependence of the discrete Hamiltonian on θ is important. Note that the Hamiltonian is a weighted measure quantifying dependence of the discrete rate vector on time interval θ.

Definition The θ-differentiable Hamiltonian function H^{n-1} is said to be weakly dependent on θ^n in a vicinity of $0+$, if for any positive number ε, a positive number η exists, such that for sufficiently small positive $\theta^n \leq \eta$ the absolute value of the product $I\theta^n\ \partial(H^{n-1})/\partial\theta^n\ I \leq \varepsilon$, for $n = 1, 2..N$. This means that in a vicinity of $0+$ the reciprocal of derivative $\partial(H^{n-1})/\partial\theta^n$ tends to zero slower than θ^n itself, or that $\lim\ (\theta^n\, \partial(H^{n-1})/\partial\theta^n) = 0$ for $\theta^n \to 0+$ regardless the form of rate functions f_k. In fact, many popular discretization schemes lead to the discrete rates and Hamiltonians weakly dependent on θ^n for positive θ in a vicinity of the point $\theta = 0$.
For Hamiltonians weakly dependent on θ^n, the following corollary holds.

Corollary *Assume fixed end states ($\mathbf{x}^0, \mathbf{x}^N$), end times ($t^0, t^N$), and an arbitrarily large number M. Observe that, for a sufficiently large total number of stages $N > M$, each free interval θ^n is sufficiently close to zero, i.e., $\lim\theta^n = 0+$ for $N \to \infty$ as the consequence of the monotonic property of time coordinate t. Then for unconstrained intervals θ^n and each Hamiltonian function $H^{n-1}(x_1^n, \ldots x_s^n, t^n, -\partial V^{n-1}/\partial x_1^{n-1} \ldots - \partial V^{n-1}/\partial x_s^{n-1}, \theta^n)$ weakly dependent on θ^n in a vicinity of $0+$, classical Hamilton-Jacobi equation holds in the limit of $N \to \infty$.*

Proof For Hamiltonians H^{n-1} weakly dependent on θ^n in a vicinity of $0+$ and a sufficiently large N, the derivative $\partial(\theta^n H^{n-1})/\partial\theta^n$ is sufficiently close to H^{n-1} for each $n = 1, \ldots N$. In the limiting case of $N \rightarrow \infty$, the sequence of free θ^n satisfies the conditions: $\lim \theta^n = 0+$ and $\lim \partial(\theta^n H^{n-1})/\partial\theta^n = \lim H^{n-1} = H(\mathbf{x}, t, \partial R/\partial \mathbf{x})$. Eq. (7.21) then goes over into the Hamilton-Jacobi equation of a continuous process

$$\partial R/\partial t + H(x_1, \ldots x_s, t, \partial R/\partial x_1 \ldots \partial R/\partial x_s) = 0 \tag{7.22}$$

The above corollary assures that the limiting Hamilton-Jacobi equation of a continuous system is obtained in unique form (7.22), regardless the discretization mode and no matter if the underlying discrete equation has form of Eq. (7.20) or (7.21). Discrete Hamiltonians weakly dependent on θ^n in a vicinity of 0+ are common in thermodynamic systems described in spaces of arbitrary state variables rather than extensive thermodynamic coordinates.

The discrete dynamics described in terms of the Hamiltonian is omitted in this chapter, where the theoretical schemes of dynamic programming leading to optimal performance functions V and R are essential. The reader interested in the first issue is referred to the literature (Sieniutycz, 2006b).

In the general case of nonsmooth solutions for the value function V, the discretization modes that satisfy the condition of the weak dependence of process rates on free θ^n in a vicinity of 0+ have solutions convergent to the viscosity solutions of continuous Hamilton-Jacobi and HJB equations.

7.4 Dynamic programming equation for maximum power from radiation

The previous section has shown that, regardless the discretizing mode, equations modeling the continuous processes (including equations of power systems considered) can be solved numerically by the discrete algorithm of dynamic programming (DP) associated with stage criterion (7.11). Consequently, one can use the DP method associated with Bellman's recurrence equation

$$R^n(\mathbf{x}^n, t^n) = \min_{\mathbf{u}^n, \theta^n} \{ l_0^n(\mathbf{x}^n, t^n, \mathbf{u}^n, \theta^n)\theta^n + R^{n-1}((\mathbf{x}^n - \mathbf{f}^n(\mathbf{x}^n, t^n, \mathbf{u}^n, \theta^n)\theta^n, t^n - \theta^n) \} \tag{7.23}$$

The enlarged Hamiltonian is not constant as in the linear, free-θ theory, but satisfies the condition (7.9), where λ is the Lagrange multiplier of the duration constraint and H_1^{n-1} is Hamiltonian of the discrete model whose rates don't involve time interval. Therefore, different models linear in θ^n (those with θ-independent rates f_k) are the primary candidates to accomplish the efficient solving of continuous equations characterized by their own Hamilton-Jacobi-Bellman equations and Hamilton-Jacobi equations.

We can now return to the difficult radiation problem described by Eqs. (7.4)–(7.6). Applying Eq. (7.23) to this problem, the following recurrence equation is obtained

$$R^n(T^n, t^n) = \min_{\mathbf{u}^n, \theta^n} \left\{ \dot{G}_c(T^n).\left(1 - \frac{\Phi T^e}{T'^n}\right)\beta \frac{T^{na} - T'^{na}}{\left(\Phi'(T'^n/T^e)^{a-1} + 1\right)T^{na-1}}\theta^n \right.$$
$$\left. + R^{n-1}\left(T^n - \theta^n \beta \frac{T'^{na} - T^{na}}{\left(\Phi'(T'^n/T^e)^{a-1} + 1\right)T^{na-1}}, t^n - \theta^n\right)\right\} \tag{7.24}$$

While the analytical treatment of Eqs. (7.1), (7.2) is a tremendous task, it is quite easy to solve recurrence equation (7.24) numerically. Low dimensionality of state vector for Eq. (7.24) assures a decent accuracy of DP solution. Moreover, an original accuracy can be significantly improved after performing the so-called dimensionality reduction associated with the elimination of time t^n as the state variable. In the transformed problem, without coordinate t^n, accuracy of DP solutions is high. Section 7.9 discusses related computational issues.

7.5 Discrete approximations and time adjoint as a Lagrange multiplier

In this section, we consider another discrete example whose DP solution converges to the viscosity solution of a continuous problem of power generation. Yet, for brevity of formulas, we restrict ourselves to systems in which nonlinearities are absent in process kinetics, although they are still present in the power expression. We consider solutions of HJB equations by discrete approximations (produced by the method of dynamic programming) in association with state dimensionality reduction (elimination of time coordinate) by using a Lagrange multiplier λ.

First, we shall outline generation of certain suboptimal costs in terms of the Lagrangian multiplier λ associated with the duration constraint. As the time adjoint, λ is constant in autonomous systems.

Consider a minimum of consumed work with constraints imposed on discrete dynamics and process duration

$$R^n(T^n, \tau^n) = \sum_{k=1}^{n} c\left(1 - \Phi' \frac{T^e}{T'^k}\right)\left(T^k - T^{k-1}\right) = \sum_{k=1}^{n} c\left(1 - \Phi' \frac{T^e}{T'^k}\right)\left(T'^k - T^k\right)\theta^k \tag{7.25}$$

$$T^k - T^{k-1} = \theta^k\left(T'^k - T^k\right) \tag{7.26}$$

$$\tau^k - \tau^{k-1} = \theta^k \tag{7.27}$$

Observe that above difference equations model, a continuous problem of minimum work subject to the kinetic constraint $\dot{T} = T' - T$ (Sieniutycz, 2007a,b).

Exploiting constancy of λ, we eliminate state variable τ by introducing a (primed) criterion of modified work

$$R'^{n}(T^n,\lambda)=\min\sum_{k=1}^{n}\left\{c\left(1-\Phi'\frac{T^e}{T'^k}\right)(T^k-T^{k-1})+\lambda\theta^n\right\} \quad (7.28)$$

or, in view of state equation (7.26)

$$R'^{n}(T^n,\lambda)=\min\sum_{k=1}^{n}\left\{c\left(1-\Phi'\frac{T^e}{T'^k}\right)+\frac{\lambda}{T'^k-T^k}\right\}(T^k-T^{k-1}) \quad (7.29)$$

In this problem, idea of parametric representations for the principal performance function, Lagrange multiplier, and process duration had proven its usefulness. While these representations are unnecessary for linear optimization problems, they are quite effective to describe solutions of nonlinear problems, where the optimal work, Lagrange multiplier, and optimal duration are obtained in terms of an optimal control variable as a parameter.

To begin with, we determine optimality conditions from Eq. (7.29). We consider two initial stages of the process, 1 and 2. A procedure leading to parametric representations is defined below.

Equation of work modified by the presence of the Lagrange multiplier λ, yet without a minimization sign,

$$R'^{1}\left(T^1,T'^1,\lambda\right)=\left\{c\left(1-\Phi'\frac{T_2}{T'^1}\right)+\frac{\lambda}{T'^1-T^1}\right\}(T^1-T^0) \quad (7.30)$$

constitutes a component of the parametric representation of $R'^1(T^1,\lambda)$ provided that the following procedure is implemented:

1. λ is determined from the extremum condition of work function R'^1 with respect to a control variable, here with respect to Carnot control

$$\lambda=c\Phi'T^e\frac{(T'^1-T^1)^2}{(T'^1)^2} \quad (7.31)$$

2. Extremum λ is substituted into the work function R'^1, and the result of this substitution

$$R'^{1}\left(T^1,T'^1\right)=\left\{c\left(1-\Phi'\frac{T^e}{T'^1}\right)+c\Phi'T^e\frac{T'^1-T^1}{\left(T'^1\right)^2}\right\}(T^1-T^0) \quad (7.32)$$

is taken together with the stationary λ.

In view of the above, parametric representation of work function $R'^1(T^1,\lambda)$ in terms of Carnot control as a parameter is given by the set of Eqs. (7.31), (7.32), or, after simplification of work equation

$$R'^{1}\left(T^{1},T'^{1}\right)=c\left(1-\Phi' T^{e}\frac{T^{1}}{\left(T'^{1}\right)^{2}}\right)\left(T^{1}-T^{0}\right)\ \ \lambda=c\Phi' T^{e}\left(1-\frac{T^{1}}{T'^{1}}\right)^{2} \tag{7.33}$$

In this example, it is possible to eliminate the parameter T'^{1} which leads to explicit function of work consumption for $n=1$

$$R'^{1}\left(T^{1},T^{0},\lambda\right)=c\left(T^{1}-T^{0}\right)-c\Phi' T^{e}\left(1-\pm\sqrt{\frac{\lambda}{c\Phi' T^{e}}}\right)^{2}\left(\frac{T^{1}-T^{0}}{T^{1}}\right) \tag{7.34}$$

Corresponding optimal control satisfies an equation

$$T'^{1}=T^{1}\left(1-\pm\sqrt{\frac{\lambda}{c\Phi' T^{e}}}\right)^{-1} \tag{7.35}$$

obtained by solving Eq. (7.31) with respect to T'^{1}.

Yet, it should be kept in mind that the elimination of the parameter is not always possible, and then parameter dependent-functions are the only representation of the solution. It is just this case when the parametric representations are inevitable and helpful.

Let us proceed further. Optimal work supply to two-stage system $R'^{2}(T^{2},\lambda)$ is described by an equation

$$R'^{2}\left(T^{2},\lambda\right)=\min_{T^{1}}\left\{c\left(T^{2}-T^{1}\right)-c\Phi' T^{e}\left(1-\pm\sqrt{\frac{\lambda}{c\Phi' T^{e}}}\right)^{2}\left(\frac{T^{2}-T^{1}}{T^{2}}\right)+c\left(T^{1}-T^{0}\right)\right.$$
$$\left.-c\Phi' T^{e}\left(1-\pm\sqrt{\frac{\lambda}{c\Phi' T^{e}}}\right)^{2}\left(\frac{T^{1}-T^{0}}{T^{1}}\right)\right\}$$

whence, after making simplifications

$$R'^{2}\left(T^{2},\lambda\right)=\min_{T^{1}}\left\{c\left(T^{2}-T^{0}\right)-c\Phi' T^{e}\left(1-\pm\sqrt{\frac{\lambda}{c\Phi' T^{e}}}\right)^{2}\left(\frac{T^{2}-T^{1}}{T^{2}}+\frac{T^{1}-T^{0}}{T^{1}}\right)\right\} \tag{7.36}$$

Optimal interstage temperature between stages 1 and 2, T^{1}, satisfies the stationarity condition for expression in the large bracket of the above equation

$$\frac{\partial}{\partial T^{1}}\left(\frac{T^{2}-T^{1}}{T^{2}}+\frac{T^{1}-T^{0}}{T^{1}}\right)=\frac{T^{0}}{\left(T^{1}\right)^{2}}-\frac{1}{T^{2}}=0 \tag{7.37}$$

Therefore, optimal interstage temperature T^{1} is the geometric mean of boundary temperatures of both considered stages.

$$T^{1}=\left(T^{0}T^{2}\right)^{1/2} \tag{7.38}$$

The minimum value of optimized expression (7.37) is

$$\min_{T^1}\left(\frac{T^2-T^1}{T^2}+\frac{T^1-T^0}{T^1}\right)=\frac{T^2-\sqrt{T^0T^2}}{T^2}+\frac{\sqrt{T^0T^2}-T^0}{\sqrt{T^0T^2}}=2\left(1-\sqrt{T^0/T^2}\right) \tag{7.39}$$

This leads to the optimum work function for $n=2$

$$R'^2(T^2,\lambda)=c(T^2-T^0)-2c\Phi'T^e\left(1-\pm\sqrt{\frac{\lambda}{c\Phi'T^e}}\right)^2\left(1-\left(\frac{T^0}{T^2}\right)^{1/2}\right) \tag{7.40}$$

For $n=3$, we apply an expression for local work

$$K'^3(T^3,T^2,\lambda)=c(T^3-T^2)-c\Phi'T^e\left(1-\pm\sqrt{\frac{\lambda}{c\Phi'T^e}}\right)^2\left(\frac{T^3-T^2}{T^3}\right) \tag{7.41}$$

which has the same structure as the one-stage function of Eq. (7.34), but the indices are shifted ahead by one. In the recurrence equation for $n=3$, explicit work function $R'^3(T^3,\lambda)$ is the result of optimization described by the following expression

$$R'^3(T^3,\lambda)=\min_{T^2}\left\{c(T^3-T^2)-c\Phi'T^e\left(1-\pm\sqrt{\frac{\lambda}{c\Phi'T^e}}\right)^2\left(\frac{T^3-T^2}{T^3}\right)\right.$$
$$\left.+c(T^2-T^0)-2c\Phi'T^e\left(1-\pm\sqrt{\frac{\lambda}{c\Phi'T^e}}\right)^2\left(1-\left(\frac{T^0}{T^2}\right)^{1/2}\right)\right\}$$

After simplifying, we obtain

$$R'^3(T^3,\lambda)=\min_{T^2}\left\{c(T^3-T^0)-c\Phi'T^e\left(1-\pm\sqrt{\frac{\lambda}{c\Phi'T^e}}\right)^2\left[\left(\frac{T^3-T^2}{T^3}\right)+2\left(1-\left(\frac{T^0}{T^2}\right)^{1/2}\right)\right]\right\} \tag{7.42}$$

Consequently, optimal interstage temperature T^2 satisfies the stationarity condition

$$\frac{\partial}{\partial T^2}\left(\frac{T^3-T^2}{T^3}+2\left(1-\left(\frac{T^0}{T^2}\right)^{1/2}\right)\right)=0 \tag{7.43}$$

Performing the differentiation with respect to T^2 one obtains in terms of T^0 and T^3

$$T^2=(T^0)^{1/3}(T^3)^{2/3} \tag{7.44}$$

and

$$T^1=(T^0T^2)^{1/2}=(T^0)^{1/2}\left[(T^0)^{1/3}(T^3)^{2/3}\right]^{1/2}=(T^0)^{2/3}(T^3)^{1/3} \tag{7.45}$$

Let us eliminate T^0 and determine T^2 in terms of T^1 and T^3. We obtain

$$T^2 = (T^0)^{1/3}(T^3)^{2/3} = (T^1)^{1/2}(T^3)^{-1/6}(T^3)^{2/3} = (T^1)^{1/2}(T^3)^{1/2} \tag{7.46}$$

Therefore, as one could expect, also optimal interstage temperature T^2 is the geometric mean of boundary temperatures of two considered stages

$$T^2 = (T^1 T^3)^{1/2} \tag{7.47}$$

Substitution of optimal temperature $T^2 = (T^0)^{1/3}(T^3)^{2/3}$ into work function (7.42) yields

$$R'^3(T^3, \lambda) = c(T^3 - T^0) - c\Phi' T^e \left(1 - \pm\sqrt{\frac{\lambda}{c\Phi' T^e}}\right)^2 \left[\left(1 - \left(\frac{T^0}{T^3}\right)^{1/3}\right) + 2\left(1 - \left(\frac{T^0}{T^3}\right)^{1/3}\right)\right]$$

which can be simplified to the form

$$R'^3(T^3, \lambda) = c(T^3 - T^0) - 3c\Phi' T^e \left(1 - \pm\sqrt{\frac{\lambda}{c\Phi' T^e}}\right)^2 \left(1 - \left(\frac{T^0}{T^3}\right)^{1/3}\right) \tag{7.48}$$

Comparing this expression with corresponding ones for $n=1$ and $n=2$, Eqs. (7.24), (7.40), leads to optimal work function for an arbitrary n

$$R'^n(T^n, \lambda) = c(T^n - T^0) - nc\Phi' T^e \left(1 - \pm\sqrt{\frac{\lambda}{c\Phi' T^e}}\right)^2 \left(1 - \left(\frac{T^0}{T^n}\right)^{1/n}\right) \tag{7.49}$$

The corresponding optimal duration is the partial derivative of optimal work function with respect to Lagrangian multiplier λ

$$\tau^n = \frac{\partial R'^n(T^n, \lambda)}{\partial \lambda} = \frac{1 - \pm\sqrt{\frac{\lambda}{c\Phi' T^e}}}{\pm\sqrt{\frac{\lambda}{c\Phi' T^e}}} n \left(1 - \left(\frac{T^0}{T^n}\right)^{1/n}\right) \tag{7.50}$$

Qualitative properties of the duration function are illustrated in Fig. 7.1. Eq. (7.50) refers to linear kinetics, in which case the curve intersects the axis of λ for $\lambda = c$. In nonlinear systems with a variable c, the intersection point may move to large values of λ, This case is also shown in Fig. 7.1. In any process, linear or not, λ is monotonically decreasing function of duration.

Knowledge of partial derivative of optimal work function with respect to Lagrangian multiplier λ is essential when one wishes to return to original work function $R^n(T^n, \tau^n)$ (without the Lagrange multiplier term). In this operation, the Legendere transformation plays an essential role (Section 7.7). Yet, consider first some properties of intensity parameter ξ.

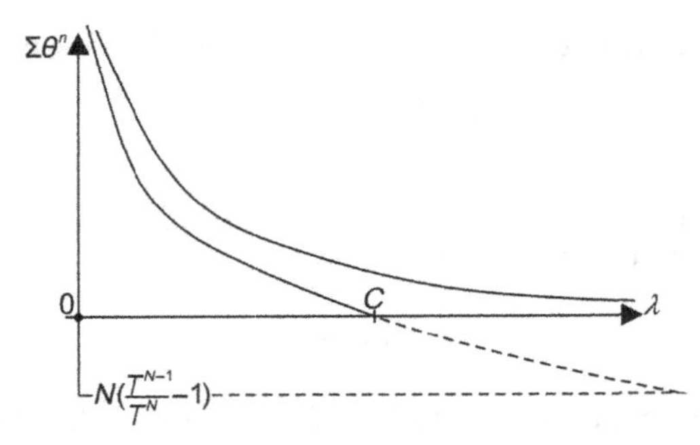

Fig. 7.1
Two cases of dependence of Lagrange multiplier λ on total optimal duration $\tau = \Sigma\Theta^n$ in a cascade of power generation systems.

7.6 Mean and local intensities in discrete processes

Further transformations are easier if the following intensity criterion is introduced

$$\xi \equiv \frac{n}{\tau^n}\left(1 - \left(\frac{T^0}{T^n}\right)^{1/n}\right) \tag{7.51}$$

To identify physical meaning of ξ, let us calculate its limiting value for n approaching infinity (continuous process). Since

$$\lim_{n\to\infty} n\left(1 - \left(T^0/T^n\right)^{1/n}\right) = \lim_{x\to 0}\frac{1-\left(T^0/T^n\right)^x}{x} = \ln\left(T^n/T^0\right)$$

we obtain from Eq. (7.51)

$$\lim_{n\to\infty}\xi \equiv \lim_{n\to\infty}\frac{n}{\tau^n}\left(1 - \left(\frac{T^0}{T^n}\right)^{1/n}\right) = \frac{\ln T^n - \ln T^0}{\tau^n} \tag{7.52}$$

Therefore ξ defined by Eq. (7.51) is a discrete counterpart of a *mean* relaxation rate of the temperature logarithm for the n-stage process.

For an arbitrary stage n, we can also introduce a local intensity of a discrete process

$$\xi^n \equiv \frac{T^n - T^{n-1}}{T^n\theta^n} \tag{7.53}$$

This quantity can be obtained after careful use of Eq. (7.51) for $n=1$ and appropriate change of symbols. Its limit for $n \to \infty$ is instantaneous logarithmic rate of state change in a continuous process

$$\lim_{n\to\infty}\frac{T^n - T^{n-1'}}{T^n\theta^n} = \frac{d\ln T}{d\tau} = \xi \tag{7.54}$$

Applying the geometric sequence property for optimal path in the considered example

$$\frac{T^0}{T^n}=\frac{T^0}{T^1}\frac{T^1}{T^2}\cdots\cdots\frac{T^{n-1}}{T^n}=\left(\frac{T^{n-1}}{T^n}\right)^n \tag{7.55}$$

we obtain

$$\frac{T^{n-1}}{T^n}=\left(\frac{T^0}{T^n}\right)^{1/n} \tag{7.56}$$

Thus, we can easily prove the equality of mean and local rates

$$\xi\equiv\frac{n}{\tau^n}\left(1-\left(\frac{T^0}{T^n}\right)^{1/n}\right)=\frac{n}{\tau^n}\left(1-\frac{T^{n-1}}{T^n}\right)=\frac{T^n-T^{n-1'}}{T^n\theta^n} \tag{7.57}$$

Local and mean intensities ξ are in general different quantities. Yet, in processes with linear kinetics, considered in the present example, there is no need to distinguish between them and the same symbol for both may be used in equations.

Writing the duration formula, Eq. (7.50), in the form

$$\xi^{-1}\equiv\frac{\tau^n}{n}\left(1-\left(\frac{T^0}{T^n}\right)^{1/n}\right)^{-1}=\pm\left(\sqrt{\frac{\lambda}{c\Phi' T^e}}\right)^{-1}-1 \tag{7.58}$$

we find an useful equality determining the Lagrange multiplier in terms of the process intensity (mean or instantaneous)

$$\lambda=c\Phi' T^e\left(\frac{\xi}{\xi+1}\right)^2 \tag{7.59}$$

Two values of ξ for a given λ correspond with heating and cooling of the resource fluid in heat-pump and engine modes (upgrading and downgrading of the resource). Both λ and ξ vanish in reversible quasistatic processes.

As the following equality is valid

$$1-\pm\sqrt{\frac{\lambda}{c\Phi' T^e}}=\frac{1}{\xi+1} \tag{7.60}$$

optimal work function in terms of ξ assumes the form

$$R'^n(T^n,\xi)=c\left(T^n-T^0\right)-\frac{c\Phi' T^e}{(1+\xi)^2}n\left(1-\left(\frac{T^0}{T^n}\right)^{1/n}\right) \tag{7.61}$$

Taking into account the limiting value of the expression

$$\lim_{n\to\infty} n\left(1-\left(T^0/T^n\right)^{1/n}\right) = \ln\left(T^n/T^0\right)$$

we find that the limiting value of function $R'^n(T^n,\xi)$ in a quasistatic ($\xi=0$) and reversible process ($\Phi=1$) represents the change of classical thermal exergy.

$$R'^n(T^n,0)=c\left(T^n-T^0\right)-cT^e\ln\left(T^n/T^0\right) \tag{7.62}$$

Therefore, optimal work function (7.61) is a finite-rate exergy of the considered discrete process. In the following section, other functions of this kind are obtained.

7.7 Legendre transform and original work function

The minimum of consumed work (7.25) is described by original principal function$R^n(T^n,\tau^n)$. This function is the Legendre transform of $R'^n(T^n,\lambda)$ with respect to λ.

$$R^n(T^n,\tau^n)=R'^n(T^n,\lambda)-\lambda\tau^n=R'^n(T^n,\lambda)-\lambda\frac{\partial R'^n(T^n,\lambda)}{\partial\lambda} \tag{7.63}$$

For our example, we obtain

$$\begin{aligned}R^n(T^n,\lambda)=&c\left(T^n-T^0\right)-nc\Phi'T^e\left(1-\pm\sqrt{\frac{\lambda}{c\Phi'T^e}}\right)^2\left(1-\left(\frac{T^0}{T^n}\right)^{1/n}\right)\\&-\lambda\frac{1-\pm\sqrt{\dfrac{\lambda}{c\Phi'T^e}}}{\pm\sqrt{\dfrac{\lambda}{c\Phi'T^e}}}n\left(1-\left(\frac{T^0}{T^n}\right)^{1/n}\right)\end{aligned} \tag{7.64}$$

which should be transformed to the space of variables T^n and τ^n. In transformations, we use intensity ξ as an intermediate variable to increase lucidity of formulas. Applying the equality

$$\frac{1-\pm\sqrt{\dfrac{\lambda}{c\Phi'T_2}}}{\pm\sqrt{\dfrac{\lambda}{c\Phi'T_2}}}=\frac{\dfrac{1}{1+\xi}}{\dfrac{\xi}{1+\xi}}=\frac{1}{\xi} \tag{7.65}$$

we obtain

$$\tau^n=\frac{\partial R'^n(T^n,\lambda)}{\partial\lambda}=\frac{n}{\xi}\left(1-\left(\frac{T^0}{T^n}\right)^{1/n}\right) \tag{7.66}$$

and

$$R^n(T^n,\xi)=c\left(T^n-T^0\right)-c\Phi'T^e\left(\frac{1}{1+\xi}\right)^2 n\left(1-\left(\frac{T^0}{T^n}\right)^{1/n}\right)-\frac{\lambda(\xi)}{\xi}n\left(1-\left(\frac{T^0}{T^n}\right)^{1/n}\right) \tag{7.67}$$

where function describing λ in terms of ξ is given by Eq. (7.50). A complementary formula expressing λ in terms of duration τ follows from Eq. (7.50)

$$\lambda = c\Phi' T^e \left(\frac{\tau^n}{n\left[1 - (T^0/T^n)^{1/n}\right]} + 1 \right)^{-2} \tag{7.68}$$

Monotonic decrease of λ withτ is a general property of both linear and nonlinear processes. Since λ is numerically equal to the Hamiltonian of the optimization problem, Fig. 7.1 shows the behavior of Lagrange multiplier λ in terms of nondimensional duration at prescribed boundary temperatures.

Using Eqs. (7.67), (7.59), we find

$$R^n(T^n, \xi) = c(T^n - T^0) - \frac{c\Phi' T^e}{1+\xi} n \left(1 - \left(\frac{T^0}{T^n} \right)^{1/n} \right) \tag{7.69}$$

This is a finite-rate exergy, yet it differs from function $R'^n(T^n, \xi)$ of Eq. (7.67) the previous section by the structure of ξ term. To single out from this equation a ξ-independent term, we write $(1+\xi)^{-1}$ as $1 - \xi/(1+\xi)$ and, then,

$$R^n(T^n, \xi) = c(T^n - T^0) - c\Phi' T^e n \left(1 - \left(\frac{T^0}{T^n} \right)^{1/n} \right) + \frac{\xi}{1+\xi} c\Phi' T^e n \left(1 - \left(\frac{T^0}{T^n} \right)^{1/n} \right) \tag{7.70}$$

Eq. (7.51) can next be applied to express the last (rate-dependent) term of Eq. (7.70) in terms of process duration. We obtain

$$\frac{\xi}{1+\xi} \equiv \frac{n\left(1 - \left(\frac{T^0}{T^n} \right)^{1/n} \right)}{\tau^n + n\left(1 - \left(\frac{T^0}{T^n} \right)^{1/n} \right)} \tag{7.71}$$

and

$$\begin{aligned} R^n(T^n, \tau^n) = {} & c(T^n - T^0) - c\Phi' T^e n \left(1 - \left(\frac{T^0}{T^n} \right)^{1/n} \right) \\ & + c\Phi' T^e \left\{ n\left(1 - \left(\frac{T^0}{T^n} \right)^{1/n} \right) \right\}^2 \left(\tau^n + n\left(1 - \left(\frac{T^0}{T^n} \right)^{1/n} \right) \right)^{-1} \end{aligned} \tag{7.72}$$

First two components of this work function are of static origin. The function describes a minimum work supplied to a resource to upgrade it from T^0 to T^n in a finite (nondimensional) time τ^n. Like in the case of primed function R', a limiting value of $R^n(T^n, \tau^n)$ in a reversible and quasistatic process ($\Phi' = 1$, $\tau^n \to \infty$) describes a change of classical thermal exergy (7.62).

This approach can also be organized in the entropy representation, where principal function R'^n_σ is the minimum entropy production modified by Lagrange multiplier term. Then

$$R'^1_\sigma(T^1, \lambda) = c\left(1 - \frac{T^0}{T^1}\right)\left(1 - \left(1 - \pm\sqrt{\frac{\lambda}{c\Phi' T^e}}\right)^2\right) \tag{7.73}$$

and

$$R'^n_\sigma(T^n, \lambda) = nc\left(1 - \left(\frac{T^0}{T^n}\right)^{1/n}\right)\left(1 - \left(1 - \pm\sqrt{\frac{\lambda}{c\Phi' T^e}}\right)^2\right) \tag{7.74}$$

Further analysis can be developed in the way analogous to that performed for work functions.

7.8 Numerical approaches applying dynamic programming

Numerical aspects of DP algorithms are only briefly mentioned here as there is a vast literature available which discusses these issues. "Curse of dimensionality" and problems of grid expansion are main difficulties (Aris, 1964; Bellman, 1961; Sieniutycz, 1991).

Optimal performance functions are direct outcome of numerical methods which apply the dynamic programming. Optimal control problems of both continuous and discrete processes with a single independent variable (time or length) can be treated in the framework of a common discrete formalism. As the computer needs require a discrete set of equations, in the continuous case prior discretizing of ordinary differential equations is required to obtain a set of difference equations. Thus, it is appropriate focus on numerical multistage optimization, so we describe generation of optimal function $V^n = \max W$.

Assume that, at the stage n (duration $\Delta t^n = \theta^n$), a profit function $D^n = D^n(\mathbf{x}^n, t^n, \mathbf{u}^n, \theta^n, n)$ is given. In maximum work problem $D^n = f_0^n \theta^n$, where D^n is the work produced at stage n. The total profit at the n-stage subprocess equals $\Sigma D^n = \Sigma f_0^m \theta^m$. Along with discrete state equations and local constraints, various data of D^n—analytic, graphic, or tabular—are sufficient to develop computational principles for cascade processes with arbitrary number of stages. It is, however, important that D^n is properly expressed at the stage n as a function of state $\mathbf{x}^n$, time t^n, and controls ($\mathbf{u}^n$, θ^n). Data of optimal work functions V^1, ...V^n,.. V^N should be generated over subprocesses composed, respectively, of the stage 1, stages 1 and 2, ...stages 1, 2 ....n ...and, finally, stages 1, 2...N.

For a given set of difference constraints (7.8), each profit function, e.g., optimal work potential V^n, is found from Bellman's equation of dynamic programming. A typical form of this equation in terms of enlarged state vector $\widetilde{\mathbf{x}}^n \equiv (\mathbf{x}^n, t^n)$ and one-stage profit D^n is

$$V^n(\widetilde{\mathbf{x}}^n) = \max_{\mathbf{u}^n, \theta^n} \left\{ D^n(\widetilde{\mathbf{x}}^n, \mathbf{u}^n, \theta^n) + V^{n-1}\left(\widetilde{\mathbf{x}}^n - \widetilde{\mathbf{f}}^n(\widetilde{\mathbf{x}}^n, \mathbf{u}^n, \theta^n)\theta^n\right) \right\} \tag{7.75}$$

Gauged profits $\widetilde{D}^n$ obtained by addition (subtraction) of a difference of a scalar function $G(\mathbf{x}^n, t^n) - G(\mathbf{x}^{n-1}, t^{n-1})$ to original one-stage profit D^n can also be considered. They are associated with Bolza functionals and necessarily nonlinear in θ^n

$$\widetilde{D}^n(\widetilde{\mathbf{x}}^n, \mathbf{u}^n, \theta^n) \equiv \widetilde{f}_0^n \theta^n \equiv \left(f_0^n(\widetilde{\mathbf{x}}^n, \mathbf{u}^n, \theta^n) + \frac{G(\widetilde{\mathbf{x}}^n) - G\left(\widetilde{\mathbf{x}}^n - \widetilde{\mathbf{f}}^n(\widetilde{\mathbf{x}}^n, \mathbf{u}^n, \theta^n)\theta^n\right)}{\theta^n} \right) \theta^n \tag{7.76}$$

Tildas over symbols refer to extended quantities; thus, a tilda over the profit symbol refers to extended profit including the effect of the gauging function G. As both profit functions D^n and $\widetilde{D}^n$ differ by the path independent increment of state function G, simultaneous generation of optimal data for D^n and $\widetilde{D}^n$ is possible. Since one has to generate computational data within a definite domain of the variables $\widetilde{\mathbf{x}}$, one can conveniently omit in Eq. (7.75) stage superscript n at the (enlarged) state vector

$$V^n(\widetilde{\mathbf{x}}) = \max_{\mathbf{u}^n, \theta^n} \left\{ D^n(\widetilde{\mathbf{x}}, \mathbf{u}^n, \theta^n) + V^{n-1}\left(\widetilde{\mathbf{x}} - \widetilde{\mathbf{f}}^n(\widetilde{\mathbf{x}}, \mathbf{u}^n, \theta^n)\theta^n\right) \right\} \tag{7.75'}$$

The solution to Eq. (7.75) or (7.75′) is obtained iteratively in the form of tables for $n = 1, 2 \ldots N$, which describe sequence of functions $V^n(\widetilde{\mathbf{x}})$, $\mathbf{u}^n(\widetilde{\mathbf{x}})$, and $\theta^n(\widetilde{\mathbf{x}})$. The iterative procedure starts with $V^0 = 0$. Potential $V^n(\widetilde{\mathbf{x}})$ is in general time-dependent even if the process is autonomous.

Organization of calculations requires a suitable computational grid. In the nodes of this grid, data of optimal functions and optimal controls are computed and stored. A total number of stages, N, is assumed. Numerical DP algorithm generates potential function $V^n(\widetilde{\mathbf{x}})$ from function $V^{n-1}(\widetilde{\mathbf{x}})$ of $n-1$ stage subprocess and state transformations, Eq. (7.8). In agreement with Bellman's optimality principle, consistent with Eq. (7.75), a computer maximizes the sum of optimal cost of all previous $n-1$ stages (optimal function V^{n-1}) and nonoptimal profit $\widetilde{D}^n$ at the stage n. To determine $V^n(\widetilde{\mathbf{x}})$ exactly for a definite n, the computer would have to numerically determine values of this function for every value of $\widetilde{\mathbf{x}}$, an impossible task. Therefore, these values are determined on a discrete subset of $\widetilde{\mathbf{x}}$, and the data are used in the way that makes possible evaluation of $V^n(\widetilde{\mathbf{x}})$ everywhere. If $< \widetilde{\mathbf{x}}_{\mathrm{I}}, \widetilde{\mathbf{x}}_{\mathrm{II}} >$ is the interval of interest, one can take $\widetilde{\mathbf{x}}_{\mathrm{I}} = \mathbf{A}\delta$ and $\widetilde{\mathbf{x}}_{\mathrm{II}} = \mathbf{B}\delta$, where δ is a small accepted value. The vectors $\mathbf{A}$ and $\mathbf{B}$ are, respectively, computed as $\widetilde{\mathbf{x}}_{\mathrm{I}}/\delta$ and $\widetilde{\mathbf{x}}_{\mathrm{II}}/\delta$. The quantity δ cannot be too large, because the results' accuracy is then poor. It cannot be too small either, as the computation time becomes very long. The discrete subset of $\widetilde{\mathbf{x}}$, for which the values $V^n(\widetilde{\mathbf{x}})$ are computed for various n, has the form.

$$\widetilde{\mathbf{x}} = \mathbf{A}\delta, (\mathbf{A}+1)\delta, \ldots\ldots\ldots\ldots, (\mathbf{B}-1)\delta, \mathbf{B}\delta \tag{7.77}$$

This refers to the linear grid of values $\widetilde{\mathbf{x}} = \mathbf{a}\delta$, where $\mathbf{a} = \mathbf{A}$, $\mathbf{A}+\mathbf{1}$, ...$\mathbf{B}$. Other values of $V^n(\widetilde{\mathbf{x}})$, e.g., those in an interval $<\mathbf{a}\delta$, $(\mathbf{a}+1)\delta>$, are defined by accepting $V^n(\mathbf{a}\delta)$ or $V^n[(\mathbf{a}+\mathbf{1})\delta]$ depending on location of $\mathbf{x}\delta$ with respect to $(\mathbf{a}+\mathbf{1/2})\delta$, or by using interpolation, e.g.,

$$V^n(\widetilde{\mathbf{x}}) = V^n(\mathbf{A}\delta) + \frac{V^n((\mathbf{A}+1)\delta) + V^n(\mathbf{A}\delta)}{\delta}(\widetilde{\mathbf{x}} - \mathbf{a}) \quad \mathbf{a}\delta \le (\mathbf{a}+1)\delta \tag{7.78}$$

The discrete subset of admissible controls $\mathbf{u}^n$ is defined in a similar way. For example, when constrains imposed on $\mathbf{u}^n$ are described by inequality $\mathbf{u}^* \le \mathbf{u}^n \le \mathbf{u}^*$, variable $\mathbf{u}^n$ may assume only the discrete values

$$\mathbf{u}^* = \mathbf{E}\gamma + (\mathbf{E}+1)\gamma, \ldots\ldots(\mathbf{F}-1)\gamma, \mathbf{F}\gamma = \mathbf{u}^* \tag{7.79}$$

for an appropriately small value γ. This refers to a linear grid of controls $\mathbf{u} = \mathbf{b}\delta$, where $\mathbf{b} = \mathbf{E}$, $\mathbf{E}+\mathbf{1}$, ...$\mathbf{F}-\mathbf{1}$, $\mathbf{F}$.

Optimal controls are determined form the formula

$$\{\mathbf{u}^n(\widetilde{\mathbf{x}}), \theta^n(\widetilde{\mathbf{x}})\} = \underset{\mathbf{u}^n\theta^n,}{\arg\max}\left\{\widetilde{D}^n(\widetilde{\mathbf{x}}, \mathbf{u}^n, \theta^n) + V^{n-1}\left(\widetilde{\mathbf{x}} - \widetilde{\mathbf{f}}^n(\widetilde{\mathbf{x}}, \mathbf{u}^n, \theta^n)\theta^n\right)\right\} \tag{7.80}$$

along with the sequence of the optimal functions V^n. The first optimal function, $V^1(\widetilde{\mathbf{x}})$, and the corresponding optimal controls for $n=1$ follow from the application of the initial condition $V^0(\widetilde{\mathbf{x}}) = 0$ in Eqs. (7.75), (7.80); this yields

$$V^1(\widetilde{\mathbf{x}}) = \max_{\mathbf{u}^1, \theta^1}\left\{\widetilde{D}^1(\widetilde{\mathbf{x}}, \mathbf{u}^1, \theta^1)\right\} \tag{7.81}$$

and

$$\{\mathbf{u}^1(\widetilde{\mathbf{x}}), \theta^1(\widetilde{\mathbf{x}})\} = \underset{\mathbf{u}^n\theta^n,}{\arg\max}\left\{\widetilde{D}^1(\widetilde{\mathbf{x}}, \mathbf{u}^1, \theta^1)\right\} \tag{7.82}$$

To find these functions, the computer choses the first point $\widetilde{\mathbf{x}} = \mathbf{A}\delta = (A_1\delta, A_2\delta)$ and compares $\widetilde{D}^1(\mathbf{A}\delta, E_1\gamma, E_2\gamma)$ with $\widetilde{D}^1(\mathbf{A}\delta, (E_1+1)\gamma, E_2\gamma)$. The larger of these values is stored and compared with $\widetilde{D}^1(\mathbf{A}\delta, (E_1+2)\gamma, E_2\gamma)$, etc. This process is continued until the whole discrete set of controls $(\mathbf{u}^1, \theta^1)$ is exhausted. The largest of the so-obtained values is always a maximum of $\widetilde{D}^1$ with respect to $(\mathbf{u}^1, \theta^1)$ for a fixed discrete point $\widetilde{\mathbf{x}}$. The coordinates of $\mathbf{u}^1$ and θ^1 which maximize $\widetilde{D}^1$ are stored. Analogous operations are next performed for $\widetilde{\mathbf{x}} = ((A_1+1)\delta, A_2\delta)$, $((A_1+2)\delta, A_2\delta)$, and so on. Again, this leads to maximum of $\widetilde{D}^1$ and optimal values of $\mathbf{u}^1$ and θ^1. The data for the same quantity differ at various points $\widetilde{\mathbf{x}}$ (different nodes of the grid). The computer outputs are DP tables which contain only optimal data: $V^1(\widetilde{\mathbf{x}})$, $\theta^1(\widetilde{\mathbf{x}})$, and $\mathbf{u}^1(\widetilde{\mathbf{x}})$.

For $n=2$ (two-stage process), and for larger n, the procedure is analogous but uses Eqs. (7.75), (7.80) in their complete form. Data of V^{n-1} are found in tables describing previous computations, for cascade with $n-1$ stages. When using these data, a difficulty can appear which is called "the danger of grid expansion." It follows from the fact that the values $V^{n-1}(\widetilde{\mathbf{x}})$ were computed within the range $\widetilde{\mathbf{x}}_{\mathrm{I}} \le \widetilde{\mathbf{x}} \le \widetilde{\mathbf{x}}_{\mathrm{II}}$, but the computation of $V^n(\widetilde{\mathbf{x}})$ from Eq. (7.75) requires the knowledge of $V^{n-1}(\widetilde{\mathbf{x}}^T)$ for the transformed state $\widetilde{\mathbf{x}}^{\mathrm{T}} \equiv \widetilde{\mathbf{x}} - \widetilde{\mathbf{f}}^n(\widetilde{\mathbf{x}}, \mathbf{u}^n, \theta^n)\theta^n$. This means that, for some forms of rate functions $\widetilde{\mathbf{f}}$, the computation of $V^n(\widetilde{\mathbf{x}})$ requires the knowledge of values $V^{n-1}(\widetilde{\mathbf{x}})$ for $\widetilde{\mathbf{x}}$ located outside of the range $\widetilde{\mathbf{x}}_{\mathrm{I}} \le \widetilde{\mathbf{x}} \le \widetilde{\mathbf{x}}_{\mathrm{II}}$. Therefore, to evaluate $V^n(\widetilde{\mathbf{x}})$ within the range satisfying $\widetilde{\mathbf{x}}_{\mathrm{I}} \le \widetilde{\mathbf{x}} \le \widetilde{\mathbf{x}}_{\mathrm{II}}$, it may be necessary to determine $V^{n-1}(\widetilde{\mathbf{x}})$ within a boundary which is larger than that described by the inequality $\widetilde{\mathbf{x}}_{\mathrm{I}} \le \widetilde{\mathbf{x}} \le \widetilde{\mathbf{x}}_{\mathrm{II}}$.

The procedure leads to the optimal values of V^n, θ^n and $\mathbf{u}^n$ stored at each node of the grid of $\widetilde{\mathbf{x}}$, for each n. These values constitute discrete representations of optimal functions $V^n(\widetilde{\mathbf{x}})$, $\theta^n(\widetilde{\mathbf{x}})$ and $\mathbf{u}^n(\widetilde{\mathbf{x}})$. Additionally, values of coordinates of the transformed state, $\widetilde{\mathbf{x}}^T \equiv \widetilde{\mathbf{x}} - \widetilde{\mathbf{f}}^n(\widetilde{\mathbf{x}}, \mathbf{u}^n, \theta^n)\theta^n$, can be stored. Data of $\widetilde{\mathbf{x}}^T$ describe optimal inlet states to the stage n in terms of outlet states from this stage, $\widetilde{\mathbf{x}}$.

Backward reading of the solution. Dynamic programming tables, which describe all computed data, can be used to find the solution of a particular (N-stage) problem in which final values of $\widetilde{\mathbf{x}}^N \equiv \widetilde{\mathbf{x}}^f$ and N are prescribed. This is a backward procedure in which we first identify in DP tables the final point $\widetilde{\mathbf{x}}^f$ for $n=N$, and, next, in these tables, we read off data of optimal controls $\theta^N\left(\widetilde{\mathbf{x}}^f\right)$ and $\mathbf{u}^N\left(\widetilde{\mathbf{x}}^f\right)$. In the tables we also find transformed outlet states $\widetilde{\mathbf{x}}^{TN} \equiv \widetilde{\mathbf{x}} - \widetilde{\mathbf{f}}^N\left(\widetilde{\mathbf{x}}, \mathbf{u}^N, \theta^N\right)\theta^N$, which are inlet states to the stage N.

Now we pass to the $N-1$ stage subprocess. It has its own outlet state $\widetilde{\mathbf{x}}^{N-1}$ which was just found as $\widetilde{\mathbf{x}}^{TN}$. By interpolating in the tables for $n=N-1$ we find all suitable data for the state $\widetilde{\mathbf{x}} = \widetilde{\mathbf{x}}^{N-1}$. We thus find optimal profit $V^{N-1}\left(\widetilde{\mathbf{x}}^{N-1}\right)$, optimal controls $\theta^{N-1}\left(\widetilde{\mathbf{x}}^{N-1}\right)$, $\mathbf{u}^{N-1}(\widetilde{\mathbf{x}}^{N-1})$, and inlet states to the stage $N-1$, $\widetilde{\mathbf{x}}^{N-2} = \widetilde{\mathbf{x}}^{TN-1}$. Continuing the procedure by the computer, we obtain an optimal solution as a sequence of optimal controls $\mathbf{u}^N$, $\mathbf{u}^{N-1}$..., $\mathbf{u}^1$ and θ^N, θ^{N-1}, ..., θ^1, and an optimal discrete trajectory, $\widetilde{\mathbf{x}}^N$, $\widetilde{\mathbf{x}}^{N-1}$..., $\widetilde{\mathbf{x}}^1$, $\widetilde{\mathbf{x}}^0$. Sequence of optimal costs for all related subprocesses V^N, V^{N-1}, ..., V^1 also follows.

The optimality principle excludes investigation of (a huge number of) all possible $n-1$ stage subprocesses, which causes extremely large savings of computational time. Another virtue of the DP method is that it always leads to absolute maximum, and, as opposed to other methods, an increase of number of constraints simplifies the computer search (fewer points to be tested). Functions $\widetilde{D}^n$ and $\mathbf{f}^n$ describing profit and state transformation need not be continuous or analytical; they may be in graphic or tabular form. Two-point boundary values do not cause problems either, as recurrence equation is not influenced by end conditions. Only evaluation of first function, $V^1(\widetilde{\mathbf{x}})$, changes depending on initial conditions. Large dimensionality of the control vector does not cause essential troubles.

It is very fortunate that many dynamical energy production problems (especially those without mass transfer) are of low state dimensionality. It is just the property that causes the easy solving procedure for Eq. (7.24).

7.9 Dimensionality reduction in dynamic programming algorithms

However, in energy problems with multicomponent mass transfer and/or chemical reactions a very serious difficulty arises connected with the use of the dynamic programming. This is the so-called "curse of dimensionality," referred to the large dimensionality of state vector, $\tilde{\mathbf{x}}$. Indeed, for $s = \dim \tilde{\mathbf{x}} = 1$, a single column of discrete set of $\tilde{\mathbf{x}}$ is sufficient, for $s=2$ the computational grid must constitute, say, a rectangle. For $s=3$, however, the grid must be cubic, for $s=4$ the data must be obtained (and stored) as a set of cubes, etc. Clearly, the number of computational points, and hence the computer memory requirements increase tremendously with state dimensionality s. Problems with $s=1$ and $s=2$ are quite easy to solve numerically, problems with higher $s=3$ are troublesome or serious, and problems with $s \geq 5$ are practically intractable if good accuracy is required. Therefore the numerical dynamic programming can effectively be applied only for problems characterized by the small dimensionality of the state vector $\tilde{\mathbf{x}}$; problems of large dimensionality, such as those encountered in the static optimization, are excluded. Fortunately, many dynamical problems of energy production are of low dimensionality. In the case of high s, other methods, especially those associated with maximum principles must be applied. These are based on canonical equations, state adjoints and a Hamiltonian function involving these variables, and are described in a number of sources (e.g., Aris, 1964; Boltyanski, 1969, 1973; Pontryagin et al., 1962; Sieniutycz, 1991). Sometimes, however, the dimensionality reduction is possible in DP problems. For the energy problems considered here dimensionality reduction is possible, among others, in autonomous systems due to the constancy of the time adjoint λ along an optimal path. This is described below.

For V^n regarded as energy production profit, net economic-like profit, or the difference between V^n and the "time penalty cost" $\lambda(t^n - t^0)$, can be defined. λ is Lagrange multiplier associated with time t. We shall designate by an asterisk subscript so-modified profits (or costs), and will focus on their properties in autonomous processes. When the discrete process is autonomous and time interval θ^n is not explicitly present in rates f_k, λ is identical with a constant Hamiltonian $H^{n-1} = H$. Under a weaker assumption of autonomous process and intervals θ^n explicit in rates f_k, the constancy property refers only to λ. To analyze both cases simultaneously we deal with modified optimal functions, net profits $V_*{}^n \equiv V^n - \lambda\,(t^n - t^0)$ or net costs $R_*{}^n \equiv R^n + \lambda\,(t^n - t^0)$, both criteria being equivalent because the second is obtained by multiplication of the first by the minus unity. The quantity λ describes the decrease of the original profit when the process duration is increased by one unit.

Local profits and costs are defined in a similar way. For single-stage profit $\widetilde{D}^n$, which appears in Eq. (7.75), a net profit is $\widetilde{D}_*^n \equiv \widetilde{D}^n - \lambda\theta^n$, where $\theta^n \equiv \Delta t^n$. Similarly total cost at the stage n is $\widetilde{K}_*^n \equiv \widetilde{K}^n + \lambda\theta^n$, where $\widetilde{K}^n \equiv -\widetilde{D}^n$.

Given net profit $\widetilde{D}_*^N$, optimal process is governed by a sequence of asterisk functions: $V_*^1, \ldots V_*^n..V_*^{N-1}$ and V_*^N. The sequence of these optimal functions obeys an equation

$$V_*^n(\mathbf{x}^n, \lambda) = \max_{\mathbf{u}^n, \theta^n} \left\{ \widetilde{D}_*^n(\mathbf{x}^n, \mathbf{u}^n, \theta^n, \lambda) + V_*^{n-1}(\mathbf{x}^n - \mathbf{f}^n(\mathbf{x}^n, \mathbf{u}^n, \theta^n)\theta^n, \lambda) \right\} \tag{7.83}$$

This differs from Eq. (7.75) by presence of vector $\mathbf{x}$ rather than $\widetilde{\mathbf{x}} \equiv (\mathbf{x}, t)$. Because of the constancy of λ along a discrete optimal path, state dimensionality of the problem described by Eq. (7.83) is decreased by one in comparison with that for Eq. (7.75). In fact, the continuous limit of Eq. (7.83) is a HJB equation

$$\max_u \left\{ \widetilde{f}_{0*}(\mathbf{x}, \mathbf{u}, \lambda) - \frac{\partial V_*(\mathbf{x}, \lambda)}{\partial \mathbf{x}} f(\mathbf{x}, \mathbf{u}, \lambda) \right\} = 0 \tag{7.84}$$

Again, this is similar to the Hamilton-Jacobi equation of classical mechanics but contains the maximizing sign before a Hamiltonian expression. Equation refers to an optimal duration $\mathcal{T} = (t^f - t^i)$ equal $\partial V_*(\mathbf{x}, \lambda)/\partial \lambda$.

Optimal functions of work production, V^n and V_*^n, preserve a number of basic qualitative properties of economic production profits and total economic profits. The same remark refers to functions of work consumption, $R^n = -V^n$ and $R_*^n = -V_*^n$. For multistage control processes, optimal data generated by DP always have form of sequence functions $V^n(\mathbf{x}, t)$ or their duals $V_*^n(\mathbf{x}, \lambda)$, where $\mathbf{x}$ is process state, t is a time variable, and n is the number of stages. Profit functions $V^n(\mathbf{x}, t)$ and $V_*^n(\mathbf{x}, \lambda)$, or cost functions $R^n(\mathbf{x}, t)$ and $R_*^n(\mathbf{x}, \lambda)$, are related by the Legendre transformation with respect to their independent variable (Leitman, 1966; Landau and Lifshitz, 1971; Sieniutycz, 1973). The limiting case of a continuous process is characterized by functions $V(\mathbf{x}, t)$ and $V_*^n(\mathbf{x}, \lambda)$, which are mathematical equivalents of Hamilton's principal action and abbreviated action in classical mechanics or related phase functions in optics (Landau and Lifshitz, 1971). The relation between the optimal cost functions generated by dynamic programming and Pontryagin's maximum principle is now well-understood (Leitman, 1966). The optimal paths of a control problem are equivalent to mechanical paths in mechanics or light rays in optics. The use of dynamic programming in constructing finite-time potentials for discrete and continuous control separation processes has been summarized (Berry et al., 2000). A computational example showing wave-path duality is available, dealing with separation process in which a volatile component is evaporated from a porous, fluidizing solid by a hot gas (Sieniutycz, 1973; Berry et al., 2000).

The mathematics of these approaches is independent of the specific applications. This is why they can be conducted first in frames of thermodynamic models and the experience gained serves to formulate and solve more involved problems of economics.

Again, in the above discussion, we have omitted procedures of discrete maximum principle (Boltyanski, 1973) as these do not generate data of optimal performance function.

7.10 Concluding remarks

In this chapter, we have presented a basic formulation for maximum power in dynamical power systems and considered convergence of discrete value functions to viscosity solutions of HJB equations. Lagrangian multipliers associated with duration constraint have been used to reduce dimensionality of some power production problems. Analytical and numerical approaches to power generation problems, applying the dynamic programming method, were described. Legendre transform was applied to recover optimal work as function $R^n(T^n, \tau)$.

In was shown that while arbitrary discretization of the original model violates in general constancy of discrete autonomous Hamiltonians H^n, one may preserve this constancy either by working in the coordinate frames in which discrete rates don't contain time intervals θ^n or by applying the Lagrange multiplier of the duration constraint which still exhibits the constancy property.

Generalized (time-dependent) work potentials can be found for nonlinear systems. They lead to the thermodynamic bounds on the power produced (consumed) with a finite rate. These are stronger than the classical bounds as they involve constraints coming from the process kinetics.

Other important application of the considered approach involves chemical energy systems, and especially, fuel cells. Fig. 7.2 above shows the comparison of optimal work production and

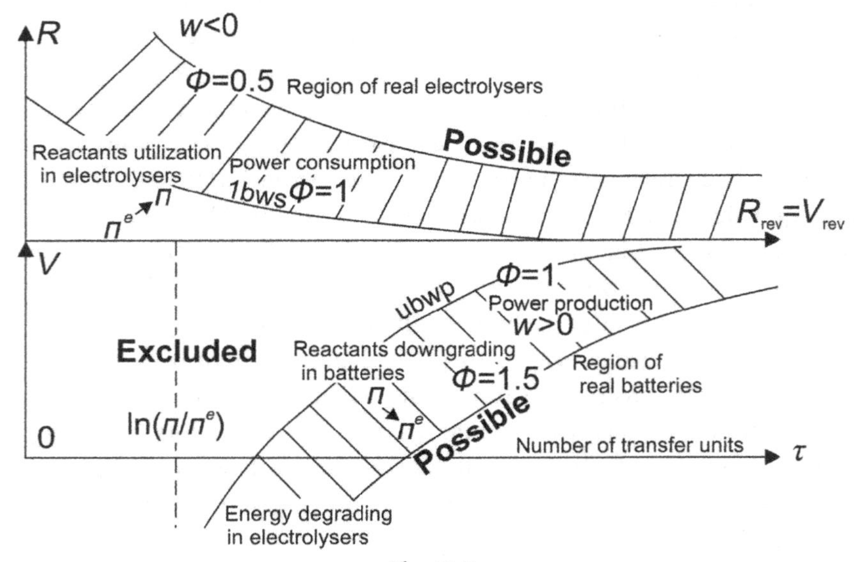

Fig. 7.2

Work limits for reversible and real electrolysers and batteries. Systems with work production are described by function $V(T,\tau)$, system with work consumption—by function $R(T,\tau)$. Endoreversible limits correspond with curves for $\Phi = 1$; weaker reversible limits are represented by the straight line $R_{rev} = V_{rev}$. *Dashed lines* mark regions of possible improvements when imperfect thermal machines are replaced by those with better performance coefficients, terminating at endoreversible limits for Carnot machines.

work consumption functions, V and R in terms of slowness variable τ, for a fixed change of system state. Reversible upper bound V_{rev} achieved in production modes is equal to reversible lower bound R_{rev} achieved in consumption modes. For the irreversible bounds, this equality does not hold, and a lower bound of R is larger than upper bound of V. Note a qualitative similarity of this plot of work limits to charts characterizing generalized exergies (Sieniutycz, 1998). This similarity is a suitable starting point to investigate energy yield in chemical and electrochemical systems.

References

Aris, R., 1964. Discrete Dynamic Programming. Blaisdell, New York.

Bardi, M., Capuzzo-Dolcetta, I., 1997. Optimal Control and Viscosity Solutions of Hamilton–Jacobi–Bellman Equations (With Appendices by Maurizio Falcone and Pierpaolo Soravia). Birkhauser, Boston.

Barles, G., 1994. Solutions de viscosite des equations de Hamilton–Jacobi. In: Collection Mathematiques et Applications de la SMAI 17. Springer Verlag, Paris.

Barles, G., Souganidis, P.M., 1991. Convergence of approximation schemes for fully nonlinear second order equations. Asymptot. Anal. 4, 271–283.

Bellman, R.E., 1961. Adaptive Control Processes: A Guided Tour. Princeton University Press, Princeton, pp. 1–35.

Berry, R.S., Kazakov, V.A., Sieniutycz, S., Szwast, Z., Tsirlin, A.M., 2000. Thermodynamic Optimization of Finite Time Processes. Wiley, Chichester, p. 117.

Boltyanski, V.G., 1969. Mathematical Methods of Optimal Control. Nauka, Moscow.

Boltyanski, V.G., 1973. Optimal Control of Discrete Systems. Nauka, Moscow.

Capuzzo-Dolcetta, I., Perthame, P., 1996. On some analogy between different approaches to first order PDE's with nonsmooth coefficients. Adv. Math. Sci. Appl. 6, 689–703.

Coclite, G.M., Risebro, N.H., 2003. Viscosity solutions of Hamilton–Jacobi equations with discontinuous coefficients. Preprint, 2003–2023. http://www.math.ntnu.no/conservation/2003/023.html.

Coclite, G.M., Risebro, N.H., 2005. Conservation laws with a time dependent discontinuous coefficients. SIAM J. Math. Anal. 36 (4), 1293–1309.

Coclite, G.M., Risebro, N.H., 2007. The research report founded by the BeMatA program of the Research Council of Norway and by the European network HYKE, funded by the EC as contract HPRN-CT-2002-00282. November 7, 2007.

Crandall, M.G., Lions, P.L., 1983. Viscosity solutions of Hamilton–Jacobi equations. Trans. Am. Math. Soc. 277, 1–42.

Crandall, M.G., Ishii, H., Lions, P.-L., 1992. User's guide to viscosity solutions of second order partial differential equations. Bull. Am. Math. Soc. 27 (1), 1–67.

Dal Maso, G., Frankowska, H., 2002. Autonomous integral functionals with discontinuous nonconvex integrands: Lipschitz regularity of minimizers, DuBois–Reymond necessary conditions, and Hamilton–Jacobi equations. Preprint, available at http://cvgmt.sns.it/papers/dalfra02/.

Falcone, M., Ferretti, R., 1994. Discrete time high-order schemes for viscosity solutions of Hamilton–Jacobi–Bellman equations. Numer. Math. 67, 315–344.

Falcone, M., Ferretti, R., 2006. Editorial. Appl. Numer. Math. 56, 1135.

Falcone, M., Makridakis, C., 2001. Numerical Methods for Viscosity Solutions and Applications. World Scientific, Singapore.

Fialho, I.J., Georgiu, T.T., 1999. Worst case analysis of nonlinear systems. IEEE Trans. Autom. Control 44, 1180–1196.

Ishii, H., 1985. Hamilton–Jacobi equations with discontinuous Hamiltonians on arbitrary open sets. Bull. Fac. Sci. Eng. Chuo Univ. 28, 33–77. See also Indiana Univ. Math J. 33, 721–749.

Ishii, H., Ramaswamy, M., 1995. Uniqueness results for a class of Hamilton–Jacobi equations with singular coefficients. Comm. Partial Differ. Equ. 20, 2187–2213.

Jin, S., Liao, X., 2006. A Hamiltonian-preserving scheme for high frequency elastic waves in heterogeneous media. J. Hyperbolic Differential Equations 3, 741–777.

Jin, S., Wen, X., 2005. Hamiltonian-preserving schemes for the Liouville equation with discontinuous potentials. Commun. Math. Sci. 3, 285–315.

Jin, S., Wen, X., 2006a. A Hamiltonian-preserving scheme for the Liouville equation of geometrical optics with transmissions and reflections. SIAM J. Numer. Anal. 44, 1801–1828.

Jin, S., Wen, X., 2006b. Hamiltonian-preserving schemes for the Liouville equation of geometrical optics with discontinuous local wave speeds. J. Comput. Phys. 214, 672–697.

Jin, S., Wu, H., Huang, Z., 2007. A hybrid phase-flow method for Hamiltonian systems with discontinuous Hamiltonians. private communication of S. Jin.

Jordan, B.W., Polak, E., 1964a. Theory of a class of discrete optimal control systems. J. Electron. Control 17, 697–713. See also Jordan, B.W., Polak, E., 1964b. Optimal control of aperiodic discrete time systems. SIAM J. Control Ser. A 2, 332–338.

Jordan, B.W., Polak, E., 1964b. Optimal control of aperiodic discrete time systems. SIAM J. Control Ser. A 2, 332–338.

Karlsen, K.H., Risebro, N.H., Towers, J.D., 2003. L1 stability for entropy solutions of nonlinear degenerate parabolic convection-diffusion equations with discontinuous coefficients. Preprint. Available at: www.math.ntnu.no/conservation.

Leitman, G., 1966. An Introduction to Optimal Control. McGraw-Hill, New York.

Landau, L.D., Lifshitz, E., 1971. Mechanics. Pergamon, Oxford.

Marsden, J.E., 1992. Lectures on Mechanics. Cambridge University Press, Cambridge.

McLachlan, R.I., 1993. Explicit Lie–Poisson integration and the Euler equations. Phys. Rev. Lett. 71, 3043.

McLachlan, R.I., Quispel, G.R.W., 1998. Generating functions for dynamical systems with symmetries, integrals and differential invariants. Physica D 112, 298–309.

McLachlan, R.I., Quispel, G.R.W., Robidoux, N., 1999. Geometric integration using discrete gradients. Philos. Trans. R. Soc. Lond. A 357, 1021–1045.

Mordukhovich, B.S., 1988. Approximate maximum principle for finite-difference control systems. Comput. Math. Math. Phys. 28, 106–114.

Mordukhovich, B.S., 1995. Discrete approximations and refined Euler–Lagrange conditions for nonconvex differential inclusions. SIAM J. Control. Optim. 33, 882–915.

Mordukhovich, B.S., Shvartsman, I., 2004. The approximate maximum principle for constrained control systems. SIAM J. Control. Optim. 43, 1037–1062.

Munos, R., Moore, A., 2001. Variable resolution discretization in optimal control. Mach. Learn. 1, 1–31.

Ostrov, D.N., 1999. Viscosity solutions and convergence of monotone schemes for synthetic aperture radar shape from-shading equations with discontinuous intensities. SJAM 59, 2060–2085.

Ostrov, D.N., 2000. Extending viscosity solutions to Eikonal equations with discontinuous spatial dependence. Nonlinear Anal. 42, 709–736.

Pontryagin, L.S., Boltyanski, V.G, Gamkrelidze, R.V., Mishchenko, E.F., 1962. The Mathematical Theory of Optimal Processes. Interscience Publishers, New York.

Quispel, G.R.W., 1995. Volume preserving integrators. Phys. Lett. A 206, 26–30.

Quispel, G.R.W., Dyt, C.P., 1998. Volume preserving integrators have linear error growth. Phys. Lett. A 242, 25–30. and references therein.

Rund, H., 1966. The Hamilton–Jacobi Theory in the Calculus of Variations. Van Nostrand, London, pp. 1–32.

Sieniutycz, S., 1973. The thermodynamic approach to fluidized drying and moistening optimization. AICHE J. 19, 277–285.

Sieniutycz, S., 1991. Optimization in Process Engineering, second ed. Wydawnictwa Naukowo Techniczne, Warsaw, pp. 151–194.

Sieniutycz, S., 1998. Hamilton–Jacobi–Bellman theory of irreversible thermal exergy. Int. J. Heat Mass Transf. 41, 183–195.
Sieniutycz, S., 2003a. A synthesis of thermodynamic models unifying traditional and work-driven operations with heat and mass exchange. Open. Syst. Inf. Dyn. 10, 31–49.
Sieniutycz, S., 2003b. Carnot controls to unify traditional and work-assisted operations with heat and mass transfer. Int. J. Appl. Thermodyn. 6, 59–67.
Sieniutycz, S., 2006a. Thermodynamic limits in applications of energy of solar radiation. Dry. Technol. 24, 1139–1146.
Sieniutycz, S., 2006b. State transformations and Hamiltonian structures for optimal control in discrete systems. Rep. Math. Phys. 49, 289–317.
Sieniutycz, S., 2007a. Hamilton–Jacobi–Bellman equations and dynamic programming for power-maximizing relaxation of radiation. Int. J. Heat Mass Transfer 50, 2714–2732.
Sieniutycz, S., 2007b. Dynamical converters with power-producing relaxation of solar radiation. Int. J. Therm. Sci. 66, 219–231.
Sieniutycz, S., Jeżowski, J., 2018. Energy Optimization in Process Systems and Fuel Cells, third ed. Elsevier, Oxford. (Especially Chapter 2).
Sieniutycz, S., Kuran, P., 2005. Nonlinear models for mechanical energy production in imperfect generators driven by thermal or solar energy. Int. J. Heat Mass Transf. 48, 719–730.
Sieniutycz, S., Kuran, P., 2006. Modeling thermal behavior and work flux in finite rate systems with radiation. Int. J. Heat Mass Transf. 49, 3264–3283.
Stromberg, T., 2003. On viscosity solutions of irregular Hamilton–Jacobi equations. Arch. Math. 81 (6), 678–688.
Szwast, Z., 1979. Discrete Algorithms of Maximum Principle with a Constant Hamiltonian and their Selected Applications in Chemical Engineering. (Ph.D. thesis). Institute of Chemical Engineering, Warsaw University of Technology.
Szwast, Z., 1988. Enhanced version of a discrete algorithm for optimization with a constant Hamiltonian. Inz. Chem. Proc. 3, 529–545.
Tourin, A., 1992. A comparison theorem for a piecewise Lipschitz continuous Hamiltonian and application to shape from shading problems. Numer. Math. 62, 75–85.

Further reading

Recio-Garrido, D., Perrier, M., Tartakovsky, B., 2016. Review: Modeling, optimization and control of bioelectrochemical systems. Chem. Eng. J. 289, 180–190.
Sieniutycz, S., Jeżowski, J., 2009a. Optimal decisions for chemical and electrochemical reactors. In: Energy Optimization in Process Systems. first ed. Elsevier, Oxford, pp. 321–366. (Chapter 9).
Sieniutycz, S., Jeżowski, J. 2009b. Energy Optimization in Process Systems and Fuel Cells, first ed. Elsevier, Oxford, 2009.
Sieniutycz, S., Jeżowski, J., 2013. Energy Optimization in Process Systems and Fuel Cells, second ed. Elsevier, Oxford.

CHAPTER 8

Chemical systems with catalyst decay and regeneration

8.1 Mathematical model for kinetics of reaction and deactivation

In this chapter, by using the maximum principle theory and its equations, an algorithm applicable to distributed parameter systems is obtained and then used to determine optimal temperature control for reactors with deactivating catalyst. The algorithm involves differential equations which define an extremum control in the case of two state variables. The state equations contain reaction extent, x, and catalyst activity, a, as the state variables and temperature, T, as a single control. The theory has been applied to obtain an analytical solution for the extremum temperature control in the fixed bed reactor with deactivating catalyst in the case of first-order reaction and deactivation. Application of the algorithm shows that the optimal profile for the effective reaction constant, $k_{\text{eff}}(t)$, is formed as the result of prescribed activity distributions, initial $w_p(\tau)$ and final $w_k(\tau)$. When the chronological time t_k increases, the effective reaction rate constant corresponds to higher reaction temperatures. Decrease of the operation time causes the time-dependent temperature profiles $T(\tau, t)$. This function may be associated with too high temperatures at the final moments of the process, which may cause excessive corrosion of the reactor material. Therefore, a generalized optimization problem with constrained temperature control should be formulated and used for short duration processes. This problem is considered along with the analysis of the cycle composed of an infinite number of deactivation and regeneration steps working sequentially in a steady cyclic operation.

Further, a cocurrent tubular reactor with temperature profile control and recycle of moving deactivating catalyst is investigated. For the temperature-dependent catalyst deactivation, the optimization problem has been formulated in which a maximum of a profit flux is achieved by the best choice of temperature profile and residence time of reactants for the set of catalytic reactions $A+B \rightarrow R$ and $R+B \rightarrow S$ with the desired product R. The rates of reactions have been described separately for every reagent by expressions containing reaction rate constants, concentrations of reagents, catalyst activity, as well as the catalyst concentration in the reacting suspension and a measure of the slip between reagents and solid catalyst particles. Algorithms of maximum principle have been used for optimization. The optimal solutions show that a shape of the optimal temperature profile depends on mutual relations between

Complexity and Complex Thermo-Economic Systems. https://doi.org/10.1016/B978-0-12-818594-0.00008-8

activation energies of reactions and catalyst deactivation. It has been proved that the optimal temperature profile is a result of the following factors: compromise between the overall production rate of desired reagent R (production rate in the first reaction minus disappearance rate in the second one), catalyst saving, and saving of reagents' residence time (reactor volume). The most important influence on the optimal temperature profile is associated with the necessity of saving the catalyst. When the catalyst recycle ratio increases, the number of catalyst particles in the reactor increases, and the optimal temperatures save the catalyst because the optimal profile is shifted in direction of lower temperatures. The same is observed when the catalyst slip increases. Despite variation in the catalyst concentration, the optimal profile is practically the same because the decay rate is affected only by the instantaneous activity of catalyst. When the reactor unit volume price decreases, catalyst residence time increases, whereas the optimal temperature profile is shifted to lower temperatures. When the economic value of unit activity of outlet catalyst increases (catalyst with a residual activity still has an economic value), catalyst saving should be more and more intense. As far as possible, the optimal profile is shifted in the direction of lower temperatures, whereas the optimal residence time is still the same. In this case, the optimal profile is isothermal at the level of minimum allowable temperature, whereas the catalyst is saved as its residence time in reactor decreases. All the above led to a condensed picture of optimal systems, the picture which still is by necessity simplified because of the inherent process complexity.

8.1.1 A quantitative description of reaction-catalyst deactivation phenomena

The catalyst usually deactivates itself, that is, its activity decreases in time. In the literature, the notion of activity is used either to compare the reaction acceleration properties of various catalysts or to compare the different states of the same catalyst. Below, the catalyst activity is considered in the latter meaning, consequently, the activity variable, a, is defined as the quotient of the reaction rate in a given moment, r, to the reaction rate r_0 in the presence of the fresh catalyst.

$$a=\frac{r}{r_0} \tag{8.1}$$

Catalyst deactivation is a compound phenomenon. Several fundamental mechanisms have been set apart. Aging of the catalyst, which especially occurs in high temperatures, causes the reduction of catalyst's active surface. Poisoning refers to the catalyst deactivation caused by the contaminated stream of reacting substances; in this case, the decay results from the deposition of nonvolatile substances. These mechanisms, which were described in Chapter 5 of a book (Sieniutycz and Szwast, 1982), may work one at a time or simultaneously. Despite the phenomenon's complexity, it was possible to obtain valuable generalizations of its modeling from the perspective of deactivation kinetics (Szepe, 1966; Szepe and Levenspiel, 1968). Some of these issues are discussed below.

Let us consider the rate at which the catalyst's activity diminishes—da/dt. Generally, it depends on operational conditions: temperature Tp, concentrations $\mathbf{c}$, and the current catalyst state a, that is,

$$-\frac{da}{dt}=\rho(T,\mathbf{c},a)\quad c=(c_1,\ c_2,\ \ldots,\ c_s) \tag{8.2}$$

Eq. (8.2) determines the deactivation process rate. The simplest function ρ has a separated structure, that is,

$$-\frac{da}{dt}=k_d(T)\phi(\mathbf{c})\psi(a) \tag{8.3}$$

It has been proved that kinetic equations of this structure provide good approximations for a large number of deactivation processes. They do not, however, include situations when activity can increase in time or deactivation is reversible (Szepe, 1966; Szepe and Levenspiel, 1968). For equations similar to (8.3), effects of temperature, concentrations, and activities may be examined independently. In Eq. (8.3), the temperature-dependent term has the Arrhenius structure, that is,

$$k_d(T)\equiv k_{d0}\exp\left(\frac{-E_d}{RT}\right) \tag{8.4}$$

where E_d is the activation energy for the deactivation process of the catalyst. In most cases, the value E_d is positive, that is, the temperature increase increases the deactivation rate.

On the ground of a large number of experiments (Szepe and Levenspiel, 1968), it has been proven that the activity term of Eq. (8.3), in the simplest but most frequent case, is described well by the power function

$$\psi(a)=a^m \tag{8.5}$$

where m is a nonnegative constant known as the *deactivation order*.

What is more, the concentration term of the Eq. (8.3) is also, in its simplest form, a power function, which generally comprises all components of the system (substrates, products, and poisons). In applications, considerable and frequent case refers to the constant function,

$$\phi(\mathbf{c})=1 \tag{8.6}$$

that is, to the case when the process temperature has a main influence on the activity changes. Deactivation is then called the *concentration independent.*

Further, we shall consider a simple form of a single nonreversible reaction $A\rightarrow B$, in which the rate constant shows the Arrhenius structure. From the activity definition, Eq. (8.1), the following equation is accepted for the catalytic reaction rate:

$$-\frac{dc}{dt}=k_0\exp\left(\frac{-E}{RT}\right)af(c) \tag{8.7}$$

By definition, the product of the reaction rate constant k and activity a is referred to as the effective reaction rate constant.

$$k_{ef} \equiv ka = k_0 \exp\left(\frac{-E}{RT}\right)a$$

Substitution of Eq. (8.4) into Eq. (8.3) yields an equation that describes changes of activity

$$-\frac{da}{dt} = k_{d0} \exp\left(\frac{-E_d}{RT}\right)\psi(a)\phi(\mathbf{c}). \tag{8.8}$$

Eqs. (8.7), (8.8) constitute the model kinetics of reaction and deactivation in single irreversible reaction $A \rightarrow B$ of any order.

8.2 Optimal temperatures for single catalytic reaction in a batch reactor and moving bed reactors

8.2.1 General remarks

In the following, optimization of a batch reactor and reactors with the moving catalyst bed—both cocurrent and countercurrent—is examined simultaneously for the purpose of the analysis shortening.

For batch reactors, optimization usually entails the following problems: (a) how to operate between regenerations (operational problem) and (b) when to regenerate catalysts and when to dispose them (regeneration problem). This section covers only the first of the listed issues, which includes finding the optimal operational temperature policy $\widetilde{T}(t)$ that maximizes the final conversion rate, maintaining fixed the following values: final time t_k and final catalyst activity a_k.

Concurrently, the optimization of an unsteady batch reactor is pursued simultaneously with the optimization of reactors with moving packing (cocurrent and countercurrent) which operate at steady state in time. In this case, the ability to operate the simple generalization of kinetic models (8.7) and (8.8), which is common for all three reactor types, has been applied. The joint nature of the model applies to the operational problem only and remains formal; in batch reactors slowly deactivating catalysts are used, whereas in flow reactors with moving packing catalysts of short half-life are exploited. Therefore, despite common elements in analysis, applications usually concern different reactions.

The present analysis focuses on a single irreversible reaction $A \rightarrow B$, accelerated by the deactivating solid catalyst and occurring in the liquid phase that contains both reactants and products of the reaction. It is assumed that the flow of fluid and catalyst in the reactor is piston-like, and total space times of fluid and solid are Θ_r and Θ, respectively. For the batch reactors, perfect mixing of the both phases is assumed. It is also assumed that in each reactor the both phases are in the thermal equilibrium.

8.2.2 Common set of equations and formulation of the problem

Now, a common system of equations for the considered processes is formulated and examined. Fig. 8.7 depicts two fundamental contacting schemes of fluid and catalyst as well as designations of basic parameters. From the mathematical perspective, batch reactor model may be regarded as a special case of a cocurrent reactor model if total space times of both moving phases in the latter are equal to the chronological time in the batch process.

All considered cases are described by the following system of equations:

$$\frac{dc}{dt} = -k_0' \exp\left(\frac{-Ey}{R}\right) af(c) = f_1(c, a, y) \tag{8.9}$$

$$\frac{da}{dt} = -k_{d0} \exp\left(\frac{-E_d y}{R}\right) \psi(a)\phi(\mathbf{c}) = f_2(c, a, y) \tag{8.10}$$

where, $y = T^{-1}$; and t, solid residence time. The constant k_0' is defined as

$$k_0' \equiv \begin{matrix} k_0\Theta_r/\Theta & \text{cocurrent} \\ -k_0\Theta_r/\Theta & \text{countercurrent} \\ k_0 & \text{batch} \end{matrix} \tag{8.11}$$

where Θ and Θ_r, total space time for solid stream and reagents stream, respectively. Clearly, $\Theta = t_k - t_p$ and $\Theta_r = t_{rk} - t_{rp}$. Functions $f(c)$ and $\phi(c)$ in Eqs. (8.9), (8.10) describe the effect of concentration, while functions $g(a) = a$ and $\psi(a)$—the effect of activity on reaction rate and deactivation rate, respectively. Considerations regarding specific forms of the discussed functions have been omitted. However, as it has already been mentioned, one of the most popular forms is the exponential function (Szepe, 1966; Szepe and Levenspiel, 1968). In the light of diversity and quite common uncertainty of these forms, our attention will be directed towards deriving general optimal relationships, which would enable us to draw some conclusions without the knowledge of the detailed data.

In the model of optimization defined in Eqs. (8.9), (8.10), state variables comprise molar concentration c of a reactant and activity of the catalyst a.

The reciprocal of the absolute process temperature $y = T^{-1}$ is the only decision variable.

Optimization of cocurrent processes (Fig. 8.1) may be formulated as the minimization of the final concentration c_k for the prescribed space time of the solid grains Θ and the initial and final catalyst activity (respectively, a_p and a_k). For batch processes, the problem formulation is identical, except that t_k denotes the chronological time. For countercurrent processes, it is more

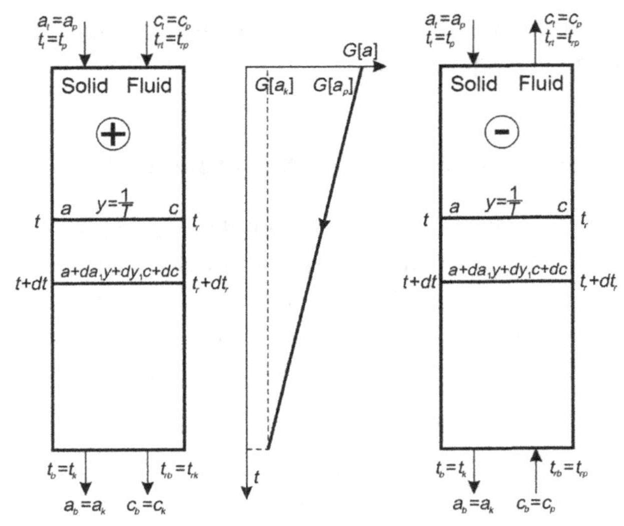

Fig. 8.1

Two ways of contacting (cocurrent and countercurrent). Function $G(a)$ as the property describing the stationary optimal activity in terms of time t. In Figs. 8.1–8.3, labels with signs + and − correspond, respectively, to cocurrent and batch processes or to the countercurrent process.

convenient to employ the alternative formulation, which assumes that the input concentration c_p is maximized, while the output concentration c_k is fixed. Variables Θ, a_p, and a_k are prescribed as in the previous case.

For optimization, the case of $E_d > 0$, analyzed in detail below, is of particular interest (if $E_d < 0$, then the deactivation rate decreases along with the temperature increase, and the maximum conversion of irreversible reactions is obtained at the highest permissible temperature; while the isothermal policy $\widetilde{T} = T^*$ is also applicable if $E_d = 0$). If $E_d > 0$, the reaction rate is accelerated by the temperature increase and retarded by the decrease of catalyst activity. The role of optimization is to find a policy that constitutes the best possible result of the two above-mentioned effects.

First, the finding of a stationary optimum is considered. Next, the results are used for the analysis of the optimization problem in the face of a temperature restriction.

8.2.3 Method of elimination of adjoint variables

The easiest way of solving the problem formulated above in its stationary version is proven to be the use of a method which leads directly to the differential equation for the stationary optimal decision, thanks to the elimination of conjugate variables in the formulation of the maximum principle (Chapters 1 and 12 in Sieniutycz and Szwast, 2018, for example). Such elimination is possible in the presence of two state variables x_1 and x_2 and one control variable u. In such a case, the system of equations for the maximum principle with Hamiltonian

$$H = z_1 f_1 + z_2 f_2$$

has the following structure:

Equations of state:

$$\frac{dx_1}{dt} = f_1(x_1, x_2, u, t) \tag{8.12}$$

$$\frac{dx_2}{dt} = f_2(x_1, x_2, u, t) \tag{8.13}$$

Adjoint equations:

$$\frac{dz_1}{dt} = -z_1 \frac{\partial f_1}{\partial x_1} - z_2 \frac{\partial f_2}{\partial x_1} \tag{8.14}$$

$$\frac{dz_2}{dt} = -z_1 \frac{\partial f_1}{\partial x_2} - z_2 \frac{\partial f_2}{\partial x_2} \tag{8.15}$$

Hamiltonian stationarity condition is

$$z_1 \frac{\partial f_1}{\partial u} + z_2 \frac{\partial f_2}{\partial u} = 0 \tag{8.16}$$

From Eq. (8.16) $z_2 = -z_1 \left(\frac{\partial f_1}{\partial u} / \frac{\partial f_2}{\partial u} \right)$, and this is substituted to (8.15). This yields

$$z_1 \frac{d}{dt} \left(\frac{\partial f_1}{\partial u} / \frac{\partial f_2}{\partial u} \right) + \frac{dz_1}{dt} \left(\frac{\partial f_1}{\partial u} / \frac{\partial f_2}{\partial u} \right) = z_1 \frac{\partial f_1}{\partial x_2} + z_2 \frac{\partial f_2}{\partial x_2} \tag{8.17}$$

Once the Eq. (8.14) is used for the purpose of eliminating the derivative dz_1/dt, Eq. (8.17) takes the following form:

$$z_1 \frac{d}{dt} \left(\frac{\partial f_1}{\partial u} / \frac{\partial f_2}{\partial u} \right) + \left(z_1 \frac{\partial f_1}{\partial x_1} + z_2 \frac{\partial f_2}{\partial x_1} \right) \left(\frac{\partial f_1}{\partial u} / \frac{\partial f_2}{\partial u} \right) = z_1 \frac{\partial f_1}{\partial x_2} + z_2 \frac{\partial f_2}{\partial x_2} \tag{8.18}$$

Substitution of z_2 from the formula (8.16) into Eq. (8.18) gives, after simplification by $z_1 \neq 0$ and rearrangement,

$$\frac{dZ}{dt} + Z \left(\frac{\partial f_2}{\partial x_2} - \frac{\partial f_1}{\partial x_1} \right) + Z^2 \frac{\partial f_2}{\partial x_1} - \frac{\partial f_1}{\partial x_2} = 0 \tag{8.19}$$

where

$$Z = Z(x_1, x_2, u, t) \equiv \frac{\partial f_1}{\partial u} / \frac{\partial f_2}{\partial u} \tag{8.20}$$

Eq. (8.19) employs only the variable: x_1, x_2, u, t. Since

$$\frac{dZ}{dt}=\frac{\partial Z}{\partial t}+\frac{\partial Z}{\partial x_1}f_1+\frac{\partial Z}{\partial x_2}f_2+\frac{\partial Z}{\partial u}\frac{du}{dt} \tag{8.21}$$

thus, substitution of the Eq. (8.21) into Eq. (8.19) produces the following differential equation for the optimal control:

$$\frac{du}{dt}=\left(\frac{\partial Z}{\partial u}\right)^{-1}\left[\frac{\partial f_1}{\partial x_2}-\frac{\partial f_2}{\partial x_1}Z^2+\left(\frac{\partial f_1}{\partial x_1}-\frac{\partial f_2}{\partial x_2}\right)Z-f_1\frac{\partial Z}{\partial x_1}-f_2\frac{\partial Z}{\partial x_2}-\frac{\partial Z}{\partial t}\right] \tag{8.22}$$

which has the implicit form: $du/dt=g(x_1,x_2,u,t)$. It is called Horn's equation (Horn, 1961).

Eq. (8.22) is a suitable tool to derive equations describing optimal temperature profile in systems of chemical reactions described by two state variables x_1 and x_2 and one control variable u. For its early applications, see Horn (1961), Sieniutycz (1976, 1991), and Szwast (1994).

8.2.4 Optimality conditions for reaction and catalyst deactivation

For the analyzed processes with catalyst deactivation, Eq. (8.22) assumes the following form:

$$\frac{dy}{dt}=\left(\frac{\partial Z}{\partial y}\right)^{-1}\left[\frac{\partial f_1}{\partial a}-\frac{\partial f_2}{\partial c}Z^2+\left(\frac{\partial f_1}{\partial c}-\frac{\partial f_2}{\partial a}\right)Z-f_1\frac{\partial Z}{\partial c}-f_2\frac{\partial Z}{\partial a}-\frac{\partial Z}{\partial t}\right] \tag{8.23}$$

while for the model that is described with the Eqs. (8.9), (8.10)

$$Z=\frac{Ek_0'}{E_d k_{do}}\exp\left(\frac{-(E-E_d)y}{R}\right)\frac{af(c)}{\psi(a)\phi(c)} \tag{8.24}$$

The use of the Eq. (8.24) in (8.23) after some transformations leads to the following equation Sieniutycz (1976)

$$\frac{dy}{dt}=\frac{-k_{d0}R}{E}\exp\left(\frac{-E_d y}{R}\right)\frac{\psi(a)\phi(\mathbf{c})}{a}-\frac{k_0'R}{E_d}\exp\left(\frac{-Ey}{R}\right)\frac{af(c)\phi'(c)}{\phi(c)} \tag{8.25}$$

The analysis of the second Hamiltonian derivative $\partial^2H/\partial y^2$ proves that Eq. (8.24) and other stationary relationships are applicable to the optimal process only if

$$E_d>E \tag{8.26}$$

If $E_d<E$ and $E_d=E$ (all policies are equivalent), then the optimal policy consists in the employment of the maximum admissible temperature $T=T^*$. This results from the fact that if $E_d<E$, the increase in temperature favors the reaction rate to the deactivation rate's disadvantage. However, if $E_d>E$, the increase of the deactivation rate prevails. However, temperature cannot be too low, since it would make the final conversion too small. That is why this condition (8.25) is the best compromise in a situation when $E_d > E$.

According to Eq. (8.25), the structure of the optimal policy $y(t)$ depends on the contacting scheme between phases (cocurrent, countercurrent batch). If deactivation is independent of concentration, the Eq. (8.25) is reduced to the form:

$$\frac{dy}{dt} = \frac{-k_{d0}R}{E} \exp\left(\frac{-E_d y}{R}\right) \frac{\psi(a)}{a} \tag{8.27}$$

which demonstrates that in this particular case the optimal policy is common for each of the three exchange types. This is directly the consequence of the qualities of optimal activity profiles, which is discussed further.

In order to find optimal trajectories $c(t)$ and $a(t)$ as well as the optimal policy $y(t)$, one must solve the system comprised of Eqs. (8.9), (8.10), (8.25). By analyzing the right-hand side of Eq. (8.25) and comparing it with Eqs. (8.9), (8.10), an observation can be made that Eq. (8.25) can be presented in the following form:

$$\frac{dy}{dt} = \frac{R}{E}(\ln a)' \frac{da}{dt} + \frac{R}{E_d}[\ln \phi(c)]' \frac{dc}{dt} \tag{8.28}$$

and as such can be easily integrated, producing

$$\exp\left(\frac{-Ey}{R}\right) a\phi(c)^{E/E_d} = C_1 \tag{8.29}$$

In case of the deactivation independent from concentration, Eq. (8.29) indicates that in optimal processes

$$k_{eff} = ka = \text{constant} \tag{8.30}$$

meaning that an optimal process is required to maintain the effective reaction rate constant k_{eff} at the unchanged level.

If deactivation and concentration are interrelated, the optimal policy consists in maintaining the constant product of the effective constant rate and the specified function of concentration, compliant with Eq. (8.29).

The discussed conditions are very simple and general. Since activity a decreases in time, in case of the concentration-independent deactivation, $\phi(c) = 1$, the optimal temperature $T(t)$ should be monotonically increasing function of time, according to Eq. (8.30). What is more, in view of the constancy of k_{eff} for $\phi(c) = 1$, concentration changes in processes with deactivation are the same as those in an isothermal reaction without deactivation. The concentration profile is then easily acquired through the integration of Eq. (8.9), in which the product of all factors preceding the function $f(c)$ is constant and equal to k_{eff}.

The analysis in case of $f(c) \neq 1$ is not necessarily more difficult. By combining Eq. (8.29) with Eqs. (8.9), (8.10), a system of equations is obtained

$$\frac{dc}{dt} = -k_0' C_1 f(c) \phi(c)^{(-E/E_d)} \tag{8.31}$$

$$\frac{da}{dt} = -k_{d0} C_1^{E_d/E} \psi(a) a^{(-E_d/E)} \tag{8.32}$$

that easily leads to solutions $c(t)$ and $a(t)$ because dependent variables are separated. In this solution, the following functions of optimal concentration and optimal activity will appear:

$$F(c) = \int_0^c \phi(\gamma)^{E/E_d} \psi(\gamma)^{-1} d\gamma \tag{8.33}$$

$$G(c) = \int_1^a g(\gamma)^{E_d/E} \psi(\gamma)^{-1} d\gamma \tag{8.34}$$

By using Eq. (8.34), one can easily found the constant C_1 after integration of Eq. (8.32) with conditions $a(t_p) = a_p$ and $a(t_k) = a_k$. Therefore

$$C_1 = \left\{ \frac{G(a_k) - G(a_p)}{-k_{do}(t_k - t_p)} \right\}^{E/E_d} \tag{8.35}$$

By employing Eq. (8.35), it is possible to integrate Eq. (8.31) between time t_p and t in cocurrent and batch processes. In countercurrent processes, integration is conducted between times t and t_k. Schemes of all these processes are shown on Fig. 8.1. The equation that describes the optimal concentration profile takes the following form:

$$F(c) = -k_0' \left[\frac{G(a_k) - G(a_p)}{k_{do}(t_k - t_p)} \right]^{E/E_d} (t - t_p) + F(c_p) \tag{8.36}$$

for the concurrent and batch processes and the form

$$F(c) = -k_0' \left[\frac{G(a_k) - G(a_p)}{k_{do}(t_k - t_p)} \right]^{E/E_d} (t - t_k) + F(c_p) \tag{8.37}$$

for the countercurrent process.

The use of an identical procedure in order to solve the Eq. (8.32) produces the optimal activity profile that is expressed as the function

$$G(a) = \frac{[G(a_k) - G(a_p)]t + G(a_p)t_k - G(a_k)t_p}{t_k - t_p} \tag{8.38}$$

which is common for all the processes considered herein.

By employing Eq. (8.35) for constant C_1 in Eq. (8.29) and converting the result, one obtains an equation for the temperature reciprocal $y(t)$

$$y(t) = \frac{1}{T(t)} = \frac{R}{E}\ln[a(t)] + \frac{R}{E_d}\{\ln\phi[c(t)]\} - \frac{R}{E_d}\ln\frac{G(a_k) - G(a_p)}{-k_{do}(t_k - t_p)} \tag{8.39}$$

that describes the optimal stationary temperature, provided the condition (8.26) is satisfied and subject to the restriction that boundary values a_p, a_k, t_p and t_k lead to positive temperatures T. Eq. (8.39) uses functions $a(t)$ and $c(t)$ appearing in Eqs. (8.36) to (8.38).

The question arises whether Eqs. (8.36), (8.37) match suitable application forms since their form enables the determination of the function $t(c)$ rather than $c(t)$. Nevertheless, this is insignificant since it is possible to operate with diagrams $t - c$, produced on the ground of these equations, which facilitate reading concentrations c for any time t. Analogical diagrams may be produced for the activity by using Eq. (8.38). Then, a diagram of the optimal temperature profile is drawn on the basis of Eq. (8.39). Despite the fact that in specific cases one may avoid using diagrams, the above-described procedure is the most general one—it is applied to any functions $f(c)$, $\phi(c)$, and $\psi(a)$ in Eqs. (8.9), (8.10).

8.2.5 Properties of stationary profiles of optimal activity, concentration and temperature

Eq. (8.38) indicates that stationary optimal activity profiles $a(t)$ do not differ between cocurrent and countercurrent arrangements. This is consistent with Fig. 8.1 where they all are characterized by the same straight line in coordinates Gt.

In order to compare the optimal concentration profiles c, it is recommended to introduce to the Eqs. (8.36), (8.37) the residence time t_r of reacting substances instead of the residence time t of solid. The analyzed equations then assume the following common form:

$$F(c) = -k_0\left[\frac{G(a_k) - G(a_p)}{-k_{do}(t_k - t_p)}\right]^{E/E_d}(t_r - t_{rp}) + F(c_p) \tag{8.40}$$

The above equation points that the shape of the curve $c(t_r)$, which describes the stationary optimal concentration profile in relation to time t_r, remains independent of the process type. This conclusion is analogical to the corresponding conclusion for the stationary optimal catalyst activity profile in relation to the time t.

In consequence, final optimal concentrations c_k are identical in cocurrent and countercurrent processes if space time of two phases remains the same in both processes. Thus, none of these two processes is preferred to the other one.

The graphical interpretation of the role of the quotient Θ_r/Θ is provided in Fig. 8.2. It applies to the situation when the total space time of a solid phase Θ is constant, and so are the initial and final activity of solid. Then, final concentrations c_k depend only on the numerical value of the quotient Θ_r/Θ.

If the quotient Θ_r/Θ is smaller than unity, the final optimal concentrations in cocurrent and countercurrent processes are lower than the final optimal concentration of a batch process ($\Theta_r/\Theta=1$). Otherwise, if the quotient Θ_r/Θ is bigger than the unity, better results are achieved in cocurrent and countercurrent processes. If $\Theta_r/\Theta=1$, then cocurrent, countercurrent, and periodic processes do not differ in the value of the final stationary optimal concentration.

As it follows from Eqs. (8.39), (8.40), the optimal temperature profile does not depend on the process type if space time t is selected as an independent variable and if catalyst deactivation is concentration-independent, $\phi(c)=1$. In such a case, temperature T increases monotonically with time ($ka=\text{const}$).

If deactivation is concentration-dependent, the term which includes the function $\phi(c)$, Eq. (8.39) influences the differences between stationary temperature profiles in various processes (Fig. 8.3, cases A–C).

In cocurrent and batch processes, concentration c decreases with time t. Therefore, if $\phi(c)$ is a monotonically increasing function of c, first two terms on the right-hand side of Eq. (8.39) decrease with time t. This means that in cocurrent and batch processes the optimal temperature

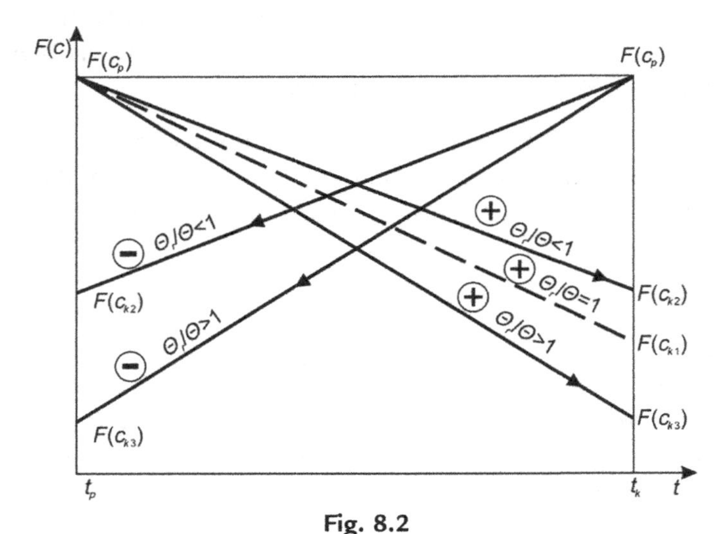

Fig. 8.2
Comparison of the performance of stationary cocurrent and countercurrent processes as well as the batch process for identical initial and final catalyst activity a_p and a_k and for the same total residence time Θ.

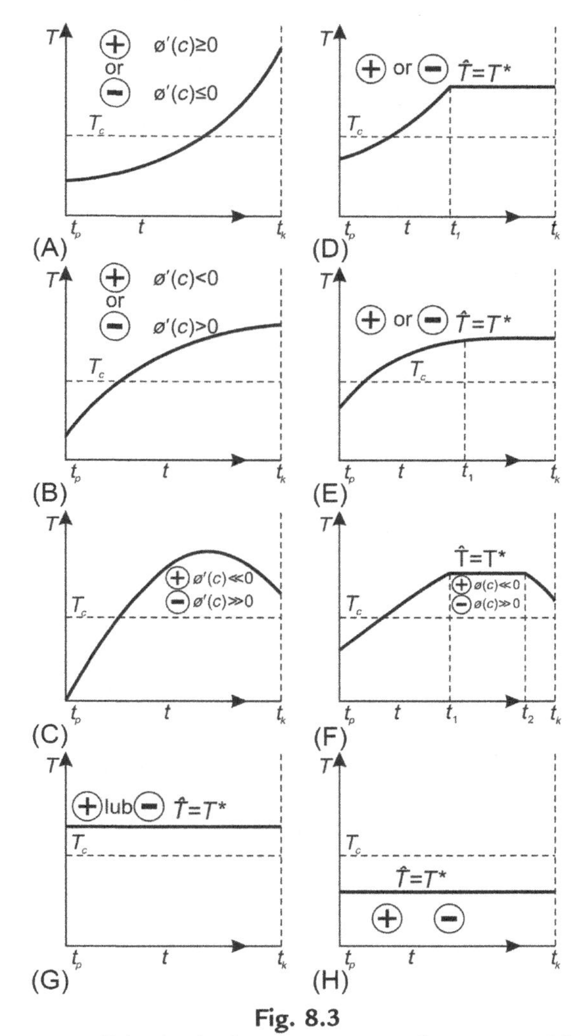

Fig. 8.3

Various optimal temperature policies in the batch reactor and reactors with the moving packing of a deactivating catalyst: (A–C) stationary policy for $E^* > E$; (D–F) policy with the restriction on temperature for $E_d < E$ if $T^* > T_c$; (G) the policy with the restriction on temperature for $E_d < E$ and if $T^* > T_c$; (H) the policy with the restriction on the temperature for $E_d < E$ and for $E_d > E$ if $T^* < T_c$.

increases monotonically with time t, as in the optimal process with the concentration-independent deactivation.

In the countercurrent process, concentration c rises along with time t. If $\phi(c)$ is a monotonically increasing function, then the concentration term of the right-hand side of Eq. (8.39) also increases in a monotonic fashion. Since the activity term decreases, local minimum of the variable y is possible, that is, maximum temperature T, may occur.

The opposite situation is observed when $\phi(c)$ is a monotonically decreasing function. Local maximum of optimal temperature may then occur in cocurrent or batch processes, while a monotonic increase of the temperature can be observed in the countercurrent process.

8.2.6 Problem with a constrained temperature

So far, only the stationary temperature policy has been considered. Let us now look into the situation when a certain maximum admissible temperature is imposed. In such a case, the optimal policy may comprise an isothermal part $T=T^*$ and one or two stationary parts (Fig. 8.3, cases D–F). Depending on threshold conditions and the admissible temperature, it is possible to have the completely isothermal optimal policy (cf. cases G and H in Fig. 8.3).

In order to illustrate the optimization procedure with the constraint imposed on the temperature as briefly as possible, only the case of concentration-independent deactivation has been considered here; it is characterized by the existence of one part of the optimal stationary policy (Fig. 8.3, case D).

A crucial point of the optimization process, which is discussed below, is the search for the activity value a_1, time t_1, and concentration c_1 for the apparatus section where two parts of the optimality policy intersect, that is, stationary and isothermal (Fig. 8.4). As it results from the optimality principle, all relations determined for the unconstrained case can be employed to describe the stationary part of the optimality policy, once adequate changes in designations are introduced to adjust them to those used in Fig. 8.4. By virtue of Eqs. (8.29), (8.25), which include, respectively, variables a_1 and t_1 instead of variables a_k and t_k for the point of intersection (if $\phi(c)=1$)

$$a_1 \exp\left(\frac{-E}{RT^*}\right) = \left[\frac{G(a_k)-G(a_p)}{-k_{do}(t_k-t_p)}\right]^{E/E_d} \tag{8.41}$$

By introducing the function

$$I(a) = \int_1^a \frac{d\gamma}{\psi(\gamma)} \tag{8.42}$$

and integrating the activity Eq. (8.10) for $T=T^*$, for all types of contacting, one obtains the following equation:

$$t = \frac{I(a_k)-I(a)}{k_d(T^*)} + t_k \tag{8.43}$$

The use of Eq. (8.118) for $t=t_1$ and $a=a_1$ in Eq. (8.41) produces the following relationship:

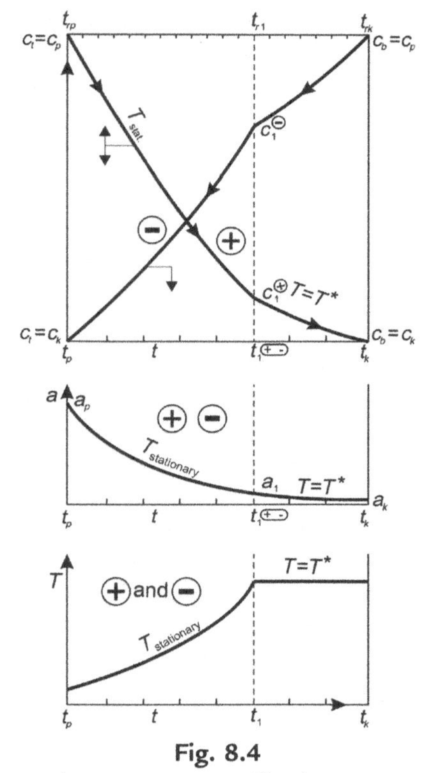

Fig. 8.4
Optimal concentration, activity, and temperature profiles in systems with the deactivating catalyst—a case with the restriction on process temperature and the catalyst deactivation independent of concentration.

$$a_1 \exp\left(\frac{-E}{RT*}\right) = \left\{\frac{G(a_1) - G(a_p)}{-k_{do}\left[\frac{I(a_k) - I(a_1)}{k_d(T*)} + t_k - t_p\right]}\right\}^{E/E_d} \tag{8.44}$$

which indicates that activity a_1 does not depend on the exchange type if deactivation remains independent of concentration. This activity is calculated as the root of Eq. (8.44) for the particular function G and I. Then, the space time t_1 is determined through the Eq. (8.43) employed for $a = a_1$. As develops from this equation, the numerical value of time t_1 is also independent of the process type. If values a_1 and t_1 are known, it is possible to calculate the optimal activity profile with the use of Eq. (8.43) and Eq. (8.38) containing a_1 instead of a_k. These equations are employed, respectively, for the intervals of space time $t_p < t < t_1$ and $t_1 < t < t_k$. Temperatures of the stationary part of an optimal process result from Eq. (8.39), with the value a_k having been replaced by a_1 and the value t_k having been replaced by t_1.

Now, final concentrations c_k for different exchange types can be compared. Eq. (8.43) leads to the conclusion that activity a must be expressed as a certain function $j(t)$, which is independent of process type. By substituting the above-mentioned function to Eq. (8.10), used for $T=T^*$ if $\phi(c)=1$, and integrating the obtained value within limits (t_1, t_k), one obtains the following result:

$$F(c_b) - F(c_1) = -k'^* \int_{t_1}^{t_k} j(t)dt \tag{8.45}$$

(cf. symbols c_b and c_t in Fig. (8.1)). The stationary part of the optimal policy is obtained from Eqs. (8.29), (8.35)–(8.37) employed for $T=T^*$, $a_k=a_1$ and $t_k=t_1$

$$F(c_1) - F(c_t) = -k'^* a(t_1 - t_p) \tag{8.46}$$

Eqs. (8.45), (8.46) can be used to calculate concentration c_1 for countercurrent processes and concurrent or batch processes, respectively. Then, profiles of the optimal concentration can be determined by employing analogical formulae, which utilize the current space time and the one that replaces one of the threshold times t_1 or t_k in formulas (8.45) or (8.46). The analyzed profiles are presented in Fig. 8.4. It can be noticed that current concentrations c depend on the gas-solid exchange type, unlike in a purely stationary optimal process, which does not feature any temperature restrictions.

By eliminating $F(c_1)$ from Eqs. (8.45), (8.46) and allowing for threshold values of c_b and c_t (Fig. 8.1), an equation common for all contacting types is produced

$$F(c) = F(c_0) - k'^* \Theta_r / \Theta \left[a(t_1 - t_p) + \int_{t_1}^{t_k} j(t)dt \right] \tag{8.47}$$

Eq. (8.47) proves that the final optimal concentration c_k for the cocurrent process is identical with the one for the countercurrent process, provided that deactivation is independent of concentration, that is, $\phi(c)=1$. The simple evidence presented above is possible thanks to the fact that for $\phi(c)=1$; both the stationary and isothermal parts of the optimal temperature may be described with the same function $F(c)$.

Finally, if $f(c)=1$, influence of admissible temperatures on the structure of the optimal policy may be analyzed.

Let us consider the first instance where the stationary part of an optimal process approaches zero. Such situations occur when temperature T^* has a specified value T_c, which depends upon threshold values a_p, a_k, t_p, t_k and corresponds to the isothermal contacting type.

$$T_c = \left\{ \frac{R}{E_d} \ln \left[\frac{-k_{d0}(t_k - t_p)}{I(a_k) - I(a_p)} \right] \right\}^{-1} \tag{8.48}$$

Now, let us look into the second instance where the isothermal part of an optimal process approaches zero. In such a case, the admissible temperature T^* is equal to the maximum stationary temperature T'. If $\phi(c) = 1$, the highest value of the stationary temperature is achieved for the final space time t_k—see Fig. 8.3, case D. Thus, Eq. (8.39) gives

$$T' = \left\{ \frac{R}{E} \ln a_k - \frac{R}{E_d} \ln \left[\frac{G(a_k) - G(a_p)}{-k_{d0}(t_k - t_p)} \right] \right\}^{-1} \tag{8.49}$$

Knowledge of formulas (8.48) and (8.49) enable the following discussion about the role of admissible temperature:

If $T^* \geq T'$, the optimal policy is described by Eq. (8.39). In a situation when the admissible temperature remains within the range

$$\left\{ \frac{R}{E_d} \ln \left[\frac{-k_{d0}(t_k - t_p)}{I(a_k) - I(a_p)} \right] \right\}^{-1} \leq T^* \leq \left\{ \frac{R}{E} \ln a_k - \frac{R}{E_d} \ln \left[\frac{G(a_k) - G(a_p)}{-k_{d0}(t_k - t_p)} \right] \right\}^{-1} \tag{8.50}$$

then, in the optimal process, there is a stationary and isothermal part, and the Eqs. (8.37)–(8.47) are applicable. Finally, if $T^* \leq T_c$, optimality is ascribed to the isothermal policy. Neither of the discussed inequalities depend on the contacting type.

We shall now refer the reader to further considerations of the deactivation optimization problems in other reactor systems. Levenspiel and Sadana (1978) and Grzesik and Skrzypek (1983) determine optimal temperature profile for heterogeneous catalytic parallel reactions. They treat fixed bed catalytic reactor with catalytic parallel reactions. Effects of transport phenomena (external or internal diffusion) are investigated. Grubecki and Wójcik (2008) determine analytically optimal temperature profiles for the reactions occurring in the presence of microorganism cells. Grubecki (2010) analyzes a biotransformation process in a batch bioreactor with parallel enzyme deactivation. He is particularly interested in effects of initial and final biocatalyst activity and conversion on the course of the optimal temperature profiles. In particular, he shows that for processes which run at any range of the substrate concentrations, in order to achieve the shortest duration time of their course, it is appropriate to conduct them at possibly the lowest concentration range.

8.3 Tubular reactor with fixed catalyst bed and optimal reaction-regeneration cycle

8.3.1 Simplest cases described by ordinary differential equations

In a general case, reaction and deactivation of the catalyst in a tubular reactor are described with partial differential equations. This is a consequence of the fact that changes in concentration and activity occur both in time and space. Still, if deactivation remains independent of concentration, $\phi(c)=1$, standard differential equations are applicable

$$\frac{da}{dt}=-k_d(T)\psi(a) \tag{8.51}$$

since spatial changes in concentration do not influence the alterations of activity. Eq. (8.51) is valid in a special but still the most realistic case in which the control T depends only on the chronological time t, similarly as in batch reactors, for instance.

In the kinetic chemical equation,

$$\frac{\partial c}{\partial \tau}=-k(T)af(c) \tag{8.52}$$

where τ is spatial time, a certain simplification has been applied, namely the quasi-static approach, and the partial derivative of concentration in relation to chronological time t was omitted.

Considerations of the optimization for models shown in Eqs. (8.51), (8.52) are rather complicated and time-consuming, so we will only analyze the first-order reaction $A \rightarrow B$, for which Eq. (8.52) may be written as follows:

$$\frac{\partial x}{\partial \tau}=-k(T)a(1-x) \quad x(0)=x_p \tag{8.53}$$

where x is a molar fraction of the product.

Integration of Eq. (8.53) within the range 0 to Θ_r yields the function

$$x_w(a,T)-x_p=1-\exp(-ka\Theta_r) \tag{8.54}$$

The subscript w refers to the outlet concentration. This function describes the relationship between concentration change in the reactor and both temperature and activity in moment t. In Eq. (8.54), $k=k(t)$ and $a=a(t)$.

The optimization problem consists in maximizing the total production of component B in a prescribed time $t_k=\bar{t}_k$ and with the defined final activity of $a_k=\bar{a}_k$. When the inflow of reacting substances is fixed, this entails the maximization of the integral

$$\int_0^{\bar{t}_k} \left[x_w(a,T) - \bar{x}_p\right] dt = \int_0^{\bar{t}_k} [1 - \exp(-ka\Theta_r)] dt \qquad (8.55)$$

with a single equation of state (8.51). In the analyzed model, Eq. (8.53) is not an equation of state—it has only been employed with the purpose of deriving the expression (8.54).

The most convenient method consists in solving the analyzed problem in its stationary variant with the calculus of variations by converting the index (8.55) so that it contains the derivative $da/d\tau$ instead of temperature T concealed in the constant k. In order to do so, one must eliminate T from Eq. (8.51) and substitute the result to Eq. (8.55). By denoting,

$$p = \frac{E}{E_d} \qquad (8.56)$$

it can be easily proven that with $k(t)$ and $k_d(t)$ satisfying the Arrhenius equation the following relationship applies:

$$k(T) = g[k_d(T)]^p; \ \gamma \equiv \frac{k_0}{k_{d0}^p} \qquad (8.57)$$

Hence, after substituting Eq. (8.51) into Eq. (8.57)

$$k(T) = \gamma \left[\frac{-da}{\psi(a)dt}\right]^p \qquad (8.58)$$

Furthermore, by substituting Eq. (8.58) into Eq. (8.55), an integral is obtained

$$\int_0^{\bar{t}_k} \left\langle 1 - \exp\left\{ -\gamma a [\psi(a)]^{-p} \left(\frac{-da}{\psi(a)dt}\right)^p \Theta_r \right\} \right\rangle dt \qquad (8.59)$$

Thus, the problem has been reduced to the maximization of an autonomous functional (8.59), the general form of which is:

$$S = \int_0^{\bar{t}_k} L\left(a, \frac{da}{dt}\right) dt \qquad (8.60)$$

Instead of analyzing the Euler-Lagrange equation for the functional (8.59), it is better to remember that in autonomous cases exists—see, for example, Eq. (4.64) in Sieniutycz (1991 &4.4, p. 137)—the first integral of this equation. For the functional (8.60), the first integral has the following form:

$$L - \dot{a}\frac{\partial L}{\partial \dot{a}} = \text{constant} \qquad (8.61)$$

On this basis, the following first integral is obtained for the functional (8.59):

$$1-(1-L)[1-p\ \ln(1-L)]=\text{constant}' \tag{8.62}$$

Since in the considered case $L=x_w-x_p$, Eq. (8.62) means that

$$x_w=\text{constant} \tag{8.63}$$

Therefore, the stationary optimal control $T(t)$ must render conversion at the reactor outlet constant in time. Simplicity of this condition and its technical facility bear particular significance. What is more, conditions (8.63) and Eq. (8.54) indicate that in a stationary optimal processes, when $E_d>E$, our earlier condition (8.30) holds again:

$$k_{\text{eff}}=k(T)a=\text{constant}\,(t) \tag{8.30}$$

This means that whenever the deactivation is independent of concentration, constancy of the effective reaction rate constant k_{eff} applies. The conclusion we arrive at is reactors with catalyst flow. It turns out that the conclusion remains valid for reactions of any order; however, proving so is more difficult (Szepe, 1966; Szepe and Levenspiel, 1968).

Validity of conditions (8.63) and (8.30) can also be proven for fluidized bed reactors, both in case of perfect and bubbling fluidization.

In a book (Sieniutycz and Szwast, 2018), the problems that have been initiated in Section 8.3.3 are continued under the assumption of catalyst deactivation

8.3.2 Problem formulations for optima of cyclic processes

Below, several problems of optimal control for a cyclic reaction-regeneration system are considered. Frequently, a cyclic reaction-regeneration system may be regarded as a complex system of hierarchical structure, with decisions of higher order (also known as coordinating decisions), such as production time, exchange time, boundary activities of the catalyst, as well as low-level (local) decisions, such as reactor temperature or regenerator temperature.

Cyclic reaction-regeneration processes, in which a chemical reaction undergoes within a specified time t_k, and then, catalyst regeneration takes place within a time t_R, belong to an important, yet peculiar, group of processes in which production and downtime alternate. One of the characteristic features of these processes is the feature of their subprocesses (stages) which differ by signs of profit functions. During the production stage (reaction stage), new value is generated, making net profit of this subprocess positive. However, the regeneration stage only brings the incurrence of costs. A sort of dialectic may be observed here: the production stage generates profit, but it also causes detrimental effects, such as catalyst deactivation, aging of the apparatus, etc. On the other hand, the regeneration stage, which eliminates these harmful effects, is itself uneconomical.

Since regeneration brings nothing but costs, and no profits are recorded at the regeneration stages, it would be economically unfeasible in the absence of the production stage. Still, maintaining production without any regeneration (apparatus maintenance or catalyst regeneration) for very long periods is also uneconomical because of inevitable aging of the apparatus, catalyst, etc. All this would eventually make the increase rate of the net profit inefficient, causing a gradual reduction of the time average of the production profit. Only the mutual link of production stages (reaction) and renovation stages (regeneration), arranged in a cyclic process in which these stages are alternating, guarantees the sufficiently high positive average net profit, which determines the economic viability of the cyclic process considered.

Processes of this sort are a common occurrence in a wide array of economic and social aspects of life, for example, replacement of a car that is damaged beyond repairability, periodic fertilization of field crops due to the deterioration of soil fertility, turnover of employees that is forced by the decreasing professional effectiveness, replacement of machinery due to progressing corrosion of their elements, etc.

General quality index (average net profit, S), or the performance criterion of the analyzed cyclic processes, usually has a structure quite complex in mathematical terms. This structure is discussed on an example of a reaction-regeneration system with catalyst circulation, subject to the assumptions that the chemical reaction undergoes in an apparatus with a stationary bed, and that the regeneration takes place in the same apparatus. The quality index S for a steady state cyclic process assumes a general structure:

$$S=\frac{\int_0^{t_k} R[c(\tau_k,t)c(0,t),\tau_k]dt}{t_k+t_R}\frac{\int_0^{t_k}\int_0^{\tau_k} Q(\mathbf{x},\mathbf{a},\mathbf{u},\tau,t)d\tau dt}{t_k+t_R} \\ \frac{\int_0^{t_k+t_R}\int_0^{\tau_k} Q(\mathbf{x},\mathbf{a},\mathbf{u},\tau,t)d\tau dt}{t_k+t_R}-\frac{I_\tau\left[a_p(0,\tau).a_k(t_k,\tau),t_R,\tau_R,\mathbf{x}_R,\mathbf{u}_R,y\right]}{t_k+t_R}-\frac{I_1[\tau_k.....]}{t_k+t_R} \tag{8.64}$$

(only changes in a single spatial direction are assumed), where R, intensity of the profit yield, resulting from the production of valuable substances, c, concentration of the substrate, Q and Q_R, functions describing the control costs, I_R and I_1, respectively, the functionals describing regeneration and capital expenditures of a single cycle, t_k, total chronological time of reaction stage, t_R, regeneration time, τ_k, contacting time, y, quantity of the fresh catalyst added per one unit of the catalyst mass in the system, $\mathbf{x}$ and $\mathbf{x}_R$, vectors composed of state variables (excluding c) for reactions and regenerations, respectively, **a**, vector of catalyst activity, **u**, $\mathbf{u}_R$, control vectors for reaction and regeneration operations, respectively.

The production stage of a packed reactor is characterized by PD equations of state:

$$\frac{\partial \mathbf{c}}{\partial t} = \mathbf{f}(\mathbf{c}, \mathbf{a}, \mathbf{u}), \quad \frac{\partial \mathbf{a}}{\partial \tau} = \mathbf{g}(\mathbf{c}, \mathbf{a}, \mathbf{u}) \tag{8.65}$$

which constitute differential constraints applied in maximization of the performance index (8.64). Other restrictions are algebraic. They can correspond, for example, to the quest for the specific profile of inlet concentrations $\mathbf{c}(0,t) = v_p(\tau)$, to the initial and final activity profile $a(\tau,0) = w_p(\tau)$ and $a(\tau, t_k) = w_k(\tau)$, as well as to the admissible changes in the control vector $\mathbf{u} \in \mathbf{U}$. However, it must be pointed out that both discussed activity profiles must generally be treated as distributed controls and thus sought as components of an optimal solution.

The decrease of instantaneous profit flux in time t_k accompanies the reduction of average regeneration cost $I_R(t_k + t_R)^{-1}$. The difference between these two values can possess a natural (stationary) maximum with respect to the duration of the reaction stage t_k, the property which substantiates both the rationale and capability of the cycle optimization. The decision t_k, which influences both the reactive and the regenerative term of the criterion S, does not resemble a local decision that is significant for only one stage; it rather approximates a systemic (coordinate) decision, which is common for both stages of reaction and deactivation. This brings a significant difference between this decision and decisions of other types. The latter are known as local (**u** and $\mathbf{u}_R$). Each of these decisions influences (for $t_k = \text{const}$) only a specific part of the performance index, related either to reaction or to regeneration. The occurrence of both, coordinating and local, decisions proves that the analyzed cyclic system constitutes, in fact, a system of hierarchical structure, characterized by the specific subordination of local decisions to the coordinating ones. The latter are also referred to as upper-level decisions and are frequently made by a certain superior authority (designer, management, etc.) for the entire system.

The analysis of the expression in Eq. (8.64) allows us to observe that, apart from the production time t_k, the role of upper-level decisions may be additionally played by such values as regeneration time t_R and reactor contact time τ_k, as well as distribution of activity: initial $w_p(0, \tau)$ and final $w_k(t_k, \tau)$. For instance, restrictions of the form, $\mathbf{u}^* < \mathbf{u} < \mathbf{u}^*$, may appear. Presence of local and coordinating decisions is shown on a scheme of Fig. 8.9. The figure also illustrates the system structure.

The optimization procedure is as follows. First, the quality index is maximized with respect to the local decisions **u**. They are temperatures $T(\tau, t,)$ and $T_R(\tau, t,)$ as well as the amount of the fresh catalyst y, for a definite set of coordinating decisions **v**. This optimization results in the acquisition of optimal local decisions and local quality indices expressed as functions of coordinating decisions (upper-level controls). Then, the optimization with respect to coordinating decisions takes place.

By denoting the vector of coordinating decisions by **v**, we assign for the reaction-regeneration cycle:

$$\mathbf{v} = \left[w_p(t), w_k(t), t_k, t_R, \tau_k\right]^T \tag{8.66}$$

and

$$\begin{aligned} S(\mathbf{v}, \mathbf{u}) &= f\left(w_p(\tau), w_k(\tau), t_k, t_R, \tau_k, T(\tau, t), T_R(\tau, t), \ldots\right) \\ &= \frac{P\left(w_p, w_k, t_k, \tau_k, T\right)}{t_k + t_R} - \frac{I\left(w_p, w_k, t_R, \tau_k, T_R, y\right)}{t_k + t_R} \end{aligned} \tag{8.67}$$

where P, net production profit, I, regeneration expenditures including investment costs expressed in terms of optimization variables. The structure of Eq. (8.67) defines basic properties of cyclical multilevel systems analyzed.

An optimization scheme is now applied to the quality index (8.67) based on the principle of the so-called parametric optimization (Sieniutycz, 1991, Section 6.3.1) in the form

$$S(\mathbf{v}, \mathbf{u}) = \overline{Z}(\mathbf{v}, \mathbf{u}) - \overline{K}_R(\mathbf{v}, \mathbf{u}_R) \tag{8.68}$$

where $\overline{Z}$ is the average net profit acquired from the reactor. $\overline{K}_R$ is an average regenerator cost corresponding to a single cycle. (These quantities include the cycle time, e.g., $K_R = I/(t_k + t_R) > 0$.)

In Eq. (8.68), **u** and $\mathbf{u}_R$ are vectors of local decisions, while **v** is the vector of coordination decisions. Since **u** and $\mathbf{u}_R$ are disjoint sets, it is possible to apply the so-called decomposition, that is, the low-level problem for $\mathbf{v} = \text{const}$ may be replaced by two independent optimization tasks involving the indices $\overline{Z}$ and $\overline{K}_R$, as the quantities characterizing reactor and regenerator.

Problems of the first level (lower level) are local optimization problems described by the formulae

$$\max \overline{Z}(\mathbf{v}, \mathbf{u}) = \overline{Z}(\mathbf{v}) \quad \mathbf{u} \in \mathbf{U}(\mathbf{v}) \tag{8.69}$$

For the regenerator, it is also possible to exploit the profit function; which is, however, negative, that is, $\overline{Z}_R < 0$, see Fig. 8.9. Assumptions ensuring convexity of functions and constraining sets are required.

$$\max\left[-\overline{K}_R(\mathbf{v}, \mathbf{u}_R)\right]_{\mathbf{u}_R} \equiv -\min\left[-\overline{K}_R(\mathbf{v}, \mathbf{u}_R)\right]_{\mathbf{u}_R} = -\overline{K}_R(\mathbf{v}) \tag{8.70}$$

where $\mathbf{v} \ni \mathbf{V}$. They emphasize parametric nature of local problems with regard to coordinating variables v. However, the problem of superior level is described by the relationship

$$\max S(v) = \max\left[\tilde{\tilde{Z}}(\mathbf{v}) - \tilde{\tilde{K}}_R(\mathbf{v})\right] \tag{8.71}$$

The optimal solution to first-level problems is, similarly to optimal indices $\tilde{\tilde{Z}}(\mathbf{v})$ and $\tilde{\tilde{K}}_R(\mathbf{v})]$, parametric with respect to $\mathbf{v}$

$$\hat{\mathbf{u}}(\mathbf{v}) \text{ and } \hat{\mathbf{u}}_R(\mathbf{v}) \tag{8.72}$$

Eq. (8.72) describe local parametric decisions that have been decomposed. A condition for this decomposition is the ability to express functions (8.72) in an explicit form. Data in such a form are acquired in the course of analysis involving individual reactors and regenerators.

The solution to the superior problem is comprised of optimal values for coordinating variables (upper-level decisions).

$$\mathbf{v} = \hat{\mathbf{v}} \tag{8.73}$$

which, once substituted into the Eq. (8.72), define the optimal values of lower-level decisions

$$\hat{\mathbf{u}}(\tilde{\mathbf{v}}) \text{ and } \hat{\mathbf{u}}_R(\tilde{\mathbf{v}}) \tag{8.74}$$

Formulas shown in (8.73) and (8.74) describe the solution to our optimization problem.
In case of homogeneous activity profiles, w_p and w_k, and when the regeneration stage includes only the removal of a certain amount of spent catalyst y and the mixing of its residue with fresh catalyst (whose quantity is not allowed to exceed a value y^*), the amount of fresh catalyst is no longer an independent decision, but becomes a quantity that is unequivocally related to activities w_p and w_k. As a result, local extremizing of the regenerative component of performance index with regard to the local decision y, and subject to fixed coordination variables $\mathbf{v}$ (including w_p and w_k), means an unequivocally specified decision y (no choice-no selection). In this calculation scheme, the function $y(w_p, w_k)$ follows from the balance of mixing.

The mass balance in Fig. 8.5 is referred to mixing of the used catalyst, of instantaneous activity $a(t_k)$, with the fresh catalyst, for which $a=1$. As the result of this operation, a catalyst possessing the activity $a_p(0)$ is reproduced, which is next supplied to the reactor.

When homogeneity of profiles w_p and w_k is assumed, we have $a_p = w_p$ and $a_k = w_k$, which gives

$$\tilde{y}(\mathbf{v}) = \frac{w_p - w_k}{1 - w_k} \tag{8.75}$$

whereas for inhomogeneous profiles $w_p(\tau)$,

$$\frac{1}{\tau_k}\int_0^{\bar{\tau}_k} w_p(\tau|\,)d\tau = 1 - \hat{y} + \left(\frac{1}{\tau_k}\int_0^{\bar{\tau}_k} w_k(\tau)d\tau\right)(1 - \hat{y}) \tag{8.76}$$

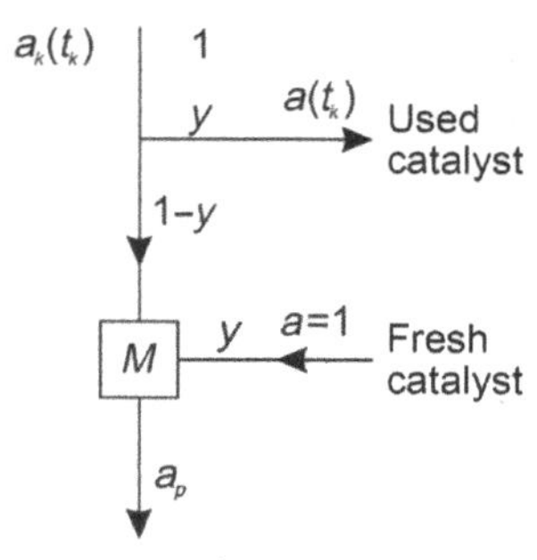

Fig. 8.5
Catalyst regeneration through mixing (M) the fraction $1-y$ of used catalyst with the fresh catalyst, y.

which involves manipulating with values averaged over the spatial distribution. Details of analysis for Eq. (8.76) are omitted here.

Optimization of the upper level, that is, with respect to the coordination variables, should be performed subject to the constraint

$$\tilde{y}(\mathbf{v}) = \frac{w_p - w_k}{1 - w_k} \tag{8.77}$$

imposed on the admissible boundary activities w_p and w_k.

Other constraints imposed on coordination variables are also possible, such as $0 < t_k < t_k{}^*$, $t_R < t_R{}^*$, $0 < \tau_k < \tau_k{}^*$, as well as constraints on total capital expenditures, I. However, the latter are not necessary since Eq. (8.71) attains the natural (i.e., stationary) maximum with respect to the variables t_k, t_K, or τ_k (Fig. 8.8). In relation to constraint (8.77), a boundary solution is often obtained, that is, w_k and w_p assume values that satisfy $y(\mathbf{v}) = y^*$ in Eq. (8.77). This occurs, for example, when costs of catalyst regeneration by mixing increase monotonically with the quantity of fresh catalyst. If the mode of regeneration is different, these costs may change in more complex fashion, and then catalyst amount y is not directly linked with boundary activities w_p and w_k, that is, Eqs. (8.75), (8.76) should be replaced by their more general counterparts.

8.3.3 Properties of optimal temperature control T(x, t)

Optimization of local decisions is performed at the lower level for a fixed set of coordinating decisions (t_k, $w_p(\tau)$, $w_k(\tau)$, τ_k). Consequently, the determination of the time-dependent optimal temperature profile in a fixed bed reactor comes down to the maximization of the integral

$$P = \int_0^{t_k} R dt = \dot{G} C \int_0^{t_k} (c_0 - c_w) dt \tag{8.78}$$

subject to the constraints represented by Eq. (8.65) and for prescribed activity profiles w_p and w_k. The integral describes the net profit from the production of a valuable reagent within the period between $t=0$ and $t=t_k$. In Eq. (8.78), $G^{\cdot}$ is the mass flow of all reacting substances, c_0 and c_w are, respectively, input and output concentrations in the reactor, while C is the unit price of the product (Sieniutycz, 1976, 1980).

A stationary solution to the maximization problem of P has been achieved at by applying the method of eliminating conjugate variables for processes with distributed parameters (Sieniutycz, 1980).

We shall now discuss a general solution for the case with a temperature constraint. As stated in the literature (Sieniutycz, 1976, 1980, 1991), the objective of our considerations is the first-order nonreversible reaction and the catalyst deactivation independent of concentration but dependent on temperature T.

General kinetic equations (8.65) assume now a special form represented by a set of following equations

$$\frac{\partial c}{\partial \tau} = -k(T)ca, \; k(T) = k_0 \exp\left(-\frac{E}{RT}\right) \tag{8.79}$$

$$\frac{\partial a}{\partial t} = -k_d(T)a, \; k_d(T) = k_{d0} \exp\left(-\frac{E_d}{RT}\right) \tag{8.80}$$

The initial concentration $c(0,t)$ as well as the initial and final activity distributions take the following form:

$$c(0,t) = c_0(t) \tag{8.81}$$

$$a(\tau,0) = w_p(\tau) \tag{8.82}$$

$$a(\tau,t_k) = w_k(\tau) \tag{8.83}$$

If $E_d < E$, an increase of temperature favors reaction rather than deactivation, so that the upper admissible temperature is optimal, that is, $T(t,\tau) = T^*$. The stationary segment of $T(t,\tau)$ is possible for $E_d > E$ since temperatures can neither be too high (otherwise, the deactivation rate would increase) nor too low (if they were, the reaction would be quenched). The optimal stationary temperature increases in time t for $\tau = \text{const}$ and satisfies the condition equivalent with Eq. (8.30), that is,

$$k_{\text{eff}} = ka = \text{constant}'(t) \tag{8.84}$$

At the same time, this condition guarantees the constancy of the outlet conversion $x = (c_0 - c_w)$ in time t.

For an imposed temperature constraint, $T \leq T^*$, the optimal temperature in a given reactor section τ=const starts to increase from a certain initial temperature $T_p(\tau)$, and it can reach the maximum admissible temperature T^*. As the time t goes on, the optimal temperature $T(\tau, t)$ tends to T^* until the end of the stage of reaction, that is, for $t=t_k$. The stationary temperature control, which fulfills the condition (8.84), is described by an equation (Sieniutycz, 1976, 1980)

$$\frac{1}{T(\tau,\ t)} = \frac{R}{E_d}\left\langle \ln k_{do}\frac{E_d}{E_0}\left\{t'_k\left[1-\left(\frac{w_k(\tau)}{w_p(\tau)}\right)^{E_d/E}\right]^{-1} - t\right\}\right\rangle \tag{8.85}$$

Here t'_k is the time required to reach the final activity profile $w_k(\tau)$. Whenever the optimal control has an isothermal part, the time t^* needed to reach temperature T^* can be determined by applying in the above result $T=T_p(\tau)$ for $t=0$ and $T=T^*$ for $t=t_k=t^*$, which gives:

$$t^*(T_p,\ \tau) = \frac{E}{k_{d0}E_d}\left[\exp\left(\frac{E}{RT_p(\tau)}\right) - \exp\left(\frac{E}{RT^*}\right)\right] \tag{8.86}$$

The above equation describes a moment of replacing the stationary strategy by an isothermal one, expressed as a function of the initial temperature $T_p(\tau)$. Further formulas will be presented in terms of this function, and only then will T_p be varied so that it can be selected optimally.

For times $t_k \leq t^*(T_p \cdot \tau)$ we obtain:

$$k_{\text{eff}}(T_p(\tau), \tau) = k_0 \exp\left(-\frac{E_d}{RT(\tau)}\right) w_p(t) \equiv k^*_{\text{eff}} \tag{8.87}$$

hence

$$a(\tau, t) = w_p(\tau)\ \exp\left[E/RT(\tau, t) - E/RT_p(\tau)\right] \tag{8.88}$$

Whereas if $t_k > t^*(T_p.\ \tau)$, then

$$k_{\text{eff}}(T_p, \tau, t) = k(T^*)a(T_p, T^*, \tau, t, t^*) \tag{8.89}$$

where the function $a(T_p, T^*, \tau, t, t^*)$ results from integrating the activity Eq. (8.80) along the isothermal line $T=T^*$, ranging from $a^*(T_p, t)$ to $a(T_p, T^*, t, t^*)$. The function $a^*(T_p, \tau)$ is calculated from Eq. (8.88) for $T=T^*$. Time t^* is determined from Eq. (8.86). Since

$$\begin{aligned} a(T_p, T^*, \tau, t, t^*) &= a^*(T_p, \tau) \exp\left[-k^*_d(t - t^*)\right] \\ &= \frac{k^*_{\text{eff}}(T_p, \tau)}{k(T^*)} \exp\left[-k^*_d(t - t^*)\right] \end{aligned} \tag{8.90}$$

and stationary rate constant $k_{\text{eff}}(T_p,\tau) = k_{\text{eff}}{}^*(T_p,\tau)$, so, eventually, based on Eqs. (8.87), (8.90)

$$a(T_p, T^*, \tau, t, t^*) = w_p(\tau)\frac{k_0}{k(T^*)}\exp\left(-\frac{E_d}{RT_p(\tau)}\right)\exp\left[-k_d^*(t-t^*)\right] \tag{8.91}$$

Thus, the effective rate constant (8.89) (for $t_k > t^*$)

$$\begin{aligned} k_{\text{eff}}(T_p\tau, t) &= w_p(\tau)k_0\exp\left[\left(-\frac{E}{RT_p(\tau)}\right) - k_{d0}\left(-\frac{E_d}{RT^*}\right)(t-t^*)\right] \\ &= k_{\text{eff}}^*\exp\left[k_d^*(t^*-t)\right] \end{aligned} \tag{8.92}$$

as well as the final activity profile

$$\begin{aligned} w_k(\tau) &= a(T_p, T^*, \tau, t_k, t^*) \\ &= w_p(\tau)\exp\left(\frac{E}{RT^*} - \frac{E}{RT_p(\tau)}\right)\exp\left[k_{d0}\exp\left(-\frac{E_d}{RT^*}\right)(t^*-t_k)\right] \end{aligned} \tag{8.93}$$

If the profile $T(\tau, t)$ is purely stationary, final activity is obtained by assuming $t=0$ and $T=T_p$ in formula (8.85), which yields

$$w_k(T_p, \tau, t_k, w_p) = w_p(\tau)\left[1 - E_d k_{d0} t_k' E^{-1}\exp\left(-E_d/RT_p(\tau)\right)\right]E/E_d \tag{8.94}$$

Integration of Eq. (8.79) for $t \le t^*(\tau)$ produces in view of the constancy $k_{\text{eff}} = k_{\text{eff}}^*$

$$\begin{aligned} c(T_p, \tau, t) &= c_0(t)\exp\left[-\int_0^\tau k_{\text{eff}}^*(\tau')d\tau'\right] \\ &= c_0(t)\exp\left[-\int_0^\tau k_0 w_p(\tau')\exp\left(-\frac{E_d}{RT_p(\tau')}\right)d\tau'\right] \end{aligned} \tag{8.95}$$

In the case of $t > t^*(\tau)$, the use of t instead of t_k in Eqs. (8.79), (8.93) yields

$$\begin{aligned} \frac{\partial \ln c}{\partial t} &= -k(T^*)a = \\ &= -k_0\exp\left(\frac{E}{RT^*}\right)w_p(\tau)\exp\left(\frac{E}{RT^*} - \frac{E}{RT_p(\tau)}\right)\exp\left[k_{d0}\exp\left(-\frac{E_d}{RT^*}\right)(t^*-t)\right] \end{aligned} \tag{8.96}$$

where, after integration and using Eq. (8.86), we obtain:

$$\begin{aligned} c(T_p, \tau, t) &= c_0(t)\exp\left\langle -\int_0^\tau k_{\text{eff}}^*(\tau, T_p)\exp\left[-k_d^*(t - t^*(\tau'))\right]d\tau'\right\rangle \\ &= c_0(t)\exp\left\langle -\int_0^\tau k_{\text{eff}}^*(\tau, T_p)\exp\left\{-k_d^*\left[t - \frac{E}{k_{d0}E_d}\left(\exp\left(\frac{E_d}{RT_p(\tau)}\right) - \exp\left(\frac{E_d}{RT^*}\right)\right)\right]\right\}d\tau'\right\rangle \end{aligned} \tag{8.97}$$

As a result, the conversion at the moment t and for the intersection τ is

$$x(T_p,\tau_p,t)=\frac{c_0-c}{c_0}=1-\exp\left(\int_0^{\tau}k^*_{\text{eff}}(\tau')d\tau'\right)\ \text{for } t\le t^* \tag{8.98}$$

$$x(T_p,\tau_p,t)=\frac{c_0-c}{c_0}=1-\exp\left\langle\int_0^{\tau}k^*_{\text{eff}}(\tau')\exp\left[-k^*_d(t-t^*(\tau'))\right]\right\rangle\ \text{for } t>t^*(\tau) \tag{8.99}$$

Abbreviation $k^*_{\text{eff}}(\tau)$ has been introduced for the function $k^*_{\text{eff}}(\tau,T_p(\tau))$.

It is also important to know the average conversion in the reactor after the time t.

$$\begin{aligned}\bar{x}_w(t_k,T_p)&\equiv\frac{1}{t_k}\int_0^{\tau_k}x(\tau_k,t_k,T_p)dt=\frac{1}{t_k}\int_0^{t^*(t_k)}x(\tau_k,t_k,T_p)dt+\\&+\frac{1}{t_k}\int_{t^*(t_k)}^{t_k}x(\tau_k,t_k,T_p)dt=\frac{1}{t_k}\int_0^{t^*(t_k)}\left[1-\exp\left(-\int_0^{t_k}k^*_{\text{eff}}(\tau')\right)d\tau'\right]dt\\&+\frac{1}{t_k}\int_{t^*(t_k)}^{t_k}\left\{1-\exp\int_0^{t_k}-k^*_{\text{eff}}(\tau')\exp\left[-k^*_d(t-t^*(\tau'))\right]\right\}d\tau'dt\end{aligned} \tag{8.100}$$

where $k^*_{\text{eff}}(\tau')$ depends on $T_p(\tau')$, in concord with Eq. (8.87).

Let us also consider the particular case of a homogeneous profile of the initial activity $w_p(\tau)=w_p$ and the final activity $w_k(\tau')=w_k$. In such a case, the initial distribution of optimal temperature $T_p(\tau)$ is also homogeneous. Eq. (8.100) reduces to the following form:

$$\begin{aligned}\bar{x}_w(t_k,T_p)&\equiv\frac{1}{t_k}\int_0^{\tau_k}x(\tau_k,t_k,T_p)dt\\&=1-\frac{1}{t_k}\int_0^{t^*}\exp\left(-k^*_{\text{eff}}\tau_k\right)dt-\frac{1}{t_k}\int_{t^*}^{t_k}\exp\left[-k^*_{\text{eff}}\tau_k\exp\left(-k^*_d(t-t^*)\right)\right]dt\end{aligned} \tag{8.101}$$

By calculating the first integral in the second line of this formula and introducing into the second integral a new variable

$$p=k^*_d(t-t^*)=k_{do}\exp\left(-\frac{E_d}{RT^*}\right)(t-t^*) \tag{8.102}$$

we obtain

$$\bar{x}_w(t_k, T_p) = 1 - \frac{t^*}{t_k}\exp\left(-k^*_{\text{eff}}\tau_k\right) - \frac{\exp(E_d/RT^*)}{k_{d0}t_k}\int_0^{p_k}\exp\left[-k^*_{\text{eff}}\tau_k\exp(-p)\right]dp \tag{8.103}$$

where $p_k = p(t_k)$.

By introducing another new variable,

$$Y = k^*_{\text{eff}}\tau_k\exp(-p) = Y^*\exp(-p) \tag{8.104}$$

we obtain

$$\bar{x}_w(t_k, T_p) = 1 - \frac{t^*}{t_k}\exp(-Y^*) - \frac{\exp(E_d/RT^*)}{k_{d0}t_k}\int_{Y^*}^{Y_k}\frac{\exp(-Y)dY}{Y} \tag{8.105}$$

for $dp = -dY/Y$. On the other hand, by denoting

$$\int_{-\infty}^{Y}\frac{\exp(-t)dt}{t} = E(Y)$$

the expression in (8.105) is written as follows

$$\bar{x}_w(t_k, T_p) = 1 - \frac{t^*}{t_k}\exp(-Y^*) - \frac{\exp(E_d/RT^*)}{k_{d0}t_k}[E(Y_k) - E(Y^*)] \tag{8.106}$$

where

$$Y_k = Y^*\exp\left(-k^*_d\left(t_k - t^*(\tau_p)\right)\right)$$

$$Y^* = k^*_{\text{eff}}\tau_k = k_0\exp\left(-\frac{E}{RT_p}\right)w_p(\tau_k)$$

Let us consider a partial optimum related to the maximum yield of the product. This is not an optimum of the cycle. Average value of outlet conversion (objective function) is closely connected with the initial optimal temperature T_p influencing values of variables t^*, Y^*, and Y_k. The conversion maximum corresponds to the vanishing derivative $d\bar{x}_w/dT_p$, which gives:

$$\exp(Y^* - Y_k) - EY^*/E_d - 1 = 0$$

Numerical solution of this equation determines Y^*. This result allows to find the optimal starting temperature T_p of the reactor providing the maximum average conversion in the period $(0, t_k)$. The problem would, therefore, be solved, but this would not be the solution to the cycle optimization problem in every case. Let us note that setting the initial temperature T_p enables us to determine the final activity w_k (as well as other quantities) for known initial activity w_p. However, the activity w_k is optimal not for the whole reactive/regenerative system, but for the reactor itself. Applying the so-obtained decision for w_k would be false since it would have been undertaken at too low management level (with respect to the partial performance criterion, not a global one). Whole cycle needs to require rather those from previous results which do not conceal assumptions of the variation T or w_k at the reactor level. This means that the previously analyzed equation should be ignored.

By analyzing Eq. (8.85) for known $w_k(\tau)$, calculated from the formula (8.94), it can be noted that temperatures increase in time t in every section of the reactor (T=const) until the admissible temperature T^* is reached. Afterwards (for longer t), temperatures remain constant in time t and equal to T^*. If the entire reactor has not yet reached T^* in every of its sections, that is, stationary profiles $T(\tau, t)$ are still operative, then, for c_0=constant (t), the optimal outlet conversion $x_w(\tau_k,t)$ remains constant in time. Else (which usually happens in later periods of the reactor's operation), optimal conversion x_w decreases in time as the consequence of decrease of effective reaction rates in isothermal parts of temperature strategies.

We have qualified the initial temperature function $T_p(\tau)$ as the decision essential for the entire cycle, not only for the reactor. Eqs. (8.95)–(8.106) are then employed for the purpose of optimizing the cycle when one of the coordinating entities is $T_p(\tau)$. It may, however, turn out that it is more convenient to operate with the final activity $w_k(\tau)$ as the coordination variable. This can be realized by introducing the function $T_p(\tau, w_k, t_k, w_p)$ to Eqs. (8.95)–(8.106). If the temperature strategy is entirely stationary (inactive constraint imposed on T), a suitable temperature function is obtained by substituting $t=0$ and $T=T_p$ into Eq. (8.85), which gives:

$$\frac{1}{T_p(\tau,\ t_k)} = \frac{R}{E_d}\ \ln\left\{\frac{k_{do}E_d t_k}{E}\left[1-\left(\frac{w_k(\tau)}{w_p(\tau)}\right)^{E_d/E}\right]^{-1}\right\} \tag{8.107}$$

On the other hand, for the limited temperature strategy, the function $T_p(\tau, w_k, t_k, w_p)$ is determined from the formula (8.93).

Summing up, we have presented above an example of the approach accomplishing the replacement of coordination variables (Sieniutycz, 1991; Sieniutycz and Szwast, 2018).

8.3.4 Analyzing the role of coordination variables in optimal cycle

Let us subsume the calculation example for the cycle optimization. The simplest regeneration has been assumed, namely the one that consists in mixing the fresh catalyst with portions of the spent catalyst (Fig. 8.5). Regeneration time constant t_R and the inactivity of temperature constraints have been assumed. Control costs have been omitted, while capital expenditures are assumed as proportional to the contact time τ_k with the coefficient μ. Possible distribution of the catalyst activity along the apparatus has been ignored. As one of the coordinating variables, w_k rather than T_p has been used.

In the analyzed case, cycle performance index (8.64) assumes the following form:

$$S = \frac{\dot{g}\int_0^{\tau_k}(c_0 - c_w)dt - \left(\lambda_R^{/}y + \mu\right)\tau_k}{t_k + t_R} \tag{8.108}$$

where $\dot{g} \equiv \dot{G}C/K$, value flux of reacting substances per unit of the contact mass; $\lambda_R^{/}$, multiplier of regeneration costs modified by inclusion of fresh catalyst's price; μ, multiplier of capital expenditures, which increase along with the contact time τ_k.

If the constraining T^* is inactive, the effective rate constant is $k_{\text{eff},}$ constant in time is t, but its magnitude depends on t_k. The same remark applies to c_w. The outlet concentration is

$$c_w = c_0 \exp\left(-k_{\text{eff}}\tau_k\right) \tag{8.109}$$

while the quality index after local optimizations

$$S = \frac{\dot{g}c_0\left\{1 - \exp\left[-k_{\text{eff}}\left(w_p, w_k, t_k\right)\tau_K\right]\right\}t_k - \lambda\tau_k}{t_k + t_R} \tag{8.110}$$

This is the performance index expressed as the function of coordinating variables. The function $k_{\text{eff}}(w_p, w_k, t_k)$ is acquired as $k(T_p)w_p$ through the application of the Arrhenius equation and Eq. (8.107), which gives

$$k_{\text{eff}}\left(w_p, w_k, t_k\right) = k_0\left(\frac{k_{do}E_d t_k}{E}\right)^{-\frac{E}{E_d}} w_p\left[1 - \left(\frac{w_k(\tau)}{w_p(\tau)}\right)^{E_d/E}\right]^{E/E_d} \tag{8.111}$$

We designated here

$$\lambda = \left(\frac{\lambda_R^{/} + \mu}{y^*}\right)y^* \equiv \lambda_R y^* \tag{8.112}$$

as the multiplier that takes into consideration both regeneration expenses and capital expenditures. Note that the condition $k_{\text{eff}} = \text{const}$ also holds for a batch reactor. The state equation for the state variable c is $d\ln c/dt = k_{\text{eff},}$ while the related conversion $\Delta c(t_k) = c_0 - c = c_0[1 - \exp\left(-k_{\text{eff}}t_k\right)]$. The unit production is $Z = Gc/K$, where G is the number of moles of reacting mixture and K defines the catalyst's quantity.

Calculations of the most significant quantities that characterize this problem have been completed (Sieniutycz, 1991), acquiring the following data: effective rate constant $k_{\text{eff},}$

Eq. (8.111), outlet substrate concentration $c_w = c_0 \exp(-k_{\text{eff}}\tau_k)$, and the production resulting from the reaction

$$P \equiv \int_0^{\tau_k} (c_0 - c_w)dt = (c_0 - c_w)t_k$$

from the regeneration cost

$$I = \lambda_R y(w_p, w_k)\tau_k = \lambda_R y^* \tau_k \text{ for } w_k = w_k(w_p, y^*)$$

(see comments at the end of Section 8.3.2; quality index S, Eq. (8.110)).

In order to acquire a qualitative picture illustrating the role of coordinating decisions, components of the vector **v** have been varied in a wide scope (wider than required in practice). The following fixed input values have been assumed:

$$k_0 = 3.3 \times 10^9, \quad k_{do} = 2.0 \times 10^{19}, \quad E_d = 126{,}060, \quad E = 64{,}851, \quad \dot{g} = 1, \quad c_0 = 1.$$

However, we will not discuss changes of activity in a detail. We shall only state that the optimal value of the initial activity w_p is lower than one, and it remains in the range $(1, y^*)$.

Let us discuss the role of production time t_k in some detail. This is illustrated in Figs. 8.6–8.8. The effective reaction rate constant k_{eff} decreases with time t_k, which implies lower reaction temperatures for larger values of t_k. Still, k_{eff} remains independent of τ_k. Substrate outlet concentration c_w rises along with the increase of t_k. At the same time, c_w decreases with the contact time τ_k and with the increase in use of the catalyst activity Δw.

What is interesting is the effect of the conversion increase for shorter t_k in the optimal process, Fig. 8.6. It is explained by the observation that optimal temperatures $T(t)$ and effective constant rates k_{eff} are higher for short production times t_k. The process is then more intensive, however noticeably shorter. As a result of the clash between two opposing factors T and t_k, with

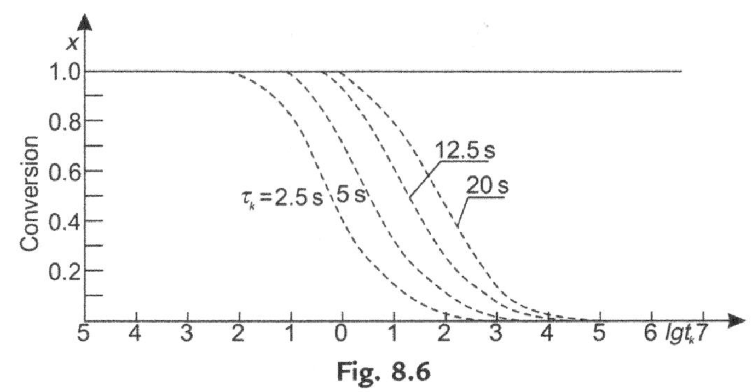

Fig. 8.6
The influence of contact time τ_k and reaction time t_k on the optimal conversion.

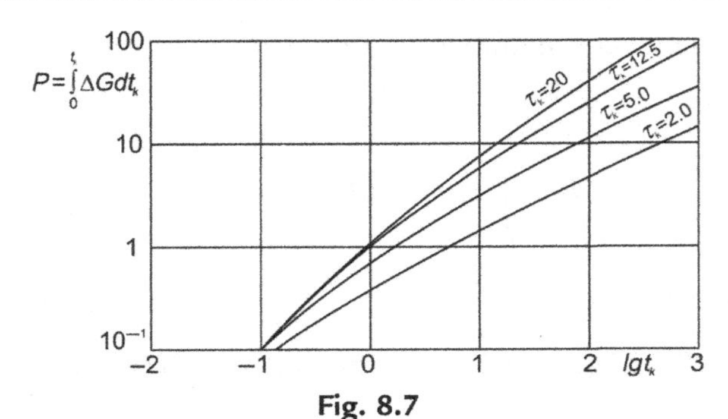

Fig. 8.7
Production (locally optimal) of the valuable component as a function of times t_k and τ_k.

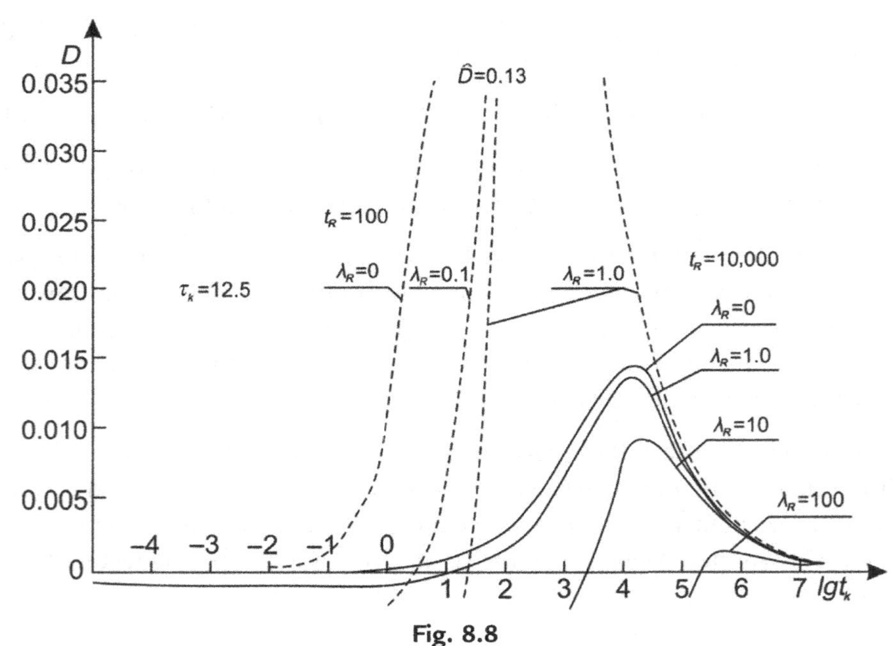

Fig. 8.8
Profit D depending on reaction time t_k, regeneration time τ_R, and price λ_R (for $\tau_k=12.5$).

the latter being stronger, the optimal production P decreases with the increase of time t_k (Fig. 8.7). Naturally, a remarkable increase in production takes place along with the increase of contacting time t_k.

The relation between the quality index $S=D(t_R,t_k,\tau_k)$ and the coordination variable t_k is illustrated in Fig. 8.8. Its data correspond to the assumption of a reasonable value of the coordinating variable t_k. However, the decrease of τ_k, caused, for example, by limiting the quantity of steel used for the reactor construction, brings dramatic drops in the profit. The effect of similar strength is observed for the downtime τ_R (Fig. 8.8). The longer τ_R corresponds with the increased production downtime, that is, in a situation when the idle duration of the reactor is caused not only by regeneration, but also by the personnel taking their break. The effects of excessive production downtimes (impact of t_R) are equally harmful as the effects of incorrect upper-level decisions regarding τ_k.

The optimal state corresponds to a suitably balanced incidence of regenerative processes, high enough to provide proper intensity of the process at the production stage, and simultaneously low enough to ensure the unprofitability of regeneration does not overly reduce the global gain from the system's operation.

8.3.5 Significance of hierarchical controls

Analyses of solutions to many complex chemical systems, for example, reaction-catalyst regeneration systems, show that in hierarchical systems the role of coordination controls is substantial. Therefore, in a typical reaction-catalyst regeneration system, global quality index is very sensitive to changes in such basic quantities as process duration, clock time t_k, contact time of phases, τ_k, and the degree of catalyst decay, Δw. In fact, practical engineers know that the best profit increase is achieved when the above quantities are upper-level rather than lower-level decisions. The goals of each upper level, which formulates and transfers its orders to lower levels, are more general than goals of lower levels and entail tasks of higher priority.

Upper-level management of the system (Fig. 8.9) cannot be based solely on the knowledge of all details of all lower-level subsystems. Every information transferred onto the higher level must be condensed or aggregated. This follows from the observation that upper-level optimizations use only a part of the total available information. On the other hand, local optimizations at lower level(s) use more detailed information involving, for example, dynamical equations, kinetic constants for the reactor and the regenerator, activation and deactivation energies, etc.

Data of aggregated information for the reactor working at the temperature $T \leq T^*$, that is, the temperature T not exceeding an admissible limit T^*, may be generated in the form of tables which describe the effective reaction rate constant, k_{eff}, the product of the standard reaction rate constant, and catalyst activity. Knowledge of these data (along with other data, such as, regeneration time t_R and its unit cost λ_R) by the upper-level staff (designers) enables the calculation of the quality index S in terms of coordinating controls as well as the optimization of the system as a whole. The aggregated data are transferred from lower levels to the upper level by communication channels (Fig. 8.9). In particular, communication may transmit aggregated information involving data of partial quality indices, at fixed coordinating decisions. More research is, however, necessary in this difficult area.

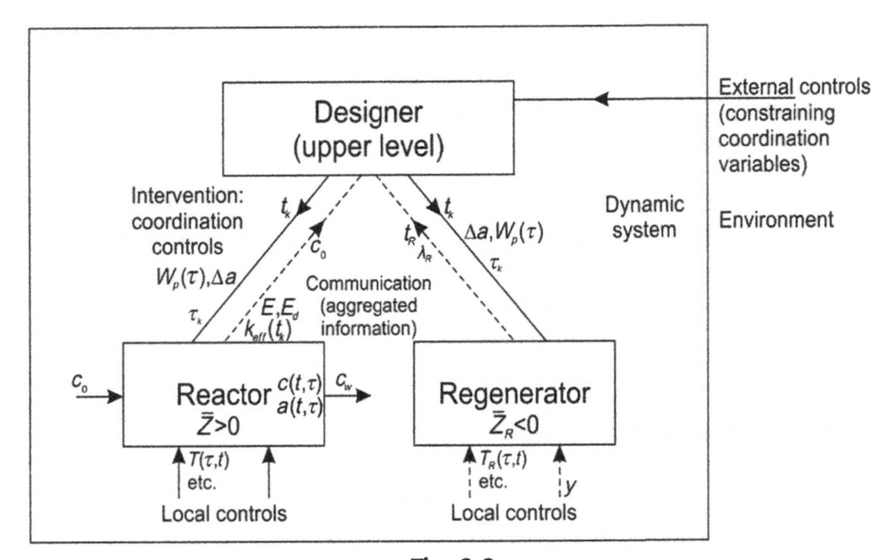

Fig. 8.9
Reactor-regenerator arrangement as a complex system of the hierarchical structure.

Exact distinction between coordinating and local decisions may be obstructed if there is sufficiently large number of decisions and management levels in the system. Difficulties may arise in case of adequate assigning of some controls to appropriate levels. The observation that upper levels operate with exceedingly high number of decisions may follow from the nature of the problem and often provides the evidence for too excessive centralization of controls. In case of the reaction-regeneration system analyzed here, excessive centralization means that the optimization problem is

$$\max_{\mathbf{v},\mathbf{u},\mathbf{u}_R} S(\mathbf{v},\mathbf{u},\mathbf{u}_\mathrm{R}) \equiv \max_{\mathbf{v},\mathbf{u},\mathbf{u}_R} (Z(\mathbf{v},\mathbf{u}) - K_R(\mathbf{v},\mathbf{u}_R)) \tag{8.113}$$

that is, that the local decisions (temperature of the reactor $T(\tau, t)$ or the one of regenerator $T_R(\tau, t)$) are undertaken simultaneously with coordinating controls at the higher level (e.g., by the system's designer). Such a scheme may produce nearly correct results in systems with large number of decisions, provided that only moderately disaggregated information package is transferred to the upper level. Yet, the situation is not normal since a partial aggregation of information at the upper-level management is in practice unavoidable. However, too excessive centralization diverts attention of the upper-level managers from the highest-priority coordinating decisions $\mathbf{v}(t_k, \Delta w, \tau_k)$. The mathematical consequence of centralization is simultaneous optimization of the system with respect to local decisions $\mathbf{u}$ and coordinating decisions $\mathbf{v}$, while the numerical consequence is usually an increase of the computation time. Shifting some decisions $\mathbf{u}$ from upper level to the more correct local level, that is, proper decentralization, is linked with decomposition. Its numerical benefit should be the reduction of the computation time. This idea is contained in algorithms of Eqs. (8.69)–(8.74).

Decentralization may be inefficient when some coordinating decisions **v** are shifted on too low level and selected from the viewpoint of a particular subsystem. Erroneous nature of this procedure is clear since the system's optimum cannot be determined directly from partial optima of subsystems. An example of such inappropriate procedure would be the shift of the coordinating decision τ_R to the regenerator level (it does not apply to the problem analyzed above, where τ_R has been prescribed). The decision regarding the reaction time τ_k may then be undertaken approximately as the optimal one from the regenerator's perspective, but not necessarily from the point of view of the entire cycle.

8.3.6 Optimization of cyclic modes of operation

In an early paper, Ogunye and Ray (1971a) formulated an optimal control problem for optimal policies in tubular reactors experiencing catalyst decay. To solve the problem, they derived a weak maximum principle for such distributed parameter systems. From this principle, an efficient computational algorithm was found which allows the simultaneous solution of the problems of both catalyst distribution along the reactor and optimal control. Detailed numerical examples have been worked out for isothermal reactors, adiabatic reactors, and adiabatic reactors with both catalyst distribution problem and an optimal control problem. For irreversible reactions, constant conversion policies are found to be optimal for a significant portion of the time. Literature analysis shows some flaws in previous solutions related to earlier works on the subject.

Ogunye and Ray (1971b) have also derived a sort of weak maximum principle as well as a computational algorithm for the optimization of cyclic plug flow tubular reactors with catalyst deactivation. They focus on reactors which are operated in the semiregenerative fashion for an on-stream time θ followed by a regeneration period θ_R. Semiregenerative reactors suffer a disadvantage that the whole process has to be shut down for the period of catalyst regeneration. This disadvantage can be overcome if the reactors are operated in a cyclic manner, where one or more reactors are always regenerating off-line. The cyclic mode of operation is a means of maintaining a continuous flow of products without the needs of shutting down the process for regeneration. In this paper, a maximum principle developed by Ogunye and Ray (1971c) is extended to take into account the cyclic nature of operation. A computational scheme is tested on examples to determine the computation required for the optimal solution of the cyclic reactors. A parametric study involves the effect of kinetic parameters and the sequence of beds. The authors discuss several earlier approaches for finding the optimum regeneration sequence. They also work out and analyze examples to illustrate the efficiency of the computational approach.

In their papers, Ogunye and Ray (1971a,b,c) considered the optimization of reactors experiencing catalyst decay without the product recycle. However, an abundant bibliography of authors exists who have considered this and also many other, different aspects. Elnashaie and Elshishini (1993) treated modeling, simulation, and optimization of industrial fixed bed catalytic reactors in which hydrodynamic imperfections can be very serious. Megiris and Butt (1990a,b) pointed out effects of poisoning on the dynamics of fixed bed reactors, respectively, in the context of isothermal and constant conversion policy of operation. Since water content in hygroscopic catalysts may be detrimental for their catalytic properties, knowledge of drying granular solids and keeping these materials in dry environment is recommended (Strumillo and Kudra, 1987).

A book by Ogunnaike and Ray (1994) is a suitable source which offers a broad context to the processes considered in this chapter and provides a modern view of process control in the context of today's technology. It contains the standard material in a coherent presentation and uses a notation that is more consistent with the research literature in process control. Especially worth reading are some unique topics that include a unified approach to model representations, process model formation and process identification, multivariable control, statistical quality control, and model-based control. This book has been designed to be used as an introductory text for undergraduate courses in process dynamics and control. In addition to chemical engineering courses, the text would also be suitable for similar courses taught in mechanical, industrial, and metallurgical engineering departments. The students will be benefited by material enhancing their experience with measurement instruments, real-time computers, process dynamics experiments, and control problems.

8.4 Discussion of basic results

The results obtained demonstrate that for a single nonreversible reaction $A \rightarrow B$, the rate of which depends on the catalyst deactivation, it is possible to acquire simple and general results without the knowledge of neither reaction order and deactivation order nor specific analytic expressions describing the reaction and deactivation rates.

Optimal cocurrent and countercurrent and batch operations deliver the product of the same final conversion, provided that total residence times of both phases are equal, and the temperature policy is stationary. If total times Θ_r and Θ are not equal, $T = T_{\text{stationary}}$ and the ratio $\Theta_r/\Theta < 1$, then the batch reactor ensures better conversion than the reactor with a moving catalyst bed, under the assumption that time Θ remains identical for various processes. Otherwise, if the ratio $\Theta_r/\Theta > 1$, the better final conversion is obtained in cocurrent or cocurrent reactors. There is no difference between final optimal concentrations acquired in cocurrent and countercurrent processes for the same values Θ_r and Θ. These results are applicable also in case of the active temperature constraint if the catalyst deactivation is not a concentration-dependent process [$\phi(c) = 1$].

8.5 Consecutive-parallel reactions with exponential catalyst decay

8.5.1 Introductory remarks

Below, problems initiated in Section 8.3.3 are continued, but under the assumption of catalyst deactivation. Considerations have been confined to systems described with difference equations or standard differential equations. Computational aspects are also discussed.

The consecutive-parallel system (Eqs. 8.114a, 8.114b) is one of the most common reaction systems, as it occurs in a large number of reactions related to technological processes (oxidation of carbon and sulfur, chlorination of benzene, etc.). Suitable examples can be found in books on nonorganic and organic chemistry technologies.

An essential computational tool is a discrete algorithm with a constant Hamiltonian (Sieniutycz, 1991, Section 5.5; Szwast, 1994), applied for the purpose of acquiring a computer program (Sieniutycz and Szwast, 2018), which solves the problem of the optimal temperature profile for a cascade of well-mixed reactors with reactions (8.114a) and (8.114b) running in the presence of a catalyst subject to exponential (first-order) deactivation.

The cascade model is employed to obtain a discrete solution for a real cascade (with finite number of stages, N) and a limiting solution for an infinite number of stages ($N \rightarrow \infty$). The latter solution refers to continuous reactors, such as batch, tubular, cocurrent, and countercurrent reactors. Optimal solutions for continuous reactors (as those described with ordinary differential equations) are obtained in the course of assuming enough steps of a cascade in order to reach the sufficient approximation of the continuous system through the discrete system.

The computational program (Fig. 8.10 and Section 8.5.5 in Sieniutycz and Szwast, 2018) is principally designed for a set of parallel-consecutive reactions, but, after easy modifications, can be used either for the purpose of optimizing simpler kinetic schemes (e.g., parallel and consecutive reactions, single reversible reactions) or schemes that are more complex (e.g., consecutive-parallel reactions with catalyst deactivation that depends on the temperature and concentration of reacting substances, which are not necessarily first-order in relation to catalyst activity).

The original formulation of the optimization problem for noncatalytic consecutive-parallel reactions (Skrzypek, 1974; Burghardt and Skrzypek, 1974), where the notion of deactivation is naturally absent, is briefly described in Chapter 7, Section 7.3.3, of the book by Sieniutycz and Szwast (2018). However, when a catalytic reaction proceeds, attention should always be directed towards the effect of catalyst deactivation as well as the significance of optimization of the total space time. The solution obtained for various reactors by placing them in the realm of broader-class processes in cascades approaches the continuous solution when the total number of cascade steps N approaches infinity (and in practice, when N is high enough).

In a special case, when total residence times of reacting substances and the catalyst are equal, the solution for the batch reactor is acquired. In most cases, the batch case requires different input data. The assumed method and results are also significant for optimization of cascades composed of only a small number of steps, that is, for multistage reactors, which are naturally discrete.

8.5.2 Problem, systems, and assumptions

We are looking at such a temperature profile which maximizes the final concentration of the valuable product c_{Rk} subject to a constraint imposed on the space time of reacting stream, t_k. If the volumetric flux of reacting substances q, the inlet molar concentration of the product $c_R(0)$, and the unit price of the product p_R are constant, the problem investigated is equivalent to maximization of the economic value of production subject to the limited capital expenditures K_i (the discussed equivalence holds provided that the cost K_i is a function rather than a functional of the final time t_k). The change in t_k at the last stage of the optimization procedure (cf. Section 8.5.7) allows to easily include the changes of K_i, although the maximization of product concentration c_{Rk} always remains the most important and most difficult stage of the solution.

The following systems of *reactions with the exponential deactivation* are considered and compared:

1. System of irreversible consecutive-parallel reactions with catalyst deactivation

$$A + B \overset{k_1, E_1}{\longrightarrow} R \tag{8.114a}$$

$$B + R \overset{k_2, E_2}{\longrightarrow} S \tag{8.114b}$$

   This reaction system remains fundamental in our considerations. Further reaction schemes serve the sole purpose of comparison of the optimization results with the results obtained for the above system.

2. Parallel system with the catalyst deactivation

$$S \overset{k_2, E_2}{\longleftarrow} B \overset{k_1, E_1}{\longrightarrow} R \tag{8.115}$$

3. Consecutive system with the catalyst deactivation

$$A \overset{k_1, E_1}{\longrightarrow} R \overset{k_2, E_2}{\longrightarrow} S \tag{8.116}$$

4. A single reversible catalytic reaction with the catalyst deactivation

$$A \underset{k_2, E_2}{\overset{k_1, E_1}{\rightleftarrows}} B \tag{8.117}$$

In all cases, R denotes the desired (valuable) product, the concentration of which should be maximized.

The constant concentration of the mixture of reacting substances and the first-order catalyst deactivation that is independent from temperature and concentrations is assumed. This deactivation is described by the differential equation,

$$\frac{da}{dt'} = -\alpha_d a \tag{8.118}$$

and the initial condition

$$a\left(t'_p\right) = a_0 \tag{8.119}$$

where t' is the residence time of the catalyst. The symbol α_d denotes the average value of the product of constant deactivation rate k_d and the concentration term $\varphi(c)$ in Eq. (8.8), within the analyzed range of temperatures and concentrations of reacting mixtures. Approximation of the catalyst aging, described by Eq. (8.118), is as simple as possible. In fact, however, some processes such as catalytic cracking do not require it to be more complex. For a constant coefficient α_d, Eq. (8.118) leads to an exponential formula

$$a = a_0 \exp(-\alpha_d t') \tag{8.120}$$

Since noncatalytic reactions (8.114a) and (8.114b) will also be considered, using the residence time of the catalyst (as in Eqs. 8.9–8.11) would be troublesome in limiting situations when catalyst concentrations tend towards zero. Therefore, as an independent variable t, we shall use this time as the residence time of reacting substances. This releases us from the necessity to introduce to the reaction kinetics equations the modified reaction rate constants k'_0, which contain the quotient of space times of reacting substances and the catalyst as in the Eq. (8.11). This quotient will, however, be present in the equations that describe the catalyst in heterophase reactor with gliding between the catalyst phase and the reacting substance phase.

Let us analyze the cocurrent process, for instance. Since $t' = t\Theta/\Theta_r$, the Eq. (8.120) will take a form

$$a = a_0 \exp\left(-\frac{\alpha_d \theta t}{\theta_r} t'\right) = a_0 \exp(-\alpha t) \tag{8.121}$$

where t is the space time of reacting substances, while α is the coefficient, which is acquired through the multiplication of α_d by the quotient of catalyst and reacting substances' space times θ/θ_r. Only for batch reactors (with either homogeneous or heterogeneous catalysts), $\alpha = \alpha_d$. Eq. (8.121) may be employed with the unchanged reaction Eqs. (8.135)–(8.138) at the stable density of the catalyst suspension. For heterogeneous flow reactors, α and α_d differ in accordance with the general definition

$$\alpha = \begin{cases} \alpha_d \theta/\theta_r & \text{cocurrent reactor} \\ -\alpha_d \theta/\theta_r & \text{countercurrent reactor} \\ \alpha_d & \text{batch reactor} \end{cases} \tag{8.122}$$

Such a value is referred to as the substitutional deactivation constant. It will constitute a convenient parameter of optimal solutions obtained with the standard process temperature restriction

$$T_* \leq T \leq T^* \tag{8.123}$$

The Arrhenius relation of constant reactions rates and temperature is assumed.

8.5.3 Problem formulation for consecutive-parallel reactions

For the reaction system (8.114) and deactivation kinetics (8.121), the following kinetics for catalytic reactions have been assumed:

$$\frac{dc_A}{dt} = -k_1 a c_A c_B \tag{8.124}$$

$$\frac{dc_B}{dt} = -k_1 a c_A c_B - k_2 a c_B c_R \tag{8.125}$$

$$\frac{dc_R}{dt} = -k_1 a c_A c_B - k_2 a c_B c_R \tag{8.126}$$

$$\frac{dc_S}{dt} = k_2 a c_B c_R \tag{8.127}$$

The initial conditions for the above system comprise the set input concentrations $c_j(0) = c_{j0}$. For products R and S

$$c_R(0) = c_S(0) = 0 \tag{8.128}$$

All final concentrations are unbound, that is they should be found as elements of an optimal solution.

The system shown in (8.114) has two reaction invariants which develop from the stoichiometry. Still, only one of them is required for further considerations.

$$c_R(t) - c_B(t) + 2c_A(t) = c_R(0) - c_B(0) + 2c_A(0) = \text{const} \tag{8.129}$$

which enables the elimination of one of the concentrations, for example, $c_R(t)$.

By taking the Eqs. (8.128), (8.129) into account, the following quantities are introduced:

$$x_i(t) = c_i(t)/c_B(0) \tag{8.130}$$

$$M = c_A(0)/c_B(0) = c_{A0}/c_{B0} \tag{8.131}$$

$$x_R = x_B - 2x_A + 2M - 1 \tag{8.132}$$

$$G_{i0} = k_{i0}a_0c_B(0) \tag{8.133}$$

$$G_i(T) = G_{i0}\exp(-E_i/RT) \tag{8.134}$$

Then, Eqs. (8.124)–(8.127) as well as boundary conditions can be written in the form

$$\frac{dx_A}{dt} = -G_1(T)x_Ax_B\exp(-\alpha t) \tag{8.135}$$

$$\frac{dx_B}{dt} = -[G_1(T)x_Ax_B + G_2(T)x_B(x_B - 2x_A + 2M - 1)]\exp(-\alpha t) \tag{8.136}$$

$$\frac{dx_R}{dt} = [G_1(T)x_Ax_B - G_2(T)x_B(x_B - 2x_A + 2M - 1)]\exp(-\alpha t) \tag{8.137}$$

$$\frac{dx_S}{dt} = G_2(T)x_B(x_B - 2x_A + 2M - 1)\exp(-\alpha t) \tag{8.138}$$

$$x_A(0) = M \tag{8.139}$$

$$x_B(0) = 1 \tag{8.140}$$

$$x_R(0) = x_S(0) = 0 \tag{8.141}$$

$$x_i(t_k) - \text{free} \tag{8.142}$$

The optimization criterion for the problem of maximization $x_R(t_k)$ results from the Eqs. (8.137), (8.141) as

$$F = \int_0^{t_k} [G_1(T)x_Ax_B - G_2(T)x_B(x_B - 2x_A + 2M - 1)]\exp(-\alpha t)dt \tag{8.143}$$

Since the Eq. (8.137) has been replaced by its equivalent shown in Eq. (8.143), we don't need to consider it anymore. Eq. (8.138) may be ignored as well, since it does not include the variable x_S and the final value of the variable $x_S(t_k)$ remains unbound. In consequence, the problem of optimization is broken down to the maximization of criterion (8.143), where only two differential restrictions (8.135) and (8.135) as well as initial conditions (8.139) and (8.140) apply. The total space time of reacting substances may be either prescribed or free. The optimal temperature control must satisfy the constraint (8.123).

The introduction of a new variable $u = (a_0 - a)/\alpha$ allows the conversion of the equation $\dot{a} = -\alpha a$ (i.e., differential Eq. 8.118) to the following form: $du/dt = a$. Then, the Eq. (8.135) assumes the form

$$dx_A/du = -G_1x_Ax_B$$

that is, the catalyst activity has been eliminated. By employing this procedure, it is possible to eliminate activity from Eqs. (8.135), (8.136), and (8.143) and reduce the problem to the optimization without catalyst deactivation, in which the variable u replaces time t. Return to the variable t takes place after the optimization completion, by integration of the equation

$$dt = \frac{du}{a} = \frac{du}{a_0 - \alpha u} \tag{8.144}$$

which results from the definition of the variable u. As a result, we obtain

$$t = \frac{1}{\alpha} \ln \frac{a_0}{a_0 - \alpha u} \tag{8.145}$$

where $u(0) = 0$ and $0 \leq u < a_0/\alpha$. By employing this approach, one can use the solution to the continuous problem without deactivation (Skrzypek, 1974; Burghardt and Skrzypek, 1974), which is based on the same idea as Horn's method of eliminating conjugate variables (at the end of Section 8.2.3). Still, such an approach cannot be extended to cover the case of three state variables, when the catalyst activity constitutes an independent dynamic variable described by its own differential or difference equation (Horn's method applies only for two-dimensional processes). Therefore, another approach has been adopted below, based on the discrete maximum principle with a constant Hamiltonian (Sieniutycz, 1991, Section 5.5 therein; Szwast, 1994), for which the above-described extension including the temperature-dependent and concentration-dependent deactivation is simple. Assuming a chemical process described by the state variables x_A, x_B, and a, the continuous limit of the maximum principle is employed for the purpose of acquiring qualitative conclusions regarding the optimal control profiles in continuous reactors. However, the quantitative results for the control in continuous and discrete processes are acquired with the aid of a discrete algorithm of the maximum principle with the constant Hamiltonian.

8.5.4 Qualitative properties of continuous temperature controls

It is worthwhile to demonstrate first how to use the continuous maximum principle for the purpose of foreseeing the qualities of optimal temperature controls in continuous systems (batch reactors in their natural unsteady state and tubular reactors—both cocurrent and countercurrent—in a steady state).

By employing Pontryagin's maximum principle for the criterion function (8.143), a Hamiltonian is defined as follows

$$\begin{aligned} H(x_A, x_B, z_A, z_B, T, t) = \{ & G_1 x_A x_B - G_2 x_B (x_B - 2x_A + 2M - 1) + \\ & - z_A G_1 x_A x_B - z_B [G_1 x_A x_B - G_2 x_B (x_B - 2x_A + 2M - 1)] \} \exp(-\alpha t) \end{aligned} \tag{8.146}$$

or, in a simplified form:

$$H = [G_1 x_A x_B (1 - z_A - z_B) - G_2 x_B (x_B - 2x_A + 2M - 1)(1 + z_B)] \exp(-\alpha t) \tag{8.147}$$

The Hamiltonian H is explicitly dependent on time, and therefore

$$\frac{dH}{dt} = \frac{\partial H}{\partial t} = -\alpha H \tag{8.148}$$

whence

$$H(t) = H(0) \exp(-\alpha t) \tag{8.149}$$

This equation indicates that the Hamiltonian maintains its sign along the optimal trajectory, or that it vanishes in the entire range of t, that is, for $0 \le t \le t_k$. Later, it will show that, in terms of physical qualities, only nonnegative values of H are of interest, thus the main attention should be directed towards the case $H \ge 0$.

Adjoint equations take the following form:

$$\frac{dz_A}{dt} = -\frac{\partial H}{\partial x_A} = -[G_1 x_B (1 - z_A - z_B) + 2G_2 x_B (1 + z_B)] \exp(-\alpha t) \tag{8.150}$$

$$\begin{aligned} \frac{dz_B}{dt} &= -\frac{\partial H}{\partial x_B} = \\ &= \{-G_1 x_B (1 - z_A - z_B) + G_2 [(x_B - 2x_A + 2M - 1) + x_B](1 + z_B)\} \exp(-\alpha t) \end{aligned} \tag{8.151}$$

and for free, unbound $x_A(t_k)$ and $x_B(t_k)$

$$z_A(t_k) = z_B(t_k) = 0 \tag{8.152}$$

From Eqs. (8.226), (8.222),

$$\frac{dz_B}{dt} = -\frac{H(t)}{z_B} + G_2 x_B (1 + z_B) \exp(-\alpha t) \tag{8.153}$$

The above equation provides the evidence that for

$$H(t) \ge 0 \tag{8.154}$$

and satisfied conditions in (8.152) within the range $0 \le t \le t_k$ the following inequality takes place:

$$(1 + z_B) > 0 \tag{8.155}$$

In order to prove the above-mentioned inequality by *reductio ad absurdum*, it sufficies to assume that $1 + z_B$ is negative or zero within the range $(0, t_k)$. Then, for the satisfied Eq. (8.154) and positive x_B and G_2, Eq. (8.153) proves that the sum $1 + z_B$ must also be negative for all $t>t'$, including t_k, and therefore, conditions (8.152) could not be fulfilled. Thus, the inequality (8.155) is valid.

Also, on the ground of Eqs. (8.147), (8.148), (8.155), the following holds

$$1-z_A-z_B\begin{cases}>0 \text{ for } 0\le t\le t_k\\ =0 \text{ for } H(t)=0 \text{ and } t=0\end{cases} \tag{8.156}$$

because in the presence of Eqs. (8.132), (8.141) one has

$$x_R(0)=x_B-2x_A+2M-1=0$$

The relation between expressions $1-z_A-z_B$ and $1+z_B$ is defined in the form of an inequality describing the difference between time derivatives of these expressions. Eqs. (8.150), (8.151), (8.154), (8.156) give

$$\frac{d(1-z_A-z_B)}{dt}-\frac{d(1+z_B)}{dt}=G_1x_B(1-z_A-z_B)\exp(-\alpha t)+\frac{2H(t)}{x_B}>0 \tag{8.157}$$

Since Eq. (8.152) shows that $1-z_A(t_k)-z_B(t_k)=1+z_B(t_k)=1$, it follows from inequality (8.157) that

$$1-z_A-z_B<1+z_B \tag{8.158}$$

The use of this inequality in Eq. (8.147) leads to the conclusion that the production rate of the product R

$$[G_1x_Ax_B-G_2x_B(x_B-2x_A+2M-1)]\exp(-\alpha t) \tag{8.159}$$

is positive for any t, only if $H>0$. Additionally, if $H=0$, this rate stays positive for any t, except t_k, and vanishes for $t=\hat{t}_k$, where $\hat{t}_k$ is an optimal time, a free variable.

We may now analyze the optimality problem of the concentration x_{Rk} with respect to the process duration, t_k. (If $x_R(0)=0$, it is also the optimality problem for the generation of the product R). Vanishing of the final Hamiltonian (and each current value of the Hamiltonian in the present problem) constitutes the necessary condition for the optimality of the final time t_k. As a result, the dependencies that have already been calculated can be interpreted as follows:

(a) for the time-optimal trajectory, concentration of the product R increases monotonically with time t (length of a tubular reactor), and the production rate is always nonnegative, reaching the zero value only for $t=t_k$,
(b) processes, for which $H>0$, are characterized by the total time t_k, which is shorter than the optimal time $\hat{t}_k$, and the production rate of R is positive for any $0\le t\le t_k$,
(c) processes with the negative final production rate R are characterized by the total time t_k that is longer than the optimal time $\hat{t}_k$; for such processes, the Hamiltonian H remains negative.

Therefore, for processes that satisfy the inequality $t_k > \hat{t}_k$, an increase of capital expenditures with t_k is accompanied by the decrease of the production. Clearly, these processes should be ignored, that is, we will only analyze the processes for which $t_k \leq \hat{t}_k$ and $H(t) \geq 0$.

Processes with $t_k = \hat{t}_k$

The subject of the present analysis is the ability to reach optimal stationary temperature along a time-optimal trajectory. Eq. (8.147) implies the simultaneous satisfaction of two following optimality conditions:

$$\frac{\partial H}{\partial T} = \frac{1}{RT^2}\left[E_1 G_1 x_A x_B (1 - z_A - z_B) - E_2 G_2 x_B (x_B - 2x_A + 2M - 1)(1 + z_B)\right] \exp(-\alpha t) = 0 \tag{8.160}$$

$$H = \left[G_1 x_A x_B (1 - z_A - z_B) - G_2 x_B (x_B - 2x_A + 2M - 1)(1 + z_B)\right] \exp(-\alpha t) = 0 \tag{8.161}$$

However, this is not possible when $E_1 \neq E_2$; what is more, the temperature does not influence the solution when $E_1 = E_2$. If $E_1 = E_2$, then various profiles $T(t)$ lead to identical values of the profit function, although final times and the production rate of R differ in most cases. Therefore, it is justified to assume the maximum allowed temperature $T = T^*$, which minimizes t_k, as the optimal temperature if $E_1 = E_2$. Provided that $E_1 \neq E_2$, Eq. (8.161) proves that the derivative $\partial H/\partial T$ in Eq. (8.160) is negative for $E_2 = E_1$ and positive for $E_1 > E_2$. This means that the optimal temperature $T(t)$ (maximizing H) is

$$\hat{T}(t) = \begin{cases} T_* \text{ if } E_2 > E_1 \text{ and } t_k = \hat{t}_k \\ T^* \text{ if } E_1 > E_2 \text{ and } t_k = \hat{t}_k \end{cases} \tag{8.162}$$

Processes with t_k shorter than $\hat{t}_k$

In this case,

$$H = \left[G_1 x_A x_B (1 - z_A - z_B) - G_2 x_B (x_B - 2x_A + 2M - 1)(1 + z_B)\right] \exp(-\alpha t) > 0 \tag{8.163}$$

For the reactor inlet, Eqs. (8.132), (8.141), (8.163) yield

$$H(t = 0) = G_{10} x_A x_B (1 - z_A - z_B) \exp(-\alpha t) \exp(-E_1/RT) > 0 \tag{8.164}$$

Thus, the continuous optimal profile starts at $T(0) = T^*$ if only $x_R(0) = 0$.

For stationary temperatures, conditions (8.160) and (8.163) must be satisfied at the same time, which is possible only if $E_2 > E_1$. Whenever $E_1 \geq E_2$, the derivative $\partial H/\partial T$, Eq. (8.160), is always positive, and that is why the optimal profile is an isotherm $\hat{T}(t) = T^*$.

Considerations regarding optimal stationary temperatures when $E_2 > E_1$ contain some important details. Eq. (8.160) can be used for the purpose of eliminating the expression $1 - z_A - z_B$ from the inequality (8.163), which gives

$$H(t) = G_2 x_B (x_B - 2x_A + 2M - 1)(1 + z_B)(E_2/E_1 - 1)\exp(-\alpha t) \tag{8.165}$$

or by using Eqs. (8.132), (8.149)

$$G_2 x_B x_R (1 + z_B)(E_2/E_1 - 1) = H(0) > 0 \tag{8.166}$$

This equation is commonly used to describe the shape of the optimal profile qualitatively. In order to foresee this shape, it should be noted first that the product $x_B x_R$ can reach a maximum for a certain t (if t is short, x_R is small, whereas if t is long, x_B is small). Obviously, it is possible that the product $x_B x_R$ would be an increasing function t within $(0, t_k)$, especially for short times t_k.

In order to characterize the time dependence of the expression $1 + z_B$, Eq. (8.160) is employed to convert the canonical Eq. (8.151) into the following form:

$$\frac{d(1 + z_B)}{dt} = G_1 x_A (1 - z_A - z_B)\left[\frac{E_1}{E_2}\frac{x_R + x_B}{x_R} - 1\right]\exp(-\alpha t) \tag{8.167}$$

Because of the temperature constraint (8.123) and the fact that x_R increases monotonically with time t, while x_B decreases with time t, Eq. (8.167) shows that the sum $1 + z_B$ may achieve a maximum value for a certain value t, similarly as the product $x_B x_R$. Thus, the expression

$$x_B x_R (1 + z_B) \tag{8.168}$$

can also reach a maximum for a certain t. Consequently, Eq. (8.166) proves that the rate constant G_2 and so the temperature can reach the minimum value. Moreover, if only expression (8.168) has an increasing part, then the stationary part of the optimal profile for $T(t)$ is limited to a decreasing curve.

Taking the above into consideration and minding that the optimal process starts at $T(0) = T^*$ (due to the absence of R in the inlet stream), it may be concluded that the general optimal profile of a continuous system is comprised of the following parts: isothermal $T = T^*$, stationary decreasing ($T = T_{stat}$), isothermal ($T = T_*$), stationary increasing($T = T_{stat}$), and isothermal $T = T^*$. The general profile is defined as the one, in which all possible parts of the temperature strategy can be reached in a single process.

In specific cases, the shape of the optimal profile may be less complex; however, it must contain the initial part of the above-mentioned general profile. For instance, a specific process may feature only the segment $T = T^*$. Although the constraint $T \leq T^*$ is always operative, the constraint $T \geq T_*$ may often be inoperative. This means that the shift from decreasing to increasing part of optimal curve $T(t)$ can be realized along a single stationary curve having the temperature minimum, such that $T_{min} > T_*$.

It needs be pointed out that the above conclusions are valid only if $x_R(0) = 0$. Otherwise, the specific optimal profile does not necessarily start from $T(0) = T^*$, and then it can constitute any part of the general profile, provided that the order of particular sections of the line $T(t)$ is secured. For example, if the initial part is comprised of an increasing section $T(t)$ (which precedes the isothermal part $T=T^*$ in the general profile), then the following isothermal part can only be the final part of an optimal process.

The reader is now prepared to develop an optimization algorithm for the purpose of calculation of optimal temperature controls.

8.5.5 Consecutive-parallel reactions in a cascade

8.5.5.1 Optimization algorithm

It is well-known that the maximum principle constitutes an effective tool for prediction of qualitative properties of optimal processes. However, finding concrete quantitative results for the functions $T(t)$, $x_R(t)$, etc. requires solving the canonical system with the decision T derived from the Hamiltonian maximum condition. Usually this system lacks the analytical solution, and a numerical procedure is required in order to make it solvable.

Discretization of differential equations is a standard tool for calculating the optimal controls, especially in cases when computers are used. It can be accomplished in several ways. Here the effect of discretization is shown in Eqs. (8.169)–(8.174) below, which serve as the discrete model used in a standardized computer program, presented in Fig. 8.10A and B of Sieniutycz and Szwast (2018). The computational procedure, which involves the discrete model, Eqs. (8.169)–(8.174), is outlined below. A discrete representation of a continuous process at the same time constitutes the difference model of a cascade with perfectly mixed reactors. Any cascade with perfectly mixed stages is discrete by nature. The discrete optimization solution may contain the continuous solution as a specific limiting case (not reversely). The discrete solution (more general than continuous) may pass into the continuous solution for $N \to \infty$ (practically for very large N).

Methods of dynamical optimization and the discrete algorithm with constant Hamiltonian (e.g., Section 1.5 in Chapter 1 and Chapter 12 in a book by Sieniutycz and Szwast, 2018) should be used in calculation of optimal trajectories and optimal controls. We assume that the optimization solution approaches the continuous solution for very large N.

In the analyzed problem, the interval of space time of reacting substances at each stage forms an additional (in comparison with the continuous system) decision Θ^n which satisfies the equation

$$\Theta^n = t^n - t^{n-1} \quad n=1,\ 2,\ \dots\ N \tag{8.169}$$

(index n is related to time and state following the stage n as well as to the decision at this stage.

The optimality criterion of a discrete process develops from Eq. (8.143) as

$$F = \sum_{n=1}^{N} \left[G_1^n(T) x_A^n x_B^n - G_2^n(T) x_B^n \left(x_B^n - 2x_A^n + 2M - 1 \right) \right] \exp\left(-\alpha t^n\right) \Theta^n \tag{8.170}$$

and the discrete equations of state based on formulas shown in (8.135) and (8.136) take the form

$$\frac{x_A^n - x_A^{n-1}}{\Theta^n} = -G_1^n x_A^n x_B^n \exp(-\alpha t^n) \tag{8.171}$$

$$\frac{x_B^n - x_B^{n-1}}{\Theta^n} = -\left[G_1^n x_A^n x_B^n + G_2^n x_B^n \left(x_B^n - 2x_A^n + 2M - 1\right)\right] \exp(-\alpha t^n) \tag{8.172}$$

where

$$G_i^n = G_{i0} \exp(-E_i/RT^n) \tag{8.173}$$

Boundary conditions are actually the relations (8.139)–(8.142), in which $x_i(0)$ has been replaced with $x_i^0 (i = A, B, R)$ and $x_i(t_k)$ has been replaced with x_i^N. In the presence of $t^0 = 0$, the fixed total time t^N satisfies the equation

$$t^N = \sum_1^N \Theta^n = \tau \tag{8.174}$$

where τ is a prescribed number.

It should be noted that the factor $\exp(-\alpha t^n)$ in equations of state and the criterion function is an approximation. In a sense, it replaces the allowance for an additional equation of state for activity

$$a^n - a^{n-1} = -\alpha a^n \Theta^n \tag{8.175}$$

The cascade model employing the above equation is more accurate, but it may be proven that, even for a small number of stages in the cascade, N, and a large value of the constant α, the error caused by the replacement of the model based on the Eq. (8.175) by the exponential model is not substantial. In most cases, it does not exceed a couple of percent. As the error increases when N decreases or α increases, in cases that require utmost precision, a model that incorporates Eq. (8.175) should be employed. This would, in turn, entail the increased number of calculations, resulting from the replacement of the two-dimensional model (state variables x_A^n, x_B^n) by the three-dimensional one (state variables x_A^n, x_B^n, a^n). To begin with, we may restrain ourselves to the two-dimensional model. However, if the catalyst deactivation is strongly dependent on temperature (and/or concentration), using the model that includes the Eq. (8.175) is necessary. Therefore, the computational algorithm (Fig. 8.10) is constructed in such a way that the shift to the algorithm including the catalyst activity a^n as an additional state variable does not cause essential difficulties.

By employing the discrete algorithm with the constant Hamiltonian, a following function is defined:

$$\begin{aligned} H^{n-1}\left(x_A^n, x_B^n, z_A^{n-1}, z_B^{n-1}, T^n, t^n\right) = \{ & G_1^n x_A^n x_B^n - G_2^n x_B^n \left(x_B^n - 2x_A^n + 2M - 1\right) + \\ & - z_A^{n-1} G_1^n x_A^n x_B^n - z_B^{n-1}\left[G_1^n x_A^n x_B^n + G_2^n x_B^n \left(x_B^n - 2x_A^n + 2M - 1\right)\right] \} \exp(-\alpha t^n) \end{aligned} \tag{8.176}$$

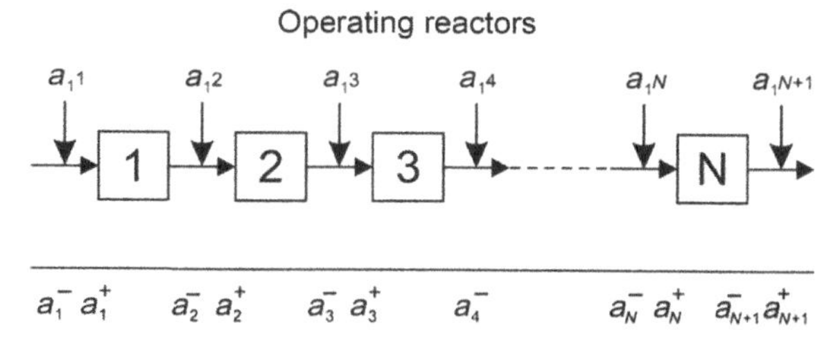

Fig. 8.10
Cyclic reactor scheme analyzed by Ogunye and Ray (1971a,b,c).

that is, a discrete counterpart of Eq. (8.146). It may be simplified to the following form:

$$H^{n-1} = \left[G_1^n x_A^n x_B^n \left(1 - z_A^{n-1} - z_B^{n-1}\right) - G_2^n x_B^n \left(x_B^n - 2x_A^n + 2M - 1\right)\left(1 + z_B^{n-1}\right)\right] \exp\left(-\alpha t^n\right) \quad (8.177)$$

For the above H^{n-1} the adjoint equations are as follows:

$$\begin{aligned}\frac{z_A^n - z_A^{n-1}}{\Theta^n} = -\frac{\partial H}{\partial x_A^n} = -[G_1^n x_B^n \left(1 - z_A^{n-1} - z_B^{n-1}\right) + \\ + 2G_2^n x_B^n \left(1 + z_B^{n-1}\right)] \exp\left(-\alpha t^n\right)\end{aligned} \quad (8.178)$$

$$\begin{aligned}\frac{z_B^n - z_B^{n-1}}{\Theta^n} = -\frac{\partial H}{\partial x_B^n} = \{-G_1^n x_A^n \left(1 - z_A^{n-1} - z_B^{n-1}\right) + \\ + G_2^n\left[\left(x_B^n - 2x_B^n + 2M - 1\right) + x_B^n\right]\left(1 + z_B^{n-1}\right)\} \exp\left(-\alpha t^n\right)\end{aligned} \quad (8.179)$$

Final conditions for conjugate variables are:

$$z_A^N = z_B^N = 0.$$

The fundamental difference between the continuous and discrete variants consists in that the equation

$$\frac{H^n - H^{n-1}}{\Theta^n} = -\frac{\partial H^{n-1}}{\partial t^n} = -\alpha H^{n-1} \quad (8.180)$$

does not result (as in the continuous case) from canonical equations, but it constitutes an independent optimality condition that entails an additional decision which is the discrete interval of time, Θ^n.

For a prescribed final time t^N, the final value of the Hamiltonian H^N is unspecified.

Stationary temperature control satisfies the equation

$$\frac{\partial H^{n-1}}{\partial T^n} = \frac{1}{R(T^n)^2}\left[E_1 G_1^n x_A^n x_B^n \left(1 - z_A^{n-1} - z_B^{n-1}\right) - E_2 G_2^n x_B^n \left(x_B^n - 2x_A^n + \right.\right. \\ \left.\left. + 2M - 1\right)\left(1 + z_B^{n-1}\right)\right] \exp\left(-\alpha t^n\right) = 0 \tag{8.181}$$

thus, after its transformation

$$L^n = E_1 G_1^n x_A^n x_B^n \left(1 - z_A^{n-1} - z_B^{n-1}\right) - E_2 G_2^n x_B^n \left(x_B^n - 2x_A^n + \right. \\ \left. + 2M - 1\right)\left(1 + z_B^{n-1}\right) = 0 \tag{8.182}$$

The variable L^n has been introduced for computational purposes. Vanishing L^n at the stage n proves the satisfaction of the necessary condition for the optimal stationary temperature.

Eqs. (8.171), (8.172) demonstrate that the expression of state before the stage in the state function after this stage does not cause difficulties, while the transformations in the opposite direction are in general not easy to obtain. This influences the forms of transformations used by the computer program.

From Eqs. (8.169), (8.171), (8.172),

$$t^{n-1} = t^n - \Theta^n \tag{8.183}$$

$$x_A^{n-1} = x_A^n + \Theta^n G_1^n x_A^n x_B^n \exp\left(-\alpha t^n\right) \tag{8.184}$$

$$x_B^{n-1} = x_B^n + \Theta^n \left[G_1^n x_A^n x_B^n + G_2^n x_B^n \left(x_B^n - 2x_A^n + 2M - 1\right)\right] \exp\left(-\alpha t^n\right) \tag{8.185}$$

Whereas, solving the conjugate Eqs. (8.178), (8.179), with respect to the adjoints z_A^{n-1} and z_B^{n-1} gives

$$z_B^{n-1} = \frac{z_B^n - B^n z_A^n + B^n\left[1 - 2G_2^n x_B^n \Theta^n \exp\left(-\alpha t^n\right)\right] - C^n}{1 + C^n + B^n\left[1 + 2G_2^n x_B^n \Theta^n \exp\left(-\alpha t^n\right)\right]} \tag{8.186}$$

$$z_A^{n-1} = \left(A^n\right)^{-1}\left\{z_A^n - \left[x_B^n \Theta^n \left(G_1^n - 2G_2^n\right) z_B^{n-1} - x_B^n \Theta^n \left(G_1^n + 2G_2^n\right)\right]\right\} \exp\left(-\alpha t^n\right) \tag{8.187}$$

where

$$A^n = 1 + G_1^n x_B^n \Theta^n \exp\left(-\alpha t^n\right)$$

$$B^n = \left[G_1^n x_A^n \Theta^n \exp(-\alpha t^n)\right]/A^n$$

$$C^n = G_2^n \Theta^n \left[\left(x_B^n - 2x_A^n + 2M - 1\right) + x_B^n\right] \exp(-\alpha t^n)]/A^n$$

Possession of Eqs. (8.186), (8.187) enables us to express Eqs. (8.177), (8.182) exclusively through the variables that contain the index n (the state after the stage n). Moreover, Eq. (8.180) may be written as:

$$H^{n-1} = \frac{H^n}{1 - a\Theta^n} \tag{8.188}$$

The above equations enable the construction of a computational program. It is available elsewhere (Sieniutycz and Szwast, 2018).

The computation program enables determination of the optimal T^n with accuracy up to KR 1/1000 (e.g., for the assumed KR $1 = 10\,\text{K}$, accuracy of calculations reaches 0.01 K).

Optimization of time intervals Θ^n ensures that the Hamiltonian values calculated from Eqs. (8.176), (8.188) are identical. The program assumes that the condition is fulfilled when the relative difference between two Hamiltonians does not exceed the certain arbitrarily assumed accuracy ε.

Assumption of the value Θ^n for the next iteration, based on the data from the previous iteration, results from the following reasoning:

As it has been already demonstrated, in continuous processes, the Hamiltonian retains the positive sign along the entire optimal trajectory, provided that the final time t_k is shorter than the optimal time $\hat{t}_k$. Also, H is constant and zero if time t_k is equal to the optimal time $\hat{t}_k$. Therefore, if the solution for a cascade with perfectly mixed stages is supposed to reflect the solution for a continuous process (e.g., for a tubular reactor) for the time shorter than optimal, then all values H^0, H^1, $\ldots H^N$ should be positive. Eqs. (8.180), (8.188) show that the decreasing of positive values H^n along the trajectory implies the following inequality:

$$\Theta^n < \alpha^{-1} \text{ for } H^n > 0 \tag{8.189}$$

which holds for each stage of the cascade if $t^N < \hat{t}^N$.

The analysis of Eqs. (8.176)–(8.179), which allows for the final conditions of conjugate variables, leads to the conclusion that values H^{n-1}, calculated from the Eq. (8.176), decrease if Θ^n increases, whereas the values H^{n-1} calculated from (8.188) increase with Θ^n. This enables us to decide whether Θ^n should be increased or decreased at the next iteration.

Great significance is attached to the fact that the prescribed time t^N shorter than the optimal time of a continuous reactor (e.g., a tubular one) can prove to be longer than the optimal time for a cascade with a finite number of stages (in tubular reactors the catalyst entails all intermediate activities, while in a single mixed reactor only final activities are employed). In consequence, this case corresponds with $H^N < 0$.

Let us now take into consideration that in continuous reactors with decaying catalysts the positive nonautonomous Hamiltonian (8.224) decreases along the optimal trajectory. The only discrete solution of the same qualitative shape as the continuous solution (with the Hamiltonian decreasing along subsequent stages) is the solution in which

(a) each H^{n-1} for $n = 1, 2, \ldots, N$ is positive and only $H^N < 0$,
(b) for each stage, except the last one, inequality (8.189) is valid,
(c) at the last stage, spatial time Θ^N as the last contribution to the prescribed duration τ, satisfies the inequality

$$\Theta^N > \tau - (N-1)\alpha^{-1} \tag{8.190}$$

This way, the surplus of time (with reference to the optimal duration defined for a constant N) contained in a prescribed value τ becomes noticeable in the last step of the cascade. Provided that N is large enough, the values Θ^N satisfy, except Eq. (8.190), also the inequality (8.189), related to $H^N > 0$.

For $H^N < 0$, that is, when $1 - \alpha\Theta^N < 0$, an increase of the Hamiltonian H^{N-1} calculated for the last iteration from Eq. (8.176) requires decreasing of Θ^N (i.e., the situation is opposite to the one when $H^N > 0$ and $1 - \alpha\Theta^N > 0$). In the first iteration, the lowest value $\Theta^n = \Theta_*$ has been assumed (most often Θ_* slightly higher than zero are used in calculations). In the second iteration, the value Θ^n is increased by DELTA = DELTAP (i.e., $\Theta^n = \Theta_* +$ DELTAP), where DELTAP is any assumed initial value of the DELTA variable. The analysis of Hamiltonian values calculated in the second iteration from Eqs. (8.176), (8.188) indicates the need to increase Θ^n or decrease Θ^n, respectively. In the presence of that, Θ^n assumes, respectively, the values $\Theta^n = \Theta_* +$ 2DELTAP or $\Theta^n = \Theta_* +$ DELTAP − DELTAP/10, where DELTAP/10 is the current value of the DELTA variable. Additional variables SP, SK, WP, and WK determine whether the value of DELTA should remain unchanged or be decreased 10 times before the next iteration. The computational accuracy of Θ^n is DELTAP $\times 10^{-i}$ where i is an accuracy coefficient. The variable ST in the program describes the total space time or duration τ. Blocks "NO SOLUTION" denote, respectively:

1. For the assumed values x_A^N, x_B^N, and H^N, there is no solution in terms of positive values of Θ^n.
2. For the assumed values x_A^N, x_B^N, and H^N, the optimal space time in the current stage is higher than the highest expected value of Θ^*.
3. For the assumed computational accuracy of Θ^n equal to DELTAP $\times 10^{-i}$ and the precision of T^n amounting to KRI $\times 10^{-3}$, it is impossible for the Hamiltonian to reach the required accuracy ε.

It is possible to conduct an analysis (similar to the one presented above for the continuous reactor), which allows a shape foreseeing for the discrete optimal temperature profile. It turns out that, qualitatively, the general discrete profile T^n has the same shape as the profile of an optimal continuous reactor (e.g., the tubular one). Still, in specific cases, the discrete optimal profile can comprise any part of the general profile (unlike in the continuous reactor), that is, the optimal T^n does not necessarily starts from T^*. This results from the fact that the optimal value T^n in the stage $n=1$ entails the nonzero concentration x_R^1, which is typical for the first-stage output state (not the input state, as it would be in the continuous reactor). It is now right to remind previous considerations concerning Eq. (8.168). If N is large enough, an optimal discrete profile of temperature has the same qualitative properties as the continuous profile, that is, it starts at $T^1=T^*$.

Calculations for the following numeric data have been conducted: c_A^0, $c_B^0=1$ mol/L, $c_R^0=c_S^0=0$, $t^0=0$, $a=1$, $k_{10}=5\times10^{10}$ L/(kmol min), $k_{20}=3\times10^{17}$ L/(mol min), $E_1=66.72$ kJ/mol, $E_2=125.1$ kJ/mol, $t^N=\tau=3$ min, $T_*=335$ K, $T^*=355$ K.

The first series of results have been found for $\alpha=0$ and various number of stages, N (Fig. 8.11). They demonstrate the influence of N on the discrete optimal profile $\hat{T}^n$ and serve the purpose of determining the particular value of N, above which the profile $\hat{T}^n$ has the same quality parameters as its continuous counterpart. As shown in Fig. 8.11, for a single perfectly mixed reactor, the optimal temperature is slightly lower than $T^*=355$ K. For a three-stage reactor, optimal temperatures form an increasing sequence that ends at $T^3=T^*$. The analogical profile is observed in six-stage cascades (in Fig. 8.11, for $N=6$, temperatures of the last two stages constitute a common segment). For $N=12$ and 30 (12- and 30-stage approximation of the tubular reactor), optimal temperatures decrease initially, only to start increasing, eventually reaching T^* at the end of the process. In both cases, initial temperatures T^n are even lower than T^*. However, for the 44-stage approximation, the discrete optimal profile shows all the qualities of a continuous profile. In particular, it starts with the temperature $T^1=T^*$.

Fig. 8.12 corresponds to the same run of calculations. Yet, it presents dimensionless concentrations A, B, and R in a stream that leaves the last step of a cascade as functions of the total number of steps N. As it follows from the illustration, beginning from $N=10$, any further increase of N exerts almost no influence on the process quality index (concentration x_R^N). Therefore, if x_R^N is the only criterion, the 10-stage cascade is the sufficient approximation of a continuous process. Nonetheless, in further calculations, $N=45$ has been assumed as the number of stages which guarantees that the discrete profile demonstrates all qualitative features of a continuous process.

The second run of calculations corresponds to $N=45$. The effective deactivation rate constant α has been modified with other values of constants. Since the value $N=45$ is in practice associated with a continuous process, continuous curves of the optimal temperature have been plotted onto Fig. 8.13. As the illustration shows, if α increases, the optimal profile shifts towards higher temperatures. For α that is high enough, the optimal profile is entirely isothermal with $T=T^*$ (not shown in the figure).

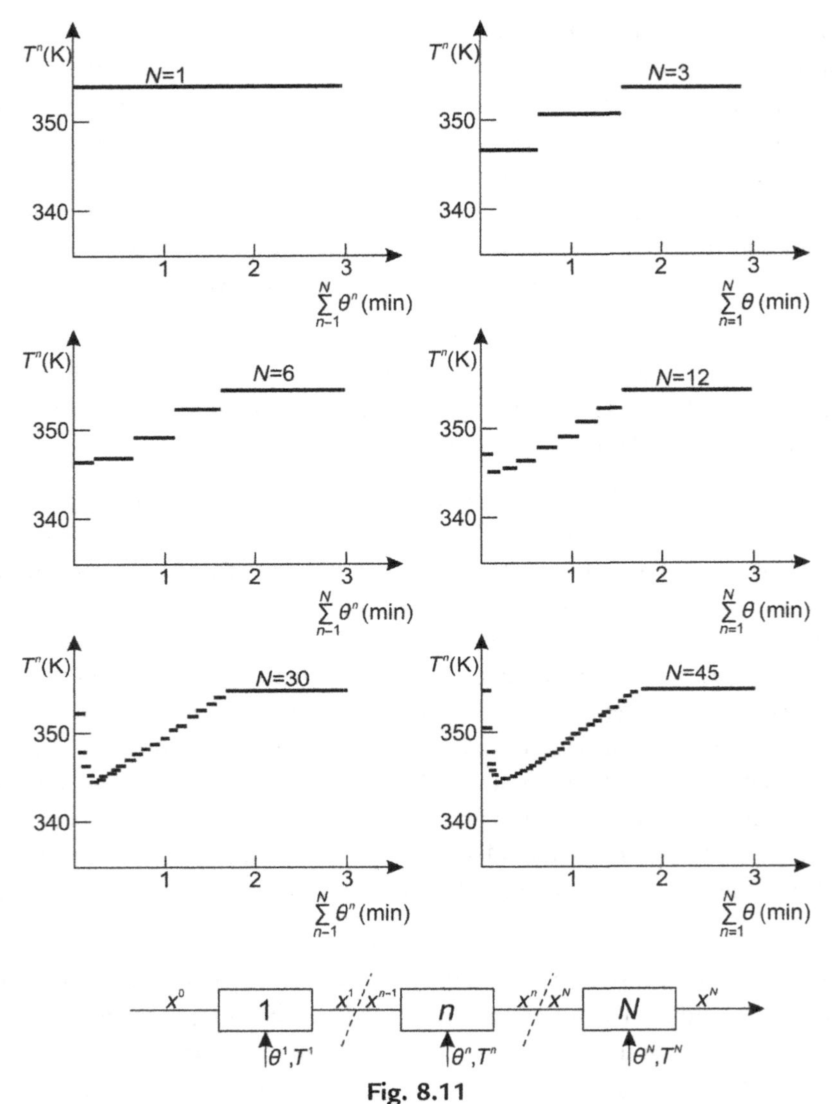

Fig. 8.11
Optimal discrete temperature profiles for consecutive-parallel reactions ($\alpha=0$, variable N).

In Fig. 8.14, the distribution of space times for cascades is presented. For $\alpha=0$, times Θ^n increase uniformly with the number of steps. For $\alpha=2\,\text{min}^{-1}$, such an increase of Θ^n occurs in first N–1 steps, and residence times for these steps satisfy the inequality (8.189), that is, $\Theta^n<\alpha^{-1}=0.5\,\text{min}$. As it has been already explained, in the last step space time is considerably longer than in preceding steps. This effect abates with the increase of N and disappears for N that is large enough.

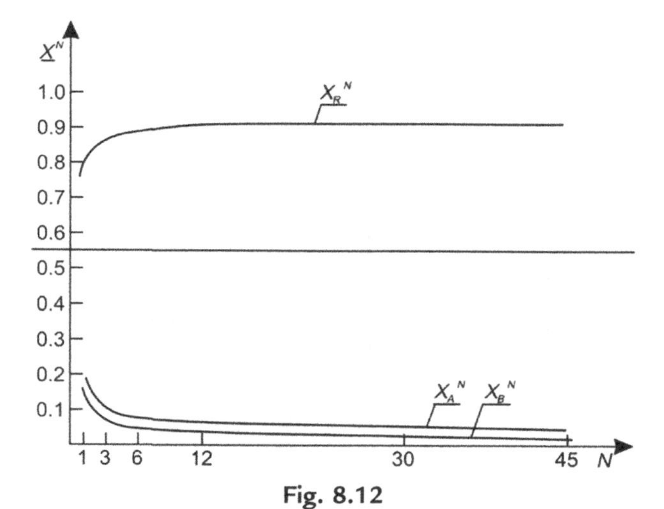

Fig. 8.12
Final optimal concentrations as a function of total number of stages N.

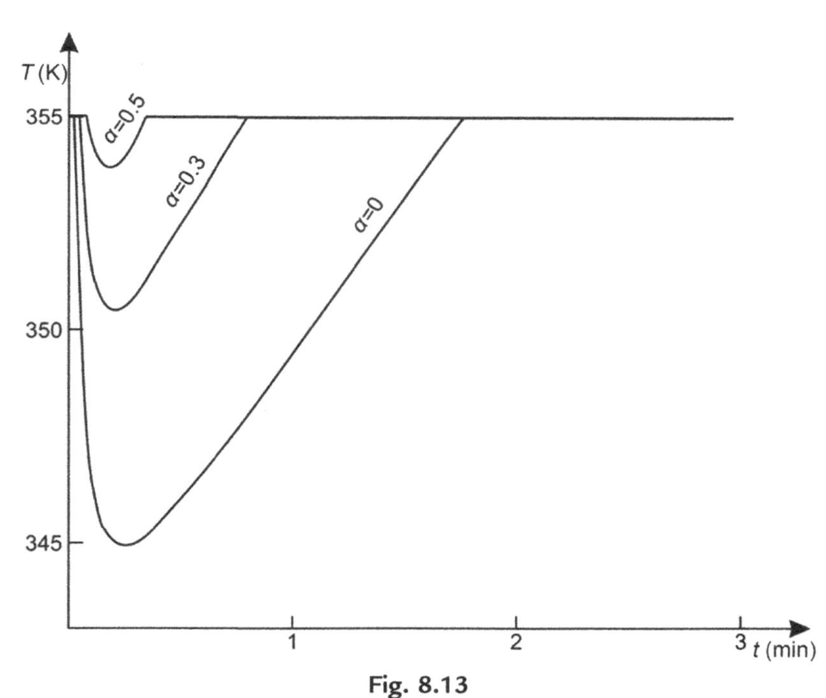

Fig. 8.13
Impact of catalyst deactivation on optimal temperature profiles for consecutive-parallel reactions in a cocurrent tubular reactor.

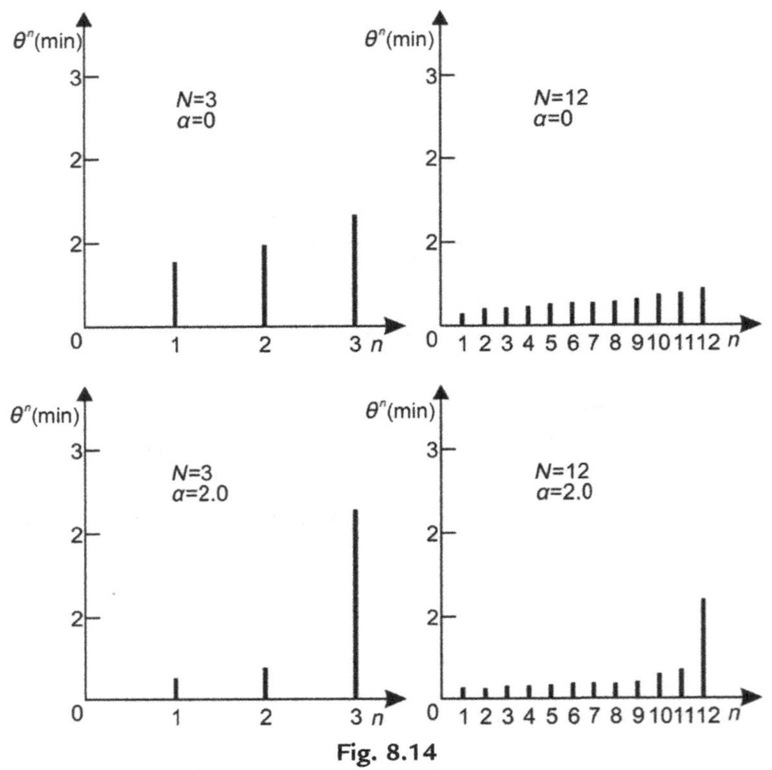

Fig. 8.14
Distribution of space times Θ^n in optimal cascades.

8.5.6 Comparison with a reversible reaction and simple irreversible systems

Let us return to the equation set (8.124)–(8.127) with the assumption that the right-hand side of the Eq. (8.125) is completed with an additional source term, which serves the purpose of an auxiliary control, with dc_B/dt that vanishes, that is, with c_B that remains constant in time t. For a constant c_B, Eq. (8.126) describes the pseudo-first-order kinetics of a single reversible reaction (8.117) with catalyst deactivation. (Reaction rate constants for such a pseudo-first-order kinetics scheme are $k_1' = k_1 c_B$ and $k_2' = k_2 c_B$). Therefore, optimal temperature profiles for a single reaction may serve as a simple, yet free, reference system for other, more complex, systems. Apart from that, one must be aware that in the strict line of reasoning kinetics of the single reversible reaction (8.117) can by no means be regarded as the boundary case of the kinetics of reactions (8.124)–(8.127).

Optimal profiles for a single reversible reaction in the absence of deactivation are described in many literature sources, see, for example Section 7.1.2 of chapter and Fig. 7.2 in Sieniutycz and Szwast (2018). We may enrich those information by considering the impact of the deactivation rate constant α upon the shape of the profile $T(t)$. Fig. 8.15 illustrates the effect of

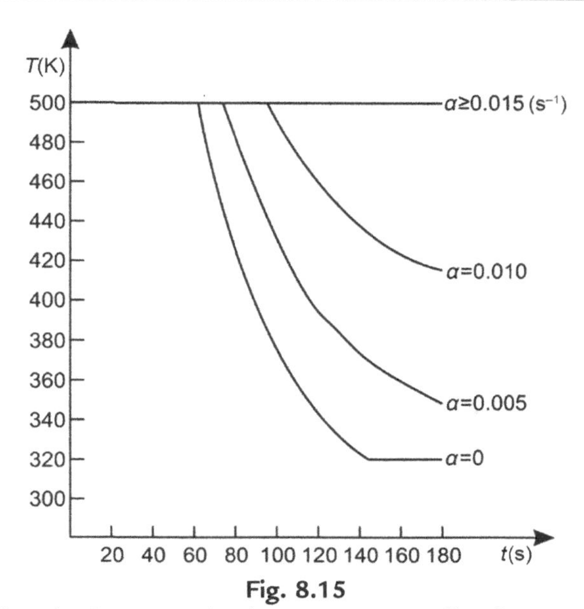

Fig. 8.15
Impact of the catalyst deactivation on optimal temperature profiles for a single reversible exothermic reaction.

α for the following data: $c_{A0}=1\,\text{mol/L}$, $c_{R0}=0$, $a_0=1$, $k_{10}=0.09\,s^{-1}$, $k_{20}=0.11\,\text{s}^{-1}$, $E_1=6\,\text{kJ/mol}$, $E_2=12\,\text{kJ/mol}$, $t_k=3\,\text{min}$, $T_*=320\,\text{K}$, and $T^*=500\,\text{K}$. Calculations based on the Pontryagin's algorithm apply to an exothermic case ($E_2>E_1$) where the general profile comprises three consecutive parts: isothermal T^*, decreasing stationary, and isothermal T_*. As Fig. 8.15 demonstrates, the increase of the deactivation constant causes the shift of the optimal function $T(t)$ towards higher temperatures. For relatively large values of α, the lower isothermal part $T=T_*$ disappears, while for the values that are high enough, the optimal solution is provided by the isothermal line $T=T^*$. On the other hand, if $E_1\geq E_2$, then $T=T^*$ is always an optimal solution. Losing any terminal part of the profile gives a different optimal profile for the shorter time t_k. This occurs for every α, regardless of the relation between E_1 and E_2.

Investigation of optimal temperatures for consecutive reactions proves that results of the same qualitative properties as those in Fig. 8.15 may be attributed to the system of two irreversible consecutive reactions $A \rightarrow R \rightarrow S$ with catalyst deactivation. In a sense, someone may say the Fig. 8.15 describes correctly the consecutive system $A \rightarrow R \rightarrow S$ as well, that is, not only a single reversible reaction for which it has been obtained.

On the other hand, for systems of parallel reactions $S \leftarrow B \rightarrow R$ with the catalyst deactivation, calculations prove that a qualitatively optimal profile $T(t)$ for this system is the final part of the optimal profile found for consecutive-parallel reactions. These approximate rules may enable memorizing properties of optimal temperature profiles for various kinetic schemes.

8.5.7 Inclusion of capital costs

Returning to the system of consecutive-parallel reactions (8.114a) and (8.114b), it is worth reminding that a simple result has been proven, namely for the case of the optimal final time $t_k \equiv \hat{t}_k$, the profile is always isothermal, and $\hat{T} = T^*$ if $E_1 \geq E_2$, as well as $\hat{T} = T_*$ if $E_2 > E_1$. The technical facility of realizing such profile makes it significant in terms of practical use. Still, one must remember that the analyzed result (acquired for $H_k = 0$) is valid only for negligible capital costs K_i.

The effect of costs K_i can be taken into account by making the optimal final Hamiltonian a function of technical and economic quantities. What is significant is the fact that the inclusion of K_i into the profit criterion takes place without any additional modifications of the original optimization problem. One only needs to note that nonzero changes in K_i are related to process times t_k shorter than optimal $\hat{t}_k$, and thus, to the nonzero final Hamiltonian. The method consists in associating changes in K_i described by the derivative $h_k = \partial K_i / \partial \hat{t}_k$ with the Hamiltonian H.

For significant changes of K_i, the optimal time $\hat{t}_k$ follows from the evident formula for the profit extremum

$$\frac{\partial}{\partial \hat{t}_k}[q c_B(0) p_R F - K_i] = 0 \tag{8.191}$$

where q, volume flux of reagents; p_R, price of 1 mol of product R; K_i, capital cost; F, criterion (8.143). Since $\partial F / \partial t_k = H_k$, the above equation yields:

$$H_k = \frac{\partial K_i / \partial t_k}{q c_B(0) p_R} \tag{8.192}$$

This is the sought-after optimality condition, which serves the purpose of determining the final optimal time when capital costs are variable and cannot be omitted. Since $h_k \equiv \partial K_i / \partial t_k > 0$, we are dealing with the case of a nonzero positive Hamiltonian (analyzed in Section 8.5.4), to which corresponds a final time shorter than $\hat{t}_k$. Only for h_k approaching zero, does an economically optimal final time correspond to the vanishing Hamiltonian H_k.

In consequence, the inclusion of capital expenditures (e.g., during design) is related to the optimal time t_k calculated for the positive Hamiltonian that satisfies Eq. (8.192). In case of a fixed h_k, the computational procedure could rather consist in assuming t_k for known $H(t_k) = H^N$ rather than assuming $H_k = H^N$ for the known t_k. Despite that, the computer program can be used in its primary form since H_k may be calculated for various t_k in order to plot the function $t_k(H_k)$. The graph of this function enables us to find the economically optimal final time t_k (shorter than $\hat{t}_k$) for H_k that satisfies Eq. (8.192). The higher the H_k (higher h_k and lower p_R), the shorter is this optimal final time in comparison with the time $\hat{t}_k$.

The optimal temperature profile for nonzero capital expenditures shows the already discussed properties related to times t_k shorter than $\hat{t}_k$. If $E_1 \geq E_2$, then $\hat{T}(t) = T^*$, whereas if $E_2 > E_1$ then the function $\hat{T}(t)$ is usually comprised of five consecutive parts: $T = T^*$, decreasing stationary, $T = T_*$, increasing stationary, and $T = T^*$. The technical realization of such complex curves can only be approximate, and the inclusion of costs incurred by this process as penalty for the variation $\hat{T}$ in time substantiates the search for simplifications based on preferential exploiting of isothermal parts. For small p_R/h_k (large h_k and short t_k) in particular, a practical optimum entails the assumption of $T = T^*$, while for large p_R/h_k (small h_k and times t_k close to $\hat{t}_k$, $H_k \to 0$), the practical optimum is $T = T_*$ when $E_2 > E_1$ and $T = T^*$, when $E_1 \geq E_2$. These conclusions are drawn from the already discussed properties of optimal profiles $\hat{T}(t)$ obtained as a result of the exact analysis.

8.5.8 Temperature-sensitive catalysts

Now, we will discuss optimal continuous temperature profiles for the same system of consecutive-parallel reactions (8.114a) and (8.114b) as before, but in the presence of the catalyst deactivation that is dependent on temperature.

Our attention will be focused on the case of no slip between the phases (finely granulated catalyst) or the homophase system of the reacting mixture.

The kinetic equations of reactions (8.124)–(8.127) and their boundary conditions remain unchanged. Once the relations (8.129)–(8.132) are used, the kinetic equations for $0 < t < t_k$ may be noted in the following form:

$$\frac{dx_A}{dt} = -k_1 a x_A x_B \tag{8.193}$$

$$\frac{dx_B}{dt} = -k_1 a x_A x_B - k_2 a x_B (x_B - 2x_A + 2M - 1) \tag{8.194}$$

$$\frac{dx_R}{dt} = k_1 a x_A x_B - k_2 a x_B (x_B - 2x_A + 2M - 1) \tag{8.195}$$

$$\frac{dc_S}{dt} = k_2 a x_B (x_B - 2x_A + 2M - 1) \tag{8.196}$$

Rate constants G_i, Eqs. (8.133), (8.134) are not introduced this time. The above equations are accompanied by boundary conditions (8.139)–(8.142). In Eqs. (8.193)–(8.196), symbols k_1 and k_2 denote modified reaction rate constants, which are products of actual rate constants$k_i^r$and the initial concentration of reacting substance B, $c_B(0)$.

$$k_i(T) \equiv k_i^r(T) c_B(0) \quad \text{for} \quad i = 1, 2 \tag{8.197}$$

The form of the kinetic equation of the catalyst deactivation shown in Eq. (8.118) is replaced with the form (8.198) that allows for the impact of temperature on the rate of deactivation.

$$\frac{da}{dt} = -k_3(T)a^m \tag{8.198}$$

where $m > 0$ defines the order of deactivation. The deactivation rate constant satisfies the structure of the Arrhenius relationship

$$k_3(T) = k_{30}\exp(-E_3/RT) \tag{8.199}$$

As previously, the initial activity is fixed, while the final one remains unbound (free).

Assuming a use of the catalyst leaving the reactor, the optimization criterion has been defined as the sum of value fluxes of the reacting substance R and the catalyst which leave the reactor, reduced by the costs K of the process.

$$Z = p_R q c_R(t_k) + SC_s(a_k) - K(t_k) \tag{8.200}$$

where s, prescribed catalyst flux; $C_s(a_k)$, price of the catalyst as a function of its activity.

Costs K may include flow values of the fresh catalyst and substrates. Variable capital expenditures, expressed in terms of t_k, constitute their important component. The use of the definition of nondimensional concentrations (8.130), allowance for $x_R(0) = 0$, and then utilization of Eq.(8.195) enable the replacement of the criterion Z with the following index

$$\widetilde{F} = \int_0^{t_k} k_1 a x_A x_B - k_2 a x_B(x_B - 2x_A + 2M - 1)dt + \frac{SC_s(a_k)}{p_R q c_B(0)} - \frac{K}{p_R q c_B(0)} \tag{8.201}$$

Let us notice that in order to unequivocally describe the state of the process and the quality index, one only needs to operate nondimensional concentrations x_A and x_B and the catalyst activity a. Therefore, these three variables are state variables, while the Eqs. (8.193), (8.194), (8.198) are state equations.

The optimization problem comes down to the selection of the optimal temperature profile and the total residence time t_k, which maximize the quality index (8.201) in the presence of equations of state (8.193), (8.194), and (8.198) as well as temperature restriction (8.123). The characteristic feature of the problem is the freedom of all coordinates of the final state, including the final time.

By using the continuous maximum principle, it is possible to note for $0 \le t \le t_k$

$$\begin{aligned} H(x_A, x_B, a, z_1, z_2, z_3, T) &= k_1 a x_A x_B - k_2 a x_B(x_B - 2x_A + 2M - 1) + \\ &- k_1 a x_A x_B z_1 - [k_1 a x_A x_B + k_2 a x_B(x_B - 2x_A + 2M - 1)]z_2 - k_3 a^m z_3 \end{aligned} \tag{8.202}$$

or after putting in order

$$H = k_1 a x_A x_B (1 - z_1 - z_2) - k_2 a x_B (x_B - 2x_A + 2M - 1)(1 + z_2) - k_3 a^m z_3 \tag{8.203}$$

The final value of the Hamiltonian is described by relationship (8.171)

$$H_k = \frac{\partial K / \partial t_k}{p_R q c_B(0)} \tag{8.204}$$

(note a link with a transversality condition linked with the variation of the final time t_k and the HJB equation). With the absence of its explicit dependence on time, the value of the Hamiltonian remains constant along the entire trajectory, this constant being equal to H_k. The only practical function is the nondecreasing function $K(t_k)$ since the increase of residence time must cause the increase of costs K or, in limiting case, it must not change its value, thus

$$H(t) = H_k = \text{constant} \geq 0 \tag{8.205}$$

In a variant of the problem that features the prescribed residence time, the economically viable Hamiltonian values satisfy the inequalities (8.205). Compare the previous discussion in Section 8.5.4.

The final value of a variable conjugated with the catalyst activity $z_3(t_k)$ is defined analogically.

$$z_3(t_k) = \frac{S \partial C_s / \partial a_k}{p_R q c_B(0)} \tag{8.206}$$

As Eq. (8.204), the above formula remains compliant with a transversality condition.

The more active the catalyst, the higher its value. If the catalyst leaving the reactor, independently of its activity, is not used, both the function C_s and its derivative are equal to zero, thus

$$z_3(t_k) \geq 0 \tag{8.207}$$

In view of the free values of $x_A(t_k)$ and $x_B(t_k)$

$$z_1(t_k) = z_2(t_k) = 0 \tag{8.208}$$

In the first place, we will analyze the course of functions z_3, $1 + z_2$, and $1 - z_1 - z_2$ that appear in Eq. (8.203). Understanding characteristics of these functions will prove useful when analyzing optimal temperature profiles.

$$\begin{aligned} \frac{dz_3}{dt} = -\frac{\partial H}{\partial a} &= -k_1 x_A x_B (1 - z_1 - z_2) + \\ &+ k_2 x_B (x_B - 2x_A + 2M - 1)(1 + z_2) + m k_3 a^{m-1} z_3 \end{aligned} \tag{8.209}$$

After transforming this result we obtain

$$\frac{dz_3}{dt} = -\frac{H + (1-m)k_3 a^m z_3}{a} \tag{8.210}$$

An analysis of Eq. (8.210) leads to the conclusion that, regardless of the sign of the expression $(1-m)$, inequality $z_3(t) > 0$ is valid for $0 \le t \le t_k$, except the case of $H=0$ and $z_3(t_k)=0$, when $z_3(t) = \text{const} = 0$.

$$\begin{aligned} \frac{d(1+z_2)}{dt} &= \frac{dz_2}{dt} = -\frac{\partial H}{\partial x_b} = -k_1 a x_\mathrm{A}(1 - z_1 - z_2) + \\ &+ k_2 a(x_\mathrm{B} - 2x_\mathrm{A} + 2M - 1)(1+z_2) + k_2 a x_\mathrm{B}(1+z_2) = \\ &= -\frac{H + k_3 a^m z_3}{x_\mathrm{B}} + k_2 a x_\mathrm{B}(1+z_2) \end{aligned} \tag{8.211}$$

Analysis of Eq. (8.211), which takes the boundary condition for the variable z_2 into consideration, brings the conclusion that for $0 \le t \le t_k$

$$1 + z_2 > 0 \tag{8.212}$$

In view of nonnegative expressions for z_3 and $1+z_2$ as well as the fact that $x_R(0)=0$ and $x_R(t \neq 0) \neq 0$, Eq. (8.203) demonstrates that

$$1 - z_1 - z_2 \begin{cases} > 0 \\ = 0 \text{ for } t=0, \text{ if } H=0 \text{ and } z_3(t_k)=0 \end{cases} \tag{8.213}$$

$$\begin{aligned} \frac{d(1-z_1-z_2)}{dt} &= -\frac{dz_1}{dt} - \frac{dz_2}{dt} = \frac{\partial H}{\partial x_a} + \frac{\partial H}{\partial x_b} = \\ &= k_1 a x_\mathrm{B}(1 - z_1 - z_2) + k_2 a x_\mathrm{B}(1+z_2) + \frac{H + k_3 a^m z_3}{x_\mathrm{B}} \end{aligned} \tag{8.214}$$

Thus, the value of the expression $1 - z_1 - z_2$ increases along the entire optimal trajectory.

Using Eqs. (8.210), (8.214), one can compare rates of changing values in expressions $1 - z_2$ and $1 - z_1 - z_2$.

$$\frac{d(1-z_1-z_2)}{dt} - \frac{d(1+z_2)}{dt} = k_1 a x_B(1 - z_1 - z_2) + 2\frac{H + k_3 a^m z_3}{x_B} \tag{8.215}$$

Eq. (8.215) shows that the increase rate of $1 - z_1 - z_2$ is for every $0 \le t \le t_k$ not lower than the increase rate of $1 + z_2$. Taking into consideration the fact that at $t = t_k$ both expressions are equal unity (as it results from Eq. 8.208), it can be stated that

$$\left.\begin{aligned} &0 < 1 - z_1 - z_2 < 1 + z_2 \text{ for } 0 \le t < t_k \\ &1 = 1 - z_1 - z_2 = 1 + z_2 \text{ for } t = t_k \end{aligned}\right\} \tag{8.216}$$

Eq. (8.203), provided that $H \geq 0$ and $z_3(t) \geq 0$, demonstrates that

$$\begin{aligned} &k_1 a x_A x_B (1 - z_1 - z_2) + \\ &- k_2 a x_B (x_B - 2x_A + 2M - 1)(1 + z_2) \begin{cases} > 0 \\ = 0 \text{ if } H = 0 \text{ and } z_3(t_k) \end{cases} \end{aligned} \tag{8.217}$$

From the two latter relationships, it follows that

$$k_1 a x_A x_B - k_2 a x_B (x_B - 2x_A + 2M - 1) > 0 \tag{8.218}$$

except the case when this expression is equal to zero for $t = t_k$ if $H = 0$ and $z_3(t_k) = 0$.

Since the left-hand side of the relationship (8.218) describes the rate of producing the required reacting substance R (Eq. 8.189), it can be concluded that the concentration of this substance increases monotonically along its entire optimal trajectory. The production rate of R vanishes only for $t = t_k$ provided that $H = 0$ and $z_3(t_k) = 0$.

For the optimal stationary policy, the following relationship is valid:

$$\begin{aligned} \frac{\partial H}{\partial T} = \frac{1}{RT^2} &[E_1 k_1 a x_A x_B (1 - z_1 - z_2) - E_2 k_2 a x_B (x_B - 2x_A + \\ &+ 2M - 1)(1 + z_2) - E_3 k_3 a^m z_3] = 0 \end{aligned} \tag{8.219}$$

which should be satisfied along with the Eq. (8.203). This proves that the shape of the optimal temperature profile must be influenced by the mutual relation of the values of the activation energy E_1, E_2, and E_3.

8.6 Possible reactions and related special cases

The following relations are possible:

(1) $E_1 < E_2 < E_3$
(2) $E_1 < E_3 < E_2$
(3) $E_2 < E_1 < E_3$
(4) $E_3 < E_1 < E_2$
(5) $E_2 < E_3 < E_1$
(6) $E_3 < E_2 < E_1$

Eqs. (8.203), (8.219) are appropriate for any moment t, for which there is a valid stationary policy of optimality. Whereas in view of the relation

$x_R(0) = x_B(0) - 2x_A(0) + 2M - 1 = 0$
for $t = 0$, the discussed equations are reduced to the form:

$$H = k_1 a x_A x_B (1 - z_1 - z_2) - k_3 a^m z_3 \tag{8.220}$$

$$\frac{\partial H}{\partial T} = \frac{1}{RT^2}[E_1 k_1 a x_A x_B (1 - z_1 - z_2) - E_3 k_3 a^m z_3] = 0 \tag{8.221}$$

Case of positive Hamiltonian. The system of Eqs. (8.220), (8.221) demonstrates that for $H > 0$:

(a) if $E_1 > E_3$ then always $T(0) = T^*$,
(b) if $E_1 < E_3$, then the optimal temperature can be the stationary temperature, or the optimum occurs for T^* or T_* (Eq. (8.221) is satisfied or the partial derivative $\partial H/\partial T$ assumes any sign).

Case of vanishing Hamiltonian; $H = 0$:

(a) if $z_3(t_k) > 0$ (consistent with $z_3(t) > 0$), then the stationary temperature at $t = 0$ is unfeasible. If $E_1 > E_3$, then $T(0) = T^*$, whereas if $E_1 < E_3$, then $T(0) = T_*$,
(b) if $z_3(t_k) = 0$, then $z_3(t) = 0$ and $1 - z_1(0) - z_2(0) = 0$, thus Eqs. (8.220), (8.221) are always satisfied. They do not determine the optimal temperature, allowing any temperature instead.

Eqs. (8.203), (8.219) simplify also for $t = t_k$ if $z_3(t_k) = 0$, assuming the form:

$$H = k_1 a x_A x_B (1 - z_1 - z_2) - k_2 a x_B (x_B - 2x_A + 2M - 1)(1 + z_2) \tag{8.222}$$

$$\frac{\partial H}{\partial T} = \frac{1}{RT^2}[E_1 k_1 a x_A x_B (1 - z_1 - z_2) - E_2 k_2 a x_B (x_B - 2x_A + 2M - 1)(1 + z_2)] = 0 \tag{8.223}$$

From the above system of equations, it results that for $H > 0$:

(a) if $E_1 > E_2$ then $T(t_k) = T^*$,
(b) if $E_1 < E_2$, then it is possible for both stationary temperature and boundary temperatures C and T^* to occur at the end of the profile.

Whereas for $H = 0$, the stationary temperature $T(t_k)$ is not possible, and then

(a) if $E_1 > E_2$ then $T(t_k) = T^*$,
(b) if $E_1 < E_2$ then $T(t_k) = T_*$

For times $t \neq 0$ and $t \neq t_k$, in case when the stationary policy is optimal, Eqs. (8.203), (8.219) must be valid. It is easily noticeable that the stationary policy for $H > 0$ is not possible if E_1 is simultaneously higher than E_2 and E_3. Then, the optimal profile is an isothermal profile; $T(t) = T^*$. This corresponds to cases 5 and 6 of the mutual relation of activation energy.

The stationary profile may occur in cases 1–4. If $H = 0$ and $z_3(t_k) = z_3(t) = 0$, the optimal profile is always an isothermal profile: $T(t) = T^*$ if $E_1 > E_2$, and $T(t) = T_*$ if $E_1 < E_2$. When $H = 0$ and $z_3(t_k) > 0$, then the isothermal profile occurs in the following cases: $T(t) = T^*$ if $E_1 > E_2$ and E_3, and $T(t) = T_*$ if $E_1 < E_2$ and E_3. In cases 3 and 4, that is, when E_1 has the value that is intermediate between E_2 and E_3, the stationary profile may occur.

In order to analyze cases 1–4 for $H>0$ as well as cases 3 and 4 for $H=0$ and $z_3(t_k)>0$, further transformations are introduced to Eq. (8.228).

Elimination of the expression $(1-z_1-z_2)$ from Eqs. (8.203), (8.219) leads to the relation:

$$E_1 H = (E_2 - E_1) k_2 a x_R x_B (1+z_1) + (E_3 - E_1) k_3 a^m z_3 \tag{8.224}$$

Due to the constancy of the Hamiltonian, the time derivative of the right-hand side of the Eq. (8.224) must be equal to zero.

$$\begin{aligned} &\frac{1}{RT^2} E_2 (E_2 - E_1) k_2 \frac{dT}{dt} a x_B x_R (1+z_2) + (E_2 - E_1) k_2 \frac{d}{dt} [a x_B x_R (1+z_2)] + \\ &+ \frac{1}{RT^2} E_3 (E_3 - E_1) k_3 \frac{dT}{dt} a^m z_3 + (E_3 - E_1) k_3 \frac{d}{dt} (a^m z_3) = 0 \end{aligned} \tag{8.225}$$

and therefore

$$\frac{1}{RT^2}\frac{dT}{dt} = -\frac{(E_2 - E_1) k_2 \frac{d}{dt} [a x_B x_R (1+z_2)] + (E_3 - E_1) k_3 \frac{d}{dt} (a^m z_3)}{E_2 (E_2 - E_1) k_2 a x_B x_R (1+z_2) + E_3 (E_3 - E_1) k_3 a^m z_3} \tag{8.226}$$

Let us examine the sign of derivatives occurring on the right-hand side of the Eq. (8.226). By converting canonical equations, Eqs. (8.227), (8.228) have been obtained

$$\begin{aligned} &\frac{d}{dt} [a x_B x_R (1+z_2)] = -k_3 a^m x_B x_R (1+z_2) + \\ &- k_1 a^2 x_A x_B x_R (1-z_1-z_2) + k_1 a^2 x_A x_B (1-z_2)(x_B - x_R) \end{aligned} \tag{8.227}$$

From Eq. (8.227), it follows that the discussed derivative has a positive value at the start of the process (x_R increases starting from zero). Later in the process, this derivative may have a negative value. This is not necessary, however, because—for instance—due to the short residence time, the process may not reach the state corresponding to the change of sign of the derivative

$$\frac{d}{dt} (a^m z_3) = -a^{m-1} (H + k_3 a^m z_3) \tag{8.228}$$

Eq. (8.228) shows that the analyzed derivative is constantly negative (except the case when $H=0$ and $z_3(t_k)=z_3(t)=0$, but it is not significant since the optimal policy has been determined for it).

Now, Eq. (8.226) can be employed for the purpose of defining shapes of optimal stationary temperature profiles. Considerations covered $H>0$ and consecutive cases 1–4, for which—as it has already been demonstrated—stationary parts of optimal temperature profiles may occur.

Case 1, $E_1 < E_2 < E_3$

Both terms in the denominator of the fraction on the right-hand side of Eq. (8.226) are positive, and so is the fraction's denominator itself. While the numerator of this fraction, in the presence of the constantly negative second term and the first term that is initially positive and then possibly negative (as it has been previously noted, the sign of the derivative $d[ax_B x_R(1+z_2)]/dt$ may be limited to a positive sign), can be constantly negative, constantly positive or initially positive and then negative. The optimal stationary temperature profile may be constantly increasing, constantly decreasing, or the decreasing part may precede the increasing part.

Fig. 8.16 shows optimal temperature profiles for various activation values in the catalyst deactivation process and identical values of other data: $E_1 = 66.7\,\text{kJ/mol}$, $E_2 = 125.1\,\text{kJ/kmol}$, $k_{10} = 5 \times 10^{10}\,\text{L/(mol min)}$, $k_{20} = 3 \times 10^{17}/(\text{mol min})$, $k_{30} = 4 \times 10^{15}\,\text{min}$, $m = 1$, $T_* = 335\,\text{K}$, $T^* = 355\,\text{K}$, $c_{A0} = c_{B0} = 1\,\text{mol/L}$, $t_k = 3\,\text{min}$, $z_3(t_k) = 0$. The curve that refers to activation energy $E_3 = 150\,\text{kJ/mol}$ applies to the above-discussed case 1.

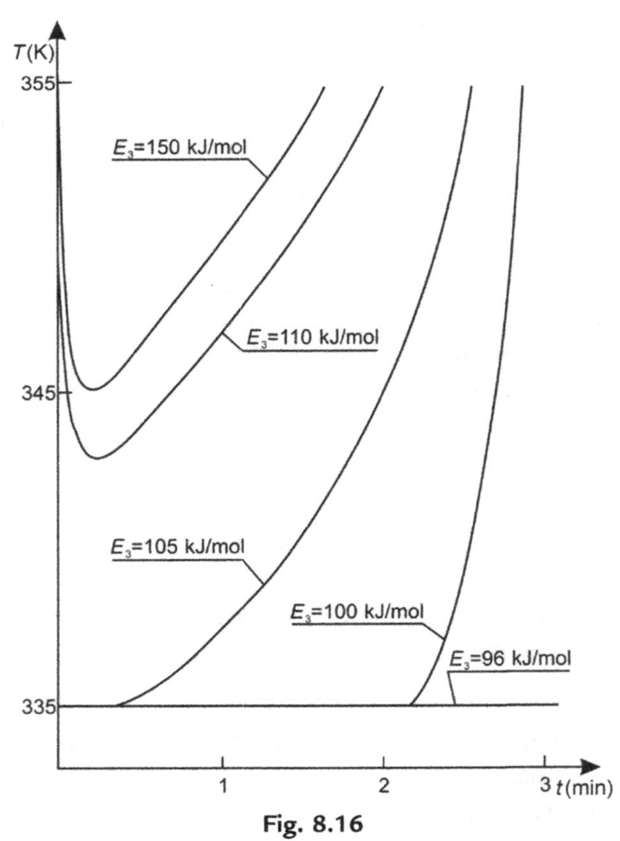

Fig. 8.16
Optimal temperature profiles in a tubular reactor with consecutive-parallel reactions for different values of deactivation energy E_3.

Case 2, $E_1 < E_3 < E_2$

Considerations and conclusions are identical as for case 1. In Fig. 8.16, all profiles except that for $E_3 = 150\,\text{kJ/mol}$ apply to this case.

Case 3, $E_2 < E_1 < E_3$

The first term on the right side of Eq. (8.224) is negative, while the second one is positive. The sum of these terms is positive, thus the absolute value of the second term is higher than the first one. Moreover, $E_3 > E_1$. Thus, the denominator of the fraction on the right side of Eq. (8.226) equation is positive. The numerator of this fraction, in the presence of the first term that is initially negative and then possibly positive and the constantly negative second term, can be constantly negative or initially negative and then positive. The optimal stationary profile may be thus constantly increasing, or its increasing part precedes the decreasing part. It is also possible for the optimal profile to begin with an isothermal section that is long enough to make the moment of the profile's entrance onto the stationary part correspond to the already positive value of the fraction's numerator ((8.226)). The stationary part of the optimal temperature profile is then decreasing. The decreasing part of the profile cannot be the final stationary part if $z_3(t_k) = 0$ since in this case, as it has been previously demonstrated, the optimal profile must end with the maximum permissible temperature.

Case 4, $E_3 < E_1 < E_2$

Analogically, it may be proved that the denominator of the fraction on the right-hand side of Eq. (8.226) is positive. The first term of this fraction's numerator is initially positive, but then it may become negative. The second term of the numerator is always positive. Therefore, the numerator is initially positive, but its value may switch the sign.

This way, the stationary temperature profile may either be decreasing or its decreasing part may precede the increasing part.

Cases 3 and 4 for $H = 0$ are left for the reader as an exercise.

The discussion enables to conclude that the shape of the optimal control curve in case of the temperature-dependent deactivation can demonstrate large variety, depending on the mutual relation of parameters E_1, E_2, and E_3, and any intuitive attempts at assessing the profile's shape and level fail. The above analysis, based on the optimality conditions, is necessary even when significant accuracy of quantity results is not required. Its negligence may produce the results that are incorrect even qualitatively.

8.7 Final remarks

For all of the analyzed reactors, and the reactors with fixed catalyst bed in particular, the condition of constancy of the effective reaction rate remains in force: $ka = \text{constant}$, for $\phi(c) = 1$. This is an interesting condition of great generality. It requires the optimal temperature policy to ensure such changes in the system's concentrations as would be observed in an isothermal process without catalyst deactivation. This is a substantial hint for controlling catalytic processes.

Treating the fixed bed catalytic reactors operating in a cyclic fashion (reaction-regeneration cycles) as hierarchical systems reveals the essential role of coordinating controls, which are also known as higher-level decisions. These controls include such quantities as the reaction stage time, contact time, and the decay degree of the catalyst. Therefore, it is possible to simplify structures of local decisions (such as T or k_{eff}) by keeping them constant along the path, for example, to limit the system's control to optimization of decision levels and resigning from profile optimizations. This facilitates the control accomplishment.

A condensed consideration has been devoted to semiregenerative reactors which suffer a disadvantage that the whole process has to be shut down for the period of catalyst regeneration. This disadvantage can be overcome if the reactors are operated in a cyclic manner, where one or more reactors are always regenerating off-line. The cyclic mode of operation is, therefore, a means of maintaining a continuous flow of products without the needs of shutting down the process for regeneration. In this book, theory and some computational results developed in papers by Ogunye and Ray (1971a,b,c) are exploited to take into account the cyclic nature of operation.

In a book (Sieniutycz and Szwast, 2018), computational aspects are discussed in detail for the problems treated in Section 8.3.3 which involve effects of catalyst deactivation. Considerations involve systems described with difference equations or ordinary differential equations. A general computer program allowing easy adaptation to current needs is also available there.

A book by Ogunnaike and Ray (1994) offers a modern view of process control in the context of today's technology. It provides the standard material in a coherent presentation and uses a notation that is more consistent with the research literature in process control. Topics that are unique include a unified approach to model representations, process model formation and process identification, multivariable control, statistical quality control, and model-based control. This book is designed to be used as an introductory text for undergraduate courses in process dynamics and control. In addition to chemical engineering courses, the text would also be suitable for such courses in mechanical, nuclear, industrial, and metallurgical engineering departments. The material is organized so that modern concepts are presented to the student, but details of the most advanced material are left to later chapters. The text material has been developed, refined, and classroom-tested over the last 15 years in the two US universities. As part of the course, a laboratory has been developed to allow the students hands-on experience with measurement instruments, real-time computers, and experimental process dynamics and control problems.

References

Burghardt, A., Skrzypek, J., 1974. Optimal temperature profiles in a tubular reactor for a system of consecutive-competing reactions. Chem. Eng. Sci. 29, 1311–1315.

Elnashaie, S.S.E.H., Elshishini, S.S., 1993. In: Hughes, E. (Ed.), Modelling, Simulation and Optimization of Industrial Fixed Bed Catalytic Reactors. In: Topics in Chemical Engineering, vol. 7. Gordon and Breach, New York.

Grubecki, I., 2010. Optimal temperature control in a batch bioreactor with parallel deactivation of enzyme. J. Process. Control 20 (5), 573–584.

Grubecki, I., Wójcik, M., 2008. Analytical determination of the optimal temperature profiles for the reactions occurring in the presence of microorganism cells. Biochem. Eng. J. 39, 362–368.

Grzesik, M., Skrzypek, J., 1983. Optimal temperature profiles for heterogeneous catalytic parallel reactions. Chem. Eng. Sci. 38, 1767–1773.

Horn, F., 1961. Optimale Temperatur- und Konzentrationsverläufe. Chem. Eng. Sci. 8, 77–89.

Levenspiel, O., Sadana, A., 1978. The optimum temperaturę policy for a dezactivating catalytic packed bed reactor. Chem. Eng. Sci. 33 (10), 1393–1394.

Megiris, C.E., Butt, J.B., 1990a. Effects of poisoning on the dynamics of fixed bed reactors, 1: Isothermal in acyclic policy of operation. Ind. Eng. Chem. Res. 29 (6), 1065–1072.

Megiris, C.E., Butt, J.B., 1990b. Effects of poisoning on the dynamics of fixed bed reactors, 2: Constant conversion policy of operation. Ind. Eng. Chem. Res. 29 (6), 1072–1075.

Ogunnaike, B.A., Ray, W.H., 1994. Process Dynamics, Modeling, and Control, first ed. Oxford University Press, Oxford.

Ogunye, A.F., Ray, W.H., 1971a. Optimal control policies for tubular reactors experiencing catalyst decay. Part 1. Single bed reactors. AICHE J. 17 (1), 43–51. https://doi.org/10.1002/aic.690170112.

Ogunye, A.F., Ray, W.H., 1971b. Optimization of cyclic tubular reactors with catalyst decay. Ind. Eng. Chem. Process. Des. Dev. 10 (3), 410–416. also ibid 9, 416.

Ogunye, A.F., Ray, W.H., 1971c. Optimization of recycle reactors having catalyst decay. Ind. Eng. Chem. Process. Des. Dev. 10 (3), 416.

Sieniutycz, S., 1976. Optymalizacja reaktorów przeciwprądowych i współprądowych z ruchomym złożem katalizatora ulegającego dezaktywacji (Optimization of concurrent and countercurrent reactors with the moving bed of deactivating catalyst). Inz. Chem. VI (2), 451–468.

Sieniutycz, S., 1980. Optymalizacja reaktorów przeciwprądowych i współprądowych z ruchomym złożem katalizatora ulegającego dezaktywacji (Optimization of concurrent and countercurrent reactors with the moving bed of deactivating catalyst). Inz. Chem. I (3), 547–556.

Sieniutycz, S., 1991. Optymalizacja w Inżynierii Procesowej, second ed. Wydawnictwa Naukowo-Techniczne, Warszawa.

Sieniutycz, S., Szwast, Z., 1982. Practice of Optimization (in Polish). Wydawnictwa Naukowo Techniczne, Warszawa.

Sieniutycz, S., Szwast, Z., 2018. Optimizing Thermal, Chemical, and Environmental Systems. Elsevier, Oxford (Chapter 8).

Skrzypek, J., 1974. Optimal temperature profiles in systems of parallel and consecutive-parallel reactions undergoing in tubular reactions type. Zesz. Nauk. Pol. Śląskiej 410, 1 (in Polish).

Strumillo, C., Kudra, T., 1987. In: Hughes, E. (Ed.), Drying: Principles, Applications and Design. In: Topics in Chemical Engineering, vol. 3. Gordon and Breach, New York.

Szepe, S., 1966. The Deactivation of Catalysts and Some Related Optimization Problems (Ph.D. thesis). Illinois Institute of Technology, Chicago.

Szepe, S., Levenspiel, O., 1968. Optimal temperature policies for reactors subject to catalyst deactivation—I. Batch reactor. Chem. Eng. Sci. 23, 881–884.

Szwast, Z., 1994. Optimization of Chemical Reactions in Tubular Reactors with Moving Catalyst Deactivation. (Habilitation thesis). Reports of Faculty of Chemical and Process Engineering at The Warsaw University of Technology XXI, pp. 5–140 (in Polish).

Further reading

Szwast, Z., Sieniutycz, S., 1987. Optimal temperature profiles for a system of parallel—Consecutive reactions with catalyst deactivation—Exponential decay. Comput. Chem. Eng. 11 (5), 441–456.

CHAPTER 9

Dynamical properties of systems

9.1 Degrees of coherence and additivity (Mynarski, 1979, p. 25)

As organized arrangements, systems are characterized by a definite degree of cohesion, or the so-called coherence. This requires a linkage of individual elements/components to ensure that a change in any/one element causes the changes in remaining elements. This is intended "to arouse/incite/induce changes in these remaining elements." Thanks to cohesion property, the system is "something more than the sum of its all parts," and sometimes constitutes a "different quality than the one expected for this sum" (Mynarski, 1979).

An opposite case is lack/absence of any links between elements, so that a change performed within any element does not cause any change in any other element. Such a change is called additivity or independence because changes are then the sum of individual changes of all elements.

Cohesion (coherence) and additivity (independence) are not two different properties of the system, but two limiting cases of the same property which quantifies the system quality. Both these limits can assume different values.

Every real system, which has to behave as an integrity, must characterize itself by a certain finite degree of cohesion (coherence); otherwise, it will be a set of functionally independent elements, for which a better name will be "a group" or "a collection."

As a rule, degrees of coherence and additivity are not constant in time, but they may change and pass from one limit to another. Whenever the system passes from coherence to additivity, then (we say that) a *progressive division* takes place in the system. The natural character of the system becomes then lost, and its separate elements no longer fit well into the proper system's structure (no longer accomplish the system's goals).

We can distinguish two sorts of the progressive division: (1) disintegration, which is a natural effect of system or aging, and (2) proliferation—an effect of continual revival of systems. Disintegration is a common phenomenon associated with slow but continual run-down of system elements, which therefore become less and less efficient and functional. This process may be faster or slower, depending on the intensity of the system functioning, but cannot be stopped or reversed. In fact, every moment of the system's birth is followed by a subsequent approach to its death, a result of the progressive division performed by disintegration.

Complexity and Complex Thermo-Economic Systems. https://doi.org/10.1016/B978-0-12-818594-0.00009-X

Even shut-off systems suffer this disintegration, not to mention accelerated disintegrations in systems of exceedingly high intensity/activity. Not surprisingly, parts of device which are in continual motion disintegrate faster. Yet, these parts can become less efficient in time if they are left without appropriate maintenance. For example, in a car left without maintenance for a longer time, metal parts will corrode, tires will wear out, and in effect, the car will no longer behave as a system. This pertains, in fact, to any device which is subject to disintegration (Mynarski, 1979).

The second sort of a progressive division is proliferation. In this case, the weakening of internal cohesion between various elements leads to their independence and creation of new systems or on distinguishing of definite functions. In opposition to disintegration, this process leads to creation of something new within the old system. An example may be the development of an embryo, which consists in the embryo's passage from the initial stage of coherence to the final stage of additivity during the formation of independent organs. When the embryo becomes an independent system, its further development and further proliferation occur only via metabolism.

During the (described above) "negative evolution" of systems, as opposed to the common tendency to pass from coherence to additivity, positive (inverse) processes may also occur called *progressive integration*. They are characterized by enhancement of existing links, increase the number of links, or inclusion of new elements (blocks/stages) to the system. In all these cases, an increase of system coherence and its efficiency is observed. This process results in an associated increase of functionality of all system elements. This may be exemplified by the development of the communication system, which, in the initial stage of its history, had a local range (reach/scope), but in the course of time become more and more integrated. Consequently, today we can speak not only about national or transcontinental communication systems, but also—thanks to telecommunication satellites—about the world communication system (Mynarski, 1979).

Another example may be the development of the system of data processing in companies, banks, business firms, etc. Initially, when suitable technical devices were not yet available, data processing systems were characterized by weak links between the different human resource and human organization cells. However, after introducing technical amendments and innovations in the field of information gathering, transfer, and transformation, a lumping process followed to merge various activities/actions, first in the framework of individual cells, then branches, companies, federations, unions, and, presently in the scale of countries, continents, or even the whole globe.

Progressive integration constitutes a process leading to an increase in the degree of organization of the system, in opposition to the progressive division, which decreases the degree of system organization. Both these processes can occur simultaneously or separately. For example, in the metabolic processes, anabolism, which is representative of the progressive integration, occurs simultaneously with the catabolism—representative of the progressive division. As a result of the simultaneous occurrence of both these processes, the bio-system is in a steady state, in which the regeneration of its activity decrease occurs due to the continuity of the metabolic processes. On the other hand, an example of the separate occurrence of progressive integration and progressive division can be provided by considering developmental stages of a human family composed of parents and children. In the initial stage, parents and children constitute a coherent family system. However, when children become mature, they usually become independent of their parents and start their own families. Then the cohesion of the original family is lost, and it is rather the cohesion within two families which come into existence.

Besides properties of cohesion and additivity, in large systems also centralization and decentralization properties should be considered, which characterize typical hierarchical systems. Both centralization and decentralization can change in time resulting in an increase of controlling and leading functions (progressive centralization) or a decrease of these functions (progressive decentralization). These processes may occur simultaneously with processes of progressive division or progressive integration, or can take place separately. Most frequently, progressive centralization accompanies progressive integration, whereas progressive decentralization accompanies progressive division.

In an integrated system, its upper level management can control the performance of lover level elements more easily than in a disintegrated system. Similarly, in a centralized system, the degree of cohesion is greater than in decentralized systems. Whenever these properties are mutually inconsistent, new and foreign forms of system functioning may appear, frequently accompanied by unexpected and false behavior, which may jeopardize the realization of original goals of the system.

9.2 Basic topological structures

One of structurally simplest systems is a cascade, a sequence of blocks or elements. Fig. 9.1 shows control application of the cascade pertaining to the use of the forward optimization algorithm of the dynamic programming method (DP). In such an algorithm, computational results (DP tables) are generated in terms of final or outlet states from stages, and the results along an optimal trajectory are read off in the unnatural (flow-against) direction of the stream flowing through the cascade. Ellipse-shaped balance areas pertain to sequential subprocesses that grow by inclusion of preceding balance areas as calculations go on.

However, practical optimizers need keep in mind that, because of stability problems, the most popular are backward DP optimization algorithms in which computational results (DP tables) are generated in terms of initial or inlet states to stages, whereas the results along

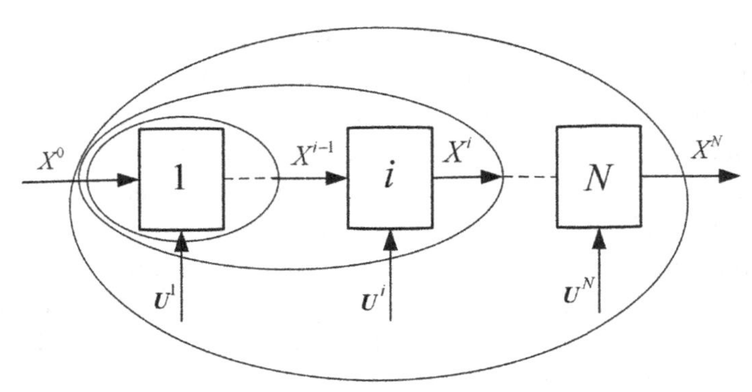

Fig. 9.1
Cascade as a complex system of the simplest structure. In the computational algorithm of the forward dynamic programming method, X^i are state variables and U^i are control variables. Ellipse-shaped balance areas comprise sequential sub-processes that grow by inclusion of preceding balance areas as calculations go on.

an optimal trajectory are read-off in the natural (with-flow) direction of the stream passing through the cascade.

See Sieniutycz and Jeżowski (2009, 2013, 2018) for more information about variety of DP optimizations.

Generalizing the cascade structure, or simple sequential combining of blocks/stages to more complex systems, we can complicate the topology of the scheme in Fig. 9.1 in many ways. The first (easiest) way works without addition of extra blocks, but by inserting certain raw streams (or adding circulation streams) into the main controlled stream or the horizontal stream of the cascade. This is described by scheme in Fig. 9.2.

In the second (more general) way, we admit addition of extra production or transformation blocks, and also, as in the first way, we admit inserting certain raw streams or circulation streams into the main controlled stream ("horizontal stream").

We begin with the listing schemes of these systems (next, they will be illustrated, analyzed, and commented) (Figs. 9.3–9.10).

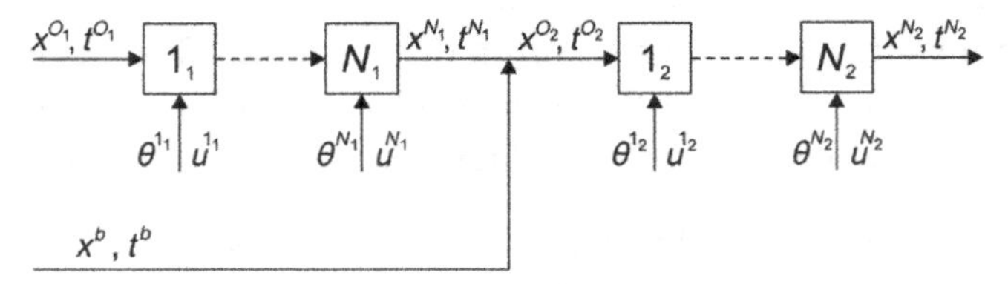

Fig. 9.2
A system with additional fresh feed fed at an intermediate point.

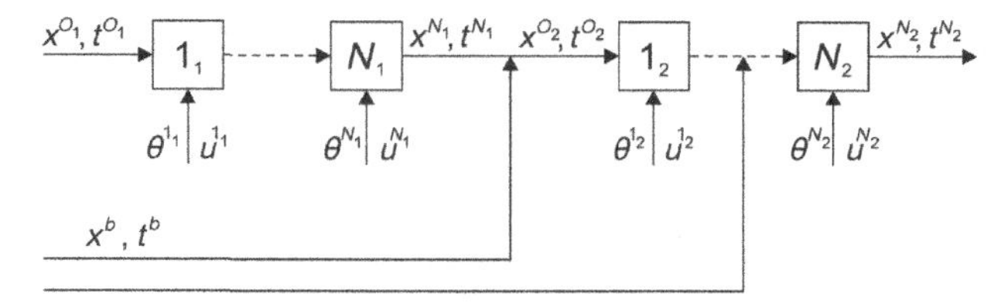

Fig. 9.3
A system with two extra fresh feeds (at two intermediate points).

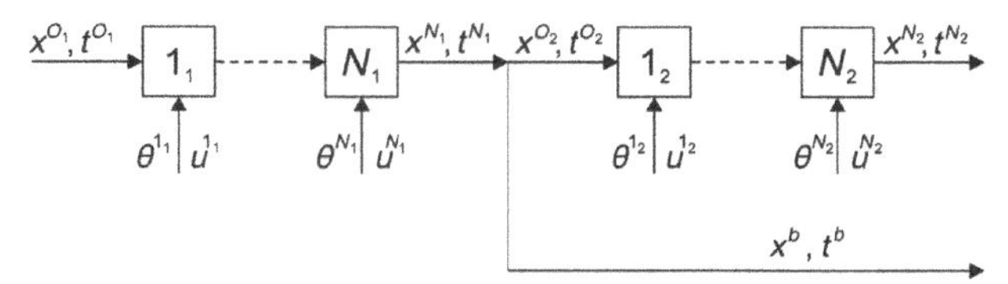

Fig. 9.4
A system with additional product withdrawn from an intermediate point.

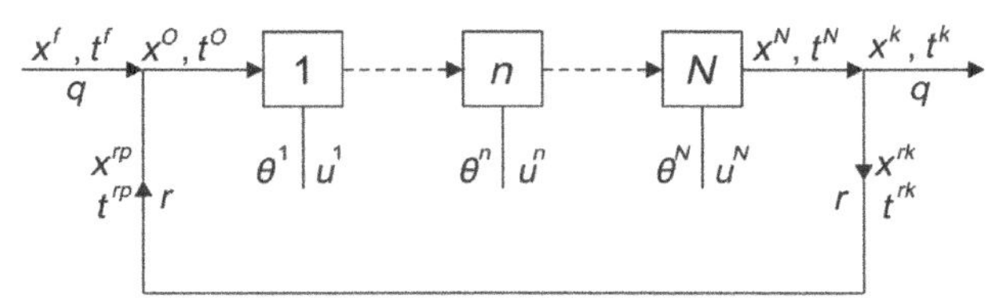

Fig. 9.5
A simple system with product recycle.

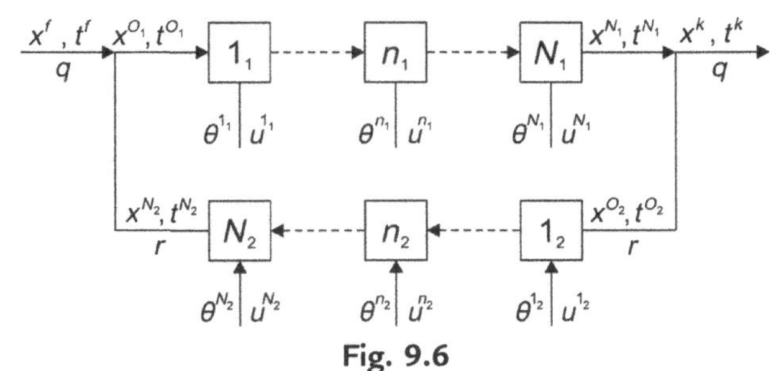

Fig. 9.6
A complex system with product recycle.

9.3 Dynamical aspects of hierarchical systems

There is no doubt that, in view of affirmative opinions about/on the role of the classical exergy in economic and thermo-economic optimizations, many readers may ask: Is there a general hint regarding the differentiation between the role played in optimization by the well-known classical exergy and the so-called finite-time exergies (sometimes called finite-rate

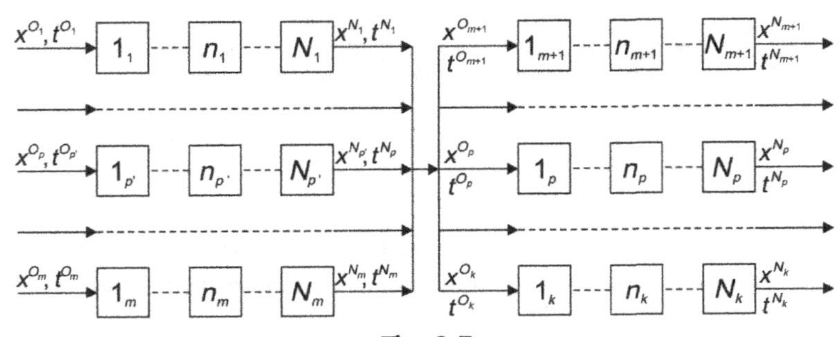

Fig. 9.7
A crossing streams system.

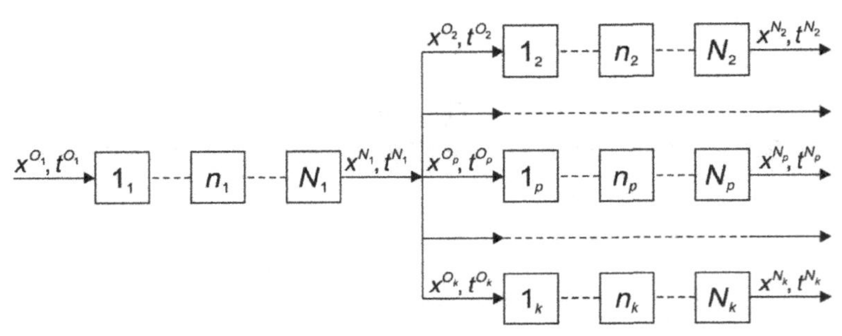

Fig. 9.8
A separating point system.

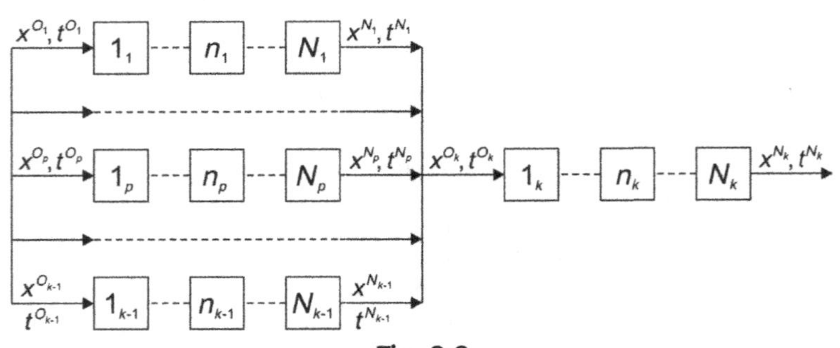

Fig. 9.9
A combining point system.

availabilities), Fig. 9.11. In this case, a following statement should be made. Classical (static) exergy, also called available energy, is the sole quantity which forms a universal base to set the scale of unit prices of chemical components produced or consumed in arbitrarily complex systems. The usefulness of the classical exergy is especially clear in systems with hierarchical structures where a universal measure of the value of each component is a highly demanded property. However, the use of nonclassical (rate sensitive) exergies is in general ineffective in

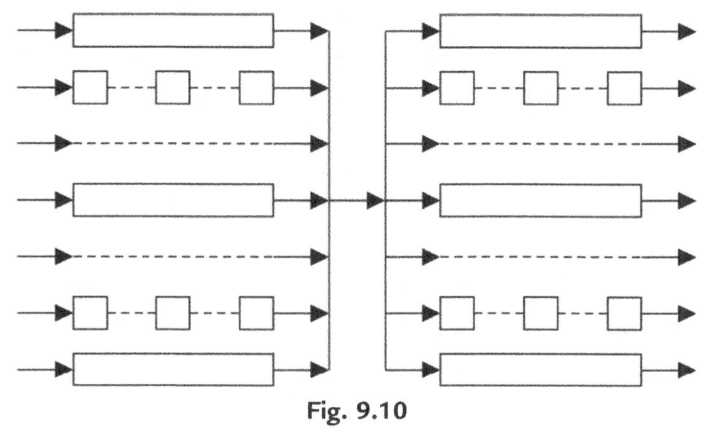

Fig. 9.10
Crossing scheme in continuous and multistage sub-processes.

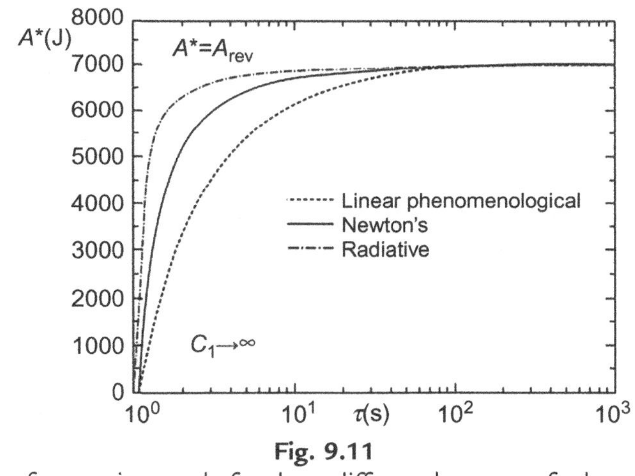

Fig. 9.11
Finite-time exergy of an engine mode for three different heat transfer laws (Xia et al., 2010).

the context of pricing and related optimizations. This is because the potential (path-independent) property is lost for nonclassical exergies due to their dissipative nature.

Fig. 8.9 of the previous chapter schematizes an arrangement of reactor and regenerator as a complex system with hierarchical structure. An analysis of optimization solutions to this and many other complex systems reveals that the role of coordination decisions is substantial. Therefore, the global quality index is most sensitive to changes in fundamental quantities, such as clock time t_k, contact time, τ_k, and the degree of catalyst decay, Δw (upper level decisions). The goals of the upper level (which governs all lower levels) are more general than the goals of lower levels and entail tasks of higher priority with regard to tasks of lower levels.

Upper level management of the system cannot be based on the knowledge of all details of all lower level subsystems. Every information transferred onto the higher level must be condensed or aggregated. This statement follows from the observation that upper level optimizations use only a part of the total available information. On the other hand, local optimizations at lower levels may use the more detailed information, i.e., that involving dynamic equations, kinetic constants of the reactor and the regenerator, activation and deactivation energies, etc.

Data of aggregated information about the reactor working at $T \leq T^*$ may be generated in the form of tables describing the effective reaction rate constant, k_{eff}. Possession of these data (along with other data, such as, e.g., those of t_R and λ_R) by the upper level staff (designer) enables the calculation of the quality index S in terms of coordinating variables, Eq. (8.66) in the previous chapter, as well as the optimization of the system as a whole. The aggregated data are transferred from lower levels to the upper level by communication channels (Fig. 8.9). In particular, communication may transmit aggregated information involving data of partial quality indices, at fixed coordinating decisions.

The downstream flow of information from the upper to the lower level is known as intervention. It may consist in feeding suitable numerical values of coordination variables from the upper level to the lower level (e.g., for testing effectiveness of coordination decisions or their influence on admissible local decisions, e.g., their influencing the limiting level of temperatures T and T_R). Proper intervention of the superior level is capable of considerably improving operation of the system due to the crucial nature of coordinating decisions. For the same reason, lack of communication brings negative consequences that are many times more serious than bad low-level decisions (e.g., optimizing the process in a reactor will not do much good if it has been incorrectly designed).

Harmfulness of actions undertaken by a superior level in the case of improper coordinating decisions is comparable to detrimental action of a foreign (external) unit, i.e., another system realizing its own goals. From the considered system's perspective, such foreign units may be capable of imposing their own decisions on our control system in a way restraining its own coordinating decisions, e.g., by reducing the quantity of steel intended for a reactor's construction (thus causing an excessive shortening of the contact time τ_k). The essence is that—by optimizing its own quality index and by influencing our reactor/regenerator performance—the foreign system may cause negative effects only. Negative influences of foreign systems can be particularly detrimental when some of their decisions are in conflict with the coordinating decisions of our system.

In systems optimization, difficulties may arise in assigning of some controls to appropriate levels. The observation of exceedingly high number of decisions at upper levels may constitute evidence for excessive centralization of controls. In the case of reaction-regeneration systems, excessive centralization means that local decisions ($T(\tau, t)$, $T_R(\tau, t)$, etc.) are undertaken simultaneously with coordinating controls at the higher level (e.g., by the system's designer). Such a scheme may produce correct results in systems with large number of decisions if only a moderately disaggregated information package is transferred to the upper level. Yet, this situation is not normal since a partial aggregation of information at the upper level management is, in practice, unavoidable. However, too excessive centralization diverts attention of the upper level managers from the highest-priority coordinating decisions, e.g., $\mathbf{v}(t_k, \Delta w, \tau_k)$. The mathematical consequence of centralization is simultaneous optimization of the system with respect to local decisions $\mathbf{u}$ and coordinating decisions $\mathbf{v}$, while the numerical consequence is an increase of the computation time. Shifting some decisions $\mathbf{u}$ from upper level to more correct local levels, i.e., proper decentralization, leads to a decomposition which leads to reduction of computational time. This idea is incorporated Eqs. (8.69)–(8.74) in Chapter 8 which are optimization algorithms applying certain controls in a hierarchical chemical system.

9.4 Aging and growth

Mynarski (1979), Bejan and Lorente (2008), and Sieniutycz (2016) describe macroscopic, microscopic, and hydrodynamic effects which accompany the aging and growth of diverse systems. General theory covers all cases: animate, inanimate, and engineered systems.

Aging of inanimate systems, such as, catalysts or enzymes, occurs in high temperatures, especially in presence of impurities and other factors that cause the reduction of the catalyst's/ enzyme active surface, thus decreasing system activity. Aging also refers to processes such as sintering, fouling, and other detrimental processes or reactions in the solid phase of catalysts. Sintering, common for metallic catalysts, damages and/or reduces the active surface of catalysts. Metallic catalysts are most often used in the form of microscopic crystals, which are distributed on the very large surface of a carrier. Excessive fast growth of these crystals is the basic cause of their textural changes. Whenever sintering is the main source of catalyst deactivation, the observed decrease in catalyst activity is fast at the process beginning, and then, a long period of slow deactivation follows. A good example of this behavior is iron catalysts of ammonia synthesis, or cuprum catalysts of methanol synthesis. Temperature increase is always detrimental here, as it significantly accelerates the sintering processes.

Aging in animate systems (such as, social systems or ecological systems) occurs mostly due to high stress conditions, bad living conditions, starvation, etc. Aging is often an effect of system disintegration, or detrimental proliferation. It constitutes, in fact, an effect of continual decay of systems. Aging occurs in time for all man-made engineering systems.

9.5 Properties of large systems

Large-scale composition systems are very diverse and full of surprises. Perhaps the most interesting is that, despite their complexity on the small scale, sometimes they crystallize into large-scale patterns that can be conceptualized rather simply, just as crazy swirls of colors crystallize into a beautiful and meaningful picture when we step back from the wall and take a broader view of a mural. These salient patterns are called emergent properties. Emergent properties manifest not so much the material base of the compound, but indicate how the materials are organized. Belonging to structural rather than material aspect, they are totally disparate from the properties of constituents, and the concepts considering them are paradoxical when applied to the constituents. Life emerges in inanimate matter, consciousness emerges in some animals; social organization emerges from individual actions. Less conspicuous, but no less astonishing, the rigidity of solids and turbulence of fluids emerge from the intangible quantum phases of elementary particles; rigidity and turbulence are as foreign to elementary particles as beliefs and desires are to neurons. Without emergent properties, the world would be dull indeed, but then we would not be there to be bored.

One cannot see the patterns of a mural with his nose on the wall; he must step back. The nature of complex compounds and our ability to adopt different intellectual focuses and perspectives jointly enable various sciences to investigate the world's many levels of organization.

9.6 Entropic theory of urban organization

Information-theoretic entropies have already found meaningful applications, for instance, for the development of entropy theory of urban modeling (Wilson, 2012). A "science of cities and regions" is critical for meeting future challenges. The world is urbanizing: huge cities are being created and are continuing to grow rapidly. There are many planning and development issues arising in different manifestations in countries across the globe. These developments can, in principle, be simulated through mathematical models which provide tools for forecasting and testing future scenarios and plans. These models can represent functioning of cities and regions, predicting spatial demography and economy, main flows such as journey to work or to services, and the mechanisms of future evolution. In Wilson's (2012) book, main principles involved in the design of such models are articulated, providing an account of the current state of the art as well as future research challenges. Wilson distilled his experience and discoveries into what serves as both an introduction to and a review of the research frontier. The topics covered include Lowry's model, the retail model, principles of account-based models, the methods rooted in Boltzmann-style statistical modeling, and the Lotka-Volterra approach to system evolution. Applications range from urban and regional planning to wars and epidemics.

9.7 Entropy principle in social and living systems

For a certain number of researches, such as Georgescu-Roegen (1970, 1971), the main importance is in stressing the relevance of the entropy law in describing and identifying limits of the economic processes. Their major errors concern the description of the long-term resource base and, thus, the possible future development of human society (Kåberger and Månsson, 2001). However, materials used in society and dissipated in nature do not need to be lost forever; future generations need not have fewer resources available to them than the present generation. Human industrial activities could be transformed into a sustainable system where the more abundant elements are industrially used and recycled, using solar energy as the driving resource. An economic theory, fit to guide industrial society in that development, must not disregard thermodynamics, nor must it overstate the consequences of the laws of thermodynamics (Kåberger and Månsson, 2001).

9.8 Disequilibrium thermodynamics and evolutionary hypothesis: Pierce work

Here we shall briefly discuss an interesting thermodynamic theory of development proposed by Pierce (2002). In this work, an alternate evolutionary hypothesis is proposed that is identified with (a component of) nonequilibrium thermodynamics. It is argued that evolution is an axiomatic consequence of organismic information obeying the second law of thermodynamics and is only secondarily connected to natural selection. As entropy increases, the information within a biological system becomes more complex or variable. This informational complexity is shaped or organized through historical, developmental, and environmental constraints of natural selection. Biological organisms diversify or speciate at bifurcation points as the information within the system becomes too complex and disorganized. These speciation events are entirely stimulated by intrinsic informational disorganization and shaped by the extrinsic environment (in terms of natural selection). Essentially, the entropic drive to randomness underlies the phenomenon of both variation and speciation and is, therefore, the ultimate cause of evolution.

Pierce (2002) further shows that like the convection of water, biological organisms are self-organizing dissipative structures; they take in and give off energy from the environment in order to sustain life processes, and in doing so, function at a state of nonequilibrium. This argument is illustrated by Fig. 9.12 which mimics original Fig. 3 in Pierce (2002).

In conclusion, although biological organisms maintain a state far from equilibrium, they are still controlled by the second law of thermodynamics. Like all physiochemical systems, biological systems are always increasing their entropy or complexity due to the overwhelming drive toward equilibrium (Fig. 9.13). But, unlike physiochemical systems,

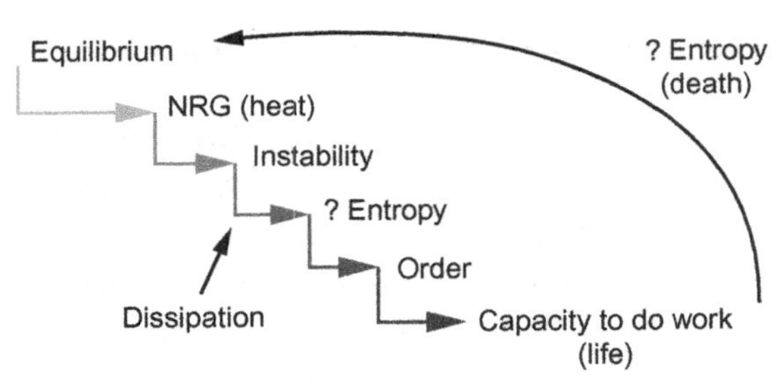

Fig. 9.12
A flow diagram illustrating Fig. 3 of Pierce (2002) which depicts the emergence of a dissipative structure from a state.

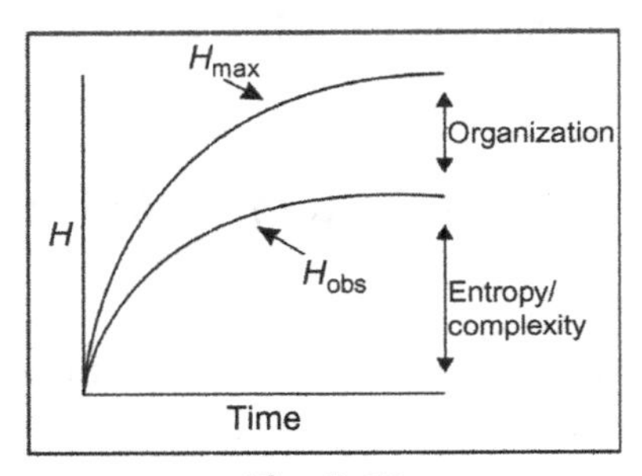

Fig. 9.13
An illustrative scheme showing the time development of the difference between the maximal entropy H_{max} and the observed entropy H_{obs}. The difference $\Delta H = H_{max} - H_{obs}$ is a measure of system organization, whereas the value H_{obs} is the measure of the actual entropy of the system. Note that our interpretation of the scheme is different from the similar interpretation of Pierce (2002).

biological systems possess "information" that permits them to self-replicate and continuously amplify their complexity and organization through time (Pierce, 2002).

Andresen an Essex (2003) give another example of the use of disequilibrium thermodynamics in characterizing biological organisms. They worked out finite-time thermodynamics optimization of mitochondria that are the fuel cells of the living body (Figs. 9.14 and 9.15).

In mitochondria, Figs. 9.14 and 9.15, the large free energy of reaction between hydrogen and molecular oxygen is harvested in a number of steps by the cytochrome chain. As reactants and products are exchanged with the matrix fluid, only those steps which produce ATP are included. This process is of great interest because mitochondria are the fuel cells of the body. The context is nonetheless foreign to contemporary fuel cell design. Thus, insights gained here are valuable for design questions generally and energetic optimization of industrial fuel cells specifically. Their work aims toward demonstrating the use of thermodynamic geometry and optimization at equipartition of thermodynamic distance through this

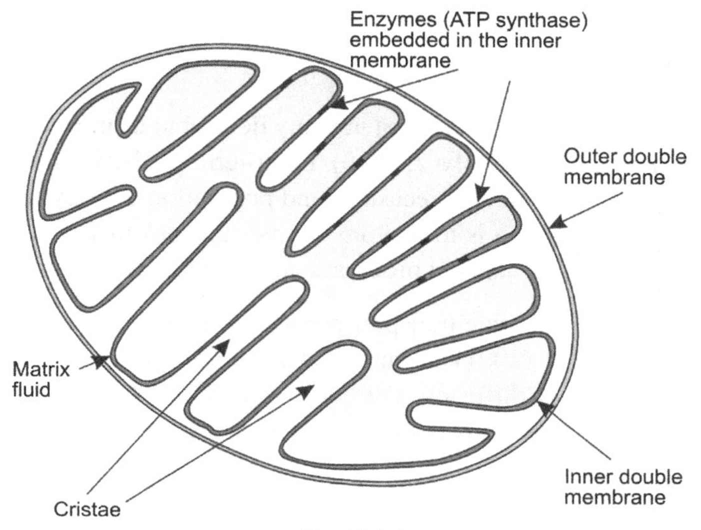

Fig. 9.14
Schematic drawing of a mitochondrion showing the two double membranes and the location of the cytochrome chain embedded in the inner membrane (Andresen and Essex, 2003).

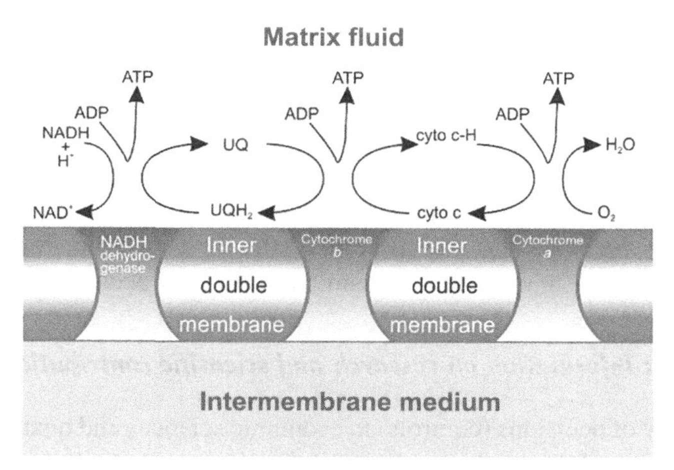

Fig. 9.15
Close-up of the enzymes of the cytochrome chain embedded in the inner membrane of the mitochondrion indicating that the reactants and products are exchanged with the matrix fluid. Only those steps which produce ATP are included. The spatial sequence of reaction sites denotes the logical sequence only. There need be no such ordering on the inner membrane (Andresen and Essex, 2003).

biochemical example. They accomplish their task through computing theoretically optimal sequence of energy degradation steps and comparing the result with the actual energy steps in the cytochrome chain. The researchers show how this problem can be formulated in a way that can be the subject of a quantitative analysis.

9.9 A brief review of various theories of dynamical (order/disorder) phenomena

Chaos is the topic which has attracted interest in many fields, but its most essential applications have to do with predictability (Sieniutycz, 2016). Long-term prediction remains an unrealized goal in many contexts from weather forecasting and population biology to stock market forecasting. The aim of a research is to explore the mechanisms that lead to failure in predictability associated with dynamical processes and to consider what may be done about it.

Gaspard and Nicolis (1990) evaluate transport properties, Lyapunov exponents, and entropy unit time. Also, Gaspard et al. (2001) subsume the scientific information on the fractality of the hydrodynamic modes of diffusion. Further, Gaspard (2004) presents an advanced theory of nonequilibrium fluctuations in chemically reacting systems.

Gaspard's (1998) book on chaos, scattering, and statistical mechanics describes recent advances in the application of chaos theory to classical scattering and disequilibrium statistical mechanics generally and to transport by deterministic diffusion in particular. Gaspard presents the basic tools of dynamical systems theory, such as dynamical instability, topological analysis, periodic-orbit methods, Liouvillian dynamics, dynamical randomness, and large-deviation formalism. These tools are applied to chaotic scattering and to transport in systems near equilibrium and maintained out of equilibrium. The book is designed for researchers interested in chaos, dynamical systems, chaotic scattering, and statistical mechanics, especially in theoretical, computational and mathematical physics, and in theoretical chemistry.

The above text is probably a shortest possible review of complex dynamical processes occurring in our physiochemical world. But, how is in the economic world?. The appendices below lead the reader to the dynamical aspects of economic processes.

9.9.1 Appendix 1: Information on research and scientific contribution of H.A. Simon

Mathematical theory of decisions (controls) in economic sciences and quantitative descriptions of effects of decision making in economic processes dates back to the seventies of 20th century. Scientific achievements performed in these years were crowned on October 16, 1978, which is the date of receiving Nobel Prize in Economics in 1978 by Herbert A. Simon "for his pioneering research into the decision-making process within economic organizations." This formulation stressed Simon's role in his application of optimization and optimal control in economic world.

The announcement (press release) of Swedish Academy of Sciences states between others: "The Royal Swedish Academy of Sciences" has decided to award the 1978 Alfred Nobel Memorial Prize in Economic Sciences to Professor *Herbert A. Simon*, Carnegie-Mellon

University, United States, for his pioneering research into the decision-making process within economic organizations. Simon's scientific output goes far beyond the disciplines in which he has held professorships—political science, administration, psychology, and information sciences. He has made contributions in, among other fields, science theory, applied mathematical statistics, operations analysis, economics, and business administration. In all areas in which he has conducted research, Simon has had something of importance to say; and, as a rule, he has developed his ideas to such an extent that it has been possible to use them as a basis for empirical studies. But he is, most of all, an economist—in the widest sense of that word—and his name is associated, most of all, with publications on structure and decision making within economic organizations, a relatively new area of economic research.

In older, traditional economic studies, no distinction was made between enterprises and entrepreneurs, and it was assumed that the entrepreneurs had only one goal: profit-maximizing. The purpose of this classic and rather rudimentary theory of the firm was primarily to serve as a basis for studies of total market behavior and not of the behavior of the individual firms. As long as these companies consisted of small, patriarchally run units, their activities remained relatively uninteresting. As companies grew in size, however, as running them became separated more and more from owning them, as employees began to form labor unions, as the rate of expansion increased, and as price competition between many was replaced by competition with regard to quality and service between few, the behavior of the individual companies attained quite another degree of interest.

Influenced by the organizational research that was being conducted in other social sciences, however, economists in the 1930s began to look at the structure of companies and at the decision-making process in an entirely new way. Simon's work was of the utmost importance for this new line of development. In his epoch-making book, *Administrative Behavior* (1947), and in a number of subsequent works, he described the company as an adaptive system of physical, personal, and social components that are held together by a network of intercommunications and by the willingness of its members to cooperate and to strive towards a common goal. What is new in Simon's ideas is, most of all, that he rejects the assumption made in the classic theory of the firm of an omniscient, rational, profit-maximizing entrepreneur. He replaces this entrepreneur by a number of cooperating decision makers, whose capacities for rational action are limited, both by a lack of knowledge about the total consequences of their decisions and by personal and social ties. Since these decision makers cannot choose the best alternative, as can the classic entrepreneur, they have to be content with a satisfactory alternative. Individual companies, therefore, strive not to maximize profits, but to find acceptable solutions to acute problems. This might mean that a number of partly contradictory goals have to be reached at the same time. Each decision maker in such a company attempts to find a satisfactory solution to his own set of problems, taking into consideration how others are solving theirs.

Simon's theories and observations about decision making in organizations apply very well to the systems and techniques of planning, budgeting, and control that are used in modern business and public administration. These theories are less elegant and less suited to overall economic analysis than is the classic profit-maximizing theory, but they provide greater possibilities for understanding and predictions in a number of areas. They have been used successfully to explain and predict such diverse activities as the distribution of access to information and decision making within companies, market adjustment to limited competition, choosing investment portfolios and choosing a country in which to establish a foreign investment. Modern business economics and administrative research are largely based on Simon's ideas.

Simon has been awarded by 1978 prize in economics for his research into the decision-making process within economic organizations, and he also made other important contributions to the science of economics. For example, his interest in simplifying and understanding complex decision-making situations led him at an early stage to the problem of breaking down complex equation systems. His studies of "causal order" in such systems have been of particular importance'.

9.9.1.1 Barros (2010) Discussion of Simon's conception of rationality

There are several texts devoted to H.A. Simon's conception of rationality. As a suitable example, we have chosen Barros (2010) paper, who discusses Simon's rationality in two of its principally general definitions: bounded rationality and procedural rationality. Simon argues that the latter is the one that better synthesizes his own view about rational behavior and that the former fills mainly a critical function. The definitions of bounded rationality and procedural rationality are complementarily used by Simon. It is argued (Barros, 2010) that the low degree of specificity of the concept of bounded rationality is one of the reasons for its relatively greater success.

Let us discuss in more detail some of the issues involved presented in Barros' (2010) paper, where, among others, he writes the following:

> First, the concept of bounded rationality is characterized, above all, by its low degree of specificity. Second, this characteristic can explain much of the (relative) current popularity of the concept.

In the third place, Barros (2010) argues that:

> Simon's remaining main contributions to the debate on rationality and economic behavior, including the ones preceding 1976, can be grouped under the term *procedures* and, therefore, his behavioral theory is based on procedural rationality. In other words, it is the case of treating the concept of procedural rationality as the one that best expresses Simon's view of rational behavior, to the detriment of bounded rationality, which mostly plays a critical role to mainstream economics.

Barros (2010) comprehensive paper is composed of an *introduction* (first section) and four working sections. The second section discusses the concept of *bounded rationality* aiming at defining it and at pointing some of its important characteristics. The third section presents the concept of *procedural rationality* so that, in the fourth section, the reader can discuss the relation between the two general concepts of rationality advanced by Simon. Special attention is dedicated to the historical chronology of the construction of these concepts, bearing in mind that Simon's chronological ordering helps to make clearer the logical relation between them. Finally, in the fifth section, some subsuming considerations are made.

Simon, writes Barros (2010), consistently strived to build a theory of human behavior throughout his work. This opinion is true, in particular, with respect to his incursions in the economic field: what he produced was, above all, a theory of economic behavior. In other words, his focus was less "the economy" than "the economic agent," though this agent does not necessarily correspond to "individual." Rationality is centrally placed in this behavioral theory: it is the main explaining element, although not the only one.

To Simon, the distance between rationality and behavior is bridged by the concept of "decision" (Barros, 2010). (Needless to say, the author of this book accepts this bridge as it is formed by decisions or controls which are basic variables in static optimization and optimal control.)

The choice is a selection of one variable, among numerous possible behavior alternatives, to be carried out. Every behavior involves a selection of this kind, be it conscious or not. The decision is a process through which this selection is performed. Rationality is a criterion used in the decision that is theoretically grounded on the presupposition that the agents are intendedly rational. In other words, the agents value rationality as a criterion of choice and it is in this sense, and by this route, that rationality is taken as an explaining principle.

Rationality is defined by Simon as a relation of conformance (efficacy) between preestablished ends and the means to reach them. To him, the specification of these ends is a question of value (or a matter of its evaluation, SS) and, hence, is beyond the scope of pure science. However, for Barros (2010), the relation between means and ends is a question of fact. The factual evaluation of this conformity involves, in theory, three "steps": (i) the listing of all possible behavioral alternatives; (ii) the determination of all the consequences that will follow, in the future, to the adoption of each of these alternatives (in a determinist way or in the form of distributions of probabilities); (iii) the comparison of the alternatives that should be evaluated by sets of consequences following each one of them, according to the preestablished ends (utility, profit, performance criterion, quality index, or any other specified pay-off function).

As Barros (2010) writes: "it is necessary to have in mind that Simon was writing a *thesis in political science, having as subject decision processes within administrative organizations.*

More specifically, he was advancing a theory of administration. And he was making use of economic theory for that, or else, he was applying economic theory to administration according to his specific perspective. It is true that such an application of the canonic economic theory to a different context demanded, to Simon, an analysis regarding conditions of validity and it is also true that he did not dispense with other explaining factors, besides rationality, to administrative behavior. These concerns forced him to discuss the 'area of rationality' and its 'limits.' However, it is important to notice that *Simon's intellectual effort was directed not towards revising economic theory, but towards applying it.* And, towards applying it to another field, stretching the theory's scope, even when this move would exact some flexibility."

Nevertheless, some years later, Simon (1955, p. 241) clearly introduced boundaries to his "rationality proper," which would become decisive: besides limited access to the several kinds of information, limits on the computational capabilities step in. On synthesizing this point, his formulation is the following:

> The alternative approach employed in these papers is based on what I shall call the principle of bounded rationality: The capacity of the human mind for formulating and solving complex problems is very small compared with the size of the problems whose solution is required for objectively rational behavior in the real world — or even for a reasonable approximation to such objective rationality.
>
> ***(Simon, 1957, p. 198, see also p. 202).***

This citation permits Barros (2010) to subsume his own discussion as follows:

> As far as it is known, this quote constitutes the first appearance in print of the term "bounded rationality." An important aspect of it, which is worth stressing, is the concept of bounded rationality built as the negative of the concept of global rationality. (Some boundaries are listed, but this is not the central point.) It is essential that the concept of bounded rationality is intended to encompass the idea of *practical impossibility of exercise of global rationality*[4] and not its logical impossibility.
>
> This carries two implications. The first is that Simon, here, is effectively speaking against global rationality: he is questioning economic theory, and speaking to the public of economists.[5] Moreover, the basis of the confrontation is precisely the lack of realism of the presuppositions sustaining global rationality, resulting in an impossibility of application without any reference to practical situations.
>
> The balance Simon tries to reach in his *Administrative Behavior* (Simon, 1976) is to include economic man's maximization as a value premise to a rational administrator, but *without rendering trivial the administrative activity*. This is the reason why it is necessary to flexibilize economic theory when applied to the administrative field: if this is not done, the administrative task becomes trivial and a theory of administration becomes useless. Not to do so would be to "solve the problem"—the problem Simon himself posed—by

> declaring the problem does not exist. A conclusion we can draw from all this is that it is an anachronism to attribute to Simon's *Administrative Behavior* the emergence of the concept of bounded rationality (Barros, 2010).

A more comprehensive analysis of remaining parts of long Barros (2010) discussion is left to the reader.

9.9.1.2 Bounded rationality: A contemporary view

H.A. Simon proposed bounded rationality as an alternative basis for the mathematical modeling of decision making, as used in economics, political science, and related disciplines. It complements "rationality as optimization," which views decision making as a fully rational process of finding an optimal choice given the information available.

Now, bounded rationality is the idea that when individuals make decisions, their rationality is limited by the tractability of the decision, the cognitive limitations of their minds, and the time available to make the decision. Decision-makers in this view act as satisficers, seeking a satisfactory solution rather than an optimal one.

Simon's analogy of a pair of scissors, where one blade represents "cognitive limitations" of actual humans and the other the "structures of the environment," illustrates how minds compensate for limited resources by exploiting known structural regularity in the environment. Some models of human behavior in social sciences assume that humans can be reasonably approximated or described as "rational" entities.

Many decision-making models assume that people are on average rational and can, in large enough numbers, be approximated to act according to their preferences. The concept of bounded rationality revises this assumption to account for the fact that perfectly rational decisions are often not feasible in practice because of the intractability of natural decision problems and the limited or insufficient computational resources available for making them.

Appendix 2 in Section 9.9.2 below provides a comprehensive account of Simon's rationality concept evaluated and extended by Morecroft (1982, 1983) who focuses on cognitive limitations on the information gathering and processing powers of human decision makers.

9.9.2 Appendix 2: System dynamics: Portraying bounded rationality by J.D.W. Morecroft

Abstract. This text, presented originally by J.D.W. Morecroft at the 1981 System Dynamics Research Conference, in The Institute on Man and Science, Rensselaerville, New York October 14–17, 1981, revised in January 1982, is based on an updated version of a conference paper (Morecroft, 1982), which examines the linkages between system dynamics and the human decision making in economics. It is argued that the structure of system dynamics models implicitly assumes bounded rationality in decision making and that recognition of this assumption would aid system dynamicists in model construction and in communication to other social science disciplines. The text begins by examining Simon's "Principle of Bounded Rationality" (Simon, 1979) where the reference on the original Morecroft's (1982) reference list included into the reference list of Chapter 9 at the end of this text. The paper draws attention to the cognitive limitations on the information gathering and processing powers of human decision makers. Forrester's "Market Growth Model" (Forrester, 1968a,b) is used to illustrate the central theme that system dynamics models are portrayals of bounded rationality. Close examination of the model formulation reveals decision functions involving simple rules of thumb and limited information content. In the final part of the paper, there is a discussion of the implications of Carnegie philosophy for system dynamics, as it affects communication, model structuring, model analysis, and future research.

Introduction

The text below shows how Morecroft (1982) examines the linkages between system dynamics school in the treatment of human decision making. Morecroft argues that the structure of system dynamics models in economics implicitly assumes bounded rationality in decision making and that the recognition of this assumption would aid system dynamicists in model construction and in communication to other social science disciplines. Morecroft's (1982) paper begins by examining Simon's "Principle of Bounded Rationality" (Simon, 1979), where the reference on the original (Morecroft's, 1982) reference list scheduled at the end of his text. The present text draws attention to the cognitive limitations on the information gathering and processing powers of human decision makers. Forrester's "Market Growth Model" (Forrester, 1968a,b) is used to illustrate the central theme that system dynamics models are portrayals of bounded rationality. Close examination of the model formulation reveals decision functions involving simple rules of thumb and limited information content. In the final part of the text, there is a discussion of the implications of Carnegie philosophy for system dynamics, as it affects communication, model structuring, model analysis, and future research.

The field of system dynamics, understood as above and focused on economical issues, has long been viewed by its practitioners as a discipline distinct from other major methodologies dealing with business and social systems. In particular, a distinction should be made between economic system dynamics and the conceptual frameworks offered by classical economics and operations research. However, to those outside the field, this system dynamics is often regarded as nothing more than a rather specialized form of simulation modeling which belongs in the general tool kit available to management scientists. Part of the reason for the persistence of these divergent viewpoints has been the inability of the system dynamics community to express its basic philosophy of social system structure in a language understandable to those outside the field. Without a clearly communicated philosophy, there is nothing to separate the subject from the simulation technique it uses (Morecroft, 1982).

Following Morecroft (1982), the essential aspect of the present considerations is examining the linkage between the philosophy of human decision making, expounded in the spirit of the Carnegie School, and the feedback structures of central concern in system dynamics. In addition, as argued by Morecroft (1982), the Carnegie way of thinking offers a language and set of concepts that may greatly improve communication between system dynamics and other fields and develop a stronger internal sense of the contribution the subject can make to the analysis of social systems. There are major exceptions to the general statement that system dynamicists have not attempted to communicate their philosophy, for example, in papers by Meadows (1980) and Andersen (1980), which compare and contrast the system dynamics paradigm to other prevalent mathematical modeling paradigms. This analysis adds a new dimension to the work of Meadows and Andersen by focusing on the implicit assumptions that system dynamics makes about human decision-making processes, the contribution that the subject can make to the analysis of social systems.

The name of Carnegie School refers to a school of thought named in recognition of the institution (Carnegie-Mellon University), where much pioneering work on human decision making was done in the 1950s and 1960s. A common and powerful theme underlying the work of the School is the notion that there are severe limitations on the information processing and computational abilities of human decision makers. As a result, decision making can never achieve the ideal of perfect (objective) rationality, but is destined to a lower level of intended rationality. The Carnegie School contends that the behavior of complex organizations can be understood only by taking into account the psychological and cognitive limitations of its human members. Such a viewpoint focuses attention on the flow of information in a complex system, the quantity and quality of information that is amenable to human judgmental processing, and the form of decision rules used to represent judgments. It is at this most fundamental level of information flow and processing that the ideas of the Carnegie School overlap with basic views of system dynamics. It is also at exactly the same fundamental level that the Carnegie School departs radically from the traditional views of economics and operations research (e.g., Simon, 1979; Allison, 1971; Eilon, 1977).

In the main body of the Morecroft's text, the ties between the Carnegie School and system dynamics are developed. In doing so, the intention is not to create the impression that the two fields are the same. Rather, they share in common a philosophy of the limitations of human decision-making philosophy that the Carnegie School has made explicit in its writings and that in system dynamics has always been implicit (Morecroft, 1982). Further, they share in common a belief that simulation is a powerful means of gaining insight into decision-making processes (see for example Dutton and Starbuck, 1971). Morecroft's development begins with a more careful look at the "principle of bounded rationality," which is the cornerstone of Carnegie philosophy. Next, the structure and behavior of "Market Growth Model" (Forrester, 1968a,b; Dutton and Starbuck, 1971) is interpreted in the light of the principle of bounded rationality, leading to the major conclusion that bounded rationality is embodied in the feedback structure of system dynamics models. Finally, there is a discussion of implications of Carnegie philosophy for system dynamics, as it affects communication, model structuring, model analysis, and future research.

Bounded rationality

The principle of bounded rationality

The principle of bounded rationality was formulated by Simon as the basis for understanding human behavior in complex systems. The principle recognizes that there are severe limitations on thinking and reasoning power of the human mind. If we wish to predict the behavior of human decision makers within the context of the systems in which they work and live, it is first necessary to take account of their psychological properties.

The principle of bounded rationality was defined in the following way (Simon, 1957): *The capacity of the human mind for formulating and solving complex problems is very small compared with the size of the problems whose solution is required for objectively rational behavior in the real world or even for a reasonable approximation to such objective rationality.*

The principle of bounded rationality provides a basis for the construction of a theory of organizational behavior (Simon, 1976). In Simon's words: "Organization theory is centrally concerned with identifying and studying those limits to the achievement of goals that are, in fact, limitations on the flexibility and adaptibility of goal striving individuals and groups of individuals themselves."

The principle of bounded rationality suggests that the performance and success of an organization is governed primarily by the psychological limitations of its members: the amount of information they can acquire and retain, and their ability to process that information in a meaningful way. These limitations in their own turn are not physiological and fixed; they depend on the organizational setting within which decision making takes place.

Bounded rationality and organizational decision making

The principle of bounded rationality leads one to expect that organizations will undertake decision making in such a way as to greatly simplify the information processing and computational load placed on the human decision makers it contains. The pioneering work of Cyert and March (1963) indicated that decision making in real business firms is indeed much simpler than one would anticipate based on classical models that assume objectively rational behavior. In the following section, the work of Cyert and March will be drawn to identify a number of empirical features of organizational decision making that can be interpreted as consequences of the principle of bounded rationality. Morecroft (1982) considers such matters as organizational goal structure and the information collecting and processing habits of human decision makers (Andersen, 1980). According to Morecroft (1982), the ultimate purpose is to show that many of these features are implicit in the structure and policy formulations of a system dynamics model. Following him, we will not consider the search-and-choice aspects of organizational decision making which seem to lie below the level of aggregation of most policy formulations in system dynamics models (Morecroft, 1982).

1. Factored decision making

Common experience with human organizations will reveal that decision making responsibility is factored or parceled out among a variety of subunits (Morecroft, 1982). For example, many business firms adopt a functional structure that divides decision making between marketing, production, pricing, finance, labor management, etc. Cyert and March (1963) point to a "division of labor" in decision making. The decision problems that an organization must solve are so complex that they cannot be handled by an individual. Separable pieces of decision making are assigned to organizational subunits in the form of subgoals. Each subunit is charged with the responsibility of meeting its own subgoal and thereby contributing to the broader objectives of the organization. Of course, this scheme will work well only if there is no inherent conflict in goals—something which cannot be guaranteed when subunits are interdependent (Morecroft, 1982). Nevertheless, factored decision making is necessary given the scope and complexity of activities in human organizations.

2. Partial and certain information

Decisions are made on the basis of relatively few sources of information that are readily available and low in uncertainty. While the above statement is not a direct prescription of a school, it can rather be inferred from comments that are made about the way information is obtained and processed by an organization. Morecroft (1982) observes that empirical observations of decision making in organizations indicate that decision makers seek only a small proportion of the information that might be considered relevant to full consideration of a given situation. Their search for information tends to be conditioned by a focus upon problem symptoms and by a desire to avoid the use of information that is high in uncertainty (Morecroft, 1982). Both these tendencies in information selection favor the use of local feedback information reflecting current conditions in the immediate operating environment of a subunit, rather than information gathered more widely whose impact upon the subunit can be only vaguely conjectured. Both Cyert and March, and Simon, comment on the frequency with which such local feedback information is used rather than more global information required for "optimal decision making" (Morecroft, 1982).

3. Rules of thumb

The organization uses standard operating procedures or rules of thumb to make and implement choices. In the short run, these procedures do not change and represent the accumulated learning embodied in the factored decision making of the organization. Rules of thumb need employ only small amounts of information of the kind that would be made available through local feedback channels. Rules of thumb process information in a straightforward manner, recognizing the computational limits of normal human decision makers under pressure of time (Morecroft, 1982).

Consider, for instance, the pricing decision of a business firm. Microeconomic theory would suggest that pricing decisions result from a sophisticated profit-maximizing computation which equates marginal cost and marginal revenue. In fact, there is evidence to suggest that computationally simpler markup pricing is common. Under this method, average variable cost is taken as a base and is increased by a fractional markup to obtain the selling price. The markup is a rule of thumb which is heavily influenced by past tradition and by feedback information on profit, return on investment, market share, etc. (For an example of rule of thumb pricing, see Mass (1977), pp. 31–35.)

Bounded rationality in a system dynamics model

In this section, Morecroft (1982) takes an existing system dynamics model and interprets its structure and behavior in the light of the principle of bounded rationality. The model selected is based on Forrester's (1968a,b) analysis and describes the policies governing the growth of sales and production capacity in a new product market. Forrester's original model resulted from a case study of an electronics-manufacturer. It represents the opinions of senior management of the company about the way that corporate growth is managed.
We will first discuss the structure of the model and show how it embodies the organizational features of factored decision making, partial information, and rules of thumb. We will then show, using simulation runs of the model, how the bounded rationality of organizational subunits can cause problems in market growth.

Factored decision making in the market growth model

In common with the Carnegie School view, the market growth model can be broken down into a number of organizational subunits, each of which is responsible for a part of the decision making that produces growth in the system as a whole.

Fig. 9.16 depicts an organization with decision making factored into four subunits. In subunit 1 on the right of the figure, customers make their ordering decisions. Ordering is influenced by the number of customer contacts made by the marketing department and by customers' perceptions of the delivery delay in obtaining the product. In subunit 2, the marketing department makes decisions on the hiring of marketing personnel. An upper limit on marketing personnel is set by a marketing budget which moves in proportion to sales volume. Hiring adjusts personnel to this budgetary limit. In subunit 3, the firm makes decisions on order filling. The rate of order filling depends on the available production capacity and its intensity of utilization. Finally, in subunit 4, the firm makes decisions on capacity management. Additional capacity is ordered whenever high delivery delay indicates there is a capacity shortage.

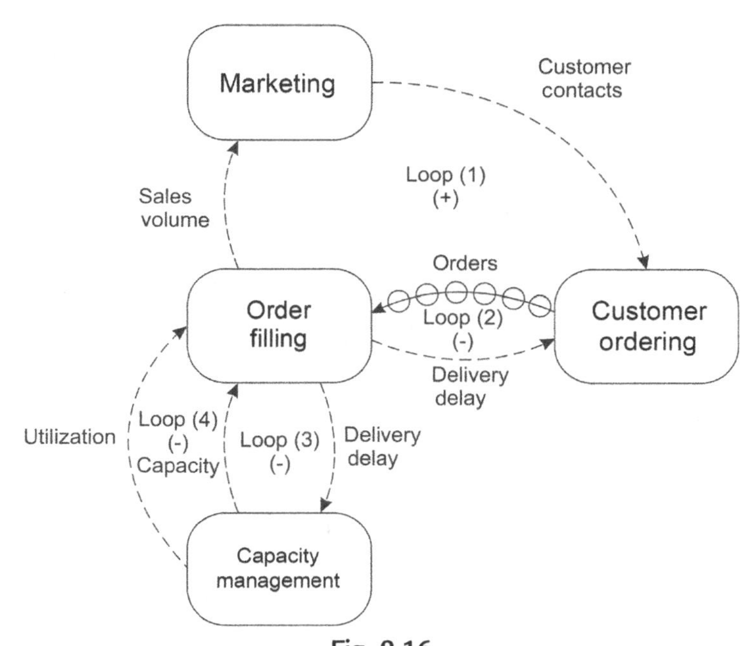

Fig. 9.16
Organizational submits in the Market Growth Model.

Partial information and rules of thumb

In this section, Morecroft (1982) considers in more detail the decision rules for capacity management, marketing, and customer ordering to illustrate examples of rules of thumb and the use of partial and certain information.

1. Capacity management

Capacity management is represented as a two-stage decision making process involving first the detection of capacity shortage and then the ordering of capacity to eliminate the shortage. During the analysis, the reader should bear in mind that the basic objective of capacity management is to adjust capacity to a level that will support demand. One could readily visualize a "rational" decision function in which future expectations of demand are generated across the lead time of capacity, and the expectations are used to drive capacity ordering. As we may see, in Morecroft (1982), the decision-making process is computationally simpler and requires far less information. Let us first look at the equations for the detection of capacity shortage:

$$\mathrm{DDC}(t) = \mathrm{DDRC}(t)/\mathrm{DDOG}(t) \tag{9.1}$$

$$\mathrm{DDRC}(t) = \mathrm{DDRC}(t_0) + \int^t (\mathrm{TDDRC})^{-1}(\mathrm{B}(t)/\mathrm{OFR}(t)) - \mathrm{DDRC}(t)\,dt) \tag{9.2}$$

$$\mathrm{DDOG}(t) = \mathrm{DDT}(t)^*\mathrm{DDW} + \mathrm{DDMG}^*(1 - \mathrm{DDW}) \tag{9.3}$$

$$\mathrm{DDT}(t) = \mathrm{DDT}(t_0) + \int^t (\mathrm{TDDT})^{-1}\,[\mathrm{DDRC}(t) - \mathrm{DDT}(t)\,dt] \tag{9.4}$$

In Eq. (9.1), delivery delay condition DDC is an index of capacity shortage based on the ratio of delivery delay recognized by company DDRC to delivery delay operating goal DDOG. When DDC is greater than capacity, shortage exists since it is not possible to fill orders at a rate that will keep delivery delay equal to the operating goal.

Eq. (9.2) states that DDRC is an exponential average of the ratio of backlog B to order fill rate OFR. Eqs. (9.3), (9.4) model the delivery delay operating goal as an adaptive goal based on a weighted average of a fixed delivery delay management goal DDMG and delivery delay tradition DDT which reflects past performance. Delivery delay tradition is formulated as an exponential average of recent delivery delay DDRC.

Consider now the cognitive and information processing assumptions of Eqs. (9.1)–(9.4). The way the company recognizes the need to expand capacity is by making a judgment on delivery delay condition. The judgment requires a comparison of current delivery delay to the operating goal. Current delivery delay is known from information on backlog and order fill rate which is readily available from concrete operating data. There is no need to do elaborate market surveys and projections of future demand, both of which are likely to involve information of an uncertain and speculative nature. If demand is growing, it will be reflected in a rising backlog. Thus, in Eqs. (9.1), (9.2), we see a clear use of partial and certain information. Furthermore, in Eq. (9.1), the information is processed in a simple rule of thumb that compares current delivery performance to an operating goal. In Eqs. (9.3), (9.4), the operating goal itself is seen to be a rule of thumb which adjusts to past traditions of performance. See Forrester (1961) for an example of how Eq. (9.2) reduces to an exponential average.

In conclusion, we see that the entire process of detecting capacity shortage uses only backlog and order filling information processed on the basis of simple judgmental criteria.

Now let us consider capacity ordering which is represented by Eqs. (9.5)–(9.7) below.

$$\mathrm{COR}\,(t) = \mathrm{C}(t)^{*}\mathrm{CEF}(t) \tag{9.5}$$

$$\mathrm{C}(t) = \mathrm{CC}(t') + \mathrm{CAR}(t)\,dt \tag{9.6}$$

$$\mathrm{CEF}(t) = f(\mathrm{DDC}(t))f(1) = 0,\;\; f' > 0,\;\; f'' > 0 \tag{9.7}$$

Eq. (9.5) states that capacity ordering rate COR is the product of capacity C and capacity expansion fraction CEF. Thus, capacity ordering takes place by a fractional expansion of existing capacity. Existing capacity in Eq. (9.5) is simply the integral of capacity arrival rate CAR (assuming no capacity discards). Eq. (9.7) states that capacity expansion fraction CEF is an increasing nonlinear function of delivery delay condition DDC. When DDC is equal to 1, the function takes a value of zero, indicating there is no pressure to expand capacity because delivery delay is in line with the operating goal. As DDC increases above 1, the function becomes positive resulting in capacity expansion in Eq. (9.7). The function in Eq. (9.7) has a second derivative greater than zero, indicating more aggressive capacity ordering as rising delivery delay condition indicates a more serious capacity shortage.

Again, consider the cognitive and information processing assumptions of Eqs. (9.5)–(9.7). The most striking feature of the equations is that nowhere is there an explicit attempt to relate capacity needs to projections of customer demand. Capacity ordering is a rule of thumb that responds to "pressure" from delivery delay condition signaling capacity shortage. Delivery delay condition is the only information entering the ordering decision. There is no information from the market or from the marketing subunit. A policy of fractional expansion is computationally simple—a judgmental process that causes capacity to change in the right direction, but without the need to explicitly compute capacity requirements.

2. Marketing expansion

Marketing expansion involves a two-stage decision making process of budgeting and hiring. Consider first budgeting as represented by Eqs. (9.8), (9.9) below.

$$\mathrm{BM}(t) = \mathrm{AOFR}(t)*\mathrm{PO}*\mathrm{FBM} \tag{9.8}$$

$$\mathrm{AOFR}(t) = \mathrm{AOFR}(t_0) + \int_{t_0}^{t} \frac{\mathrm{OFR}(t) - \mathrm{AOFR}(t)dt}{\mathrm{TAOFR}} \tag{9.9}$$

The budget to marketing BM is defined as the product of average order fill rate AOFR, price of output PO, and fraction of budget to marketing FBM. The average order fill rate is defined in Eq. (9.9) as an exponential average of current order fill rate OFR. Together Eqs. (9.8), (9.9) represent a simple budgeting process in which a fixed fraction FBM of the total budget AOFR*PO is allocated to marketing.

The hiring of marketing personnel is represented by Eqs. (9.10)–(9.12).

$$\mathrm{MP}(t) = \mathrm{MP}(t_0) + \int_{t_0}^{t} \mathrm{PH}(t)dt \tag{9.10}$$

$$\mathrm{PH}(t) = (\mathrm{IMP}(t) - \mathrm{MP}(t))/\mathrm{TAMP} \tag{9.11}$$

$$\mathrm{IMP}(t) = \mathrm{M}(t)/\mathrm{MS} \tag{9.12}$$

In Eq. (9.10), marketing personnel MP is defined as the integral of personnel hiring PH. In Eq. (9.10), personnel hiring is formulated as a goal adjustment process which eliminates the discrepancy between indicated marketing personnel IMP and marketing personnel MP over a time period TAMP, a time to adjust marketing personnel. Finally, in Eq. (9.12), IMP is defined as the personnel that can be supported at a marketing salary MS by a budget BM.

Consider now the information processing assumptions of Eqs. (9.8)–(9.12). Budgeting is a simple rule of thumb involving a fixed fractional allocation to marketing. Such a "frozen" budgetary process is computationally simpler for an organization than one in which allocation fractions are derived from a zero base. Hiring is a goal adjustment process. It uses only information that is specific to the marketing subunit: the current level of marketing personnel MP and the authorized target IMP. Hiring does not include information about capacity or the delivery delay operating goal of the organization, both of which could conceivably be of relevance in a "fully informed" hiring decision.

3. Customer ordering

An important feature of the market growth model is that customer ordering is entirely endogenously generated. The initiative for growth rests with the company. The company must contact customers in the market and persuade them to buy the product. Customer ordering (here called the product order rate POR) is represented by Eqs. (9.13)–(9.16) below.

$$\mathrm{POR}(t) = \mathrm{CC}(t)^{*}\mathrm{EPA}(t) \tag{9.13}$$

$$\mathrm{CC}(t) = \mathrm{MP}(t)^{*}\mathrm{NCR} \tag{9.14}$$

$$\mathrm{EPA}(t) = f(\mathrm{DDRM}(t)) \quad f' < 0 \tag{9.15}$$

$$\mathrm{DDR}_m(t) = \mathrm{DDRM}(t_0) + \int_{t_0}^{t} \frac{\mathrm{DDRC}(t) - \mathrm{DDRM}(t)}{\mathrm{TDDRM}}\, dt \tag{9.16}$$

In Eq. (9.13), product order rate POR is formulated as the number of customer contacts CC multiplied by the effect of product attractiveness EPA (we assume that each contact can generate no more than one order). Thus, a customer will order only if contacted, and only then if the product seems attractive. In Eq. (9.14), customer contacts are expressed as a fixed multiple NCR of marketing personnel MP. NCR, the normal contact rate, represents the average number of contacts made by marketing personnel during a month. In Eq. (9.15), the effect of product attractiveness EPA is formulated as a decreasing nonlinear function of delivery delay recognized by the market DDRM. Customers are assumed to be sensitive to delivery delay: they will be discouraged from ordering as delivery delay grows. In Eq. (9.16), DDRM is formulated as an exponential average of delivery delay recognized by the company DDRC, to represent the customers' perception of delivery delay.

Consider the information processing assumptions of Eqs. (9.13)–(9.16). Customers need only two pieces of information to make their decision: they need to be made aware of the product and they need to judge its delivery delay. They need to know nothing at all about the detailed condition of the company (such as its marketing and capacity plans). Even their knowledge of delivery delay is local to the market and need not be the same as actual delivery delay the company is currently achieving.

Feedback structure in the market growth model

Morecroft's (1982) analysis of the market growth model shows that decision making can readily be interpreted in the light of the principle of bounded rationality. Decision making is factored into subunits, each striving for separate goals: marketing is striving for a personnel goal dictated by the budget; capacity management is striving to maintain delivery delay at a value dictated by the delivery delay operating goal. All the decision functions use partial and certain information which is but a small fraction of the total available to describe the state of the system. There are numerous examples of rule of thumb decision making. The decision functions, the subunits, and their linkages together form the feedback structure of the system, which is the central concern of a system dynamics analysis. Attention focuses on closed paths of cause and effect (feedback loops) that exist in the system. How do these closed paths result from the information requirements of microlevel decision-making processes? What are the implications of the existence of the closed paths for behavior and performance of the organization? Are decision makers in the system aware of the feedback structure that their own decision making creates and of the limitations that this structure imposes on organizational performance?

Returning to Fig. 9.16, Morecroft identifies four of the important feedback loops in the market growth model. Linking the customer ordering, order filling, and marketing subunits (and their internal decision-making processes), there is a positive feedback loop (1), which by itself can produce growth in orders, marketing personnel, and customer contacts. Linking the customer ordering and order filling subunits, there is a negative feedback loop (2) that regulates the order rate to equal available capacity. Finally, between the order filling and capacity management subunits, there are two negative feedback loops (3 and 4) that adjust capacity and utilization to equal the order rate.

Below, we will use simulation analysis to examine the implications of the feedback structure (Morecroft, 1982).

Bounded rationality underlies problem behavior

In this section, three simulation experiments will be presented to show how well-intentioned policies (intendedly rational) can lead to problem behavior in a complex organizational setting. We start by showing that our three policies for marketing, capacity expansion, and ordering are intendedly rational. In other words, they are capable of producing reasonable behavior when taken in isolation or in simple combination. A demonstration of intended rationality is important because it indicates that there is a rationale to support the existing policies. We then bring all three policies together in a "complex organizational system" and show that together they can fail to bring about market growth, even though the marketing policy is striving for growth and there is no inherent limit to market size.

Experiment 1: Interaction of customer ordering and marketing (positive loop)
In Experiment 1, the interaction of customer ordering and marketing subunits is examined in isolation from capacity constraints on order filling. With no capacity constraints, delivery delay is constant so product order rate FOR in Eq. (9.13) is directly proportional to marketing personnel. Fig. 9.2 shows a simulation run of the simplified system over a time period of 80 months. Product order rate and marketing personnel both display unlimited exponential growth. Expansion of marketing personnel leads to an increase in product order rate which in turn leads to an increase in the budget for marketing, thereby justifying further marketing expansion. Growth is limited only by the delays in personnel hiring and by the sales efficiency of marketing personnel. We can interpret Fig. 9.2 to mean that the marketing policy is intendedly rational in the sense that it is able to bring about market growth under conditions of perfect supply (Fig. 9.17).

Experiment 2: Capacity expansion in isolation (negative loops 3 and 4)
In Experiment 2, Morecroft (1982) looks at the behavior of the capacity expansion policies in isolation from the market. He looks at the simple question of whether the capacity expansion policies are able to bring about an increase in supply shortages in response to an exogenous increase in demand. He is, therefore, concerned not with how the demand increase is generated, but merely with the ability of capacity management to make a "rational response."

Fig. 9.18 shows the results of a simulation run over a time period of 120 months. The run starts with the system in the state of equilibrium. Order fill rate is equal to the product order rate of 1000 units per month. Capacity is steady at 2000 units per month with a utilization

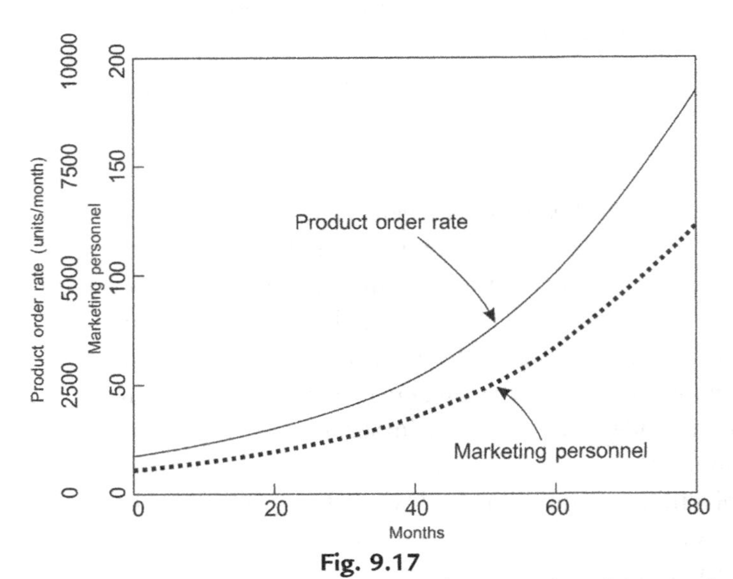

Fig. 9.17
Marketing and product ordering with no supply constraints (Morecroft, 1982).

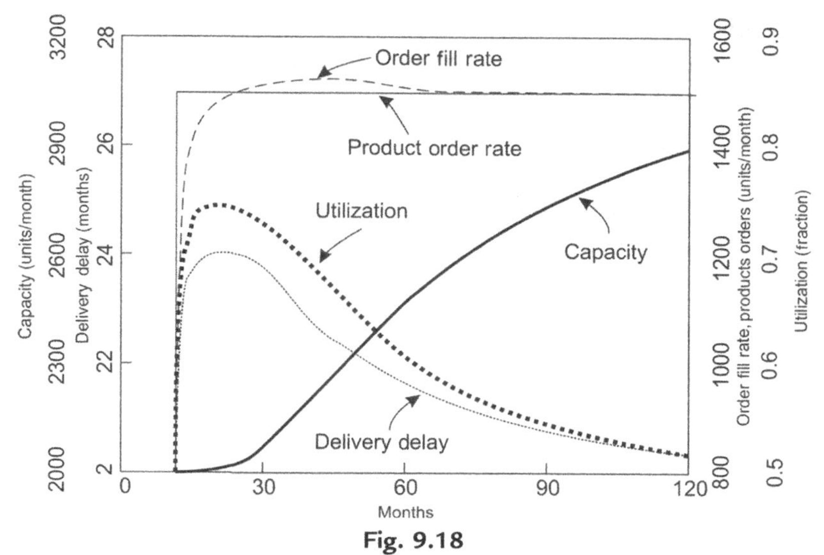

Fig. 9.18

Capacity management with exogenous product order rate. Results of simulation run over the time period of 120 months (Morecroft, 1983).

of 50%. Delivery delay is equal to the operating goal which is set at 2 months. In the 10th month of the simulation, product order rate is increased by 50%. The author traces the adjustment of the manufacturing system over time. The simulation run shows that the demand change is smoothly accommodated. Shortly after the increase in product order rate, capacity utilization increases, thereby allowing order fill rate to rise before any permanent change in the level of capacity has taken place. By month 24, order fill rate is equal to product order rate, meaning that the demand increase has been satisfied. Rising delivery delay after week 10 begins to set in motion a long-term capacity adjustment. As we might expect, capacity rises only gradually, reflecting long delays in capacity acquisition and a reluctance to commit to expansion before there is solid evidence in terms of delivery delay to justify expansion. As capacity arrives in month 24, utilization and delivery delay gradually return to their starting values.

We can interpret Fig. 9.18 to mean that the capacity expansion policy of Eqs. (9.1)–(9.7) is intendedly rational. It brings about a gradual expansion of capacity in response to reliable and conclusive delivery delay information, indicating that expansion is justified.

Experiment 3: Bounded rationality in a complex system (full set of feedback loops)
In the final experiment, Morecroft (1982) put all three policies together starting from the knowledge that the marketing policy can generate growth and that the capacity expansion policy can follow increases in demand. There is good reason to expect a system that will generate continual growth in orders and capacity. In fact, continual growth need not occur, and under extreme conditions, stagnation and decline can set in.

The final simulation run shows the extreme that results in decline. To appreciate this run, the precise conditions under which it was generated should first be described. Customer ordering and marketing perform according to exactly the same rules used in experiment 1. Capacity expansion is slightly modified in two respects in relation to experiment 2. First, delivery delay operating goal DDOG of Eq. (9.3) is made entirely a function of past tradition by setting delivery delay weight DDW equal to 1.

In experiment 2, DDOG was equal to a fixed management goal DDMG. The change is a subtle one that in no way alters the apparent "logic" of the capacity management policy. A condition of excess delivery delay will still elicit capacity expansion, but continued failure to meet the operating goal will result in goal deterioration. Second, capacity expansion fraction CEF in Eq. (9.17) is modified to include a delivery delay bias DDB as shown below:

$$\mathrm{CEF}(t) = f(\mathrm{DDC}(t)) - \mathrm{DDB}) \;\; f(1) = 0, \;\; f' > 0, \;\; f'' > 0 \tag{9.17}$$

In the modified equation, DDE plays the role of a management attitude toward capacity expansion. When DDB is >0, management has a conservative attitude toward capacity expansion, preferring initially to overcome supply shortages by increasing the utilization of capacity rather than ordering extra capacity. Yet, if delivery delay condition rises sufficiently high, capacity will be expanded in the "normal" way. Again, the underlying logic of capacity policy has not changed. Rising delivery delay will still elicit expansion, but the evidence from the delivery must be more compelling than in the case when DDB = 0.

Fig. 9.19 shows the behavior of the complete market growth model over an interval of 120 months. The run starts from a condition in which delivery delay is 4 months—twice the initial operating goal, and a clear sign that capacity expansion is required (even with DDE = 0.3). Capacity expands for the first 18 months of the run. In addition, marketing personnel expand, bringing about a growth in product ordering. The initial pattern of growth is the one we might anticipate from the previous two experiments. However, beginning in month 18, growth in the system begins to falter and eventually declines in both capacity and marketing personnel sets in. This behavior can be explained as a consequence of bounded rationality in both the capacity management and marketing policies. Capacity expansion requires convincing and solid evidence of capacity shortage in the form of high delivery delay—a requirement that is reasonable in isolation. High delivery delay depresses ordering and ultimately inhibits growth of the total budget. With a fixed budget allocation (that worked well in isolation), growth of marketing is restricted, thereby eliminating the primary driving force behind growth. The complete set of policies interacts in a way that fails to bring forth the growth potential of the market. Failure arises because the consequences of a well-intended (intendedly rational) policy within one

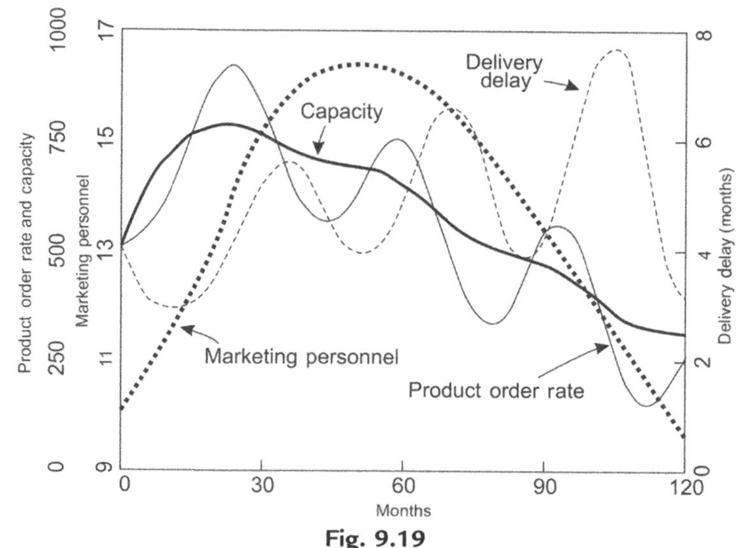

Fig. 9.19
Interaction of marketing, product ordering, and capacity management (Morecroft, 1982).

subunit radiate unintended effects elsewhere in the system. It is this failure due to unintended consequences that is the hallmark of bounded rationality.

To summarize, in this section Morecroft (1982) attempts to show how a system dynamics model can be interpreted in the light of the principle of bounded rationality. He argues that the marketing, capacity expansion, and customer ordering policies of the "Market Growth Model" are formulated recognizing implicitly the information processing limitations of human decision makers. Using simulation runs, Morecroft (1982) has shown that the policies are intendedly rational: the marketing policies promote exponential growth and the capacity policies bring about a conservative adjustment of capacity to demand changes. He has also demonstrated with a final simulation run that intendedly rational policies can produce unintended consequences which are characteristic of bounded rationality. In the market growth model, the unintended consequences lead to market stagnation and decline.

Implications for system dynamics

The principle of bounded rationality is a powerful general law underlying all social systems. Its consequences have been studied and developed in the widely known literature of the Carnegie School. The principle is an implicit part of the structure of system dynamics models. Morecroft (1982) claims that these three statements strongly suggest that the ideas of the Carnegie School have some important implications for the field of system dynamics. There are implications for the communication of the field to other social science disciplines, for the structuring and testing of system dynamics models and for research on generic structures. Each of these themes will be developed in more detail below.

Improved communication to other disciplines

System dynamics is not well-understood or accepted outside the system dynamics community (Morecroft, 1982). Part of the reason for this lack of acceptance is that system dynamics models are not clearly differentiated from the dominant mathematical modeling methods of the social sciences—economics and operations research. System dynamics often deals with areas of application similar to economics and operations research. When analysis yields the results conflicting with the more conventional approaches, the discrepancy is often explained by appeals to the importance of a "feedback approach" or "systems philosophy," neither of which conveys much meaning to those outside the field. According to Morecroft (1982), the Carnegie School offers a language in which the fundamental difference of system dynamics models is explained in terms of the models' treatment of information flow and information processing in decision making.

System dynamics models are built implicitly on the principle of bounded rationality. They portray the bounded rationality of human decision makers and human organizations. They show the distributed responsibility for decision making that is characteristic of real organizations. They contain multiple goals. They use local feedback information in decision making rather than global information drawn widely from the organization and its environment. Decision functions are portrayed as rules of thumb, requiring limited information input and limited computation of that information.

The realization that system dynamics models are portraying bounded rationality provides a basis for explaining why they should be different from models of other dominant social science disciplines. For example, classical microeconomics focuses on portrayals of efficient and rational decision making. In contrast, system dynamics focuses on portrayals of bounded rationality, with the intention of identifying the information structures that are consistent with bounded rationality. Therefore, as a theory of decision making, system dynamics differs sharply from classical economics by assuming that nonrational behavior is both likely to occur and likely to be sustained over time. As a tool for normative analysis, system dynamics differs from classical optimization techniques by setting out to explain why inefficiencies exist, and seeking decision functions that improve on existing behavior, rather than striving for optimal decision functions regardless of the existing decision-making structure of the system (Morecroft, 1982).

Conceptualization: A focus on the organization

System dynamics offers a number of structuring principles (Forrester, 1968a,b) to guide model formulation. These principles are of most value when the boundary of the model has already been set and the major interacting elements already identified. There is very little guidance for the earliest and sometimes most challenging step of initial conceptualization. Morecroft (1982) asks: What features of a situation make it suitable for analysis with system dynamics? Are there patterns of structure that one can anticipate in the construction of a system dynamics model?

In Morecroft's opinion, the process of conceptualization would be greatly aided if system dynamicists clearly recognized that they are building models of human organizations and that those organizations are governed by the principle of bounded rationality. It would then be possible to anticipate both the general form and specific structural features of a model.

In general form, system dynamics models are likely to portray a breadth of organizational structure. They will involve multiple sectors or subunits with divided decision-making responsibility. It is within the complex structure of a multisector system that bounded rationality is most likely to produce major problems in overall system behavior. Problems that are posed within the setting of a single organizational subunit are unlikely to be suitable for analysis with system dynamics.

Specifically, we would expect decision making within subunits to reflect the limits of human rationality. The set of decision functions in a model should embody the multigoal feature of organizations. Subunits will be responsible for different goals. The weight attached to information at different points in the organization will be contingent upon the goals at those points. As a consequence, there is likely to be inconsistency in the weighting schemes of different decision functions which use common sources of information. Decision functions in a given subunit are likely to employ locally available information and to ignore large quantities of information that are conceivably of relevance to decision making, but too difficult or time-consuming to collect.

At this point, let us stress Morecroft's remark that models of biological and ecological systems have also been constructed (Morecroft, 1983). He is suggesting that these should be distinguished from social systems on the grounds that their principles of organization are not necessarily the same as social systems, and the principle of bounded rationality may not apply so strongly here.

As a result, the network of influential information flows in an organization will be sparse and incomplete, relative to the network that would be required for optimal control of the organization. Decision functions are likely to take the form of simple rules of thumb, which place heavy reliance on historical precedent, rather than more fundamental information-intensive planning methods in which a decision is logically and formally tied to specific objectives and constraints of the organization. The observations above are indicative of the structuring aids that can be obtained from a Carnegie perspective. Much work could undoubtedly be done in this area building on the fine structure of decision making that has been described in Carnegie writing, but which is not covered in Morecroft's work. New structuring aids would complement, not contradict, the existing principles of formulation.

Behavior analysis: Making use of intended rationality

The analysis of system dynamics models is traditionally broken into partial and whole model tests. Partial model tests usually perform a purely technical function, enabling the modeler to eliminate formulation errors in a small model rather than unravel the same errors in the more complex setting of the complete model. The explanation of the behavior of the system is made in terms of whole model tests (Morecroft, 1983).

The Carnegie type approach to organizations suggests there may be powerful insights to be derived from contrasting partial and whole model tests. Partial model tests can be viewed as demonstrations of the intended rationality of decision making in organizational subunits. Partial model tests often reveal that the policies of a subunit make perfect sense when the subunit is free to act, independent of other organizational constraints. Whole model tests indicate how intendedly rational policies can break down and produce problem behavior in a sufficiently complex organizational setting. Contrasting partial and whole model tests, to show that rational policies can in fact produce problem behavior, is a powerful method of generating understanding of complex system behavior. Understanding is created by building upon the intuitively clear behavior of a subunit or small group of subunits. As additional subunits are added, a clear explanation can be generated of why policies begin to fail in the more complex setting. An example of the contrast of partial and whole model tests was presented in an earlier paper, in the analysis of the market growth model (Morecroft, 1983). Partial model tests of the marketing and capacity management policies indicated reasonable behavior of the two policies taken in isolation. A full model test involving customer ordering, marketing, and capacity management reveals nonrational behavior in which the firm failed to grow when faced with a limitless market for its product.

Research on generic structures

The ideas of the Carnegie School are likely to be valuable in providing methodological support for the concept of generic structures. In common with other disciplines, system dynamics is seeking order and general structure in social systems with which it is dealing. There is, of course, already a structure that is common to all system dynamics models. They all use the same basic building blocks of levels, physical flows, information flows, and decision functions. However, beyond a common rate level structure, the question remains whether there are larger groupings of basic building blocks that might occur repeatedly in social and economic systems. These larger groupings are described as generic structures (Morecroft, 1982).

To adopt a Carnegie perspective, the question of whether generic structures exist is similar to the question of whether common forms of organization exist. The principle of bounded rationality tells us that organization should evolve around the cognitive limitations of its members. It is probable (though by no means certain) that specific organizational structures have evolved to cope with these fundamental limitations of human decision makers. It is also probable that common problems are generated by these organizational structures (Morecroft, 1983).

The ideas of the Carnegie School lead to a research method for finding generic structures and testing empirically whether they are indeed generic. Morecroft suggests that generic structures should be defined by the distribution of responsibilities within different organizational subunits, by the channels of communication between subunits and by the mental shortcuts embodied in rules of thumb for decision making. If in an application a piece of structure responsible for the problem behavior is observed, we might then dissect the structure to ask whether it can be explained as a consequence of bounded rationality. What is it about the structure, and in particular the assumed complexity of the information network, that limits rationality of decision making? What changes in the information network would be compatible with more rational decision making, and why do they not currently exist? Answers to these questions could form the basis of a refutable empirical study of other organizations similar to the one that yielded the generic structure (Morecroft, 1983).

Conclusion

In Morecroft's (1983) paper, the author sets out to show the parallels between the (Carnegie) philosophy of human decision making and the philosophy of decision making implicit in system dynamics models. He argues that the key Carnegie concept of bounded rationality finds direct expression in the feedback structures that are of concern in system dynamics. He discusses three aspects of bounded rationality in decision making: factored decision making, the use of partial and certain information, and standard operating procedures. He examines the structure of the Market Growth Model (Forrester, 1968a,b) to illustrate that these aspects of bounded rationality are implicit in the formulation of many decision functions in the model.

The final part of (Morecroft, 1982) work is devoted to a discussion of the implications of (Carnegie) philosophy of bounded rationality for system dynamics. He argues that, by explaining feedback structure as a consequence of bounded rationality in the information network, it should be possible to communicate the purpose and results of a system dynamics analysis to a wider audience. He also suggests that a closer study of Carnegie concepts by system dynamicists may add more discipline to the model formulation and new dimensions to the analysis of model behavior.

9.9.3 Appendix 3: System's theory and cybernetics: A quantitative mathematical treatment by S. Mynarski

Stefan Mynarski (1935–2006) is the first scientist who systematically treated theory of systems and cybernetics by quantitative and rigorous mathematical methods. His tools included: functional analysis, methods of stability theory, probability theory, and some aspects of optimal control theory, both deterministic and stochastic.

Mynarski was born on December 1, 1935 in Stara Wieś near Oświęcim, Poland. In 1960, he obtained a master's degree in economics at the Faculty of Commerce of the Higher School of Economics in Kraków, Poland. He was then employed there, initially as an assistant and later, after obtaining in 1966 his Ph.D. degree in economic sciences. The next academic degree—Sc.D. (habilitation) Mynarski received in 1973, and in January 1978, he became a professor of economic sciences. At that time, he headed the Department of Market Analysis of the University of Economics in Kraków, where he published a number of papers on managerial functions.

His publications are of fundamental importance for market and marketing research. They include:

- Cybernetic aspects of market analysis (1972–87)
- Methods of analysis and optimization of trade in goods (1976–87)
- Marketing research methods (1990)
- Market research in competitive conditions (1995)
- Market research in an enterprise (2001)

Mynarski has a special place in the development of marketing and marketing research in Poland in the last 40 years of his scientific achievements. This place results from ontological, methodological and analytical topics related to both the cybernetic and systemic approach to the market as well as to market and marketing research. This view, initiated in the research methodology at the end of the 1960s, is currently experiencing its renaissance in the use of entropy models of market structures, hierarchical multilevel models, or modeling of market interactions in the context of relationship between marketing and service-dominant logic.

An important influence on the career and development of scientific thoughts of S. Mynarski had his research stay in the United States (scholarships granted by the Kosciuszko Foundation from October 1968 to June 1969 at three American universities: the University of California in Berkeley, the University of Pennsylvania in Philadelphia, and The Northwestern University in Chicago).

Mynarski's scientific achievements can be analyzed from the viewpoint of their ontological, methodological, and analytical aspect. From the point of view of ontology, they are close to mechanistic theories of organisms. Both the market and the economy are treated as a mechanism in which (as in biological systems) basic regulatory processes are processes of adaptation and selection in the triadic system of material, energy, and information.

In the methodological perspective, the systemic approach dominates in which the analyzed market systems are examined as hierarchical and interactive systems. This applies to both the subject of research and approaches, research methods, and techniques, under which there is a hierarchical, bi-directional dependency system.

From the analytical point of view, Mynarski's achievements are multidimensional analyzes covering both the level of structural macromodelling of the economy and the level of detailed analytical methods and techniques, such as: multivariate regression models, logistic regression (a method of three sums), factor analysis, or multidimensional correspondence analysis.

Mynarski's (1979) book titled *Elementy Teorii Systemów i Cybernetyki* (*Elements of Systems Theory and Cybernetics*) has provided the present book with information on disintegration (a natural effect of system aging), as well as on proliferation (growth) and progressive integration of systems.

In this writing, the present author exploits a part of internet information authored by Adam Sagan.

9.9.4 Appendix 4: Selten's view of bounded rationality

Modern mainstream economic theory is largely based on an unrealistic picture of human decision making (Selten, 1998, 1999). Economic agents are portrayed as fully rational Bayesian maximizers of subjective utility. This view of economics is not based on empirical evidence, but rather on simultaneous axiomization of utility and subjective probability. However, following Savage, Selten states that "the axioms are consistency requirements on actions with actions defined as mappings from states of the world to consequences" (Savage, 1954). In Selten's opinion, the imposing structure built by Savage has a strong intellectual appeal as a concept of ideal rationality. However, it is incorrect to assume that human beings conform to this ideal. The split of the person into multiple selves with conflicting goals in itself is a bound of rationality for the person as a whole, even if it is not cognitive but motivational. Not only cognitive, but also motivational bounds of rationality must be taken into account by a comprehensive theory of bounded rationality (Selten, 1999).

The further reasoning of Selten adduces Becker (1967) and Becker and Leopold (1996) who have proposed a theory of household behavior which extends aspiration adaptation theory to the right context. The household divides its monthly income into a number of funds for different kinds of expenditures like a fund for food, a fund for clothing, a fund for entertainment, etc. The goal variables are the fund sizes and upper price limits for wants, like the desire for a pair of shoes or an excursion advertised by a travel agency. Such wants are produced by a want generator, modeled as a random mechanism. When a want is generated by the want generator, another instance, the administrator, checks whether there is still enough money in the appropriate fund and whether the want remains under the price limit for such desires.

If the price limit is violated, the want is rejected. If the want remains under the price limit but there is not enough money in the fund, then the want will still be granted if transfer rules permit the transfer of the missing amount from another fund. (The structure of these transfer rules is not explained.) If such a transfer is not permissible, then the want is rejected. At the end of the spending period, a new aspiration level for the next one is formed by aspiration adaptation in the light of recent experience. If the household theory of Becker (1967) is applied to the spending behavior of a single person, then want generator and administrator are different personality components. Conflicts between them are not modeled by the theory, but it may be possible to extend the theory in this direction. Everyday experience suggests that sometimes wants are realized against the will of the administrator. The split of a person into a mechanistically responding want generator and a boundedly rational administrator seems to be a promising modeling approach not only to household theory, but also for other areas of decision making.

In conclusion, Selten (1999) hopes that he succeeded in conveying the essential features of bounded rationality. In the introductory part, he argued that rational decision making within the cognitive bounds of human beings must be nonoptimizing. The exposition of aspiration adaptation theory served the purpose of demonstrating the possibility of a coherent modeling approach to nonoptimizing but, nevertheless, systematic and reasonable boundedly rational behavior.

Finally, Selten (1999) writes: "What is bounded rationality? A complete answer to this question cannot be given in the present state of the art. However, empirical findings put limits to the concept and they indicate in which direction further inquiry should go." For other approaches and opinions see, in particular, Pattillo (1977).

The issue of bounded rationality is still vital. Read current books and the most recent papers on newest writings pro- and contra-bounded rationality after the publication date of the present book.

References

Allison, G.T., 1971. Essence of Decision. Little, Brown and Company, Boston.
Andersen, D.F., 1980. How differences in analytic paradigms can lead to differences in policy conclusions. In: Randers, J. (Ed.), Elements of the System Dynamics Method. MIT Press, Cambridge, MA (see also: Rangers, J. (Ed.), 1980. Elements of the System Dynamics Method. Pegasus Communications, Cambridge, MA).
Andresen, B., Essex, C., 2003. In: Finite-Time Thermodynamics Optimization of Mitochondrial Chemistry and Fuel Cells.Proceedings of ECOS 2003 Copenhagen, Denmark, June 30–July 2, 2003.
Barros, G., 2010. Herbert A. Simon and the concept of rationality: boundaries and procedures. Braz. J. Polit. Econ. 30 (3), 455–472.
Becker, O., 1967. Die wirtschaftliche Entscheidungen des Haushalt. Duncker and Humblot, Berlin.
Becker, O., Leopold, U., 1996. The bounds & likelihood procedure, a simulation study concerning the efficiency of visual forecasting techniques. Cent. Eur. J. Oper. Res. Econ. 4, 2–3.
Bejan, A., Lorente, S., 2008. Design With Constructal Theory. Wiley, Hoboken, NJ.
Cyert, R.M., March, J.G., 1963. A Behavioral Theory of the Firm. Prentice Hall, New Jersey.
Dutton, J.M., Starbuck, W.H., 1971. Computer Simulation of Human Behavior. Wiley, New York.
Eilon, S., 1977. More against optimization. Omega 5 (6), 627–633.
Forrester, J.W., 1961. Industrial Dynamics. Appendix E. MIT Press, Cambridge, MA.
Forrester, J.W., 1968a. Market growth as influenced by capital investment. Sloan Manag. Rev. 9 (2), 83–105.
Forrester, J.W., 1968b. Principles of Systems. MIT Press, Cambridge, MA.
Gaspard, P., 1998. Chaos, Scatterning and Statistical Mechanics. Cambridge University Press, Cambridge, UK.

Gaspard, P., 2004. Fluctuation theorem for nonequilibrium reactions. J. Chem. Phys. 120, 8898.
Gaspard, P., Nicolis, G., 1990. Transport properties, Lyapunov exponents, and entropy unit time. Phys. Rev. Lett. 65, 1693–1696.
Gaspard, P., Claus, I., Gilbert, T., Dorfman, J.R., 2001. The fractality of the hydrodynamic modes of diffusion. Phys. Rev. Lett. 86, 1506–1509.
Georgescu-Roegen, N., 1970. The entropy law and the economic problem. In: Georgescu-Roegen, N. (Ed.), Energy and Economic Myths: Institutional and Analytical Economic Essays. Pergamon, New York, NY, pp. 53–60.
Georgescu-Roegen, N., 1971. The Entropy Law and the Economic Process. Harvard University Press, Cambridge, MA.
Kåberger, T., Månsson, B., 2001. Entropy and economic processes—physics perspectives. Ecol. Econ. 36, 165–179.
Mass, N.J., 1977. Introduction to the Production Sector of the National Model. System Dynamics Group working paper D-2737-1, July, Sloan School of Management. MIT Press, Cambridge, MA.
Meadows, D.H., 1980. The unavoidable a priori. In: Randers, J. (Ed.), Elements of the System Dynamics Method. MIT Press, Cambridge, MA.
Morecroft, J.D.W., 1982. System dynamics: portraying bounded rationality. In: Presented at the 1981 System Dynamics Research Conference, The Institute on Man and Science Rensselaerville, New York, October 14–17 (Revised January 1982).
Morecroft, J.D.W., 1983. System dynamics: portraying bounded rationality. Omega: Int. J. Manag. Sci. 11 (2), 131–142.
Mynarski, S., 1979. Elementy Teorii Systemów i Cybernetyki (Elements of Systems Theory and Cybernetics). Państwowe Wydawnictwo Naukowe, Warszawa.
Pattillo, J.W., 1977. Zero-Base Budgeting. National Assoc. Accountants, New York.
Pierce, S.E., 2002. Nonequilibrium thermodynamics: an alternate evolutionary hypothesis. Crossing Boundaries 1 (2), 49–59.
Savage, L., 1954. The Foundation of Statistics. J. Wiley & Sons, New York.
Selten, R., 1998. Features of experimentally observed bounded rationality. Eur. Econ. Rev. 42, 413–436.
Selten, R., 1999. What is bounded rationality? SFB discussion paper B-454. In: Prepared for the Dahlem Conference 1999.
Sieniutycz, S., 2016. Thermodynamic Approaches in Engineering Systems. Elsevier, Oxford.
Sieniutycz, S., Jeżowski, J., 2009. Energy Optimization in Process Systems, first ed. Elsevier, Oxford (alternatively, see second ed. in 2013 and third ed. in 2018).
Simon, H.A., 1947. Administrative Behavior: A Study of Decision-Making Processes in Administrative Organization, first ed. The Macmillan Company, New York.
Simon, H.A., 1955. A behavioral model of rational choice. Q. J. Econ. 69 (1), 99–118 (compiled by Barros in 2010 and quoted after Simon (1957: 241–260)).
Simon, H.A., 1957. Rationality and decision making. In: Models of Man, Social and Rational: Mathematical Essays on Rational Human Behavior in a Social Setting. John Wiley, New York, p. 198 (see also p. 202 therein).
Simon, H.A., 1976. Administrative Behavior, third ed. Free Press, New York.
Simon, H.A., 1979. Rational decision making in business organizations. Am. Econ. Rev. 69(4).
Wilson, A., 2012. The Science of Cities and Regions: Lectures on Mathematical Model Design. Springer Briefs in Geography, Springer, New York.
Xia, S., Chen, L., Sun, F., 2010. Finite-time exergy with a finite heat reservoir and generalized radiative heat transfer law. Rev. Mex. Fis. 56, 287–296.

Further reading

Argenti, J., 1976. Corporate Col, Colapse. Halsted Press, John Wiley, New York.
Barros, G., 2004. Racionalidade e organizacoes: Um estudo sobre comportamento economico na obra de Herbert A. Simon (MA dissertation). FEA-USP, Sao Paulo.

CHAPTER 10

Some important applications of systems theory ☆

10.1 Is life a manifestation of the second law?

10.1.1 Introduction

In this introductory section, we outline Schneider and Kay's (1994) view of thermodynamic evolution for various developing systems, from primitive physical systems to complex living systems. All these systems involve processes similar to or identical with those which satisfy classical phenomenological formulations of the second law of thermodynamics. In their considerations, however, Schneider and Kay (1994) employ the second law of thermodynamics reformulated by Hatsopoulos and Keenan (1965) and Kestin (1966). Moreover, they extend this reformulation to disequilibrium regimes, where disequilibrium is represented and measured by gradients maintaining a system at some distance away from equilibrium. The reformulated second law implies that as systems are moved away from equilibrium, they will take advantage of all available means to resist externally applied gradients (Schneider and Kay, 1994). When highly ordered complex systems emerge, they develop and grow at the expense of increasing the disorder at higher levels in the system's hierarchy. Schneider and Kay (1994) assure that this resistive behavior, strongly resembling that known in LeChatelier's principle, appears universally in both physical and chemical systems. On this ground, they present a paradigm which provides a thermodynamically consistent explanation of why there is life, including the origin of life, biological growth, the development of ecosystems, and patterns of biological evolution observed in the fossil record. The use of this paradigm is illustrated through a discussion of ecosystem development; as ecosystems grow and develop, they increase their total dissipation, develop more complex structures with more energy flow, increase their cycling activity, develop greater diversity, and generate more hierarchical levels, all to help energy degradation. This view is supported by observation of species which can survive in ecosystems because such species funnel energy into their own production and reproduction and contribute to autocatalytic processes which increase total degradation (occasionally called "dissipation"

☆ Research of E. Schrödinger, J. Kay, E. Schneider, M. Gell-Man, et al.

Complexity and Complex Thermo-Economic Systems. https://doi.org/10.1016/B978-0-12-818594-0.00010-6

rather than degradation) of the energy input to the ecosystem. Summing up, ecosystems develop in ways which systematically increase their ability to degrade the incoming energy, usually solar energy. Schneider and Kay (1994) believe that their thermodynamic paradigm facilitates the study of ecosystems developed from descriptive science to predictive science based on general macroscopic physics. The reader may also note that a suitable and correct description of the above issues can be achieved in terms of information theory, in view of the link between nonequilibrium thermodynamics and Fisher information (Frieden et al., 2002).

Schrödinger (1944) points out that at first glance, living systems seem to defy the second law of thermodynamics as it insists that, within closed systems, entropy should be maximized and disorder should reign. Living systems, however, are the antithesis of such disorder. They display excellent levels of order created from disorder. For instance, plants are highly ordered structures, which are synthesized from disordered atoms and molecules found in atmospheric gases and soils. By turning to nonequilibrium thermodynamics, Schrödinger recognizes that living systems exist well in a world of energy and material and flows (fluxes). An organism stays alive in its highly organized state by taking energy from outside, from a larger encompassing system, and processing it to produce, within itself, a lower entropy, a more organized state. Schrödinger recognizes that life constitutes a far-from-equilibrium system that maintains its local level of organization at the expense of the larger global entropy budget. He proposes that to study living systems from a nonequilibrium perspective would reconcile biological selforganization and thermodynamics. Furthermore, he expects such a study to yield new principles of physics.

The paper by Schneider and Kay (1994) takes on the task proposed by Schrödinger and expands on his thermodynamic view of life. Schneider and Kay explain that the second law of thermodynamics is not an impediment to the understanding of life, but rather is necessary for a complete description of living processes. As they say, they further expand thermodynamics into the causality of living process and assert that the second law is a necessary but not a sufficient cause for life itself. In short, reexamination of thermodynamics shows that the second law underlies and determines the direction of many of the processes observed in the development of living systems. The work of Schneider and Kay (1994) links physics and biology at the macroscopic level and shows that biology is not an exception to physics, thus implying that to date we have probably misunderstood action of some rules of physics in biological context. Regardless of the ultimate opinion of the reader, it would be difficult to overlook a fresh look at thermodynamics in the research of Schneider and Kay (1994, 1995).

Since the time of Boltzmann and Gibbs, there have been major advances in thermodynamics, especially by Caratheodory, Hatsopoulos and Keenan, Kestin, Jaynes, and Tribus. Schneider and Kay (1994) assume the restated laws of thermodynamics of

Hatsopoulos and Keenan (1965) and Kestin (1966) and extend them so that disequilibrium processes and systems can be described in terms of gradients maintaining systems away from equilibrium. In their formulation, the second law states that as systems are moved away from equilibrium, they will use all available avenues to resist externally applied gradients. Schneider's and Kay's 1994 extension of the second law directly applies to complex systems with disequilibrium situations unlike classical statements which are limited to equilibrium or near equilibrium conditions. Away from equilibrium, highly ordered stable complex systems can emerge, develop, and grow at the expense of more disorder at higher levels in the system's hierarchy. These systems show the relevance and utility of these restatements of the second law by considering the classic example of dissipative structures, Benard cells. Their behavior proves that this paradigm can be applied to both physical and chemical systems, and that it allows for a thermodynamically consistent explanation of the development of far-from-equilibrium complex systems including life.

As a case study, Schneider and Kay (1994) select applications of their thermodynamic principles in the science of ecology. They view ecosystems as open thermodynamic systems with a large gradient impressed on them by the sun. The thermodynamic imperative of the restated second law is that these systems strive to reduce this gradient by action of all available physical and chemical processes. Consequently, ecosystems develop structures and functions selected to most effectively degrade or "dissipate" the imposed gradients while allowing for the continued existence of the ecosystem. In particular, they show that the unstressed ecosystem has structural and functional attributes that lead to more effective degradation of the energy provided to the ecosystem. Patterns of ecosystem growth, cycling, trophic structure, and efficiencies are explained by this paradigm.

A rigorous test of Schneider and Kay's hypothesis is the measurement of reradiated temperatures from terrestrial ecosystems. Schneider and Kay (1994) then argue that more mature ecosystems degrade incoming solar radiation into lower quality exergy, that is, have lower reradiated temperatures. Their data and arguments prove that not only are more mature ecosystems better degraders of energy (cooler final effect), but that airborne infrared thermal measurements of terrestrial ecosystems may offer a major breakthrough in providing measures of ecosystem health or integrity.

10.1.2 Classical thermodynamics and the second law

As Schneider and Kay's basic ideas are built on the principles of original thermodynamics, their work starts with a brief discussion of classical thermodynamics. They believe that their approach to the theoretical issues of nonequilibrium thermodynamics is original, and this is also the opinion of the present author. In fact, their approach permits a more satisfactory discussion of the observed phenomena than the classical nonequilibrium thermodynamics.

The second law dictates that if there are any physical or chemical processes underway in a system, then the overall quality of the energy in that system will degrade, although the total quantity of energy in a closed system will remain unchanged. The second law can be stated as: Any real process can only proceed in a direction which ensures a nonnegative entropy source (De Groot and Mazur, 1984). It is the property of the nonnegativeness of entropy source rather than entropy itself which draws the "arrow of time" in nature (Eddington, 1958) and defines the extent to which nature becomes more disordered or random (Schneider and Kay, 1994). For other formulations and discussion of the significance of so-called natural variables, see Callen (1988) and Sieniutycz (2016).

All natural processes can be viewed in light of the second law and in all cases this one-sided aspect of nature is observed. Schneider and Kay (1994) recall some elementary examples, e.g., spontaneous flow of heat from a hotter reservoir to a colder reservoir until there is no longer a temperature difference or gradient. Another example is gas flow from a high pressure to a low pressure until there is no longer a pressure difference or gradient. If one mixes hot and cold water, the mixture comes to a uniform temperature. The resulting lukewarm water will not spontaneously unmix itself into hot and cold portions. Boltzmann would have restated the above example as: it is highly improbable that water will spontaneously separate into hot and cold portions, but it is not impossible (Schneider and Kay, 1994).

Boltzmann recasts thermodynamics in terms of energy microstates of matter. In this context, entropy reflects the number of different ways microstates can be combined to give a particular macrostate. The larger the number of microstates for a given macrostate, the larger the entropy. Consider a 10 compartment box with 10,000 marbles in one of the 10 compartments and the rest of the compartments being empty (Schneider and Kay, 1994). If doors are opened between the compartments and the box is subjected to a pattern of random shaking, one would expect, over time, to see a distribution of about 1000 marbles per compartment, the distribution which has the largest number of possible microstates. This randomization of the marbles to the equiprobable distribution corresponds to the macrostate with the maximum entropy for the closed system. If one continued the shaking, it would be highly improbable but not impossible for all the marbles to reseparate themselves into the low-entropy configuration with 10,000 marbles in one compartment. The same logic is applied by Boltzmann to explain the macroscopic phenomena of thermodynamics in terms of microstates of matter. Systems tend to the macrostate which has the largest number of corresponding accessible microstates.

10.1.3 Extended laws of thermodynamics

In 1908, thermodynamics was moved a step forward by the Caratheodory's work (Caratheodory, 1976) when he developed a proof that the law of "entropy increase" is not the generalization of the second law. The more encompassing statement of the second law of thermodynamics is that: "In the neighbourhood of any given state of any closed system, there exist states which are inaccessible from it along any adiabatic path, reversible or irreversible." This statement of the second law, unlike earlier ones, does not depend on the concepts of entropy or temperature and applies equally well in the positive and negative temperature regimes.

More recently, Hatsopoulos and Keenan (1965) and Kestin (1966) put forward a principle which subsumes the 0th, 1st, and 2nd laws: "When an isolated system performs a process after the removal of a series of internal constraints, it will reach a unique state of equilibrium: this state of equilibrium is independent of the order in which the constraints are removed." This statement is called the Law of Stable Equilibrium by Hatsopoulos and Keenan and the Unified Principle of Thermodynamics by Kestin (Schneider and Kay, 1994).

The importance of the statement is that, unlike all earlier assertions that all real processes are irreversible, it dictates a direction and an end state for all real processes. All previous formulations of the second law tell us what systems cannot do (Schneider and Kay, 1994). The present statement tells us what systems will do. An example of this phenomena are two flasks, connected with a closed stopcock (Schneider and Kay, 1994). One flask holds 10,000 molecules of a gas. Upon removing the constraint (opening the stopcock), the system will come to its equilibrium state of 5000 molecules in each flask, with no gradient between the flasks.

The principles described so far hold for closed isolated systems. However, a more interesting class of phenomena belong to systems that are open to energy and/or material flows and reside at stable states some distance from equilibrium.

These much more complex thermodynamic systems are the ones investigated by Prigogine and his collaborators (Nicolis and Prigogine, 1977, 1989). These systems are open and are removed from equilibrium by fluxes of material and energy across their boundary. They maintain their form or structure by continuous dissipation of energy and, thus, are known as dissipative structures. Prigogine claimed that nonequilibrium systems, through their exchange of matter and/or energy with the outside world, can maintain themselves for a period of time away from thermodynamic equilibrium in a locally reduced-entropy steady-state. This is done at the cost of increasing the entropy of the larger "global" system in which the dissipative structure is imbedded; thus, following the mandate of the second law that overall entropy must increase. Nonliving organized systems (like convection cells, tornados, and lasers) and living systems (from cells to ecosystems) are dependent on outside energy fluxes to maintain their organization in a locally reduced entropy state.

Nonetheless, as Schneider and Kay (1994) rightly state, Prigogine's description of dissipative structures is *formally limited to the neighborhood of equilibrium*. This is because Prigogine's analysis depends on a linear expansion of the entropy function about equilibrium. This is a severe restriction on the application of his theory, which, in particular, precludes its formal application to living systems.

To deal with the thermodynamics of nonequilibrium systems, Schneider and Kay (1994) propose the following corollary that follows from the proof by Kestin of the Unified Principle of Thermodynamics; his proof shows that a system's equilibrium state is stable in the Lyapunov sense. Implicit in this "concluding corollary" is that a system will resist being removed from the equilibrium state. The degree to which a system has been moved from equilibrium is measured by gradients imposed on the system. According to Schneider and Kay (1994): *The thermodynamic principle which governs the behaviour of systems is that, as they are moved away from equilibrium they will utilize all avenues available to counter the applied gradients. As applied gradients increase, so does the system's ability to oppose further movement from equilibrium.*

In the discussion that follows Schneider and Kay (1994), we shall refer to the above statement for the thermodynamic principle as the "restated second law." (The pre-Caratheodory statements, i.e., limited-value claims that entropy should increase, are referred to as the classical second law.)

A simple example of the relevant phenomenon is the Benard cell (Chandrasekhar, 1961). The experimental apparatus for studying the Benard cell consists of a highly instrumented insulated container enclosing a fluid. The bottom of the container is a heat source and the top is a cold reservoir. When the fluid is heated from below, it resists the applied gradients (ΔT) by dissipating heat through conduction. As the gradient is increased, the fluid develops convection cells. These structures increase the rate of dissipation. They are illustrated on Fig. 10.1 and called Benard cells (Chandrasekhar, 1961).

There are many graphs in the literature illustrating instability phenomena of this sort. For an example the reader is referred to Fig. 2c in Schneider and Kay (1994) that shows a plot of a gradient (represented by the Rayleigh number, Ra), which is proportional to the temperature gradient ΔT against the available work expended in maintaining this gradient. The dynamics of the system are such that it becomes more and more difficult to move the system away from equilibrium. That is, proportionally more available work must be spent for each incremental increase in gradient as the system gets further from equilibrium, i.e., ΔT increases.

In chemical systems, LeChatelier's principle is another example of the restated second law. In his lectures on thermodynamics, Fermi (1956) points out that the effect of a change in external conditions on the equilibrium of a chemical reaction is prescribed by LeChatelier's principle. "If the external conditions of a thermodynamic system are altered, the equilibrium of the system will tend to move in such a direction as to oppose the change in the external conditions." Fermi (1956) stresses that if a chemical reaction were exothermal (i.e., $A+B = C + D+\text{heat}$), an increase in temperature will shift the chemical equilibrium to the left hand side. Since the reaction from left to right is exothermal, the displacement of the equilibrium toward the left results in the absorption of heat and opposes the rise in

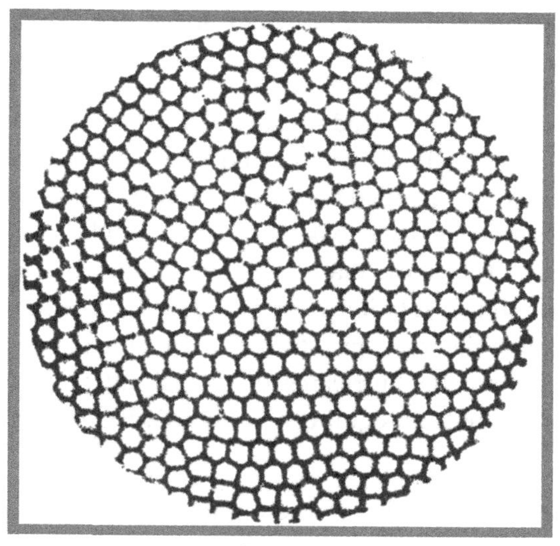

Fig. 10.1
The spontaneous organization of water due to convection: once convection begins and the dissipative structure forms a pattern of hexagonal Bénard cells appear (compare with Schneider and Kay, 1994, Fig. 1; Capra, 1996, Fig. 5.1, p. 87, Sieniutycz, 2016, Fig. 5.9, p. 273).

temperature. Similarly, a change in pressure (at a constant temperature) results in a shift in the chemical equilibrium of reactions which opposes the pressure change.

Thermodynamic systems exhibiting temperature, pressure, and chemical equilibrium resist movement away from these equilibrium states. When moved away from their local equilibrium state, they shift their state in a way which opposes the applied gradients and moves the system back toward its local equilibrium attractor. The stronger the applied gradient, the greater the effect of the equilibrium attractor on the system. The reason that the restatement of the second law is a significant step forward for thermodynamics is that it tell us how systems will behave as they are moved away from equilibrium. Implicit in this is that this principle is applicable to nonequilibrium systems, something which is not true for classical formulations of the second law (Schneider and Kay, 1994).

In fact, their "restated second law" avoids the problems associated with state variables such as entropy which are only defined for equilibrium. Their restatement of the second law sidesteps the problems of defining entropy and entropy production in nonequilibrium systems, an issue that has plagued nonequilibrium thermodynamics for years. By focusing on gradient destruction, Schneider and Kay avoid the problems encountered by Prigogine (1955), and more recently Swenson (1989), who use extremum principles based on the concept of entropy to describe self-organizing systems. Nonequilibrium systems can be described by their forces and requisite flows using the well-developed methods of network thermodynamics (Katchalsky and Curran, 1965; Peusner, 1986; Mikulecky, 1984).

10.1.4 Dissipative structures, degraders, and related problems

Prigogine and his coworkers have shown that dissipative structures self-organize through fluctuations, small instabilities which lead to irreversible bifurcations and new stable system states (Glansdorff and Prigogine, 1971; Nicolis and Prigogine, 1977, 1989). In this sense, the future states of such systems are regarded as not deterministic. Dissipative structures are stable over a finite range of conditions and are sensitive to fluxes and flows from outside the system. Glansdorff and Prigogine (1971) show that these thermodynamic relationships are best represented by coupled nonlinear relationships, i.e., autocatalytic positive feedback cycles, many of which lead to stable macroscopic structures which exist away from the equilibrium state. Convection cells, hurricanes, autocatalytic chemical reactions, and living systems are all examples of far-from-equilibrium dissipative structures which exhibit coherent behavior.

The transition between conduction and convection in a heated fluid is a striking example of emergent coherent organization in response to an external energy input. A thorough analysis of these simple physical systems allows to develop a number of general thermodynamic principles applicable to the development of complex systems as they emerge at some distance away from equilibrium (Schneider and Kay, 1994). Benard and Rayleigh (Chandrasekhar, 1961; Silveston, 1957; Brown, 1973) conducted carefully designed experiments to study this transition.

The lower surface of the experimental apparatus is heated and the upper surface is kept at a cooler temperature; see, e.g., Fig. 1 in Schneider and Kay (1994). Hence, a temperature gradient is induced across the fluid. The initial heat flow through the system is controlled by conduction; the energy is then transferred by molecular interaction. When the heat flux reaches a critical value of the temperature gradient, the system becomes unstable and convective overturning emerges. These convective structures result in highly structured coherent hexagonal surface patterns (Benard cells) in the fluids. This coherent kinetic structuring increases the rate of heat transfer and gradient destruction in the system.

As shown by Schneider and Kay (1994), this transition between noncoherent, molecule to molecule, heat transfer to a coherent structure results in excess of 10 molecules acting together in an organized manner. This seemingly improbable occurrence is the direct result of the applied temperature gradient and is the system's response to attempts to move it away from equilibrium.

Schneider and Kay (1994) studied the Benard cell phenomena in detail, using the original data sets collected by Silveston (1957) and Brown (1973). Their analyses are significant, in that they calculated for the first time the entropy production, exergy drop, and available work destruction, resulting from these organizing events (see Fig. 2 in Schneider and Kay, 1994). This analysis shows that as the gradient increases, it becomes harder to increase this gradient.

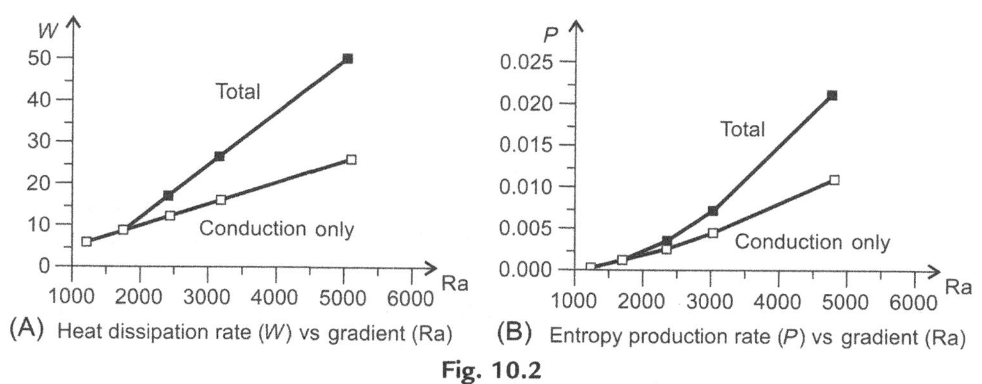

Fig. 10.2
(A) The heat dissipation rate W(J/s) and (B) entropy production rate P(J/s/K) versus Rayleigh number Ra as nondimensional measure of the temperature gradient. The sketch is based on experimental data of Silveston (1957) and Brown (1973).

Initially, the temperature gradient in the apparatus is accommodated solely by random conductive activity. When the gradient is raised, a combination of factors including surface tension effects and gravitational fluid instability converts the system to a mixed conductive-convective heat transfer system. The transition to coherent behavior occurs at $Ra = 1760$. See Table 1 and Figs. 1 and 2 in Schneider and Kay (1994) and Fig. 10.2 below, which show behavior of heat dissipation rate (W) and power as important part of these results.

As shown in Fig. 10.2 and in Fig. 2 of Schneider and Kay (1994), with the onset of convection there is a dramatic increase in the heat transfer rate across the system. From the literature, especially Chandrasekhar (1961), and also from Schneiders and Kay's comprehensive analysis of the Brown's and Silveston's data, Brown (1973) and Silveston (1957), we can observe the following behavior for these systems:

1. Heat dissipation rate (transfer of heat between the plates, W) is a linearly increasing function of the gradient ΔT (see Fig. 2a in Schneider and Kay (1994), recalling that Ra is proportional to ΔT).
2. Entropy production rate, P, vs ΔT increases in a nonlinear way (see Fig. 2b therein).
3. Destruction rate of the classical exergy Θ vs gradient ΔT increases in a nonlinear way, the shape of the curve being the same as P vs ΔT.
4. Points 2 and 3 imply that with the gradient increase, it is harder (requires more available work) to incrementally increase the gradient. The further from equilibrium the system is, the more it resists being moved further from equilibrium. In any real system, there is an upper limit to the gradient which can be applied to the system.
5. Once convection occurs, the temperature profile within the fluid is vertically isothermal outside the boundary layer, i.e., the temperature in the convection cells is constant, thus effectively removing the gradient through most of the fluid.

6. As the gradient increases, further critical points are reached. At each critical point, the boundary layer depth decreases.
7. Point 1 is valid because the rate of heat transfer is controlled by the rate of heat flow across the boundary layer, i.e., by conduction which is a linear process. This process is also responsible for most of the entropy production, as there will be little production due to convection. The slope change is caused by the decrease in the boundary layer depth at a mode change (critical point). (Recall that $Q = k\Delta T/l$, thus as l decreases, slope k/l increases.)
8. Nusselt number, Nu, equals $Q/Q_c = P/P_C = \Theta/\Theta_C$, that is, in Benard cells, the increase in dissipation at any point due to the emergent process is equal to the increase in degradation at any point due to the emergent process. This is true for any process which involves only heat transfer. Otherwise degradation differs from dissipation. (As noted by Schneider and Kay (1994), Prigogine at times mistakenly used these terms interchangeably).
9. The principle governing these systems is not one of maximum entropy production, but rather one of entropy production change being positive semidefinite as you increase the gradient. See Point 7 above. The interesting question is, how much structure emerges for a given gradient, and how much resistance exists to increasing the gradient further.

As the temperature difference increases, there are a number of further transitions at which the hexagonal cells reorganize themselves so that the cost of increasing the temperature gradient escalates even more quickly. Ultimately, the system becomes chaotic and the dissipation is maximum in this regime.

The point of this example is that in a simple physical system, new structures emerge which better resist the application of an external gradient. The Benard cell phenomenon is an excellent example of the nonequilibrium restated second law (Schneider and Kay, 1994). Other physical, chemical, and living systems exhibit similar behavior.

The more a system is moved from equilibrium, the more sophisticated are its mechanisms for resisting being moved from equilibrium. This behavior is not sensible from a classical second law perspective, but this behavior is predicted by the restated second law. No longer is the emergence of coherent self-organizing structures a surprise, but rather it is an expected response of a system as it attempts to resist and degrade externally applied gradients which would move the system away from equilibrium. See Schneider and Kay (1994) for a more comprehensive discussion which, in particular, mimics meterological phenomena.

Until now, we have focused the discussion on simple physical systems and how thermodynamic gradients drive selforganization. The literature is replete with similar phenomena in dynamic chemical systems. Ilya Prigogine and many other representatives of the Brussels School (Glansdorff and Prigogine, 1971; Nicolis and Prigogine, 1977, 1989) have documented the thermodynamic behavior of these chemical reaction systems. Chemical gradients arise in dissipative autocatalytic reactions, examples of which are found in simple inorganic chemical systems, in protein synthesis reactions, or phosphorylation, polymerization, and hydrolytic autocatalytic reactions (Schneider and Kay, 1994).

On the other hand, Månsson and Lindgren (1990) stress the role of the transformed information for creation processes by showing how destructive processes of entropy production are related to creative processes of formation. Nonlinear molecular mechanism gives rise to critical points, instabilities, bifurcations, chaotic behavior, and oscillations. It leads, in general, to various forms of organization (Ebeling, 1985; Ebeling and Klimontovich, 1984; Ebeling and Feistel, 1992; Ebeling et al., 1986; Nicolis and Prigogine, 1977; Zainetdinov, 1999).

A theoretical justification for maximum entropy production (MEP) has recently been derived from information theory (Dewar, 2003) which applied Jaynes' information theory formalism of statistical mechanics to nonequilibrium systems in steady state to show that, out of all possible macroscopic stationary states compatible with the imposed constraints (e.g., external forcing, local conservation of mass and energy, global steady-state mass and energy balance, etc.), the state of maximum entropy production is selected because it is statistically the most probable, i.e., it is characteristic of the overwhelming majority of microscopic paths allowed by the constraints.

Berezowski (2003) presents three sorts of fractal solutions to the model of recirculation nonadiabatic tubular chemical reactors with mass or thermal feedback. (This feedback is accomplished either by recycling a mass stream or by an external heat exchanger.) The first sort of fractal solutions concerns the structure of Feigenbaum's diagram at the limit of chaos. The second and third sorts concern the effect of initial conditions on the dynamic solutions of models. In the course of computations, two types of recirculation are considered: the recirculation of mass (return of a part of product stream) and recirculation of heat (heat exchange in the external heat exchanger). Maps of fractal images are obtained for this complex system.

Systems of autocatalytic reactions are a form of positive feedback, where the activity of the system or reaction augments itself in the form of self-reinforcing reactions. An example is a reaction where a compound A catalyzes the formation of compound B and B accelerates the formation of A; then the overall set of reactions is an autocatalytic or positive feedback cycle. Ulanowicz (1986) notes that, in autocatalysis, activity of any element in the cycle engenders greater activity in all the other elements, thus stimulating an aggregate activity of the whole cycle. Such self-reinforcing catalytic activity is self-organizing and is an important way of increasing the dissipative capacity of the system. Cycling and autocatalysis are fundamental in nonequilibrium systems (Schneider and Kay, 1994).

The notion of dissipative systems as gradient dissipators holds for disequilibrium physical and chemical systems and describes processes of emergence and development of complex systems (Schneider and Kay, 1994). Not only are the processes of these dissipative systems consistent with the restated second law, it is expected that they exist wherever there are gradients.

For a model of transport phenomena and chemical reaction, Burghardt and Berezowski (2003) provide analysis and numerical simulations which reveal the possibility of a generation of homoclinic orbits, the result of homoclinic bifurcation. To this end, they propose a method for the development of special type of diagrams—the so-called bifurcation diagrams. The diagrams comprise the locus of homoclinic orbits with lines of limit points bounding the region of multiple steady states, as well as the locus of points of Hopf's bifurcation. The researchers also define a parameter domain for which the homoclinic bifurcation can take place and facilitate determining conditions under which the homoclinic orbits arise. Two kinds of homoclinic orbits are observed: semistable and unstable orbits. The nature of the homoclinic orbit depends on the stability properties of the limit cycle linked with the saddle point. Interesting dynamics, illustrated in the solution diagrams and phase diagrams, are associated with two kinds of homoclinic orbits.

The unstable homoclinic orbit surrounding stable stationary point appears as a result of the merger between an unstable limit cycle and a saddle point surrounding stable stationary point. This orbit plays the role of a closed separatrix, dividing the region of state variables in the phase diagram into two subregions with two attractors. Two stable stationary points or a stable stationary point inside the orbit and a stable limit cycle outside the homoclinic orbit can be these attractors. On the other hand, merger of a stable limit cycle and a saddle point leads to the formation of a semistable homoclinic orbit which is stable on the inside and unstable on the outside. The internal stability of the orbit leads to the conclusion that the trajectories originated from the region surrounded by a semistable homoclinic orbit tend to this orbit and, consequently, to the unstable saddle point (this takes, obviously, an infinitely long time). A region of state variables is thus created with the special property: the trajectories issued from this region can reach an unstable stationary point—a phenomenon impossible without the existence of a homoclinicorbit. The authors develop stability analysis of the steady-state solutions for porous catalytic pellets in which they test the role of shape of the pellet.

Nicolis and Prigogine (1977) approach to *Self-Organization in Nonequilibrium Systems* diffuses quickly to the ecology and soon receives its own characteristic picture. Prigogine's name is best known for extending the second law of thermodynamics to systems that are far from equilibrium and implying that new forms of ordered structures could exist under such conditions. Prigogine calls these forms "dissipative structures," pointing out that they cannot exist independently of their environment. Nicolis and Prigogine (1977) show that the formation of dissipative structures allows order to be created from disorder in nonequilibrium systems. These structures have since been used to describe many phenomena of biology and ecology. It has been shown that the competition of dissipation and damping with restoring forces leads to oscillations and limit cycles. In numerous references, the reader can study the early history and the debate surrounding oscillating reactions (Glansdorff and Prigogine, 1971; Nicolis and Prigogine (1977, 1979); Nicolis and Rouvas-Nicolis (2007a,b), and many others). Applying Lyapounov functions, Nicolis and Prigogine (1979) develop a stability approach to irreversible processes with disequilibrium steady states. They argue that the competition of dissipation and damping with restoring forces lead to oscillations and limit cycles surrounding unstable fixed points.

Nicolis and Rouvas-Nicolis (2007a,b) show that the same basic approach may be used for molecular collisions, hydrodynamics, and chemical or electrochemical processes. Nonlinear molecular mechanism gives rise to critical points, instabilities, bifurcations, chaotic behavior, and oscillations. It leads, in general, to various forms of organization (Ebeling, 1985, 2002; Ebeling and Klimontovich, 1984; Ebeling and Feistel, 1992; Ebeling et al., 1986; Klimontovich, 1999; Nicolis and Prigogine, 1977; Zainetdinov, 1999).

Klimontovich (1986, 1991) compares entropies of laminar and turbulent motions with respect to the velocity of a laminar flow and the velocity of an averaged turbulent flow, for the same values of the average kinetic energies. He shows that when transition occurs from laminar flow to turbulent flow, the entropy production and entropy itself decrease. This indicates that the disequilibrium phase transition from laminar flow to turbulent flow transfers the system to a more ordered state. He also proves that the turbulent motion should be thought as motion having a lower temperature than laminar motion. Klimontovich's (1999) selforganization treatment using his "norm of chaoticity" is carried out on a number (classical and quantum) of examples of physical systems. An example of a medical-biological system is considered.

Ebeling (2002) studies a special class of nonequilibrium systems which allows to develop ensemble theory of some generality. For these so-called canonical-dissipative systems, driving terms are determined by invariants of motion. Firstly, he constructs systems ergodic on certain surfaces in the phase space. These systems, which are described by a nonequilibrium microcanonical ensemble, correspond to an equal distribution on the target surface. Next, he constructs and solves Fokker-Planck equations. In the last part, he discusses a special realization: systems of active Brownian particles. In conclusion, this work is devoted to the study of canonical-dissipative systems which are pumped from external sources with free energy. Special canonical-dissipative system is constructed whose solution converges to the solution of a conservative system with given energy or other prescribed invariant of motion. In this way, he is able to generate disequilibrium states characterized by prescribed invariants of mechanical motion. Then he postulates the existence of distributions analogous to the equilibrium microcanonical ensemble. Further, he finds solutions of the Fokker-Planck equation which may be considered as analogues of canonical equilibrium ensembles. He proposes calling these distributions canonical-dissipative ensembles. With the help of explicit disequilibrium solutions, he constructs for canonical-dissipative systems a complete statistical thermodynamics including an evolution criterion for the nonequilibrium.

Ebeling (1983) compares with experiments the theory of hydrodynamic turbulence of Klimontovich (1982), which yields effective turbulent viscosity as a linear function of the Reynolds number. This theory seems to describe well the observed effective viscosity of vortices and, after modifications, the fluid flow in tubes up to very high Reynolds numbers.

Ebeling's and Feistel's (2011) updated book on selforganization and evolution includes a review of relevant research. It retains the original fascination surrounding thermodynamic systems far from equilibrium, synergetics, and the origin of life. While focusing on the physical and theoretical modeling of natural selection and evolution processes, the book covers in detail experimental and theoretical fundamentals of self-organizing systems as well as such selected features as random processes, structural networks, and multistable systems. The authors take examples from physics, chemistry, biology, and social sciences. The result is a resource suitable for readers in physics and related interdisciplinary fields, including mathematical physics, biophysics, information science, and nanotechnology.

Entropy and entropy lowering, in comparison to their equilibrium values, are good indicators/ measures of disequilibrium. Ebeling and Engel-Herbert (1986) prove that entropy decreases with excitation under the condition of fixed energy. They also show that entropy lowering refers to the contraction of the occupied part of the phase space due to the formation of attractors. Excitation of oscillations in solids and turbulence in liquid flows serve as examples. The entropy statement formulated by Klimontovich is discussed in the turbulence context. Two examples are given: (1) self-oscillations of nonlinear oscillators, and (2) laminar and turbulent flows in tubes.

Applying his turbulence theory (Klimontovich, 1982), Klimontovich (1991) examines the link between the turbulent motion and the structure of chaos. Klimontovich (1999) considers entropy, information, and relative criteria of order for states of open systems. He points out that two meanings of the word "information" are known in the theory of communication. The first meaning coincides in form with Boltzmann's entropy. The second refers to difference between unconditional and conditional entropies. Two kinds of open systems are considered. Systems of first kind, with zero value of a controlling parameter C, are in equilibrium. Systems of second kind, with a finite value of C, are in disequilibrium. In selforganization, associated with escaping from equilibrium, information is increased. For open systems and all values of C, a conservation law holds for the sum of information and entropy with all values of the controlling parameter.

Zainetdinov (1999) attempts to attribute informational neg-entropy to selforganization processes in open systems. Zalewski and Szwast (2007) find chaos and oscillations in selected biological arrangements of the prey-predator type. They consider two models of the system.

Dependence of the average concentrations on the parameters of the models is discussed. The areas of the maximal values of rate constants in particular models are determined. Zalewski and Szwast (2008) consider chaotic behavior in a stirred tank reactor with first-order exothermic chemical reaction ($A \rightarrow B$) running in a mixture with deactivating catalyst particles. The reactor is continuous with respect to reacting mixture and periodic with respect to catalyst particles. A temperature-dependent catalyst deactivation, subject to the Arrhenius-type equation describing catalyst deactivation rate, is assumed. The direction of displacement areas in which chaos is generated is studied. Graphs show that, for older catalysts, the areas in which chaos is generated move toward higher temperatures.

More information about diversity of chemical phenomena can be found in a book on thermodynamic approaches in engineering systems (Sieniutycz, 2016, Chapters 5 and 9). Chapter 5 deals mainly with Lapunov functions and stability of paths rather than fields, whereas a part of Chapter 9 (Sections 9.7–9.9) provides an account of stabilities, instabilities, and chaos in fields, i.e., distributed chemical systems.

10.1.5 Living systems as special gradient dissipators

We shall now return to Schneider and Kay (1994) paper for a brief discussion of their results on living systems as gradient dissipators, and then will pass to their description as energy degraders. We shall terminate with conclusions regarding living systems and life in the contemporary molecular-genetic research program. Life will be understood as a composition of processes sustained by a balance between the imperatives of survival and energy degradation, supporting the "order from disorder" premise and linking biology with physics, as defined by Schneider and Kay (1994).

Our attention will be focused first on the role of thermodynamics in the evolution of living systems. Boltzmann, the creator of statistical thermodynamics, recognized the apparent contradiction between the thermodynamically predicted randomized cold death of the universe and the existence of life in nature by which systems grow, complexify, and evolve, all of which reduce their internal entropy. He observed in 1886 that the thermal gradient on Earth, caused by the energy provided by the sun, drives the living process and suggested a Darwinian-like competition for entropy in living systems:

> Between the earth and sun, however, there is a colossal temperature difference; between these two bodies, energy is thus not at all distributed according to the laws of probability. The equalization of temperature, based on the tendency towards greater probability, takes millions of years, because the bodies are so large and are so far apart. The intermediate forms assumed by solar energy, until it falls to terrestrial temperatures, can be fairly improbable, so that we can easily use the transition of heat from sun to earth for the performance of work, like the transition of water from the boiler to the cooling instillation. The general struggle for existence of animate beings is therefore not a struggle for raw materials-these, for organisms,

> are air, water and soil, all abundantly available - nor for energy which exists in plenty in any body in the form of heat (albeit unfortunately not transformable), but a struggle for entropy, which becomes available through the transition of energy from the hot sun to the cold earth. In order to exploit this transition as much as possible, plants spread their immense surface of leaves and force the sun's energy, before it falls to the earth's temperature, to perform in ways yet unexplored certain chemical syntheses of which no one in our laboratories has so far the least idea. The products of this chemical kitchen constitute the object of struggle of the animal world.
>
> ***(Boltzmann, 1886)***

As noted by Schneider and Kay (1994), Boltzmann's ideas were further explored by Schrödinger in his famous *What is Life?* book (Schrödinger, 1944). Similarly, Schrödinger was perplexed because he also noted that some systems, like life, seem to defy the classical second law of thermodynamics, yet, he recognized that living systems are open systems of classical thermodynamics. An organism stays alive in its highly organized state by taking energy from outside itself, that is, from a larger encompassing system, and processing it to produce a lower entropy state within itself. Therefore, *life can be viewed as a far-from-equilibrium dissipative structure that maintains its local level of organization at the expense of producing entropy in the larger system it is part of* (Schneider and Kay, 1994). If the Earth is viewed as an open thermodynamic system with a large gradient impressed on it by the sun, the thermodynamic command of the restated second law is that the system will strive to reduce this gradient by using all physical and chemical ways available to it (Schneider and Kay, 1994). In fact, self-organizing processes are effective means of reducing gradients. Consequently, a large fraction of scientists tends to support the opinion that *life exists on earth as another means of dissipating the solar induced gradient and, as* such, *is a manifestation of the restated second law* (Schneider and Kay, 1994). Living systems are far-from-equilibrium dissipative structures which have great potential of reducing radiation gradients on Earth (Ulanowicz and Hannon, 1987). Much of this special dissipation is accomplished by plants ($<1\%$ of it through photosynthesis).

The above features are not far from those observed for solar radiation engines. For these systems, useful conclusions are obtained via sequential systems theory (Sieniutycz and Jeżowski, 2009). This theory attributes system motions to influxes of solar and substantial streams, which drive engines as stages of whole power system. The final outcome is the transformation of flowing matter or radiation into mechanical power. The mechanical movements may be parts in various forms of motions (Shiner and Sieniutycz, 1994), as in, e.g., muscle contraction problems in living organisms (Shiner and Sieniutycz, 1997).

Consider various aspects of mechanical power yield in thermodynamic engines and chemical reactors, such as: synthesis of maximum power problems (Sieniutycz, 2012), Lapunov functions and stability of paths and fields (Sieniutycz, 2016), selection of controls in reactors (Sieniutycz, 2016), optimization in process systems (Sieniutycz and Jeżowski, 2009), optimization of process systems and fuel cells (Sieniutycz and Jeżowski, 2013), and optimization of unit operations, fuel cells, and heterogeneous chemical reactors (Sieniutycz and Jeżowski, 2018).

10.1.6 The origin of life: A brief introduction

The origin of prebiotic life may be regarded as the development of another route for the dissipation of induced energy gradients. Life with its requisite ability to reproduce insures that these dissipative pathways continue having evolved strategies to maintain the dissipative structures in the face of a fluctuating physical environment. Schneider and Kay (1994) suggest that living systems are dynamic dissipative systems with encoded memories, the gene with its DNA, which allow dissipative processes to continue without having to restart the dissipative process via stochastic events. They see living systems as sophisticated mini-tornados, with a memory (its DNA), whose Aristotelian "final cause" may be the second law of thermodynamics. However, the role of thermodynamics in living processes should not be exaggerated. In their summarizing considerations on the origin of life, Schneider and Kay (1994) maintain that "the restated second law is a necessary but not a sufficient condition for life," and, in addition, "Life should be viewed as the most sophisticated (until now) end in the continuum of development of natural dissipative processes from physical to chemical to autocatalytic to living systems."

Life should not be seen as an isolated event. Rather it is the emergence of yet another class of processes aimed at the degradation/consumption of energy attributed to thermodynamic gradients. Life should be viewed as the most sophisticated (until now) end in the continuum of development attributed to natural dissipative structures from physical to chemical to autocatalytic to living systems.

Schneider and Kay (1994) believe that autocatalytic chemical reactions are the backbone of chemical systems leading to origins of life. They assess papers of Eigen (1971) and Eigen and Schuster (1979) as contributions which rightly link autocatalytic and self-reproductive macromolecular species with thermodynamic vision of the origin of life. Ishida (1981) summarizes the essence of Eigen's work (Eigen, 1971; Eigen and Schuster, 1979).

10.2 Ways of generation of order from disorder

10.2.1 Introduction

In this section, the second basic paper by E.D. Schneider and J.J. Kay (Schneider and Kay, 1995) constitutes our basic subject of analysis. Their text considers creation of order from disorder and aims to support the thesis that this creation is the source of thermodynamic complexity in biology.

In the middle of the nineteenth century, two basic views for evolution of natural systems in time attract the scientists. Thermodynamics, as refined by Boltzmann, views nature as decaying toward a death of random disorder, in accordance with the second law. This pessimistic view of the natural evolution contrasts with Darwin's paradigm of increasing specialization, organization, and complexity of biological systems in time. In fact, the observation of natural systems shows that much of the world is inhabited by coherent disequilibrium structures, such as autocatalytic chemical reactions, convection cells, tornadoes, and life itself. Living systems exhibit a transition away from disorder and equilibrium, into organized structures that exist some distance from equilibrium (Schneider and Kay, 1994, 1995).

There is no macroscopic science able to describe natural systems better than thermodynamics. Boltzmann was not a sole great scientist who highly respected thermodynamics as the theory of structure-independent assumptions, thus it is right to recall the role of Planck and his thermodynamic reasoning in setting the theory of black body radiation. Also, Einstein, very fluent, like Planck, in thermodynamic theory, published several papers on important thermodynamic topics. When in 1905 Einstein explained the photoelectric effect, he not only used the results of Planck's discrete energy packets to describe the black body radiation, but also acknowledged Boltzmann's work, calling the expression $S = k \ln W$ "the principle of Boltzmann" (Bushev, 2000). How highly Einstein regarded thermodynamics can be seen from his famous quote: "Thermodynamics is the only physical theory of universal content, which I am convinced, that within the framework of applicability of its basic concepts will never be overthrown."

In this section, thermodynamics is used to study complex systems which exhibit transitions from disorders and equilibria into excellently organized disequilibrium structures like convection cells, tornados, lasers, etc. to living systems at end. These complex systems always reside some distance from the equilibrium, the dilemma which motivates Schrödinger to treat together the fundamental processes of biology, physics, and chemistry in his seminal book *What is Life?* (Schrödinger, 1944). He notes therein that life is comprised of two fundamental processes, one "*order from order*" and the other "*order from disorder*", and concludes that the gene generates *order from order* in a species, that is, the progeny inherits the traits of the parent.

However, Schrödinger's equally important but less understood observation was his *order from disorder* premise. This premise reports his effort to link biology with basic theorems of thermodynamics (Schneider, 1987; Schneider and Kay, 1995). Schrödinger notes that living systems seem to defy the second law of thermodynamics which states that, among all possible closed systems, the entropy of a stable equilibrium system must exhibit maximum. "Living systems, however, are the antithesis of such maximal disorder. They display marvelous levels of order created from disorder" (Schneider and Kay, 1995). "For instance, plants are highly ordered structures, which are synthesized from disordered atoms and molecules in atmospheric gases and soils" (Schneider and Kay, 1995).

Schrödinger solves his dilemma by turning to disequilibrium thermodynamics. He recognizes that living systems must be open, i.e., exist in a world of energy and material fluxes. An organism stays alive in its highly organized state by taking high-quality energy (of higher exergy) from outside itself and processing it to produce, within itself, a more organized state. Life is a far from equilibrium system that maintains its local level of organization at the expense of the larger global entropy budget. Schrödinger hopes that the study of living systems from a nonequilibrium perspective would reconcile biological selforganization and thermodynamics. Furthermore, he expects that such a study would yield new principles in physics. Indeed, less than a decade later, biology receives the research agenda that explains the molecular structure of nucleic acids, DNA, and identifies gene as one of the most important findings of the last 70 years (Watson and Crick, 1953).

This section examines the *order from disorder* research program proposed by Schrödinger and expanded on his thermodynamic view of life. Following Schneider and Kay (1994, 1995), we explain that the second law of thermodynamics is not an impediment to the understanding of life, but rather a necessary ingredient for a complete description of living processes. Using their own wording, Schneider and Kay "expand thermodynamics into the causality of the living process" and show that "the second law underlies processes of selforganization and determines the direction of the processes observed in the development of living systems."

10.2.2 Thermodynamic preliminaries

Thermodynamics applies to all work and energy systems including the classic temperature-volume-pressure systems, diverse chemical kinetics systems, and electromagnetic and quantum systems. According to Schneider and Kay (1995), thermodynamics can be viewed as the field addressing the behavior of systems in three different situations: (1) equilibrium (classical) systems, governed by the law of large numbers for molecules in a closed system, (2) systems that are some distance from equilibrium and will equilibrate, i.e., molecules in two bottles connected with a closed stopcock; (one flask holds more molecules than the other, and upon opening the stopcock, the system will come to its equilibrium state with an equal number of molecules in each flask), and (3) systems that have been moved away from equilibrium and are constrained by gradients to be at some distance from the equilibrium state, i.e., two connected flasks with a pressure gradient holding more molecules in one flask than the other.

Exergy is a basic concept in the discussion of *order from disorder* whenever energy varies in its quality or capacity to do useful work. In an arbitrary chemical or physical process, a part of the quality index or capacity of energy to perform work is irretrievably lost. Exergy is a measure of the maximum capacity of an energy system to perform useful work when a reversible benchmark process proceeds to equilibrium with a standardized environment (Szargut and Petela, 1965; Brzustowski and Golem, 1978; Ahern, 1980; Szargut, 2002, 2005; Szargut et al., 1988). See Section 11.4.4 for the exergy efficiency of plants vegetation in Szargut and Petela (1965).

The first law of thermodynamics is the result of efforts toward understanding the relation between heat and work. This law says that the total energy is a sourceless quantity, i.e., that it cannot be spontaneously created or destroyed, and that the total energy within a closed or isolated system remains unchanged. However, the quality of the energy in the system (or its exergy content) may change. The second law of thermodynamics requires that whenever there are any processes running in the system, the quality of the energy in that system will degrade. Quantitatively, this degradation is ensured by the positiveness of the quantity called entropy source in any correct entropy balance.

Alternatively, the energy degradation law (the second law) can also be stated in terms of entropy balance and the entropy source as the quantitative measure of irreversibility. For this entropy balance, the entropy source is greater than zero in any real process.

Occasionally, the second law is stated thus: any real process can only proceed in a direction which results in an entropy increase (Schneider and Kay, 1995), but this statement is in general incorrect. A correct statement reads: any real process can only proceed in a direction which ensures a nonnegative entropy source. This latter formulation is consistent with Carathéodory's formulation of the second law.

In 1908, thermodynamics progresses by the work of Carathéodory (Kestin, 1966) when he develops a proof which shows that the law of "entropy increase" is not the general statement of the second law. The more encompassing statement of the second law of thermodynamics is that "In the neighbourhood of any given state of any closed system, there exists states which are inaccessible from it, along any adiabatic path reversible or irreversible." Unlike earlier statements, this does not depend on the nature of the system, nor on concepts of entropy or temperature.

Hatsopoulos and Keenan (1965) and Kestin (1968) subsumed the 0th, 1st, and 2nd laws into a Unified Principle of Thermodynamics: "When an isolated system performs a process after the removal of a series of internal constraints, it will reach a unique state of equilibrium: this state of equilibrium is independent of the order in which the constraints are removed." This describes systems of some distance from equilibrium, but not constrained to be in a disequilibrium state. This also dictates direction and an end state for real processes and says that a system will come to a local equilibrium as constraints permit (Schneider and Kay, 1995).

10.2.3 Dissipative systems and dissipative structures

Some of principles outlined above hold for closed isolated systems. However, a truly intriguing class of phenomena belongs to the class of systems open to energy and/or material flows and reside at quasi-stable states some distance from equilibrium (Nicolis and Prigogine, 1977, 1989; Schneider and Kay, 1995). Nonliving organized systems (convection cells, tornados, lasers, etc.) and living systems (from cells to ecosystems) depend on outside energy fluxes which maintain their organization and dissipate energy gradients to carry out these self-organizing processes. According to Schneider and Kay (1995), this organization is maintained at the cost of increasing the entropy of a surrounding supersystem in which an open system is imbedded. In this situation, the total entropy change in the open system is the sum of the internal production of entropy in this system (which is always greater or equal than zero) and the entropy exchange with the environment which may be positive, negative, or zero. For the system to maintain itself in a disequilibrium steady state, the entropy exchange must be negative and larger than the entropy produced by internal processes (such as heat sources, metabolism, etc.).

Dissipative structures stable over a finite range of conditions are best represented by autocatalytic positive feedback cycles (Schneider and Kay, 1995). Convection cells, hurricanes, autocatalytic reactions, and living systems are all examples of far-from-equilibrium, coherent dissipative structures which exhibit ordered behavior. The transition in a heated fluid between conduction and convection (Bénard cells) is a clear example of emergent coherent organization in response to an external energy input (Chandrasekhar, 1961). In the Bénard experiments, the lower surface of a fluid is heated and the upper surface is kept at a cooler temperature. When the heat flux reaches a critical value, the system becomes unstable and the convective overturning results in structured coherent hexagonal to spiral surface patterns (Bénard cells). These structures increase the heat transfer rate and gradient destruction. This transition between noncoherent molecule-to-molecule heat transfer, to the coherent structure represents the system's response to attempts to move it away from equilibrium (Schneider and Kay, 1995).

To manage this class of disequilibria, Schneider and Kay (1995) propose a corollary to Kestin's (1968) Unified Principle of Thermodynamics, whose proof shows that a system's equilibrium state is stable in the sense of Lyapunov. Implicit in this corollary is that a system will resist being removed from the equilibrium state. The degree to which a system moves from equilibrium to disequilibrium is measured by gradients imposed on the system.

Schneider and Kay (1995) state: *As systems are moved away from equilibrium, they will utilize all avenues available to counter the applied gradients. As the applied gradients increase, so does the system's ability to oppose further movement from equilibrium.* Thus, following Schneider and Kay (1994, 1995), we should refer to the above formulation as to the "restated second law" and the pre-Carathéodory statements as the classical second law. For the chemically reacting systems, it is LeChatelier's principle that constitutes an example of the restated second law in the chemical world.

Thermodynamic systems resist being moved away from the equilibrium states. When this movement occurs, they shift their state in a way which opposes the applied gradients and attempt to move the system back toward its equilibrium attractor. The stronger the applied gradient, the greater the effect of the equilibrium attractor on the system. The more a system is moved from equilibrium, the more sophisticated are its mechanisms for resisting. If current conditions permit, selforganization processes arise that abet the gradient dissipation. This behavior is not sensible from a classical perspective, but is expected given the restated second law (Schneider and Kay, 1994). No longer are coherent self-organizing structures a surprise, but rather it is an expected response of a system as it attempts to resist and dissipate externally applied gradients which would move the system away from equilibrium. Consequently, we observe *order emerging from disorder* in the formation of dissipative structures (Schneider and Kay, 1994).

So far the present discussion has focused on simple physical systems and the way how thermodynamic gradients drive selforganization. Chemical gradients result additionally in autocatalytic reactions in inorganic systems, in protein synthesis reactions, and in phosphorylation, polymerization, and hydrolytic reactions (Schneider and Kay, 1995). Autocatalytic chemical systems are examples of positive feedback where the activity of system or reaction augments itself in the form of self-reinforcing reactions. Autocatalysis stimulates the aggregate activity of the whole cycle. Such self-reinforcing catalytic activity is self-organizing and is a basic way of increasing the dissipative capacity of the system (Schneider and Kay, 1995).

The property of systems as gradient dissipators holds for disequilibrium physical and chemical systems and characterizes processes of emergence and development in complex systems. Not only are processes in these systems consistent with the restated second law, but, possibly, these systems will emerge in the presence of gradients. Schrödinger's notion of *order from disorder* pertains to the emergence of these dissipative systems.

10.2.4 Living systems as consumers of matter

As ensured by Schneider and Kay (1995), Boltzmann recognized the apparent contradiction between the heat death of the universe and the existence of life in which systems grow, complexify, and evolve. He realized that the sun's energy gradient drives living processes and suggested a Darwinean like competition for entropy in living systems:

> The general struggle for existence of animate beings is therefore not a struggle for raw materials - these, for organisms, are air, water and soil, all abundantly available - nor for energy which exists in plenty in any body in the form of heat (albeit unfortunately not

> transformable), but a struggle for entropy, which becomes available through the transition of energy from the hot sun to the cold earth.
>
> ***(Boltzmann, 1886)***

Boltzmann's ideas were further explored by Schrödinger who noted that some systems, like life, seem to defy the classical second law of thermodynamics (Schrödinger, 1944). Yet, he understood that living systems are open, i.e., they are not adiabatically isolated systems of classical thermodynamics. An organism lives in its organized state due to the consumption of the high-quality energy from its local outside environment, which is degraded to sustain the systemic organization. Schrödinger described this physical situation in terms of the so-called neg-entropy (the notion of which is currently used quite seldom as unnecessary):

> the only way a living system stays alive, away from maximum entropy or death is to be continually drawing from its environment negative entropy. Thus the devise by which an organism maintains itself stationary at a fairly high level of orderliness (= fairly low level of entropy) really consists in continually sucking orderliness from its environment....plants of course have their most powerful supply in negative entropy in sunlight.
>
> ***(Schrödinger, 1944)***

Life can be viewed as a far-from-equilibrium dissipative structure that maintains its local level of organization at the expense of producing entropy in the environment (Schneider and Kay, 1995). If we treat the Earth as an open thermodynamic system with a large gradient impressed on it by the sun, the restated second law implies that the system should reduce this gradient by using all physical and chemical processes available to it. Schneider and Kay (1994, 1995) ensure that life exists on Earth to dissipate the solar-induced gradient, and as such, life is a manifestation of the restated second law. However, in a common, but rather different, language, one can rather say that life exists because our terrestrial system is capable of absorbing solar energy which induces birth processes. Living systems are far from equilibrium dissipative structures that have sophisticated potential for reducing radiation gradients (induce birth processes) on Earth (Kay, 1984; Ulanowicz and Hannon, 1987).

Consequently, the origin of life is the development of an alternative road for the dissipation of induced energy gradients (absorption of energy at flow). Life ensures that its dissipative pathways continually evolve strategies to maintain it in the fluctuating physical environment. With genes as encoded memories, living systems are dynamic dissipative structures which allow to continue life as a response to the thermodynamic imperative of dissipating gradients (Kay, 1984; Schneider, 1988). Biological growth occurs when the system adds more of the same types of pathways for degrading imposed gradients. Biological development occurs when new types of pathways for degrading imposed gradients emerge in the system.

Plant growth is an attempt to capture solar energy and dissipate usable gradients (or related driving fluxes). Plants of many species arrange themselves into assemblies to increase leaf area so as to maximize driving energy capture and its degradation. The gross energy budgets of terrestrial plants show that the majority of their energy use is for evapotranspiration

(200–500 g of water transpired per gram of fixed photosynthetic material). This mechanism is a very effective energy degrading process with c. 2500 J/g of water transpired (Gates, 1962; Schneider and Kay, 1995). Evapotranspiration is the major dissipative pathway in terrestrial ecosystems. Currie (1991) shows that the large-scale biogeographical distribution of species richness is strongly correlated with potential annual evapotranspiration. Schneider and Kay (1995) see such exemplary relationships between species richness and available exergy as a proof of a link between biodiversity and dissipative processes. Yet, his dissipation is understood here not as the positiveness of entropy production, but as the degradation of the driving energy (usually solar energy flux).

The more energy available to be distributed among species, the more pathways are available for total energy degradation. Trophic levels and food chains are based upon photosynthetic fixed material and further dissipate these gradients by making more highly ordered structures. Thus, we would expect greater diversity of species to occur where there is more available energy. Species diversity and trophic levels are vastly greater at the equator, where 5/6 of the Earth's solar radiation occurs, and there is more of a gradient to reduce (Schneider and Kay, 1995).

10.2.5 Thermodynamics of ecosystems as energy degraders

Ecosystems are biotic, physical, and chemical components of nature acting together as disequilibrium dissipative processes. Ecosystem development should increase energy degradation whenever it follows from the restated second law (Schneider and Kay, 1995). This hypothesis can be tested by observing the energetics of ecosystem development during the successional process or when ecosystems are stressed. As ecosystems develop or mature, they should increase their total dissipation and should develop more complex structures with greater diversity and more hierarchical levels to assist in energy degradation (Schneider, 1988; Kay and Schneider, 1992). Successful species are those that funnel more energy into their own production and reproduction and contribute to autocatalytic processes, thereby increasing the total dissipation in the ecosystem.

Lotka (1922) and Odum and Pinkerton (1955) have suggested that biological systems which survive are those developing the most power in flow and use this power to best satisfaction of their needs for survival. Schneider and Kay (1995) believe that a better alternative of these "power laws" may be that biosystems develop so as to increase their energy degradation rate and that biological growth, ecosystem development, and evolution represent the development of new dissipative pathways. In other words, ecosystems develop in a way which increases the amount of exergy they capture and utilize. As a consequence, as ecosystems develop, exergy of the outgoing energy decreases. It is in this sense that ecosystems develop the most power, that is, they make the most effective use of the exergy in the incoming energy, while at the same time increasing the amount of energy they capture.

Ecologists have developed analytical methods that allow analysis of material-energy flows through ecosystems (Kay et al., 1989). With these methods, it is possible to determine the energy flow and assess how the energy is partitioned in the ecosystem. Schneider and Kay (1995) analyze a data for carbon-energy flows in two aquatic tidal marsh ecosystems adjacent to a large power generating facility on the Crystal River in Florida (Ulanowicz, 1997). The ecosystems in question are a "stressed" and a "control" marsh. The "stressed" ecosystem is exposed to hot water effluent from the nuclear power station. The "Control" ecosystem is otherwise exposed to the same environmental conditions. All the flows drop in the stressed ecosystem. It follows that the stress results in the ecosystem shrinking in size, in terms of biomass, its consumption of resources, in material and energy cycling, and its ability to degrade and dissipate incoming energy.

Overall, the impact of the effluent from the power station heating water is to decrease the size of the "stressed" ecosystem and its consumption of resources while influencing its ability to retain the resources it has captured. This analysis suggests that the function and structure of ecosystems develop along the path predicted by the behavior of nonequilibrium thermodynamic structures and the application of these behaviors to ecosystem development patterns.

The energetics of terrestrial ecosystems provides another test of the thesis that ecosystems develop so as to degrade energy more effectively. It turns out that more developed dissipative structures degrade more energy. Thus, we expect a more mature ecosystem to degrade the exergy content of the energy it captures more completely than a less developed ecosystem.

Luvall and Holbo (1989, 1991) measure surface temperatures of various ecosystems using a thermal infrared multispectral scanner (TIMS). Their data shows a solid trend: when other variables are constant, the more developed the ecosystem, the colder is its surface temperature and the more the degradation of its reradiated energy. The data analysis shows that ecosystems develop structure and function that degrades imposed energy gradients more and more effectively (Schneider and Kay, 1994, 1995). Their study of energetics treats ecosystems as open systems with high-quality energy pumped into them. Clearly, an open system with high-quality energy pumped into it can be moved away from equilibrium. But nature resists movement away from equilibrium. So ecosystems, as open systems, respond, whenever possible, with the spontaneous emergence of organized behavior that consumes the high-quality energy in building and maintaining the newly emerged structure. This decreases the ability of the high-quality energy to move the system further away from equilibrium. Such selforganization is characterized by abrupt changes that occur as a new set of interactions and activities by components and the whole system emerges. This emergence of organized behavior, the essence of life, is now understood to be expected by thermodynamics. As more high-quality energy is pumped into an ecosystem, more organization emerges to dissipate (degrade) the energy. This way, Schneider and Kay (1995) find *order* emerging from *disorder* in the situation of causing even more disorder.

10.2.6 Order from disorder and order from order

Complex systems stretch from ordinary complexity (Prigoginean systems, tornadoes, Bénard Cells, autocatalytic reaction systems) to emergent complexity perhaps including human socioeconomics systems. Living systems are at the more sophisticated end of the continuum: they must function within the system and environment they are part of. If a living system does not respect the circumstances of the supersystem it is part of, it will be selected against this supersystem (Schneider and Kay, 1995). The supersystem imposes a set of constraints on the behavior of the system and those living systems which are evolutionarily successful and are prepared to live within them.

When a new living system is generated after the death of an earlier one, it would make the self-organization process more efficient if it is constrained to variations which have a high probability of success. Genes play this role in constraining the selforganization process to those options which have a high probability of success. They are a record of successful selforganization. Genes are not the mechanism of development; the mechanism is selforganization. Genes bound and constrain the process of selforganization. At higher hierarchical levels, other devices constrain the selforganization. The ability of an ecosystem to regenerate is a function of the species available for the regeneration process (Schneider and Kay, 1995).

Given that living systems go through a constant cycle of many processes (birth/development/regeneration/death), preserving information about what works and what does not is crucial for the continuation of life (Kay, 1984). This is the role of the gene and, at a larger scale, biodiversity to act as information data bases about selforganization strategies that work. This is the connection between the *order from order* and *order from disorder* themes of Schrödinger. Life emerges because thermodynamics mandates *order from disorder* whenever sufficient thermodynamic and environmental conditions exist. But if life is to continue, the same rules require that it is able to regenerate, that is to create *order from order*. Life cannot exist without both processes, *order from disorder to generate life* and *order from order to ensure the continuance of life* (Schneider and Kay, 1995).

Life represents a balance between the imperatives of survival and energy degradation. We could add: degradation by consumption of energy and matter to sustain life. To quote Blum (1968) following (Schneider and Kay, 1995):

> I like to compare evolution to the weaving of a great tapestry. The strong unyielding warp of this tapestry is formed by the essential nature of elementary non-living matter, and the way in which this matter has been brought together in the evolution of our planet. In building this

> warp the second law of thermodynamics has played a predominant role. The multi-colored woof which forms the detail of the tapestry I like to think of as having been woven onto the warp principally by mutation and natural selection. While the warp establishes the dimensions and supports the whole, it is the woof that most intrigues the aesthetic sense of the student of organic evolution, showing as it does the beauty and variety of fitness of organisms to their environment. But why should we pay so little attention to the warp, which is after all a basic part of the whole structure? Perhaps the analogy would be more complete if something were introduced that is occasionally seen in textiles, the active participation of the warp in the pattern itself. Only then, I think, does one grasp the full significance of the analogy.

As modestly explained by Schneider and Kay (1995), in their research they "tried to show the participation of the warp in producing the tapestry of life." Their frequent citing of Schrödinger accepts his assumption that life is comprised of two processes, order from order and order from disorder. They also believe that it is the work of Watson and Crick and others who described the gene and solved the order from order mystery. Schneider and Kay's (1995) paper supports Schrödinger's order from disorder premise, and as they imply, provides enhanced link between macroscopic biology and physics.

A significant contribution to understanding order and disorder in nonliving and living systems was also made by a group of Italian researchers gathered around Prof. Giacomo Bisio. Their topic is focused on thermodynamics of disequilibrium steady states, exergy degradation, and system development in the light of the nonequilibrium thermodynamics, often with particular reference to Raleigh-Benard phenomena. For details, the reader is referred to publications of this group (Bisio, 1994, 1997; Bisio et al., 1989, 1999).

Terminating this review, often saturated with thoughts of various researchers, the author of this book needs nonetheless to stress an occasional flaw in some publications: lack of clear terminologies and/or quantitative properties of considered energies or energy fluxes in terms of common, measurable quantities. Clarification and/or improved presentation of these properties should amend quantitative criteria in investigations and to increase the useful information.

10.3 Life theories within physics of self-replication

We shall begin with introducing a recent theory of life with the help of a text: *A new physics theory of life* authored by Natalie Wolchover, Katheriie Taylor, and Emily Singer (reporting), which was published in Quannta Magazine (Wolchover and Taylor, 2014). Read also England's (2013) work for physics and mathematics of self-replication.

It is the matter of continual discussions and disagreements whether science progresses more thanks to bright individuals or due to an united effort of well-organized groups. Undoubtedly, these two ways of progress efficiently contributed to the scientific evolution observed through years. Professor M. Brenner, US National Academy of Sciences, specialized in applied mathematics and physics at Harvard, and his collaborators have developed theoretical models and simulations of microstructures that self-replicate (England, 2013; Wolchover and Taylor, 2014). These clusters of specially coated microspheres dissipate energy by roping nearby spheres into forming identical clusters. "This connects very much to what Jeremy (England) is saying," Brenner says (Wolchover and Taylor, 2014). Besides self-replication, greater structural organization is another means by which strongly driven systems ramp up their ability to dissipate energy. A plant, for example, is much better at capturing and routing solar energy through itself than an unstructured heap of carbon atoms. Thus, England argues that, under certain conditions, matter will spontaneously self-organize (Wolchover and Taylor, 2014). This tendency could account for the internal order of living things and of many inanimate structures as well. "Snowflakes, and dunes and turbulent vortices all have in common that they are strikingly patterned structures that emerge in many-particle systems driven by some dissipative process," he said. Condensation, wind, and viscous drag are the relevant processes in these particular cases.

"He (England) is making me think that the distinction between living and nonliving matter is not sharp," says Carl Franck, a biological physicist at Cornell University, in a mail. And adds: "I'm particularly impressed by this notion when one considers systems as small as chemical circuits involving a few biomolecules" (Wolchover and Taylor, 2014).

If the new theory is correct, the same physics it identifies as responsible for the origin of living things could explain the formation of many other patterned structures in nature. Snowflakes, sand dunes, and self-replicating vortices in the protoplanetary disk may all be examples of dissipation-driven adaptation. England's bold idea will likely face close scrutiny in the coming years. He is currently running computer simulations to test his theory that systems of particles adapt their structures to become better at dissipating energy. The next step will be to run experiments on living systems.

Prentiss, a researcher who runs an experimental biophysics lab at Harvard, says: England's theory could be tested by comparing cells with different mutations and looking for a correlation between the amount of energy the cells dissipate and their replication rates (Wolchover and Taylor, 2014). And adds: "One has to be careful because any mutation might do many things. But if one kept doing many of these experiments on different systems and if (in dissipation and replication success) they are indeed correlated, that I would suggest this is the correct organizing principle." Brenner said he hopes to connect England's theory to his own microsphere constructions and determine whether the theory correctly predicts which self-replication and self-assembly processes can occur to answer "a fundamental question in science" (Wolchover and Taylor, 2014).

An overarching principle of life and evolution would give researchers a broader perspective on the emergence of structure and function in living things, which many of the researchers said. "Natural selection doesn't explain certain characteristics," said Ard Louis, a biophysicist at Oxford University, in an email. These characteristics include a heritable change to gene expression called methylation, increases in complexity in the absence of natural selection, and certain molecular changes Louis has recently studied (Wolchover and Taylor, 2014).

If England's approach stands up to more testing, it could further liberate biologists from seeking a Darwinian explanation for every adaptation and allow them to think more generally in terms of dissipation-driven organization. They might find, for example, that "the reason that an organism shows characteristic X rather than Y may not be because X is more fit than Y, but because physical constraints make it easier for X to evolve than for Y to evolve," Louis said.

"People often get stuck in thinking about individual problems," Prentiss said. Whether or not England's ideas turn out to be exactly right, she said, "thinking more broadly is where many scientific breakthroughs are made." (Wolchover and Taylor, 2014).

The above is a part of the information which comes with the article by Natalie Wolchover and Katherine Taylor for Quanta Magazine which was reprinted on ScientificAmerican.com and BusinessInsider.com. See more content details and figures in the complete text at: https://www.quantamagazine.org/a-new-thermodynamics-theory-of-the-origin-of-life-20140122/.

10.4 England's physics and an outline of his work

> England's New Physics? Theory of Life within his "Statistical Physics of Self-Replication" Chem. Phys. 139, 121,923 (2013); https://doi.org/10.1063/1.4818538.

10.4.1 Introduction to statistical physics of self-replication

Self-replication is an ability common to every species of the living world, and physical intuition dictates that such a process must invariably be fueled by the production of entropy. In his ambitious paper, England (2013) undertakes to make this physical feeling "rigorous and quantitative" by deriving a lower bound for the amount of heat that is produced during a process of self-replication in a system coupled to a thermal bath. He finds that the minimum value for the physically allowed rate of heat production is determined by the growth rate, internal entropy, and durability of the replicator, and he discusses implications of this finding for bacterial cell division, as well as for the prebiotic emergence of self-replicating nucleic acids (England, 2013). We will be able to discuss here only selected aspects of his methodology and results.

Every species of living thing can make a copy of itself by exchanging energy and matter with its surroundings. One feature common to all such examples of spontaneous "self-replication" is their statistical irreversibility: clearly, it is much more likely that one bacterium should turn into two than two somehow spontaneously reverting back into one. From the standpoint of physics, this observation contains an intriguing hint of how the properties of self-replicators must be constrained by thermodynamic laws, which dictate that irreversibility is always accompanied by an increase of entropy. Nevertheless, it has long been considered challenging to speak in universal terms about the statistical physics of living systems because they invariably operate very far from thermodynamic equilibrium (England, 2013), and therefore, need not obey a simple Boltzmann probability distribution over microscopic arrangements. Faced with so-described, constrained diversity of organization, it is quite reasonable to be afraid that the particular mechanistic complexity of each given process of biological self-replication might overwhelm our ability to say much in terms of a general theory (England, 2013).

For the beginnings of a way forward, the reader should consider a system of fixed particle number N and volume V in contact with a heat bath of inverse temperature β. If we label the microstates of this system by i, j, etc., and associate energies E_i, E_j, etc., respectively with each microstate, then the underlying time-reversal symmetry of Hamiltonian dynamics assures that a well-defined, detailed balance relation holds at thermal equilibrium,

$$\frac{\exp(-\beta E_i)}{Z(\beta)}\pi(i\rightarrow j;\tau)=\frac{\exp(-\beta E_j)}{Z(\beta)}\pi(j^*\rightarrow i^*;\tau) \tag{10.1}$$

which is given by the first mathematical formula of England's theory, i.e., it is identical with Eq. (1) in England (2013).

The result (Eq. 10.1) is available in the literature of stochastic methods (Gardiner, 2003). In the equation in question, there appear system microstates i, correspondingly with i^*, momentum-reversed partners of i, and $Z(\beta)$—the canonical partition function of the system. The transition matrix element π $(i \rightarrow j;\ \tau)$ is the conditional probability that the system is found to be in microstate j at time $t=\tau>0$, given that it started off in microstate i at an earlier time $t=0$. The relation obtained in this way states that when the system has relaxed to thermal equilibrium and achieved a Boltzmann distribution over microstates, the probability currents connecting i to j in the forward and time-reversed directions are equal.

Progress comes from recognizing by England (2013) that the heat expelled into the bath over a transition from i to j is given by

$$\Delta Q_{i=j}=E_i-E_j \tag{10.2}$$

and, moreover, that the quantity $\beta\Delta Q_{i\rightarrow j}$ is the amount by which the entropy of the heat bath changes over the course of this transition. Thus, the reader may arrive at the second England's equation

$$\frac{\pi(j^* \to i^*; \tau)}{\pi(j \to i; \tau)} = \exp\left[-\Delta S_{\text{bath}}^{i \to j}\right] \tag{10.3}$$

Eq. (10.3) sets up a microscopically detailed relationship between a irreversibility of a transition and the amount of entropy increased in the surroundings over the course of forward trajectory. This formula shows how much more likely it is to happen in the forward direction than in the reverse direction. Importantly, while the result (10.3) is derived from a statement of detailed balance which only holds at equilibrium, the result itself is valid for arbitrary transition between two microstates. Thus, Eq. (10.3) applies to the relaxation dynamics of undriven systems arbitrarily far from equilibrium. England's principal aim is to show that the microscopically detailed relationship between irreversibility and entropy production has general thermodynamic consequences for strongly disequilibrium macroprocesses such as biological self-replication. This allows England to derive a generalized Second Law for irreversible macroprocesses, and then use such a result to show the existence of a lower bound on heat output of a self-replicator in terms of its size, internal entropy, growth rate, and durability. Further on, by analysis of empirical data, England (2013) is able to show that his lower bound operates on a scale relevant to the functioning of real microorganisms and other self-replicants.

10.4.2 Consequences of macroscopic irreversibility

In order to determine thermodynamic constraints on a macroscopic transition between two complex coarse-grained states, we have to find first how probability, heat, and entropy are related at the microscopic level. Eq. (10.3) establishes such a suitable relation in one special case, but the linkage turns out to be far more general. In fact, it has been shown in the literature (Crooks, 1999) that for all heat bath-coupled, time-symmetrically driven disequilibrium systems whose dynamics are dominated by purely diffusive motions (at absence of ballistic effects), any trajectory starting at $x(0)$ and going through microstates $x(t)$ over time τ satisfies the equation

$$\beta \Delta Q[x(t)] = \ln\left[\frac{\pi[x(t)]}{\pi[x(\tau - t)]}\right] \tag{10.4}$$

Here $\beta = T^{-1}$ sets the bath temperature T in "natural" units, π is the probability of the trajectory $x(t)$ given $x(0)$, and ΔQ is the heat released into the bath over the course of $x(t)$. The formula (10.4) involves the main microscopic relationship between heat and irreversibility: the more likely a forward trajectory is than its time-reverse, the more the entropy of the universe is increased through the exhaustion of heat into the surrounding bath. Furthermore, it should be emphasized that this result holds for systems subject to time-symmetric external driving fields. This means that a steady-state probability distribution for the system need not be the Boltzmann distribution, nor even a distribution whose steady probability flux satisfies detailed balance, in order for this fundamental relationship between irreversibility and entropy production to hold (England, 2013).

By fixing the starting and ending points of the trajectory ($x(0)=i$, $x(\tau)=j$), we can average the exponential weight of the forward heat over all paths from i to j and obtain $\pi(j\rightarrow i;\tau)/\pi(i\rightarrow j;\tau)=\langle\exp[-\beta\Delta Q^{\tau}_{i\rightarrow j}]\rangle_{i\rightarrow j}$. This microscopic rule (of which Eq. (10.3) is the special, undriven case) must have macroscopic consequences, and to investigate these, we have to formalize what it means to talk about the system of interest in macroscopic terms. In order to do so, we first suppose there is some coarse-grained, observable condition **I** in the system, such as the criterion "The system contains precisely one healthy, exponential-growth-phase bacterium at the start of a round of cell division." If we prepare the system under some set of controlled conditions and then find that the criterion for **I** is satisfied, we can associate with **I** a probability distribution $p(i\,|\,\mathbf{I})$ which is the implicit probability that the system is in some particular microstate i, given that it was prepared under controlled conditions and then observed to be in macrostate **I** (England, 2013).

We shall now omit some further equations and their derivations. Yet, we shall refer the reader to England's paper (England, 2013) who arrives at his Eq. (8) or our Eq. (10.5), in the following form:

$$\beta\langle\Delta Q\rangle_{\mathbf{I}\rightarrow\mathbf{II}} + \ln\left[\pi(\mathbf{II}\rightarrow\mathbf{I})/\pi(\mathbf{I}\rightarrow\mathbf{II})\right] + \Delta S_{\text{tot}} \geq 0 \qquad (10.5)$$

At this point, England makes an important statement, which, as we quote here, reads:

"If **I** and **II** corresponded to identical groups that each contained all microstates i, we would have $\pi(\mathbf{II}\rightarrow\mathbf{I})=\pi(\mathbf{I}\rightarrow\mathbf{II})=1$, and the above relation, (10.5), would reduce to a simple statement of the Second Law of Thermodynamics: on average, the change in the entropy of the system ΔSint plus the change in the entropy of the bath $\beta\langle\Delta Q\rangle_{\mathbf{I}\rightarrow\mathbf{II}}$ must be greater than or equal to zero, that is, the average total entropy change of the universe must be positive. What we have shown here using Crooks' microscopic relation, however, is that the *macroscopic* irreversibility of a transition from an arbitrary ensemble of states $p(i\,|\,\mathbf{I})$ to a future ensemble $p(j\,|\,\mathbf{II})$ sets a stricter bound on the associated entropy production: the more irreversible the macroscopic process (i.e., the more negative the member

$$\ln\left[\pi(\mathbf{II}\rightarrow\mathbf{I})/\pi(\mathbf{I}\rightarrow\mathbf{II})\right])$$

is, the more positive must be the minimum total entropy production. Moreover, since the formula was derived under very general assumptions, it applies not only to self-replication but to a wide range of transitions between coarse-grained starting and ending states. In this light, the result in Eq. (10.5), or England's Eq. (8), is closely related to past bounds set on entropy production in information-theoretic terms (Blythe, 2008; Gomez-Martin et al., 2008; Verley et al., 2012; Verley, 2012), as well as to the well-known Landauer's bound for the heat generated by the erasure of a bit of information (Landauer, 1961)."

10.4.3 General constraints on self-replication

The generalization of the Second Law presented in England's (2013) paper applies in a wide range of far-from-equilibrium, heat-bath-coupled scenarios, from the operation of computers to the action of molecular motors. The result turns out to be particularly valuable, however, as a lens through which to take a fresh look at the phenomenon of biological self-replication. Interest in the modeling of evolution long ago gave rise to rich literature exploring the consequences of self-replication for population dynamics and Darwinian competition. In such studies, the idea of Darwinian "fitness" is frequently invoked in comparisons among different self-replicators in a noninteracting population: the replicators that interact with their environment in order to make copies of themselves fastest are more "fit" by definition because successive rounds of exponential growth will ensure that they come to make up an arbitrarily large fraction of the future population. The first thing to notice here is that exponential growth of the kind just described is a highly irreversible process: in a selective sweep where the fittest replicator comes to dominate in a population, the future almost by definition looks very different from the past. Happily, we are now in the position to be quantitative about the thermodynamic consequences of this irreversibility. Thus, let us suppose there is a simple self-replicator living at inverse temperature β whose population $n \gg 1$ obeys a master equation

$$dp_n(t)/dt = gn[p_{n-1}(t) - p_n(t)] - \delta n[p_n(t) - p_{n+1}(t)] \tag{10.6}$$

where $p_n(t)$ is the probability of having a population of n at time t, and $g > \delta > 0$. For simplicity, England (2013) assumes that a decay event mediated by the rate parameter δ amounts to a reversion of the replicator back into the exact set of reactants in its environment out of which it was made, and also assumes that forming one new replicator out of such a set of reactants changes the internal entropy of the system by an amount Δs_{int}, and on average puts out an amount of heat Δq into the surroundings. Furthermore, he ignores in his discussion the additional layer of complexity that comes with considering effects of spontaneous mutation in evolving populations (Nowak, 1992).

For $n(t=0) \gg 1$, we expect the behavior of the system to follow a deterministic exponential growth path of the form $n(t) = n(0)\exp(g - \delta)t$. Whatever the exact value of n, gives birth $\pi(\mathbf{I} \rightarrow \mathbf{II})$ should be gdt, while the probability of a newly born copy decaying back whence it came in the same-length interval time ($\pi(\text{II} \rightarrow \text{I})$) should be δdt. Thus, plugging into Eq. (10.5) (i.e., into England's Eq. 8), we have.

$$\Delta s_{\text{tot}} \equiv \beta \Delta q + \Delta s_{\text{int}} \geq \ln[g/\delta] \tag{10.7}$$

As an aside, it should be noted here that we have avoided including in the above a spurious multiplicative factor from the number of particles in the system. As an example, one might have thought in the case of converting one particle into two that the probability rate of reverse back to one particle ought to be 2δ, with a resulting bound on entropy production of the form $\ln(g/\delta) - \ln 2$. To see why the $\ln 2$ ought not to be included, it is important to recognize that particles in a classical physical system have locations that distinguish them from each other. Thus, we either can think of the process in question as bounding the entropy production of self-replication plus mixing of particles (in which case the mixing entropy term cancels the factor of 2) or else we can define our coarse-grained states in the transition in question so that we only consider the probability current for the reversion of the replicator that was just born. Either way, the relevant bound comes out to $\ln(g/\delta)$.

The above result is consistent with the expectation that, in order for this self-replicator to be capable of net growth, we have to require that $g > \delta$, which in turn sets a positive lower bound on the total entropy production associated with self-replication. More can be seen, however, by rearranging the expression at fixed δ, Δs_{int}, and Δq, to obtain.

$$g_{\max} - \delta = \delta\left(\exp\left[\beta \Delta q + \Delta s_{\text{int}}\right] - 1\right) \tag{10.8}$$

In other words, the maximum net growth rate of a self-replicator is fixed by three things: its internal entropy (Δs_{int}), its durability ($1/\delta$), and the heat that is dissipated into the surrounding bath during the process of replication Δq.

Several comments are in order. First of all, let us consider comparing two different replicators with fixed Δ_{int} and δ that have different heat values Δq and $\Delta q'$. If $\Delta q > \Delta q'$, then clearly $g_{\max} > g_{\max}$; the replicator that dissipates more heat has the potential to grow accordingly faster. Moreover, we know by conservation of energy that this heat has to be generated in one of two different ways: either from energy initially stored in reactants out of which the replicator gets built (such as through the hydrolysis of sugar) or else from work done on the system by some time-varying external driving field (such as through the absorption of light during photosynthesis). In other words, basic thermodynamic constraints derived from exact considerations in statistical physics tell us that a self-replicator's maximum potential fitness is set by how effectively it exploits sources of energy in its environment to catalyze its own reproduction. Thus, the empirical, biological fact that reproductive fitness is intimately linked to efficient metabolism now has a clear and simple basis in physics.

A somewhat subtler point to make here is that self-replicators can increase their maximum potential growth rates not only by more effectively fueling their growth via Δq, but Δq also by lowering the cost of their growth via δ and Δs_{int}. The less durable or the less organized a self-replicator is, all things being equal, the less metabolic energy it must harvest at minimum in order to achieve a certain growth rate. Thus, in a competition among self-replicators to dominate the population of the future, one strategy for "success" is to be simpler in construction and more prone to spontaneous degradation. Of course, in the limit where two self-replicators differ dramatically in their internal entropy or durability (as with a virus and a rabbit), the basis for comparison in terms of Darwinian fitness becomes too weak. Nevertheless, in a race between competitors of similar form and construction, it is worthwhile to note that one strategy for reducing the minimal "metabolic" costs of growth for a replicator is to substitute new components that are likely to "wear out" sooner.

10.4.4 Self-replicating polynucleotides

A simple demonstration of the role of durability in replicative fitness comes from the case of polynucleotides, which provide a rich domain of application for studying the role of entropy production in the dynamics of self-replication. One recent study has used in vitro evolution to optimize the growth rate of a self-replicating RNA molecule whose formation is accompanied by a single backbone ligation reaction and the leaving of a single pyrophosphate group. It is reasonable to assume that the reverse of this reaction, in which the highly negatively charged pyrophosphate would have to attack the phosphate backbone of the nucleic acid, would proceed more slowly than a simple hydrolysis by water. The rate of such hydrolysis has been measured by carrying out a linear fit to the early-time (20 days) exponential kinetics of spontaneous degradation for the RNAase. A substrate UpA, and the half-life measured, is on the order of 4 years (Thompson et al., 1995). Thus, with a doubling time of 1 h for the self-replicator, we can estimate for this system that $\ln[g/\delta] \geq \ln[(4\,\text{years})/(1\,\text{h})]$. Since the ligation reaction trades the mixing entropy of substrate for that of pyrophosphate at comparable ambient concentrations, we can also assume in this case that the change in internal entropy for the reaction is negligible. Thus, we can estimate the heat bound as.

$$(\Delta Q) \geq RT\ \ln[(4\,\text{years})/(1\,\text{h})] = 7\,\text{kcal}\,\text{mol}^{-1} \tag{10.9}$$

Since experimental data indicate an enthalpy for the reaction in the vicinity of $10\,\text{kcal}\,\text{mol}^{-1}$, (Minetti et al., 2003; Woo et al., 2012) it would seem this molecule operates quite near the limit of thermodynamic efficiency set by the way it is assembled. To underline this point, we may consider what the bound might be if this same reaction were somehow achieved using DNA, which in aqueous solution is much more kinetically stable against hydrolysis than RNA. In this case, we would have $(\Delta Q) \geq RT\ \ln[(3 \times 10^7\,\text{years})/(1\,\text{h})] = 16\,\text{kcal}\,\text{mol}^{-1}$, which exceeds the estimated enthalpy for the ligation reaction and is therefore prohibited thermodynamically. This calculation illustrates a significant difference between DNA and RNA, regarding each molecule's ability to participate in self-catalyzed replication reactions fueled by simple triphosphate building blocks: the far greater durability of DNA demands that a much higher per-base thermodynamic cost be paid in entropy production in order for the growth rate to match that of RNA in an all-things-equal comparison.

The key point here is that if a self-replicating nucleic acid catalyzes its own growth with a rate constant g and is degraded with rate constant δ, then the molecule should be capable of exhibiting exponential growth through self-replication with a doubling time proportional to

$1/(g-\delta)$. However, thermodynamics only sets a bound on the ratio g/δ, which means that $g-\delta$ can be made arbitrarily large while the "metabolic cost" of replication remains fixed. Thus, surprisingly, the greater fragility of RNA could be seen as a fitness advantage in the right circumstances, and this applies even if the system in question is externally driven, and even still if the replicators maintain non-Boltzmann distributions over their internal degrees of freedom. Moreover, we would expect that the heat bound difference between DNA and RNA should increase roughly linearly in the number of bases l ligated during the reaction, which forces the maximum possible growth rate for a DNA replicator to shrink *exponentially* with l in comparison to that of its RNA competitor in an all things equal comparison. This observation is certainly intriguing in light of past arguments made on other grounds that RNA, and not DNA, must have acted as the material for the prebiotic emergence of self-replicating nucleic acids (Woo et al., 2012; Gilbert, 1986). The particular example of a single nucleic acid base ligation considered here is instructive because it is a case where the relationship between irreversibility and entropy production further simplifies into a recognizable form. It is reasonable to define our coarse-graining of states so that the starting and ending products of this reaction are each in a local, thermal equilibrium, such that all of their degrees of freedom are Boltzmann-distributed given the constraints of their chemical bonds. In such a circumstance, detailed balance alone requires that the ratio of the forward and reverse rates be locked to the Gibbs' free energy change for the reaction via the familiar formula $\ln[k_f/k_r] = -\beta\Delta G \equiv -\beta\Delta H + \Delta S_{\text{int}} = \beta\Delta Q + \Delta S_{\text{int}}$. In this light, the relationship between durability and growth rate has an elegant description in terms of transition state theory: an activation barrier that is lower in the forward direction will be lower in the reverse direction as well. The key point here, however, is that whereas the relationship between free energy and reaction rates is obtained only under local equilibrium assumptions, the inequality we have derived here bounding entropy production in terms of irreversibility applies even in cases where many degrees of freedom in the system start and end the replication process out of thermal equilibrium.

On a related note, it may also be pointed out that a naïve objection might be raised to the above account of nucleic acid growth on the grounds that its conclusions about the maximum possible growth rate of DNA would seem to disagree with the empirical fact that DNA is obviously capable of undergoing replication much more rapidly than one phosphodiester linkage per hour; the processive holoenzyme DNA polymerase III is well-known for its blazing catalytic speed of $\sim$1000 base-pairs per second. The resolution of this puzzle lies in realizing that in the protein-assisted DNA replication.

Scenario being raised here, the polymerase assembly is first loaded onto DNA in an ATP-dependent manner that irreversibly attaches the enzyme to the strand using a doughnut-shaped clamp composed of various protein subunits. Thus, while the polymerase can catalyze the elongation of the DNA chain extremely rapidly, one also must take into account that the *reverse* reaction of DNA hydrolysis should happen much more readily with an enzyme tethered to the strand than it does for isolated DNA in solution. This example, therefore, underlines the care that must be shown in how the coarse-grained states **I** and **II** are defined in the computation of irreversibility.

Next, we refer the reader to a discussion of an interesting section in England's paper on the bacterial cell division (following his section on self-replicating polynucleotides).

10.4.5 Final remarks

The process of cellular division, even in a creature as ancient and streamlined as a bacterium, is so bewilderingly complex that it may come as some surprise that physics can make any binding pronouncements about how fast it all can happen (England, 2013). The reason this becomes possible is that nonequilibrium processes in constant temperature baths obey general laws that relate forward and reverse transition probabilities to heat production. Previously, such laws had been applied successfully in understanding thermodynamics of copying "informational" molecules such as nucleic acids. In those cases, however, the information content of the system's molecular structure could more easily be taken for granted, in light of the clear role played by DNA in the production of RNA and protein.

What England (2013) glimpsed is that the underlying connection between entropy production and transition probability has a much more general applicability, so long as we recognize that "self-replication" is only visible once an observer decides how to classify the "self" in the system: only once a coarse-graining scheme determines how many copies of some object are present for each microstate can we talk in probabilistic terms about the general tendency for that type of object to affect its own reproduction, and the same system's microstates can be coarse-grained using any number of different schemes. Whatever the scheme, however, the resulting stochastic population dynamics must obey the same general relationship entwining heat, organization, and durability. We may hope that this insight spurs future work that will clarify the general physical constraints obeyed by natural selection in nonequilibrium systems.

10.5 Kleidon's hierarchical machinery

10.5.1 Introduction to life and earth system

Consideration of the Earth's history shows that life evolved in time with its abundance, diversity, and complexity (Kleidon, 2010a,b, his Section 1.1).

Simultaneously, life substantially influenced the Earth's environment, evolving its state vector to states further and further away from thermodynamic equilibrium. In particular, concentrations in atmospheric oxygen increased throughout Earth's history, resulting in an increased chemical disequilibrium in the atmosphere as well as an increased red-ox gradient between the atmosphere and the reducing crust of the Earth. This behavior seems to contradict the second law of thermodynamics, which states for isolated systems, that their fluxes (gradients) and free energies are dissipated (degraded) over time, resulting in a state of thermodynamic equilibrium. This apparent contradiction is currently resolved by considering the Earth as a coupled, hierarchical, and evolving disequilibrium system that has substantially been changed by the input of free energy generated by photosynthesis.

Taking these observations into account, Kleidon (2010a,b) presents his "hierarchical thermodynamic theory of the Earth system." His considerations show that systemic thermodynamic variables are driven off thermodynamic equilibrium by transfer of power from some foreign process(es), and that the resulting state of disequilibrium reflects the past net work done on the variable. This position is taken by Kleidon to processes of the Earth to characterize the generation and transfer of free energy and its degradation (dissipation), from radiative gradients to gradients of temperature and chemical potentials. The latter gradients result in chemical, kinetic, and potential free energies and associated dynamics of the climate system and geochemical cycles.

The maximization of power transfer among the processes within the accepted hierarchy yields thermodynamic efficiencies much lower than the Carnot efficiency of equilibrium thermodynamics and is closely related to the principle of Maximum Entropy Production (MEP) proposed earlier by Kleidon and some others as the result of rectifications and amendments of many earlier formulations. See a review of Martyushev and Seleznev (2006) for a comprehensive discussion of this topic.

The role of life is discussed in terms of a photochemical process that generates substantial amounts of chemical-free energy which essentially skips the limitations and inefficiencies associated with the transfer of power within the thermodynamic hierarchy of the planet. This perspective allows to view life as a means to transform many aspects of the Earth to states even further away from thermodynamic equilibrium than is possible by purely abiotic means. In this perspective, pockets of low-entropy life emerge from the overall trend of the Earth system to increase the entropy of the universe at the fastest possible rate. The implications of the theory are discussed concerning fundamental deficiencies in Earth system modeling, applications of the theory to reconstructions of Earth system history, and regarding the role of human activity for the future of the planet.

10.5.2 Theory of life within the hierarchical earth system

See Kleidon's Section 1.3.

Kleidon (2010a,b) relates life and the thermodynamic state of the planet Earth by developing a theory of the whole Earth system. The theory is based on disequilibrium thermodynamics and is immersed in the geological context and his personal view of thermodynamics.

He develops a hierarchical view of the Earth-system processes which generate free energy, dissipate, and—most importantly—transfer power from one process to other processes higher in the hierarchy, e.g., from radiative exchange to motion of geochemical cycling.

Thermodynamics, specifically the first and the second laws, sets the rules for: operating/controlling power processes, maximizing rates of power generation and transfer, and admitting definite types of process interactions. Characteristic maximum transfer rates, which can be deduced from the maximum power principle or corresponding principle of maximum entropy production (MEP principle), define upper bounds on power transferred between processes.

In the present book, Kleidon's hierarchical view is represented by Fig. 10.3. Heat engine, which acts like a motor (indicated by the symbol "M" in Fig. 10.3), can drive various processes off equilibrium, as indicated by the "belt" around the motor. Their connection is realized within the global hydrologic cycle. The atmospheric motor drives the dehumidifier and thereby the hydrologic cycle. The hydrological motor drives continental runoff. This, in turn, acts as a motor to transport rock material in suspended and dissolved form from land to ocean.

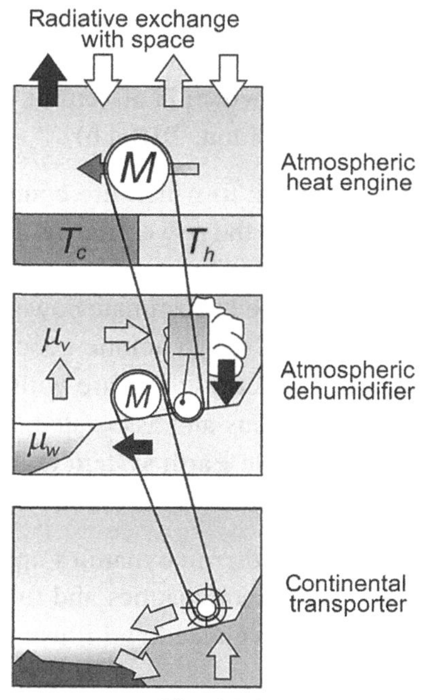

Fig. 10.3
Illustration of the formation of a hierarchy based on power transfer. A radiative gradient drives the atmospheric heat engine (top of figure). Compare this drawing with a similar illustration of the hierarchy shown in Fig. 1 of Kleidon (2010a,b).

Radiative fluxes at the Earth-space boundary lead to spatial and temporal differences in radiative heating and cooling. The resulting temperature differences fuel the heat engine that generates motion within the atmosphere, as symbolized by the engine symbol in Fig. 10.3. The kinetic energy inherent in the flow lifts moist air, cools it, brings it to saturation, condensation, and precipitation, thereby acting to remove water from the atmosphere. The atmospheric heat engine drives the atmospheric dehumidifier as indicated by the "belt" around the engine symbol in Fig. 10.3. This dehumidifier acts as another engine that drives the global cycling of water. The hydrologic cycle brings water to land at a higher elevation than the sea level, which enables continental runoff to drive the transport of dissolved ions and sediments to the sea floor. This "transporter of continental mass" interacts with interior processes of the global rock cycle and results in the geochemical cycling of rock-based elements (Kleidon, 2010a,b).

The emergent dynamics of this hierarchy, and specifically the resulting extent of chemical disequilibrium, is strongly governed by power (work flux) transferred among each of processes since dissipative processes will inevitably slow down engines and result in inefficiencies. Because of inevitable inefficiencies at each of the transfer processes, abiotic processes can only drive and maintain states of chemical disequilibrium to a relatively small extent. Photosynthetic life, in contrast, is able to directly tap into the power contained in the flux of solar radiation, skip the myriad of inefficiencies contained in abiotic power transfer processes, and is thus able to drive and maintain substantial chemical disequilibrium states. Therefore, "it is life which is of central importance in driving and maintaining planetary disequilibrium" (Kleidon, 2010a,b).

Thermodynamics lets us quantify the extent to which life contributes to disequilibrium. For this purpose, we need first to understand the power transfer in the dominant hierarchy of the Earth system processes which is associated with geochemical cycling as well as with power that drives this hierarchy. We then need to estimate power associated with biotic activity and compare this power to the power involved in abiotic geochemical processes. All this will result in a holistic thermodynamic description of life within the Earth system, in which thermodynamic nature of process interactions and associated states of maximum power play central role in functioning of the whole Earth system and for developing this hierarchical theory.

The basis for this theory is disequilibrium thermodynamics applied to the Earth system processes. The background on thermodynamics and the natural trend toward states of thermodynamic equilibrium are explained and illustrated further on in the text.

In Section 3 of his review, Kleidon shows that basic thermodynamics can be extended to practically all types of variables that characterize gradients in the Earth system. In his

Section 4, he shows how gradients are created by transferring power into the system, thereby achieving work generation within the system. The import and transfer of power (energy flux) secures maintenance of system states away from thermodynamic equilibrium and driving processes that act to deplete overall driving gradients at a faster rate. Hence, the overall effect of this hierarchy is to enhance the overall rate of entropy production, thereby resulting in the fastest approach to thermodynamic equilibrium possible at the planetary scale. Furthermore, characteristic maximum rates exist with which power can be extracted from a gradient to create other gradients. The maximization of power is related to the principle of Maximum Entropy Production (MEP).

Kleidon uses his conceptual view to describe appropriate processes of Earth system including life in a hierarchical view (his Section 6). The resulting rates at which work is performed at different levels of this hierarchy are estimated to provide a global work budget. Together with the global energy and entropy budget, this global work budget completes the description of the Earth system in terms of the first and second laws. The implications of maximum power transfer for good functioning of the Earth system is discussed. The role of life, particularly photosynthetic life, is described in terms of altering power transfer and thereby the structure and strength of this hierarchy. The resulting view of the Earth system is then compared to the Gaia hypothesis (Lovelock, 1972a,b). This hypothesis is associated with common opinions that life maintains the Earth in a homeostatic state, that conditions are optimal to life, and that the Earth system responds to change predominantly, with negative, stabilizing feedbacks.

Importantly, Kleidon reformulates the Gaia hypothesis on the basis of the requirement of chemical-free energy generation to propel and maintain atmospheric chemical equilibrium as follows:

> Hypothesis: The Earth system has evolved further and further away from a state of thermodynamic equilibrium in time mostly by the increased generation rate of chemical free energy by life.
>
> ***(Kleidon, 2010a,b)***

Kleidon realizes that his 2010 theory is a first sketch admitting some limitations and missing details, but also expects its profound implications. These implications potentially range from more basic views of understanding the Earth system feedbacks and reconstructing the Earth's history to more applied topics indicating how we should model the Earth system and how the future of humans should look like. Also, Kleidon discusses some aspects of the role of human activity in the global work budget. These issues are mostly discussed in Section 7 of his 2010 paper. He closes with a brief summary and conclusion.

Below, the reader will find a discussion of several details of Kleidon's theory of special interest to mechanical, chemical, and environmental engineers.

10.5.3 Maintaining thermodynamic disequilibrium

(*Re*: Kleidon's Section *2.3. Maintaining thermodynamic disequilibrium*)

When entropy and matter exchanges are allowed for at the system's boundary, the system can be maintained away from a state of thermodynamic equilibrium. Fig. 3a of Kleidon (2010a, b) paper, a two-box model that is shown here in Fig. 10.4, depicts a setup with entropy exchanges of two boxes with the surroundings.

The setup of Fig. 10.4 describes the differential heating of the Earth between the tropics (warm box) and the poles (cold box). To account for energy and entropy exchange across the system boundary, we need to consider the (generalized) set of Eqs. (7), (8) in Kleidon (2010a,b). Kleidon's generalized set is rewritten here in the form of Eqs. (10.10), (10.11) below.

$$c\frac{dT_h}{dt} = J_{\text{in},h} - J_{\text{out},h} - J_{\text{heat}}, \quad c\frac{dT_c}{dt} = J_{\text{in},c} - J_{\text{out},c} + J_{\text{heat}} \tag{10.10}$$

and

$$\frac{dS_h}{dt} = \frac{J_{\text{in},h}}{T_{\text{in}}} - \frac{J_{\text{out},h}}{T_h} - \frac{J_{\text{heat}}}{T_h} + \sigma_{\text{mix},h}, \quad \frac{dS_c}{dt} = \frac{J_{in,c}}{T_{in}} - \frac{J_{out,c}}{T_c} - \frac{J_{heat}}{T_c} + \sigma_{mix,c} \tag{10.11}$$

where the entropy of the incoming energy is associated with a characteristic temperature T_{in}. In this setup, entropy is produced by mixing of incoming energy fluxes $J_{\text{in},h}$ and $J_{\text{in},c}$ in the two boxes at temperatures T_h and T_c as shown in our Fig. 10.4.

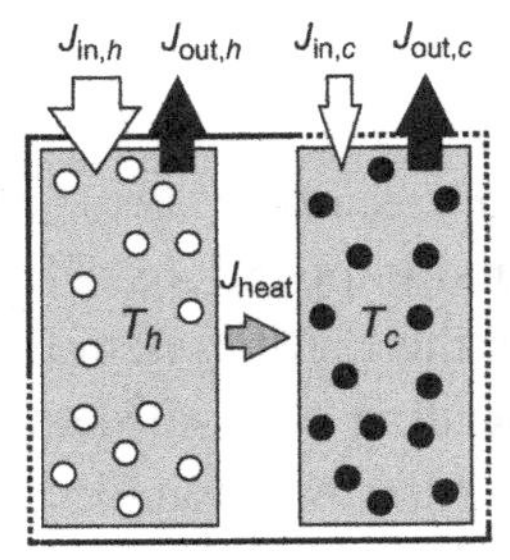

Fig. 10.4
Kleidon's setup used to describe the differential heating of the Earth between the tropics (warm box) and the poles (cold box). This is also a setup that resembles a case with entropy exchanges of two/both boxes with the surroundings. Compare this drawing with a similar illustration 3a in Kleidon (2010a,b).

The entropy budget of the whole system is given by the following equations:

$$\frac{dS_{\text{tot}}}{dt} = \frac{dS_h}{dt} + \frac{dS_c}{dt} = \sigma_{\text{mix},h} + \sigma_{\text{mix},h} + \sigma_{\text{heat}} - \text{NEE} \tag{10.12}$$

where

$$\text{NEE} = \left(\frac{J_{\text{out},h}}{T_h} + \frac{J_{\text{out},c}}{T_c}\right) - (J_{\text{in},h} + J_{\text{in},c})/T_{\text{in}} \tag{10.13}$$

is the net entropy exchange across the system boundary.

What we learn from Kleidon's Fig. 3 describing the above model is that the initial gradient in temperature is depleted, but it does not vanish in the steady state. Instead, a nonzero heat flux transports heat from warm to cold. This acts to deplete the temperature gradient that is continuously built up by the differential heating $J_{\text{in},h} - J_{\text{in},c}$. This heat flux produces entropy by the depletion of the temperature gradient, but instead of increasing the system entropy to the maximum, the entropy produced in steady state is exported by the enhanced export of entropy associated with the outgoing heat flux $J_{\text{out},h} + J_{\text{out},c}$. The enhanced entropy export results from the overall lower temperature at which the total amount of received heat is exported to the surroundings.

We shall now briefly outline contents and results of several (next) sections of Kleidon's paper, the ones which seem the most interesting to chemical and mechanical engineers.

10.5.4 Power extraction and Carnot efficiency

In his Section 4.3, Kleidon develops examples demonstrating how power is extracted from thermodynamic gradients and explains how they can be used to drive a thermodynamic variable out of equilibrium. He starts with the common derivation of the Carnot efficiency of a thermodynamic process driven by heat extraction from a gradient. While the Carnot efficiency is well-known, Kleidon's brief derivation is important in that it illustrates the inherent assumptions that result in the expression for the Carnot efficiency (Kleidon, 2010a,b). This, in turn, is the basis to understand why the maximum efficiency of natural processes is generally much smaller than the Carnot efficiency.

10.5.5 Maximum extraction of power from a gradient

Applying his basic model outlined in our Section 10.5.3 (his Section 2.3), Kleidon proves that both, maximum extraction of power and maximum entropy production (MEP), are associated with approximately the same optimum flux. Also, he notes that the maximum of extracted power is only a quarter of the power associated with the Carnot efficiency. See Chapter 10, p. 499 of a book (Sieniutycz, 2016) for the derivation of a similar result obtained in a more traditional way in the context of electrochemical power yield.

10.5.6 A hierarchy of power transfer

The two largest sources for power generation for surface and interior processes—absorption of solar radiation and heat in the interior—are the natural starting points for a planetary hierarchy of power generation and transfer. (This hierarchy is summarized in Kleidon's Fig. 11.) The planetary hierarchy is used to: (a) illustrate the importance of the hierarchical view in understanding power generation and transfer, (b) to place interactions among Earth system processes in the context of power transfer, and (c) to describe power transfer for the most important processes that drive geochemical cycles and directly relate them to life (Kleidon, 2010a,b).

10.5.7 Discussion of limitations and implications

10.5.7.1 Limitations

There are several limitations and aspects of the results, as, according to Kleidon's own words, his theory "should be only seen as a first sketch that needs improvement" (Kleidon, 2010a,b). The estimates of the global work budget can be improved by more detailed considerations, and simple models to illustrate the hierarchy of power transfer can be fully coupled and implemented in complex Earth system models to reproduce the postulated behavior. Moreover, not all power sources are included in the schematic diagram of Kleidon's Fig. 11. Other hierarchies exist in addition to the one shown, for instance, regarding the transfer of power from atmospheric motion to the oceanic circulation.

Yet the fundamental importance of thermodynamic limits on power generation and transfer (requiring to understand disequilibrium and the emergent dynamics within the Earth system) is unaffected by relatively minor limitations. What is not entirely clear is which thermodynamic variable should be maximized. While the focus in Kleidon's paper is on maximizing power generation, this optimality is very close to the proposed principle of Maximum Entropy Production and other proposed principles related to thermodynamics, such as those implying fastest depletion of gradients (Schneider and Sagan, 2005) or maximizing "access" (Bejan and Lorente, 2006). One aspect that stands out regarding the proposed MEP principle is that substantial progress has been made in recent years to develop the theoretical foundation for this principle (Dewar, 2003, 2005a,b, 2010). This theoretical foundation uses the MaxEnt inference method from information theory

(Jaynes, 1957a,b), which is a well-established basis for equilibrium statistical mechanics and is used to derive equilibrium thermodynamics from this assumption. In simple terms, the MaxEnt procedure yields the best, least biased prediction for a given set of constraints. Since for Earth system processes energy and mass balances always apply, the MaxEnt approach yields the usual Lagrange multipliers for temperature and chemical potentials of equilibrium thermodynamics, so that in practice MaxEnt should translate in maximizing thermodynamic entropy production (Dyke and Kleidon, 2010).

When dealing with Newtonian mechanics, it may well be that the inclusion of the constraints of momentum conservation into the MaxEnt algorithm yields maximum power and dissipation as a result (Dewar's personal communication to Kleidon). This aspect would need further investigation (Kleidon, 2010a,b).

Kleidon's review also addresses some of the criticisms against the proposed MEP principle (Rodgers, 1976; Volk, 2007; Caldeira, 2007; Volk and Pauluis, 2010). This includes the claim that MEP assumes teleology (i.e., that the process would need to "know" what it has to do to maximize, just like the most common objection to the Gaia hypothesis (see above), and that MEP does not tell us which entropy production need to be maximized.

To respond to the first criticism, Kleidon shows by simple means that the power generation is limited by maximum efficiencies, and these limits are fundamental and constrain the dynamics associated with the power generated. The existence of maximum power generation does not necessarily imply that these dynamics in fact utilize the maximally extractable power. The energy and mass balances and other balance constraints are compatible with a full range of possible dynamics that do not need to maximize power extraction. This is where the MaxEnt algorithm comes in. The MaxEnt algorithm tells us that the different sets of possible dynamics that are compatible with the balance constraints are not all equally probable. It is then not teleology, but a matter of highest probability that the generalities observed in the macroscopic dynamics are associated with the maximum in power generation.

The second criticism is clarified to some extent by the hierarchical view of power generation and transfer in that it puts an order in the understanding of which thermodynamic variable causes which other variable to evolve away from equilibrium, and what the influence of that disequilibrium is on the driver. It is through this hierarchy of power generation and transfer that the planetary rates of entropy production can be enhanced. The extent to which planetary entropy production can be enhanced is then limited by the maximal thermodynamic efficiencies along the hierarchy.

Last, but not the least, Kleidon's work corrects and clarifies some earlier notions on the relationship between Maximum Entropy Production and states of minimum entropy (Kleidon, 2010b,c; Paltridge, 2009).

10.5.7.2 Implications

The hierarchical view of the Earth system has implications for how we should model the Earth system, how we can get a holistic description of it that should help us to reconstruct past environments and its evolution in time, and how we should view and evaluate future change due to human activity within the system (Kleidon, 2010a). While energy and mass balances are the basis of practically all Earth system models, the work balance is typically not considered and neither is the hierarchical transfer of power among processes. The latter aspect is likely where the largest shortcomings are found in Earth system models. To give an example, Kleidon (2010a) states: Atmospheric models typically simulate dissipation of kinetic energy in the boundary layer by using semiempirical parameterizations with no-slip boundary conditions. When we think about wind blowing over the ocean or over the forest floor in autumn, then some of this wind is not dissipated in the turbulent cascade, but rather work is extracted that generates waves or lifts leaves off the ground, thereby generating potential energy. This transfer of power can be quite considerable: It is estimated that of the approximately 450 TW of kinetic energy that is dissipated near the surface (Peixoto and Oort, 1992) about 60 TW is used to generate ocean waves (Ferrari and Wunsch, 2009). It is likely that such power transfer processes are not adequately represented in current boundary layer parameterization schemes, resulting in biases in the turbulent heat fluxes (which are linked to the extent of momentum dissipation and their sensitivity to change). The same may hold true in general terms about the coupling of all kinds of thermodynamic variables within Earth system models, leading to the suspicion that these models may not be thermodynamically consistent in the very relevant areas at which interactions take place. The hierarchy shown in Kleidon's Fig. 11 describes the interrelationship of different variables of the Earth system as a function of the ability to generate power. By this, the hierarchy essentially provides a holistic description of the planetary environment. This basis should help us to get a fuller description and understanding of past environments since it shows that many of the variables are not independently driven, but strongly connected by hierarchical power generation and transfer. Hence, this should possibly form the basis to better understand past environments and relate evolutionary trends to changes in power generation (Kleidon, 2010a) and his references (Kleidon, 2009, 2010b).

10.6 Gell-Mann's view of complexities

Contents: What is Complexity? Definition of Complexity, Computational Complexity, and Information Measures in Gell-Mann's publications: Gell-Mann, M., 1995a, Gell-Mann, M., 1995b, Gell-Mann, M., 1995c: *The Quark and the Jaguar*. Below we discusss a slightly abbreviated version of: Gell-Mann, M., 1995a, *Complexity*, A reprint from John Wiley and Sons, Inc.: *Complexity*, Vol.1 no 1, 1995.

10.6.1 Can we define complexity?

What is complexity? In his considerations of the nature of complexity definition, Gell-Mann (1995a,b) states that, in fact, many different measures would be required to capture all our intuitive ideas about what is meant by complexity or by its opposite, simplicity.

Some quantities, like computational complexity (CC), are time or space measures. They are concerned with how long it would take, at a minimum, for a standard universal computer, to perform a prescribed task. CC itself is related to the least time or the smallest number of stages needed to carry out a certain computation.

10.6.2 Link with information measures

Other possible quantities are information measures, pertaining to the length of the shortest message transmitting information. For instance, the algorithmic information content (AIC) of a string of bits is defined as the length of the shortest program which will cause a standard universal computer to print out the string of bits and then stop (halt).

As measures of quantities like complexity, all such quantities are to some extent context-dependent or even subjective. They depend on the coarse graining (level of detail) of the description of the entity, on the previous knowledge and understanding of the world that is assumed, on the language employed, on the coding method used for conversion from that language into a string of bits, and on the particular ideal computer chosen as a standard (Gell-Mann, 1995c).

However, if one is considering sequences of similar entities of increasing size and complexity and is interested only in how the complexity measure behaves as the size becomes large, then, of course, many of arbitrary features become negligible. Thus, students of computational complexity are typically concerned with whether a sequence of larger and larger problems can be solved in time that grows as a polynomial in the problem size (rather than an exponential or something worse). It is probably safe to say that any measure of complexity is most useful for comparisons between things, at least one of which has high complexity by that measure (Gell-Mann, 1995a,c).

10.6.3 Difficulties, trials, and hesitations

Many of the candidate quantities are not computable. For example, the algorithmic information content (AIC) of a long bit string can readily be shown to be less than or equal to some value. But for any such value, there is no way to exclude the possibility that the AIC could be further reduced by some as yet undiscovered theorem revealing a hidden regularity in the string. A bit string that is incomprehensible has no such regularities and is defined as "random." A random bit string has maximal AIC for its length, since the shortest program that will cause the standard computer to print it out and then halt is just the one that says PRINT followed by the string. This property of AIC, which leads to its being called, on occasion, "algorithmic randomness," reveals the unsuitability of the quantity as a measure of complexity, since the works of Shakespeare have a lower AIC than random gibberish of the same length than would typically be typed by the proverbial roomful of monkeys (Gell-Mann, 1995c).

A measure that corresponds much better to what is usually meant by complexity in ordinary conversation, as well in scientific discourse, pertains not to the length of the most concise description of an entity, which is roughly what algorithmic information content (AIC) is, but to the length of a concise description of a set of the entity's regularities. Thus, something almost entirely random, with practically no regularities, would have effective complexity near zero. So would something completely regular, such as a bit string consisting entirely of zeroes. Effective complexity can be high only in a region intermediate between total order and complete disorder (Gell-Mann, 1995a,b,c).

10.6.4 Are identifications of regularity classes helpful?

There can exist no procedure for finding the set of all regularities of an entity. But classes of regularities can be identified. Finding regularities typically refers to taking the available data about the entity, processing it in some manner into, say, a bit string, and then dividing that string into parts in a particular way and looking for mutual AIC among the parts. If a string is divided into two parts, for example, the mutual AIC can be taken to be the sum of the AICs of the parts minus the AIC of the whole. An amount of mutual algorithmic information content above a certain threshold can be considered diagnostic of a regularity. Given the identified regularities, the corresponding effective complexity is the AIC of a description of those regularities.

More precisely, any particular regularities can be regarded as embedding the entity in question in a set of entities sharing the regularities and differing only in other respects. In general, the regularities associate a probability with each entity in the set. The probabilities are in many cases all equal, but they may differ from one member of the set to another. The effective complexity of the regularities can then be defined as the AIC of the description of the set of entities and their probabilities (Gell-Mann, 1995a,c). (Specifying a given entity, such as the original one, requires additional information.)

10.6.5 Further play with algorithmic information

Some authors try to characterize complexity by exploiting the *amount* of mutual algorithmic information rather than the length of a concise description of the corresponding regularities. Such a choice of measure does not agree very well, however, with what is usually meant by complexity. Take, as a simple example, any string of bits consisting entirely of pairs 00 and 11. Such a string possesses an obvious regularity, but one that can be very briefly described: the sequences of odd-numbered and even-numbered bits are identical. The quantity of mutual AIC between those sequences is enormous for a long string. Evidently, the complexity here is better represented by the length of the brief description than by the amount of mutual algorithmic information.

10.6.6 Who or what determines the class of regularities

Since it is impossible to find all regularities of an entity, the question arises as to who or what determines the class of regularities to be identified. One answer is to point to a most important set of systems, each of which functions precisely by identifying certain regularities in the data stream reaching it and compressing those regularities into a concise package of information. The data stream includes information about the system itself, its environment, and the interaction between the environment and the behavior of the system. The package of information or "schema" is subject to variation, in such a way that there is competition among different schemes. Each "schema" can be used, along with some of the data, to describe the system and its environment, to predict the future and to prescribe behavior for the system, (Gell-Mann, 1995a,b,c).

10.6.7 Examples of complex adaptive systems

However, the description and prediction can be checked against further data, with the comparison feeding back to influence the competition schemata. Likewise, behavior conforming to a prescription has real world consequences, which can also affect the competition. In this way, the schemata evolve with a general tendency to favor better description and prediction as well as behavior conforming more or less to the selection pressures in the real world. Examples on Earth of the operation of complex adaptive systems include biological evolution, learning and thinking in animals (including people), the functioning of immune system in mammals and other vertebrates, the operation of human scientific enterprise, and the behavior of computers that are built or programmed to evolve strategies by means of neural nets or genetic algorithms. Clearly, complex adaptive systems have a tendency to give rise to other complex adaptive systems.

10.6.8 A comment on terminology

It is interesting to note that J. Holland, for example, uses a different set of terms to describe some of the same ideas. He uses "adaptive agent" for a complex adaptive system as defined above, reserving the name "complex adaptive system" for a composite complex adaptive system (such as an economy or an ecological system) consisting of many adaptive agents making predictions of one or another's behavior. What Gell-Mann calls a schema, Holland calls an internal model. As Gell-Mann states: "both of us are conforming to the old saying that a scientist would rather use someone else's toothbrush than another scientist's nomenclature" (Gell-Mann, 1995c).

Any complex adaptive system can, of course, make mistakes in spotting regularities. We human beings, who are prone to superstition and often engage in denial of the obvious, are all too familiar with such errors (Gell-Mann, 1995a,b,c).

Besides the possibility of error, we should also consider difficulty of computation. How much time is involved in deducing practical predictions from a highly compressed schema, say a scientific theory, together with some specific additional data such as boundary conditions? Here we encounter time measures of "complexity," for instance logical depth, which for a big string is related to the time required for a standard universal computer to compute the string, print it out, and then halt. That time is averaged over the various programs that will accomplish the task, with an averaging procedure that weights shorter programs more heavily. We can then consider the logical depth of any entity if a suitably coarse-grained description of it is encoded into a bit string.

A kind of inverse concept to logical depth is crypticity, which measures the time needed for a computer to reverse the process and go from a bit string to one of the shorter programs that will generate it. In the human scientific enterprise, we can identify crypticity roughly with the difficulty of constructing a good theory from a set of data, while logical depth is a crude measure of the difficulty of making predictions from the theory.

10.6.9 Is a well-known fractal set really complex?

Fractals in nature originate from self-organized critical dynamical processes (Bak and Chen, 1989; Mandelbrot, 1982). It is often hard to tell whether something that is apparently complex possesses a great deal of effective complexity or reflects instead underlying simplicity combined with a certain amount of logical depth.

Faced with a diagram of Mandelbrot's famous fractal set, for example, we may attribute to it a high effective complexity until we learn that it can be generated from a very simple formula (Gell-Mann, 1995c). It has logical depth (and not even a gigantic amount of that) rather than effective complexity. Another example of this sort is the well-known Sierpiński monofractal (Fig. 10.5).

In contemplating natural phenomena, we frequently have to distinguish between effective complexity and logical depth (Gell-Mann, 1995c; Bak and Chen, 1989). For example, the apparently complicated pattern of energy levels of atomic nuclei might easily be misattributed to some complex law of the fundamental level, but it is now believed to follow

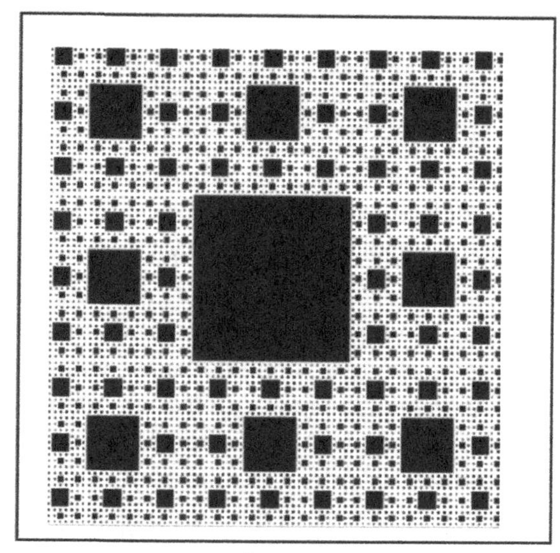

Fig. 10.5
A popular draft of monofractal image of the Sierpinski carpet.

from a simple underlying theory of quarks, gluons, and photons, although lengthy calculations would be required to deduce the detailed pattern from the basic equations. Thus, the pattern has a good deal of logical depth and very little effective complexity. (Gell-Mann, 1995c).

10.6.10 Some further comments and conclusions

It now seems likely that the fundamental law governing the behavior of all matter in the universe—the unified quantum field theory of all the elementary particles and their interactions—is quite simple. (In fact, we already have a plausible candidate in the form of superstring theory.) It also appears that the boundary condition specifying the initial condition of the universe around the beginning of its expansion may be simple as well. If both of these propositions are true, does that mean that there is hardly any effective complexity in the universe? Not at all, because of the relentless operation of chance.

Given the basic law and the initial condition, the history of the universe is by no means determined, because the law is quantum-mechanical, thus yielding only probabilities for alternative histories. Moreover, histories can be assigned probabilities only if they are sufficiently coarse-grained to display incoherence (the absence of interference terms between them). Thus, quantum mechanics introduces a great deal of indeterminacy, going far beyond the rather trivial determinacy associated with Heisenberg's uncertainty principle (Gell-Mann, 1995a,b,c).

Of course, in many cases, the quantum-mechanical probabilities are very close to certainties, so that deterministic classical physics is a good approximation. But even in the classical limit and even when the laws and initial condition are exactly specified, indeterminacy can still be introduced by lack of knowledge of previous history. Moreover, the effects of such ignorance can be magnified by the phenomenon of chaos in nonlinear dynamics, whereby future outcomes are arbitrarily sensitive to time changes in present conditions (Gell-Mann, 1995a).

We can think of the alternative possible coarse-grained histories of the universe as forming a branching tree, with probabilities at each branching. Note, these are a priori probabilities rather than statistical ones, unless we engage in the exercise of treating the universe as one of a huge set of alternative universes, forming a "multiuniverse." Of course, even within a single universe, cases arise of reproducible events (such as physics experiments), and for those events, the a priori probabilities of the quantum mechanics of the universe yield conventional statistical probabilities.

Any entity in the world around us, such as an individual human being, owes its existence not only to the simple fundamental laws of physics and the boundary condition on the early universe, but also to the outcomes of an unconceivably long sentence of probabilistic events, each of which could have turned out differently.

Now a great many of those accidents, for instance most cases of bouncing of a particular molecule in a gas to the right rather than to the left in a molecular collision, have few ramifications for the future coarse-grained histories. Sometimes, however, an accident can have widespread consequences for the future, although those are typically restricted to particular regions of space and time. Such a "frozen accident" produces a great deal of mutual algorithmic information among various parts of aspects of a future coarse-grained history of the universe, for many such histories and for various ways of dividing them up.

10.6.11 Regularities and growing universe

A situation in which there is a great deal of mutual algorithmic information generated corresponds precisely to what is called a regularity (Gell-Mann, 1995c). Thus, as time goes by in the history of the universe, accidents (with probabilities for various outcomes) accumulate, and so do frozen accidents, giving rise to regularities. Most of the effective complexity of the universe lies in the AIC of a description of those frozen accidents and their consequences, while only a smart part comes from the simple fundamental laws of the universe (the law of the elementary particles and the condition at the beginning of the expansion). For a given entity in the universe, it is, of course, for frozen accidents leading up to its own regularities that contribute, along with the basic laws, to its effective complexity (Gell-Mann, 1995a,c).

As the universe grows older and frozen accidents pile up, the opportunities for effective complexity to increase keep accumulating as well. Thus, there is a tendency for the envelope of complexity to expand, even though any given entity may either increase or decrease its complexity during a given time period. The appearance of more and more complex forms is not a phenomenon restricted to the evolution of complex adaptive systems, although for these systems the possibility arises of a selective advantage being associated under certain circumstances with increased complexity (Gell-Mann, 1995c).

10.6.12 Potential complexity for future predictions

The second law of thermodynamics, which requires average entropy (or disorder) to increase, does not in any way prevent local order from arising through various mechanisms of self-organization, which can turn accidents into frozen ones producing extensive regularities. Again, such mechanisms are not restricted to complex adaptive systems.

Different entities may have different potentialities for developing higher complexity. Something that is not particularly distinguished from similar things by its effective complexity can, nevertheless, be remarkable for the complexity it may achieve in the future. Therefore, it is important to define a new quantity, "potential complexity," as a function of future time, relative to a fixed time, say the present. The new quantity is the effective complexity of the entity at each future time, averaged over the various coarse-grained histories of the universe between the present and that time, weighted according to their probabilities.

The era may not last forever in which more and more complex forms appear as time goes on. If, in the very distant future, virtually all nuclei in the universe decay into electrons and positrons, neutrinos and antineutrinos, and photons, then the era characterized by fairly well-defined individual objects may draw to an end, while selforganization becomes rare and the complexity envelope begins to shrink.

The above considerations, (Gell-Mann, 1995a,b,c), summarize a considerable part of the material found in the book *The Quark and the Jaguar* (Gell-Mann, 1995c), which is intended for the lay reader interested in science. Gell-Mann offers therein a personal, unifying vision of the relationship between the fundamental laws of physics and the complexity and diversity of the natural world. The book explains the connections between nature at its most basic level and natural selection, archeology, linguistics, child development, computers, and other complex adaptive systems.

10.7 Proposals of other authors

Unifying approaches to complex systems are continued. Oscillatory reactions are analyzed in a unifying approach showing explosive, conservative, or damped behavior.

A synthetic review of chemical stability properties has been presented by Farkas and Noszticzius (1992) for the three basic groups of the systems showing explosive, conservative, or damped behavior, depending on parameters of a kinetic model. Their unified treatment of chemical oscillators considers these oscillators as thermodynamic systems far from equilibrium. The researchers start with the theory of the Belousov-Zhaboutinski (B-Z) reaction pointing out the role of positive and negative feedback (Zhabotinsky, 1974). Then, they show that consecutive autocatalytic reaction systems of the Lotka-Volterra type, Fig. 2.1 in Chapter 2, play an essential role in the B-Z reaction and other oscillatory reactions.

Consequently, Farkas and Noszticzius (1992) introduce and then analyze certain (two-dimensional) generalized Lotka-Volterra models similar to those in Fig. 2.1. They show that these models can be conservative, dissipative, or explosive, depending on the value of a parameter. The transition from dissipative to explosive behavior occurs via a critical Hopf bifurcation. The system is conservative at certain critical values of the parameter. The stability properties can be examined by testing the sign of a Liapounov function selected as an integral of the conservative system. By adding a limiting reaction to an explosive network, a limit cycle oscillator surrounding an unstable singular point is obtained. Some general thermodynamic aspects are discussed pointing out the distinct role of oscillatory dynamics as an inherently far-from-equilibrium phenomenon. The existence of the attractors is recognized as a consequence of the second law. The paper of Farkas and Noszticzius (1992) gives a pedagogical perspective on the broad class of problems involving the time order in homogeneous chemical systems.

References

Ahern, J.E., 1980. The Exergy Method of Energy Systems Analysis. Wiley, New York.

Bak, P., Chen, K., 1989. The physics of fractals. Physica D 38, 5–12.

Bejan, A., Lorente, S., 2006. Constructal theory of generation of configuration in nature and engineering. J. Appl. Phys. 100 (4), 041301.

Berezowski, M., 2003. Fractal solutions of recirculation tubular chemical reactors. Chaos Soliton. Fract. 16, 1–12.

Bisio, G., 1994. Non-equilibrium steady states and metastable equilibrium states in some particular systems. In: Tsatsaronis, G. (Ed.), Proceedings of Second Biennial European Joint Conference on Engineering System Design and Analysis, London, July 4–7. In: vol. 3. ASME, New York, pp. 109–116.

Bisio, G., 1997. Exergy degradation and system evolution in the light of non-equilibrium thermodynamics with particular reference to Rayleigh-Benard phenomena. In: Proceedings of 4th World Conference on Experimental Heat Transfer, Fluid Mechanics and Thermodynamics, Brussels, June 2–6 (Invited Lecture). vol. 1. Edizioni ETS, Pisa, pp. 443–450.

Bisio, G., Guglielmini, G., Tanda, G., 1989. Experimental study on Rayleigh-Benard convection in a high Prandtl number fluid. In: Presented at the Eurotherm Seminar No. 11, Natural Convection in Enclosures, Harvell Laboratory, England, 7–8 December, pp. 1–11.

Bisio, G., Pistoni, C., Schiaparelli, P., 1999. Order and disorder in non-living and living matters. Private Communication of G. Bisio to the author of the present book.
Blum, H.F., 1968. Time's Arrow and Evolution. University Press, Princeton.
Blythe, R.A., 2008. Reversibility, heat dissipation, and the importance of the thermal environment in stochastic models of nonequilibrium steady states. Phys. Rev. Lett. 100, 010601.
Boltzmann, L., 1886. Der zweite Hauptsatz der mechanischen Wärmetheorie. Alm. Akad. Wiss. 36, 225–259.
Brown, W., 1973. Heat-flux transitions at low Rayleigh number. J. Fluid Mech. 69, 539–559.
Brzustowski, T.A., Golem, P.J., 1978. Second law analysis of energy processes, part 1: exergy—an introduction. Trans. Can. Soc. Mech. Eng. 4 (4), 209–218.
Burghardt, A., Berezowski, M., 2003. Periodic solutions in a porous catalyst pellet—homoclinic orbits. Chem. Eng. Sci. 58, 2657–2670.
Bushev, M., 2000. A note on Einstein's Annus Mirabilis. Ann. Fond. Louis de Broglie 25 (3), 1–2. http://www.phys.uni-sofia.bg/upb/old/C22.PDF.
Caldeira, K., 2007. The maximum entropy principle: a critical discussion. Clim. Chang. 85, 267–269.
Callen, H., 1988. Thermodynamics and an Introduction to Thermostatistics. Wiley, New York.
Capra, F., 1996. The Web of Life: A New Scientific Understanding of Living Systems. Anchor Books, New York.
Caratheodory, C., 1976. Investigations into the foundations of thermodynamics. In: Kestin, J. (Ed.), The Second Law of Thermodynamics. In: Benchmark Papers on Energy, vol. 5. Dowden, Hutchinson, and Ross, Stroudsburg, PA, pp. 229–256.
Chandrasekhar, S., 1961. Hydrodynamic and Hydromagnetic Stability. Clarendon Press, Oxford.
Crooks, G.E., 1999. Entropy production fluctuation theorem and the nonequilibrium work relation for free energy differences. Phys. Rev. E 60, 2721.
Currie, D., 1991. Energy and large-scale patterns of animal-and-plant species-richness. Am. Nat. 137, 27–48.
De Groot, S.R., Mazur, P., 1984. Nonequilibrium Thermodynamics. Dover, New York.
Dewar, R.C., 2003. Information theory explanation of the fluctuation theorem, maximum entropy production, and self-organized criticality in non-equilibrium stationary states. J. Phys. A 36, 631–641.
Dewar, R.C., 2005a. Maximum entropy production and non-equilibrium statistical mechanics. In: Kleidon, A., Lorenz, R.D. (Eds.), Non-equilibrium Thermodynamics and the Production of Entropy: Life, Earth, and Beyond. Springer Verlag, Heidelberg, Germany.
Dewar, R.C., 2005b. Maximum entropy production and the fluctuation theorem. J. Phys. A 38, L371–L381. https://doi.org/10.1088/0305-4470/38/21/L01.
Dewar, R.C., 2010. Maximum entropy production as an inference algorithm that translates physical assumptions into macroscopic predictions: don't shoot the messenger. Entropy 11 (4), 931–944.
Dyke, J., Kleidon, A., 2010. The maximum entropy production principle: its theoretical foundations and applications to the earth system. Entropy 12 (3), 613–630.
Ebeling, W., 1983. Discussion of the Klimontovich theory of hydrodynamic turbulence. Ann. Phys. 40 (1), 25–33.
Ebeling, W., 1985. Thermodynamics of selforganization and evolution. Biomed. Biochim. Acta 44, 831–838.
Ebeling, W., 2002. Canonical dissipative systems and applications to active Brownian motion. Nonlinear Phenom. Complex Syst. 5 (4), 325–331.
Ebeling, W., Engel-Herbert, H., 1986. Entropy lowering and attractors in phase space. Acta Phys. Hung. 66 (1–4), 339–348.
Ebeling, W., Feistel, R., 1992. Theory of selforganization and evolution: the role of entropy, value and information. J. Non-Equilib. Thermodyn. 17, 303–332.
Ebeling, W., Feistel, R., 2011. Physics of Self-Organization and Evolution. Wiley-VCH, Berlin. ISBN: 978-3-527-40963-1.
Ebeling, W., Klimontovich, Y.L., 1984. Selforganization and Turbulence in Liquids. Teubner Verlagsgesellschaft, Leipzig.
Ebeling, W., Engel-Herbert, H., Herzel, H., 1986. On the entropy of dissipative and turbulent structures. Annal. Phys. (Leipzig) 498 (3–5), 187–195.
Eddington, A., 1958. The Nature of the Physical World. University of Michigan Press, Ann Arbor, MI.

Eigen, M., 1971. Authoatalytic view in the theory of life. Naturwissenschaften 58, 465 (in German).
Eigen, M., Schuster, P., 1979. The Hypercycle: A Principle of Natuml Self-Organization. Springer-Verlag.
England, J.L., 2013. Statistical physics of self-replication. J. Chem. Phys. 139, 121923. https://doi.org/10.1063/1.4818538.
Farkas, H., Noszticzius, Z., 1992. Explosive, stable and oscillatory behavior in some chemical systems. In: Flow, Diffusion and Transport Processes. Advances in Thermodynamics Series, vol. 6. Taylor & Francis, New York, pp. 303–339.
Fermi, E., 1956. Thermodynamics. Dover Publications, New York.
Ferrari, R., Wunsch, C., 2009. Ocean circulation kinetic energy: reservoirs, sources, and sinks. Annu. Rev. Fluid Mech. 41, 253–282. following A. Kleidon's citation.
Frieden, B.R., Plastino, A., Plastino, A.R., Soffer, B.H., 2002. Schrödinger link between nonequilibrium thermodynamics and fisher information. Phys. Rev. E 66, 046128.
Gardiner, W., 2003. Handbook of Stochastic Methods, third ed. Springer, New York.
Gates, D., 1962. Energy Exchange in the Biosphere. Harper and Row, New York.
Gell-Mann, M., 1995a. Complexity. Complexity 1 (1), 1995 (A reprint from John Wiley and Sons, Inc.).
Gell-Mann, M., 1995b. What is complexity? Remarks on simplicity and complexity by the Nobel prize-winning author of *The Quark and the Jaguar*. Complexity 1 (1), 16–19. https://doi.org/10.1002/cplx.6130010105.
Gell-Mann, M., 1995c. The Quark and the Jaguar: Adventures in the Simple and the Complex. W. H. Freeman, New York, ISBN: 0716727250.
Gilbert, W., 1986. Origin of life: The RNA world. Nature 319, 618. https://doi.org/10.1038/319618a0.
Glansdorff, P., Prigogine, I., 1971. Thermodynamic Theory of Structure Stability and Fluctuations. Wiley, New York.
Gomez-Marin, A., Parrondo, J.M., Van den Broeck, C., 2008. On bounds set on entropy production in information-theoretic terms. Phys. Rev. E. 78, 011107.
Hatsopoulos, G., Keenan, J., 1965. Principles of General Thermodynamics. John Wiley, New York.
Ishida, K., 1981. Nonequilibrium thermodynamics of the selection of biological macromolecules. J. Theor. Biol. 88, 257–273.
Jaynes, E.T., 1957a. Information theory and statistical mechanics. Phys. Rev. 106 (4), 620–630.
Jaynes, E.T., 1957b. Information theory and statistical mechanics. II. Phys. Rev. 108 (2), 171–190.
Katchalsky, A., Curran, P.F., 1965. *Nonequilibrium Thermodynamics* in *Biophysics*. Harvard University Press, Cambridge, MT.
Kay, J.J., 1984. Self-Organization in Living Systems (Ph.D. thesis). University of Waterloo. Systems Design Engineering. Waterloo, Ontario.
Kay, J., Schneider, E., 1992. Thermodynamics and measures of ecosystem integrity. In McKenzie, D., Hyatt, D., McDonald, J. (Eds.), Ecological Indicators. Elsevier, New York, pp. 159–181.
Kay, J.J., Graham, L., Ulanowicz, R.E., 1989. A detailed guide to network analysis. In: Wulff, F., Field, J.G., Mann, K.H. (Eds.), Network Analysis in Marine Ecosystems. Coastal and Estuarine Studies, vol. 32. Springer-Verlag, New York, pp. 16–61.
Kestin, J., 1966. A Course in Thermodynamics. Blaisdell, New York.
Kestin, J., 1968. A Course in Thermodynamics. Hemisphere Press, New York.
Kleidon, A., 2009. Maximum entropy production and general trends in biospheric evolution. Paleontol. J. 43, 130–135.
Kleidon, A., 2010a. Life, hierarchy, and the thermodynamic machinery of planet earth. Phys. Life Rev. 7 (2010), 424–460.
Kleidon, A., 2010b. Non-equilibrium thermodynamics, maximum entropy production, and earth system evolution. Philos. Trans. R. Soc. Lond. 368, 181–196.
Kleidon, A., 2010c. A basic introduction to the thermodynamics of Earth system far from equilibrium and maximum entropy production. Philos. Trans. R. Soc. Lond. B 365, 1303–1315.
Klimontovich, Y.L., 1982. Kinetic Theory of Nonideal Gases and Nonideal Plasmas. Pergamon, Oxford.
Klimontovich, Y.L., 1986. Statistical Physics. Harwood Academic Publishers, Chur.

Klimontovich, Y.L., 1991. Turbulent Motion and the Structure of Chaos. Kluwer Academics, Dordrecht.

Klimontovich, Y.L., 1999. Entropy, information, and criteria of order in open systems. Nonlinear Phenom. Complex Syst. 2 (4), 1–25.

Landauer, R., 1961. Irreversibility and heat generation in the computing process. IBM J. Res. Dev. 5, 183.

Lotka, A., 1922. Contribution to the energetics of evolution. Proc. Natl. Acad. Sci. USA 8, 148–154.

Lovelock, J.E., 1972a. Gaia as seen through the atmosphere. Atmos. Environ. 6, 579–580.

Lovelock, J.E., 1972b. Gaia: A New Look at Life on Earth. Oxford University Press, Oxford, UK.

Luvall, J.C., Holbo, H.R., 1989. Measurements of short term thermal responses of coniferous forest canopies using thermal scanner data. Remote Sens. Environ. 27, 1–10.

Luvall, J.C., Holbo, H.R., 1991. Thermal remote sensing methods in landscape ecology. In: Turner, M., Gardner, R.H. (Eds.), Quantitative Methods in Landscape Ecology. Springer-Verlag (Chapter 6).

Mandelbrot, B.B., 1982. The Fractal Geometry of Nature. W.H. Freeman and Company, New York.

Månsson, B.Å.G., Lindgren, K., 1990. Thermodynamics, information and structure. In: Nonequilibrium Theory and Extremum Principles. Advances in Thermodynamics Series. vol. 3. Taylor and Francis, New York, pp. 95–128.

Martyushev, L.M., Seleznev, V.D., 2006. Maximum entropy production principle in physics, chemistry and biology. Phys. Rep. 426 (1), 1–45.

Mikulecky, D.C., 1984. Network thermodynamics: a simulation and modeling method based on the extension of thermodynamic thinking into the realm of highly organized systems. Math. Biosci. 72, 157–179 (1984).

Minetti, C.A., Remeta, D.P., Miller, H., Gelfand, C.A., Plum, G.E., Grollman, A.P., Breslauer, J.K., 2003. The thermodynamics of template-directed DNA synthesis: base insertion and extension enthalpies. Proc. Natl. Acad. Sci. U. S. A. 100, 14719.

Nicolis, G., Prigogine, I., 1977. Self-Organization in Noneguilibrium Systems. John Wiley, New York.

Nicolis, G., Prigogine, I., 1979. Irreversible processes at nonequilibrium steady states and Lyapounov functions. Proc. Natl. Acad. Sci. USA 76 (12), 6060–6061.

Nicolis, G., Prigogine, I., 1989. Exploring Complexity. W.H. Freeman, New York.

Nicolis, G., Rouvas-Nicolis, C., 2007a. Foundations of Complex Systems. World Scientific, Singapore.

Nicolis, G., Rouvas-Nicolis, C., 2007b. Complex systems. Scholarpedia 2 (11), 1473.

Nowak, M.A., 1992. What is a quasispecies? Review. Trends Ecol. Evol. 7 (4), 118–121. https://doi.org/10.1016/0169-5347(92)90145-2.

Odum, H.T., Pinkerton, R.C., 1955. Time's speed regulator. Am. Sci. 43, 321–343.

Paltridge, G.W., 2009. A story and a recommendation about the principle of maximum entropy production. Entropy 11, 945–948.

Peixoto, J.P., Oort, A.H., 1992. Physics of Climate. American Institute of Physics, New York.

Peusner, I., 1986. Studies in Network Thermodynamics. Elsevier, Amsterdam.

Prigogine, I., 1955. Thermodynamics of Irreversible Processes. John Wiley, New York.

Rodgers, C.D., 1976. Minimum entropy exchange principle—reply. Q. J. R. Meteorol. Soc. 102, 455–457.

Schneider, E.D., 1987. Schrodinger shortchanged. Nature 328, 300.

Schneider, E., 1988. Thermodynamics, information, and evolution: new perspectives on physical and biological evolution. In: Weber, B.H., Depew, D.J., Smith, J.D. (Eds.), Entropy, Information, and Evolution: New Perspectives on Physical and Biological Evolution. MIT Press, Boston, pp. 108–138.

Schneider, E.D., Kay, J.J., 1994. Life as a manifestation of the second law of thermodynamics. Math. Comput. Model. 19 (6–8), 25–48.

Schneider, E.D., Kay, J.J., 1995. Order from disorder: the thermodynamics of complexity in biology. In: Murphy, M.P., O'Neill, L.A.J. (Eds.), What Is Life: The Next Fifty Years. Reflections on the Future of Biology. Cambridge University Press, pp. 161–172.

Schneider, E.D., Sagan, D., 2005. Into the Cool: Energy Flow, Thermodynamics and Life. University of Chicago Press.

Schrödinger, E., 1944. *What is Life*? The Physical Aspect of the Living Cell. Cambridge University Press, Cambridge, UK.

Shiner, J.S., Sieniutycz, S., 1994. The chemical dynamics of biological systems: Variational and extremal formulations. Prog. Biophys. Mol. Biol. 62, 203–221.

Shiner, J.S., Sieniutycz, S., 1997. The mechanical and chemical equations of motion of muscle contraction. Phys. Rep. 290, 183–199.

Sieniutycz, S., 2012. Maximizing power yield in energy systems—a thermodynamic synthesis. Appl. Math. Model. 36, 2197–2212.

Sieniutycz, S., 2016. Thermodynamic Approaches in Engineering Systems. Elsevier, Oxford/Cambridge, MA.

Sieniutycz, S., Jeżowski, J., 2009. Energy Optimization in Process Systems. Elsevier, Oxford, UK.

Sieniutycz, S., Jeżowski, J., 2013. Energy Optimization in Process Systems and Fuel Cells, second ed. Elsevier, Oxford, UK.

Sieniutycz, S., Jeżowski, J., 2018. Energy Optimization in Process Systems and Fuel Cells, third ed. Elsevier, Oxford, UK.

Silveston, P.L., 1957. Warmedurchange in horizontalen Flassigkeitschichtem (Ph.D. thesis). Techn. Hochsch., Muenchen, Germany.

Swenson, R., 1989. Emergent attractors and the law of maximum entropy production: Foundations to a theory of general evolution. Syst. Res. 6 (3), 187–197.

Szargut, J., 2002. Application of exergy for the determination of the proecological tax replacing the actual personal taxes. Energy 27, 379–389.

Szargut, J., 2005. Exergy Method: Technical and Ecological Applications. WIT Press, Southampton.

Szargut, J., Petela, R., 1965. Egzergia. Wydawnictwa Naukowo Techniczne, Warszawa (in Polish).

Szargut, J., Morris, D.R., Steward, F., 1988. Exergy Analysis of Thermal, Chemical and Metallurgical Processes. Hemisphere, New York.

Thompson, J.E., Kutateladze, T.G., Schuster, M.C., Venegas, F.D., Messmore, J.M., Raines, R.T., 1995. Limits to catalysis by ribonuclease. Bioorg. Chem. 23, 471.

Ulanowicz, R.E., 1986. Growth and Development: Ecosystem Phenomenology. Springer-Verlag, New York.

Ulanowicz, R.E., 1997. Ecology, the Ascendent Perspective. Columbia University Press, New York.

Ulanowicz, R.E., Hannon, B.M., 1987. Life and the production of entropy. Proc. R. Soc. Lond. B 232, 181–192.

Verley, G., 2012. Fluctuations et réponse des systèmes hors de l'équilibre. Mécanique statistique [condmat.stat-mech]. Université Pierre et Marie Curie-Paris VI, Français.

Verley, G., Chétrite, R., Lacoste, D., 2012. Inequalities generalizing the second law of thermodynamics for transitions between nonstationary states. Phys. Rev. Lett. 108, 120601.

Volk, T., Pauluis, O., 2010. It is not the entropy you produce, rather, how you produce it. Philos. Trans. R. Soc. Lond. B 365, 1317–1322.

Volk, T., 2007. The properties of organisms are not tunable parameters selected because they create maximum entropy production on the biosphere scale: A by-product framework in response to Kleidon. Clim. Ch. 85, 251–258.

Watson, J.D., Crick, F.H.C., 1953. Molecular structure of nucleic acids. Nature 171 (4356), 737–738.

Wolchover, K., Taylor, K., 2014. A new physics theory of life. Quanta Magazine. January 22, https://www.quantamagazine.org/a-new-thermodynamics-theory-of-the-origin-of-life-20140122/.

Woo, H.J., Vijaya Satya, R., Reifman, J., 2012. PLoS Comput. Biol. 8, e1002534.

Zainetdinov, R.I., 1999. Dynamics of informational negentropy associated with self-organization process in open system. Chaos, Solitons Fractals 10, 1425–1435 (Chapter 11).

Zalewski, M., Szwast, Z., 2007. Chaos and oscillations in chosen biological arrangements of the prey-predator type. Chem. Process Eng. 28, 929–939.

Zalewski, M., Szwast, Z., 2008. Chaotic behaviour of chemical reaction in a deactivating catalyst particles. In: Proceedings of the 18th International Congress of Chemical and Process Engineering, CHISA, August 24–28, Praha, Abstr. P722, pp. 1–10 (on CD). http://www.chisa.cz/2008/.

Zhabotinsky, A.M., 1967. Oscillatory Processes in Biological and Chemical Systems. Science Publishing, Moscow.

Further reading

Aharony, A., Feder, J., 1989. Fractals in Physics: Essays in honour of Benoit B. Mandelbrot. In: Proceedings of the International Conference, Vance, France. North-Holland, Amsterdam.

Andrieux, D., Gaspard, P., 2009. Kinetics and thermodynamics of living copolymerization processes. Proc. Natl. Acad. Sci. U. S. A. 105, 9516.

Bejan, A., Lorente, S., 2010. The constructal law of design and evolution in nature. Philos Trans R Soc Lond Ser B 365, 1335–1347.

Bejan, A., Marden, J.H., 2009. The constructal unification of biological and geophysical design. Phys. Life Rev. 6, 85–102.

Bisio, G., Bisio, A., 1998. La non-linearita: dato fundamentale del mondo fisico ed applicazioni in campi notevolmente diversi. In: Atti del 53o Congresso Nazionale ATI, Firenze 14–18 Septembre, I. SGE, Padova, pp. 409–420.

Boltzmann, L., 1974. The second law of thermodynamics. In: McGinness, B. (Ed.), Ludwig Boltzmann, Theoretical Physics and Philosophical Problems. D. Reidel Publishing Co., Dordrecht.

Checkland, P., 1981. System Thinking, Systems Practice. Wiley, New York. Foundation of an area called soft systems methodology for social systems management.

Devine, S.D., 2016. Understanding how replication processes can maintain systems away from equilibrium using Algorithmic Information Theory. Biosystems 140, 8–22 (Replication can be envisaged as a computational process that is able to generate and maintain order far-from-equilibrium.).

Fasham, M.J.R. (Ed.), 1984. Flows of *Energy and Materials in Marine* Ecosystems. Plenum, London, pp. 23–47. A thesis applications.

Joslyn, C., 1990. Basic Books on Cybernetics and Systems Science. Materials of Ccourse SS-501: Introduction to Systems Science. Systems Science Department of SUNY, Binghampton.

Keizer, J., 1987. Statistical Thermodynamics of Non-equilibrium Processes. Springer, New York.

Kleidon, A., Fraedrich, K., Kunz, T., Lunkeit, F., 2003. The atmospheric circulation and states of maximum entropy production. Geophys. Res. Lett. 30 (23), 2223. https://doi.org/10.1029/2003GL018363.

Lincoln, T.A., Joyce, G.F., 2009. Self-sustained replication of an RNA enzyme. Science 323, 1229–1232. https://doi.org/10.1126/science.1167856.

McLean, A.R., Nowak, M.A., 1992. Competition between zidovudine sensitive and resistant strains. Trends Ecol. Evol. 7, 118–121.

Peixoto, J.P., Oort, A.H., de Almeida, M., Tome, A., 1991. Entropy budget of the atmosphere. J. Geophys. Res. 96 (10), 981–988.

Sieniutycz, S., Farkas, H. (Eds.), 2005. Variational and Extremum Principles in Macroscopic Systems. Elsevier, Oxford, pp. 497–522.

Turner, J.S., 1979. Nonequilibrium thermodynamics, dissipative structures and self-organization: Some implications for biomedical research. In: Scott, G.P., Ames, J.M. (Eds.), Dissipative Structures and Spatiotemporal Organization Studies in Biomedical Research. Iowa State University Press.

Glossary

$\mathbf{A}$	vector of chemical affinities ($\mathrm{J\,mol^{-1}}$)
A, A^{class}	generalized and classical availability or exergy (J, $\mathrm{kJ\,kg^{-1}}$)
A_i	amount of ith component in a reacting mixture (mol, kmol, kg)
a	power exponent of temperature in energy exchange equations (–)
a, b, c	coefficients of Lorentz attractor, Fig. 2.2
a_1, a_2, b_1, b_2, p, q	coefficients of Lotka-Volterra model (–), Fig. 2.1
$a(p_0, p_1)$	deviation of a length from the Euclidean length (–), Eq. (3.4)
B	exergy content in the system (J, kJ)
$\mathbf{C}$	vector of molar concentrations in the system ($\mathrm{mol\,m^{-3}}$)
C	control parameter varying along direction of horizontal axis, Fig. 1.2
CC	number of customer contacts in market growth model, Chapter 9
c_i	molar concentration of ith component ($\mathrm{mol\,m^{-3}}$)
c_0	propagation speed of thermal waves ($\mathrm{m\,s^{-1}}$)
D	disorder (–), diffusion coefficient ($\mathrm{m^2\,s^{-1}}$), profit variable ($)
DDC	delivery delay condition in market growth model, Chapter 9
DDOG	delivery delay operating goal of Eq. (9.3) in market growth model, Chapter 9
d^2S	second differential of entropy ($\mathrm{J\,K^{-1}}$)
E	total energy or energy type quantity (J)
E	activation energy for chemical reaction ($\mathrm{J\,mol^{-1}}$)
E_d	activation energy for catalyst deactivation ($\mathrm{J\,mol^{-1}}$)
$\mathbf{E}$	electric field vector ($\mathrm{kg\,m\,A\,s^{-3}}$)
f	distribution function in Boltzmann's equation of one-component system (Chapter 5)
$f\,i(\mathbf{r}, \mathbf{c}_i, t)$	distribution functions for a mixture, $i = 1, 2, \ldots n$ (Chapter 5)
$F(c)$	function describing optimal concentration, Eq. (8.36)
$f(c), \phi(c)$	functions describing effect of concentration in Eqs. (8.9), (8.10)
$f_b(u_i)$	bipolar sigmoidal function described by Eq. (4.1)
$f_1, f_2, \ldots$	rate functions in Pontryagin's maximum principle
$F^n(\mathbf{x}^n)$	optimal performance function of dynamic programming
F_n	function vanishing for states of complexity-maximizing entropy (Chapter 3)
G	gas mass flux ($\mathrm{kg\,s^{-1}}$), molar flux ($\mathrm{mol\,s^{-1}}$)
$\mathbf{g}$	matrix of covariant metric tensor in Chapter 3
g	Reynolds flux ($\mathrm{kg\,m^{-1}\,s^{-1}}$), gravity acceleration ($\mathrm{m\,s^{-2}}$)
$g(a) = a, \psi(a)$	functions describing activity effect on reaction and deactivation
$G(a)$	function describing optimal activity profile of catalyst, Eq. (8.38)
g_1, g	partial and overall conductances ($\mathrm{J\,s^{-1}\,K^a}$)

H	information theoretic entropy (–)
H (x, u, p, t)	Hamiltonian function of continuous process
H^{n-1} ($\mathbf{x}^n$, $\mathbf{u}^n$, $\mathbf{p}^{n-1}$, t^n)	Hamiltonian function of discrete process at stage n
$\mathcal{H}$	nondimensional coefficient of homeostasis (–)
I	flux of inert component (mol s^{-1}), fluid flux passing through the system (kg s^{-1})
J	length functional, Eq. (3.4), minimized vs slope controls dp_1/dp_0 (m)
$\mathbf{i}$	molar flux density of electric current (mol m^{-1} s^{-1})
K	chemical equilibrium constant (–), kinetic energy (J), cost flux ($ s^{-1})
k	reaction rate constant for chemical reaction (mol s^{-1} for the first-order reaction)
k_0'	overall effective reaction rate constant in Eq. (8.11)
$k_{eff}=k(T)a$	effective reaction rate constant for catalyst decay problems (e.g., mol s^{-1})
$k^*_{eff}(\tau)$	abbreviated expression for effective reaction rate constant of Eq. (8.99)
k	thermal conductivity coefficient (W K^{-1} m^{-1})
k_B	Boltzmann constant (1.381×10^{-23} J K^{-1} = 1.3803×10^{-16} erg K^{-1})
k_d	deactivation rate constant (units mol s^{-1} for first-order deactivation)
$\mathbf{L}=[L_{ik}]$	Onsager's matrix of phenomenological coefficients
$L_q=\lambda T^2$	Onsager coefficient for heat transfer
l	length coordinate (m)
M_i	molar mass of ith species (kg $kmol^{-1}$)
m	mass of body, mass of particle (g, kg)
N	number of moles, number of particles (–)
N	total number of stages, number of transfer units (–)
N_A	Avogadro number equal 6.024×10^{23} (mol^{-1})
n	number of states, stage number (–), number density (m^{-3}), molar flux (mol s^{-1})
$\mathbf{P}$	pressure tensor (Pa)
P	entropy production rate in Silveston's and Brown's experiments, Chapter 9 (J s^{-1})
P, p	cumulative and local power output, respectively (J s^{-1})
P	profit integral to be maximized, Eq. (8.78) ($)
p	thermodynamic pressure (Pa)
$\mathbf{p}$	probability vector (–)
p_i	probabilities, costate variables in maximum principle (–)
$p_n(t)$	probability of having a population of n at time t, Eq. (10.6)
Q	total heat flux including mass flux (J s^{-1}), heat power (J s^{-1})
$\Delta Q_{i=j}$	heat expelled into bath over transition from microstate i to microstate j
$\mathbf{q}$	vector of heat flux density (J s^{-1} m^{-2})
Ra	Rayleigh number or product of Grashof number and Pr is the Prandtl number (–) $Ra=\frac{g\beta'\Delta T x^3}{\nu\kappa}=Gr\,\mathrm{Pr}$ where: g—acceleration due to gravity, β'—coefficient of thermal expansion of the fluid, ΔT—temperature difference, x—length, ν—kinematic viscosity, and κ—fluid's thermal diffusivity. Gr is the Grashof number and Pr is the Prandtl number
R	universal gas constant = 8.3144598 J K^{-1} mol^{-1} (1.985 cal K^{-1})
r	recycled stream (mol s^{-1})
r_j	rate of jth reaction (mol m^{-3} s^{-1})
$\mathbf{r}$	vector of reaction rates (mol m^{-3} s^{-1}), radius vector (m)
$R^n(\mathbf{x}^n, t^n)$	optimal function of cost type in terms of state, time, and number of stages
S	entropy (J K^{-1}), information-theoretic entropy (–)
S	quality index (8.64) for a steady cyclic process ($ h^{-1})

S_σ	total dissipated entropy ($J K^{-1}$)
s	specific entropy ($J K^{-1} g^{-1}$)
T	temperature of fluid mixture, catalyst temperature (K, °C)
T_g	temperature of gaseous mixture (K, °C)
T^n	temperature of phase leaving stage n (K)
T^*	upper level or limit of permissible temperature (K)
T_1, T_2	bulk temperatures of reservoirs 1 and 2 (K)
$T_{1'}, T_{2'}$	active temperatures of fluid circulating in engine (K)
T^e	constant equilibrium temperature of environment (K)
T_h, T_c	temperatures of hot and cold boxes in Kleidon's model (K)
T_c	temperature of optimal isothermal contacting, Eq. (8.48)
t	time, chronological time, space time, residence time (s)
$\hat{t}_k$	optimal final time for a process which terminates at t_k (s)
t^N	residence time of a stream in all stages N stage of multistage process (s)
$\mathbf{U}^n, U^{ni}$	control vector at stage n and its ith coordinate
$\mathbf{u}$	control vector, hydrodynamic (barycentric) velocity ($m s^{-1}$)
u_j	jth control variable as component of control vector $\mathbf{u}$
V	volume (m^3), voltage (V), potential, maximum work function ($J mol^{-1}$)
$V^n(\mathbf{x}^n, t^n)$	optimal profit function in terms of state, time, and number of stages
$v = \rho^{-1}$	specific volume ($m^3 kg^{-1}$)
W, W	work and cumulative work, respectively (J)
W	heat dissipation rate in Silveston and Brown experiments, Chapter 9 ($J s^{-1}$)
$w_p(\tau)$	initial activity distribution of catalyst or enzyme (–)
$w_k(\tau)$	final activity distribution of catalyst or enzyme (–)
$\mathbf{X}$	vector of independent thermodynamic forces, set of variables
$\mathbf{X}^n, X^{ni}$	state vector at stage n and its ith coordinate
X	concentration of active component in fuel, moles per mole of inert ($mol\, mol^{-1}$)
X	state variable in the bifurcation diagram, Fig. 1.2
X_1, X_2	state variables of Lotka-Volterra model (–), Fig. 2.1
$\mathbf{x} = (x_1, x_2, \ldots x_i \ldots x_s)$	state vector of general dynamical process
$\mathbf{x} = (x_1, x_2, \ldots x_i \ldots x_s, t)$	enlarged state vector of general dynamical process
x	molar fraction of active component in the fuel ($mol\, mol^{-1}$)
x, y, z	state variables of Lorentz attractor, Fig. 2.2
$y = T^{-1}$	reciprocal of optimal temperature (K^{-1}), Eqs. (8.23), (8.39)
$Z\ (x_1, x_2, u, t)$	Horn's variable, Eq. (8.20)
$Z(\beta)$	canonical partition function of the system, Chapter 10 (–)
$\overline{Z}$	average net profit acquired from the reactor ($\$ h^{-1}$)
z	adjoint variable, vertical coordinate (m)

Greek symbols

a	overall heat transfer coefficient ($J m^{-2} s^{-1} K^{-1}$)
α_d	average value of the product of k_d and $\phi(\mathbf{c})$ in Eq. (8.8)
ϕ	electric potential (V)
Γ	complexity, complexity potential (–)
γ	cumulative conductance of the system ($J s^{-1} K^{-1}$)
Δ	disequilibrium correction, increment, deviation (–)

Φ	internal irreversibility factor (–), dissipation function ($J K^{-1} s^{-1}$)
η	first-law thermal efficiency $p(q_1)^{-1}$ (–)
λ	heat conductivity ($J m^{-1} K^{-1} s^{-1}$), Lagrange multiplier, time adjoint
λ	overall multiplier of both regeneration expenses and capital, Eq. (8.112)
Θ	destruction rate of classical exergy ($J s^{-1}$)
θ	interval of the time type variable (s, –)
θ^n	time interval at stage n of a multistage process (s, –)
μ^k	chemical potential of kth component ($J g^{-1}$)
ν	stoichiometric coefficient (–)
ω	frequency constant (s^{-1})
ρ	mass density ($kg m^{-3}$)
ρ_e	energy density ($J m^{-3}$)
ρ_s	entropy density ($J K^{-1} m^{-3}$)
σ	entropy production ($J K^{-1} s^{-1}$)
σ_s	density of entropy production ($J K^{-1} m^{-3} s^{-1}$)
τ	relaxation time, average time between collisions (s)
$\tau = \omega t$	nondimensional time based on frequency constant ω (–)
τ	nondimensional time as the number of transfer units $xHTU^{-1}$ (–)
τ	space time variable in distributed catalyst deactivation problems (s, –)
τ_{ik}	viscous stress (Pa)
Ξ	imperfection factor (–)
ξ	process intensity factor (–)
Ω	order variable (–)
∇	differential operator

Subscripts

b	backward
C	Carnot state (open circuit)
c	cold
e	energy, equilibrium
F	feed
g	gas bulk
h	hot
i	ith state variable
in	incoming
j	reaction number
k	kth component
m	molar quantity
mix	mixing
out	outcoming
p	product
s	entropy, solid
w	water
v	per unit volume
σ	dissipative quantity
0	inlet state, reference state
1, 2	first and second, initial and final

Superscripts

a	power exponent
e	environment, equilibrium
f	final state and time
i	initial state and time
int	internal effect
n	stage number
T	transpose matrix, transform
‘	Carnot variable, modified quantity
0	ideal state, equilibrium state

Abbreviations and acronyms

AIC	algorithmic information content
CC	computational complexity
CIT	classical irreversible thermodynamics
CNCA	Chambadal-Novikov-Curzon-Ahlborn engine
DMFC	direct methanol fuel cells
DP	dynamic programming
EGM	entropy generation minimization
EIT	extended irreversible thermodynamics
FC	fuel cell
FTT	finite-time thermodynamics
GA	genetic algorithms
GEM	Gibbs equilibrium manifold
HEN	heat exchanger network
HJ	Hamilton-Jacobi equation
HJB	Hamilton-Jacobi-Bellman equation
LP	linear problem/programming
MEN	mass exchanger network
MGM	market growth model
NEE	net entropy exchange across a boundary
NLP	nonlinear problem/programming
OCT	optimal control theory
PEMFC	polymer electrolyte membrane fuel cell

Index

Note: Page numbers followed by *f* indicate figures.

B

C

D

P

Q

R

T

U

V

W

Z

Printed in the United States
By Bookmasters